Perspectives in Particles and Fields

Cargèse 1983

NATO ASI Series

Advanced Science Institutes Series

A series presenting the results of activities sponsored by the NATO Science Committee, which aims at the dissemination of advanced scientific and technological knowledge, with a view to strengthening links between scientific communities.

The series is published by an international board of publishers in conjunction with the NATO Scientific Affairs Division

A Life Sciences **B Physics**	Plenum Publishing Corporation New York and London
C Mathematical and Physical Sciences	D. Reidel Publishing Company Dordrecht, Boston, and Lancaster
D Behavioral and Social Sciences **E Engineering and Materials Sciences**	Martinus Nijhoff Publishers The Hague, Boston, and Lancaster
F Computer and Systems Sciences **G Ecological Sciences**	Springer-Verlag Berlin, Heidelberg, New York, and Tokyo

Recent Volumes in this Series

Volume 121—Electronic Structure, Dynamics, and Quantum Structural Properties of Condensed Matter
edited by Jozef T. Devreese and Piet Van Camp

Volume 122—Quarks, Leptons, and Beyond
edited by H. Fritzsch, R. D. Peccei, H. Saller, and F. Wagner

Volume 123—Density Functional Methods in Physics
edited by Reiner M. Dreizler and Jõao da Providência

Volume 124—Nonequilibrium Phonon Dynamics
edited by Walter E. Bron

Volume 125—Supersymmetry
edited by K. Dietz, R. Flume, G. v. Gehlen, and V. Rittenberg

Volume 126—Perspectives in Particles and Fields: *Cargèse 1983*
edited by Maurice Lévy, Jean-Louis Basdevant, David Speiser, Jacques Weyers, Maurice Jacob, and Raymond Gastmans

Perspectives in Particles and Fields

Cargèse 1983

Edited by

Maurice Lévy and Jean-Louis Basdevant

Laboratory of Theoretical Physics and High Energies
Université Pierre et Marie Curie
Paris, France

David Speiser and Jacques Weyers

Institute of Theoretical Physics
Université Catholique de Louvain
Louvain-la-Neuve, Belgium

Maurice Jacob

Theory Division
C.E.R.N.
Geneva, Switzerland

and

Raymond Gastmans

Institute of Theoretical Physics
Katholieke Universiteit Leuven
Leuven, Belgium

Plenum Press
New York and London
Published in cooperation with NATO Scientific Affairs Division

Proceedings of a NATO Advanced Study Institute/1983
Cargèse Summer Institute on
Particles and Fields,
held July 6-22, 1983,
in Cargèse, France

Library of Congress Cataloging in Publication Data

Cargèse Summer Institute on Particles and Fields (1983)
Perspectives in particles and fields, Cargèse 1983.

(NATO ASI series. Series B, Physics; v. 126)
"Proceedings of a NATO Advanced Study Institute/1983 Cargèse Summer Institute on Particles and Fields, held July 6-22, 1983, in Cargèse, France"—T.p. verso.
Includes bibliographies and index.
1. Particles (Nuclear physics)—Congresses. 2. Gauge fields (Physics)—Congresses. I.Lévy, Maurice, 1922- . II. North Atlantic Treaty Organization. Scientific Affairs Division. III. NATO Advanced Study Institute. IV. Title. V. Series.
QC793.C37 1983 539.7′21 85-12163

ISBN 978-1-4757-0371-9 ISBN 978-1-4757-0369-6 (eBook)
DOI 10.1007/978-1-4757-0369-6

A Division of Plenum Publishing Corporation
233 Spring Street, New York, N.Y. 10013

Softcover reprint of the hardcover 1st edition 1985

PREFACE

The 1983 Cargèse Summer Institute on Particles and Fields was organized by the Université Pierre et Marie Curie, Paris (M. LEVY and J.-L. BASDEVANT), C.E.R.N. (M. JACOB), the Université Catholique de Louvain (D. SPEISER and J. WEYERS), and the Katholieke Universiteit Leuven (R. GASTMANS). After 1975, 1977, 1979, and 1981, it was the fifth time they joined their efforts for organizing this Summer Institute.

This school was characterized by simultaneous progress in the theory of elementary particles and by impressive experimental advances. On the theoretical front, one witnessed the new developments in lattice gauge theories, which explore the world of strongly interacting particles in a non-perturbative way, and progress in a better understanding of the unity of all interactions based on supersymmetry. The experimentalists were proud to announce the discovery of the intermediate vector bosons, $W^{\pm}$ and Z^0, at C.E.R.N., while physicists working with e^+e^- colliding beams continued to probe more deeply the validity of the theoretical models of strong, weak and electromagnetic interactions.

We owe many thanks to all those who have made this Summer Institute possible!

Thanks are due to the Scientific Committee of NATO and its President for a generous grant and especially to the head of the Advanced Study Institute Program and his collaborators for their constant help and encouragement.

We also thank the National Science Foundation (USA) for their financial assistance.

Special thanks are due to the Université de Nice for having put at our disposal the facilities of the Institut de Cargèse.

We wish to thank Ms. M.-F. HANSELER, Ms. C. DUCHALAIS, Mr. D. BERNIA, and the inhabitants of Cargèse for their kind collaboration.

We thank Ms. M.-T. BERTONI for her patience and diligence in typing this manuscript.

Finally, the financial contribution of the N.V. Kredietbank (Belgium) is gratefully acknowledged. It helped to give this Summer Institute a broader international audience.

Mostly, however, we would like to thank all lecturers and participants who came from over 20 countries : the willingness of the former to answer all questions and the keen interest of the latter provided the stimulus which made (we hope) this Institute a success.

M. LEVY

J.-L. BASDEVANT

M. JACOB

D. SPEISER

J. WEYERS

R. GASTMANS

CONTENTS

ALGEBRAIC CONSTRUCTION OF GAUGE INVARIANT THEORIES

Laurent Baulieu

Laboratoire de Physique Théorique et Hautes Energies
Université Pierre et Marie Curie
F-75230 Paris, France

ABSTRACT

We give an algorithm for building gauge theories. It is based on the enlargment of space time by adjunction of additional but unphysical coordinates. The method enlights the relationship between the generating functional of anomalies in d-dimension space time and the topological invariants in d+2 dimension. We illustrate the general method with specific examples and show cases in which anomalies can occur for an even as well as for an odd number of space-time dimensions.

Given a symmetry group, all relations which are relevant for constructing an associated gauge invariant local field theory can be obtained in a remarkably direct way. We shall expose the general method, and afterwards illustrate it with examples. The construction does not depend on the space-time dimension d and uses the most convenient formalism of differential forms. It reproduces the known results in the cases of gravitation and Yang Mills theories. It has been shown to solve also puzzling cases of theories involving skew tensor gauge fields. It is very likely that this method provides an algorithm for the classification of anomalies in any gauge theory.

The main idea[1] consists in enlarging the usual space-time $\{x^\mu\}$ into a space $\{x^\mu, y^\alpha\}$. The latter is defined locally as the direct product of space time $\{x^\mu\}$ with the space $\{y^\alpha\}$. The y^α coordinates can be choosen as real as well as Grassmanian. Indeed all algebraic manipulations which we shall perform are equivalent for both choices. A deeper understanding of the global structure of the space $\{x^\mu,y^\alpha\}$, which has not been reached up to now, should allow for distinguishing between these two possibilities.

Then, one introduces differential forms ω_p^i of rank p. They are defined in the tangent plane over each point $\{x^\mu, y^\alpha\}$. Let G be the symmetry group from which we want to build a gauge theory. In general ω_p^i is a p-form which takes its values in a given representation R^i of G. In local coordinates ω_p^i can be expanded on the basis of p-forms $dx^{\mu_1}\wedge\ldots\wedge dx^{\mu_{p-g}}\wedge dy^{\alpha_1}\wedge\ldots\wedge dy^{\alpha_g}$ as follows

$$\tilde{\omega}_p^i = \sum_{0\leq g\leq p} \omega^i_{[\mu_1\ldots\mu_{p-g}]\alpha_1\ldots\alpha_g} dx^{\mu_1}\wedge\ldots\wedge dx^{\mu_{p-g}}\wedge dy^{\alpha_1}\wedge\ldots\wedge dy^{\alpha_g}. \tag{1-a}$$

Here and elsewhere the antisymmetrization on Lorentz indices acts as an idempotent and $\wedge$ denotes the wedge product.

By definition we call the following component of $\tilde{\omega}_p^i$

$$\omega^{i,g}_{[\mu_1\ldots\mu_{p-g}]} \equiv \omega^i_{[\mu_1\ldots\mu_{p-g}]\alpha_1\ldots\alpha_g} dy^{\alpha_1}\wedge\ldots\wedge dy^{\alpha_g} \tag{1-b}$$

a skew tensor gauge field of Lorentz rank p-g and ghost number g. We also define

$$\omega^{i,g}_{p-g} \equiv \omega^{i,g}_{[\mu_1\ldots\mu_{p-g}]} dx^{\mu_1}\wedge\ldots\wedge dx^{\mu_{p-g}}. \tag{1-c}$$

In field theory, the system of fields made up from the R^i-valued classical skew field $\omega_p^{i,0}$ and the ghosts $\omega_{p-g}^{i,g}$ $(p \geq g \geq 1)$ allows one generally to build up the quantum Lagrangian for a (formally) gauge independent theory depending of skew tensors $\omega_p^{i,0}$ [2]. In order to reach this result, one must assign a physical (unphysical) statistics to any field with an even (odd) ghost number*.

As a second step one introduces a field strenght $\tilde{G}^i$ as assiciated with the field form $\tilde{\omega}_p^i$. $G^i(\tilde{\omega})$ must satisfy the 3 following criteria.

(i) $\tilde{G}^i$ is linear in the exterior derivative of $\tilde{\omega}_p^i$. Thus $\tilde{G}^i$ is a (p+1) form which takes its values in R^i. In general one can write $\tilde{G}^i$ as follows :

$$\tilde{G}^i_{p+1} \equiv \tilde{d}\,\tilde{\omega}_p^i + P^i_{p+1}(\tilde{\omega}_{p'}^{i'}) \quad , \tag{2}$$

where P^i_{p+1} is a R^i-valued (p+1) form made from the $\tilde{\omega}_{p'}^{i'}$'s. $\tilde{d}$ is the exterior differential operator in the enlarged space $\{x^\mu, y^\alpha\}$. We

*The proof amounts to counting the modes which contribute to the unitarity equation.[2] It necessitates the construction of a Lagrangian which is quadratic in the field derivatives. Since the latter must have ghost number zero, it requires the introduction of anti-ghosts. It can be done trivially by doubling the unphysical space $\{y^\alpha\}$[3,4].

separate $\tilde{d}$ into two terms

$$\tilde{d} = d + s \ , \tag{3-a}$$

with

$$d = dx^{\mu}\,\partial_{\mu} \ , \qquad s = dy^{\alpha}\,\partial_{\alpha} \ . \tag{3-b}$$

From the exterior calculus rules one has identically

$$d^2 = sd + ds = s^2 = 0 \ . \tag{3-c}$$

(ii) $\tilde{G}^i$ can be set as horizontal according to the following definition

$$\tilde{G}^i_{p+1} = G^i_{p+1} = G^i_{[\mu_1 \ldots \mu_{p+1}]} dx^{\mu_1} \wedge \ldots \wedge dx^{\mu_{p+1}} \tag{4-a}$$

$$\equiv \text{horizontal.}$$

Here G^i_{p+1} is the truncation of $\tilde{G}^i_{p+1}$ in the usual space $\{x^{\mu}\}$. In other words one requires that the (p+1) form $\tilde{G}^i$ be such that all its components but the pure horizontal one (i.e. that along the physical directions), can be set equal to zero

$$G^i_{[\mu_1 \cdots, \mu_{p+1-g}], \alpha_1 \ldots \alpha_g} = 0 \ , \quad p+1 \geq g \geq 1 \ . \tag{4-b}$$

(iii) It exists a "covariant" derivative operator $\tilde{D}$ in the enlarged space

$$\tilde{D} = \tilde{d} + K(\tilde{\omega}) = D + S \ , \tag{5}$$

such that eq. (4) is compatible with

$$\tilde{D}\,\tilde{G}^i_{p+1} = D\,G^i_{p+1} \equiv \text{horizontal.} \tag{6-a}$$

Here $K(\tilde{\omega})$ is a 1-form operator acting on all representations of G. Eq. (6-a) should be interpreted as the integrability condition of eq. (4). In eq. (5) D is defined as the horizontal projection of $\tilde{D}$, $D = D_{\mu}\,dx^{\mu}$. Then, one may interpret S as the parallel transport operator along the unphysical directions y^{α}. Interestingly enough, one can rewrite eq. (6-a) as follows

$$S\,G^i_{p+1} = 0 \ , \tag{6-b}$$

since $\tilde{D} = D + S$ and $\tilde{G}^i = G^i$. Eq. (6-b) allows one to interpret the integrability condition (6-a) as characterizing the intuitive property that the classical field strenghts be left invariant under parallel transport along unphysical directions.

The following integrability conditions must be satisfied also for consistency

$$(\tilde{D}\ldots\tilde{D})\,\tilde{G}^{i}_{p+1} = (D\ldots D)\,G^{i}_{p+1} \equiv \text{horizontal} \; . \tag{7}$$

It is clear that the system of constraints (5,6,7) restricts severely enough the possible forms of a priori unknown polynomials P^{i}_{p+1} and K in eqs. (2) and (5). However, one can verify that they can be enforced in Yang Mills theories (see example 1), in theories with skew tensors[2] and in gravitation[5]. Moreover, for theories involving non Abelian skew tensors of rank higher than two the above method has allowed for the construction of yet unknown systems of gauge transformations[6].

Eqs. (4,6,7) determine in general the action of the operator $s = dy^{\alpha}\partial_{\alpha}$ on all field components $\omega^{i,g}_{p-g}$. Indeed, by expanding eq. (4) in local coordinates, and by projecting it on the basis of polynomials with a given ghost number, one gets

$$s\,\omega^{i,g}_{p-g} = -\,d\,\omega^{i,g+1}_{p-g-1} - P^{i,g}_{p-g}(\tilde{\omega}) \quad , \quad 0 \leqq g \leqq p \; . \tag{8}$$

Up to now, in all cases which have been worked out in practice, it has been verified that the definition (4,6,7) of s is compatible with $s^2 = sd + ds = 0$.

Among all equations which determine the form of s, that with ghost number 1 is of special interest

$$s\,\omega^{i,0}_{p} = -\,d\,\omega^{i,1}_{p-1} - P^{i,1}_{p}\,(\omega^{i',0}_{p'} \; , \; \omega^{i',1}_{p'-1}) \; . \tag{9}$$

This equation identifies itself with the usual definition of infinitesimal gauge transformations whenever one substitutes infinitesimal local parameters $\epsilon^{i}_{p-1} \equiv \epsilon^{i}_{[\mu_1\ldots\mu_{p-1}]}dx^{\mu_1}\wedge\ldots\wedge dx^{\mu_{p-1}}$ in place of ghosts $\omega^{i,1}_{p-1}$. Then

$$\delta\omega^{i,0}_{p} = -\,d\,\epsilon^{i}_{p-1} + P^{i,1}_{p}(\omega^{i',0}_{p'} \; , \; \epsilon^{i'}_{p'}). \tag{10}$$

The parameters ϵ^{i}_{p-1} have the same tensorial properties as the ghosts $\omega^{i,1}_{p-1}$ but the opposite (i.e. physical) statistics.

One can verify that the infinitesimal gauge transformations (10) build up a Lie algebra as a direct consequence of the nilpotency property of s and the commutation properties of ghosts $\omega^{i,g}_{p-g}$ [7].

To enlight the above formalism, we shall give two examples of this reverse construction of gauge transformations associated with a symmetry group. Before going to that, we will show a direct application of our construction. It concerns the fundamental problems which one encounters in field theory, the determination of an invariant Lagrangian* and of possible anomalies.

An invariant Lagrangian in d dimension space-time $\mathcal{L}_d^{inv}$ is by definition a d-form function of classical fields $\omega_p^{i,0}$ which is left invariant under gauge transformation, that is to say under the action of s

$$s\ \mathcal{L}_d^{inv}\ (\omega_p^{i,0}) = 0\ . \tag{11}$$

On the other hand, an anomaly in d dimension is a group invariant form Δ_d^1 with Lorentz rank d and ghost number 1. It is by definition the general solution of the so-called Wess and Zumino[9] consistency condition

$$s\ \Delta_d^1 + d\ \Delta_{d-1}^2 = 0\ . \tag{12}$$

Indeed, whenever Δ_d^1 is solution of eq.(12), it is a possible obstruction to the Ward identity associated with the s invariance of the Lagrangian[10,13]. We shall generalize eq. (12) into

$$s\ \Delta_d^g + d\ \Delta_{d-1}^{g+1} = 0\ , \tag{13}$$

where $g \geqq 1$.

For solving eq. (11), the fundamental equation (6-b), $SG^i = 0$, suggests that the s invariant d-forms $\mathcal{L}_d^{inv}(\omega_p^{i,0})$ only depend upon $\omega_p^{i,0}$ through combinations of field strenghts G^i and of their Hodge transforms $*G^i$

$$\mathcal{L}_d^{inv} \equiv P_d^{inv}\ (\ G_{p+1}^i, (*G^i)_{d-p-1})\ . \tag{14}$$

Here P_d^{inv} is a polynomial which is invariant under rigid group transformations. The constraint that $\mathcal{L}_d^{inv}$ is a d-form yields severe restriction on the possible forms of P_d^{inv}. One also excludes the possibility that $\mathcal{L}_d^{inv}$ be a pure differential.

For solving the anomaly equation (13), we shall assume the validity of Poincaré Lemma in all the directions of the enlarged space. In other words, we assume that if the equation

$$s\ X^g(\tilde{\omega}) = 0 \tag{15-a}$$

is locally satisfied, it implies

*Once a classical invariant Lagrangian $\mathcal{L}_d^{inv}(\omega_p^{i,0})$ has been built, a systematic method for building a gauge fixed, but s invariant method can be found in refs. 7,8. Such a Lagrangian leads formally to gauge independent physics.

$$X^g(\tilde{\omega}) = s\, X^{g-1}(\tilde{\omega}) \,. \qquad (15\text{-b})$$

One should note that eq. (15-b) only makes sense when $g \geq 1$. For $g = 0$ eq. (15-a) is identical to eq. (11).

We start from the anomaly equation (13). From now on we increase implicitely the space time dimension d to a large enough number such that forms with Lorentz rank d+1,..., d+g+1 do not vanish trivially. By applying on both sides of eq. (13) the d operator, and using $d^2 = sd + ds = 0$, one gets

$$s\, d\, \Delta^g_d = 0 \,. \qquad (16\text{-a})$$

It implies the existence of a form with ghost number g-1 and Lorentz rank d+1 such that

$$s\, \Delta^{g-1}_{d+1} + d\, \Delta^g_d = 0 \,. \qquad (16\text{-b})$$

One can iterate the process until one reaches a form Δ^0_{d+g} with ghost number 0 and Lorentz rank d+g such that

$$s(d\, \Delta^0_{d+g}) = 0 \,. \qquad (16\text{-c})$$

It means that the form $d\Delta^0_{d+g}(\tilde{\omega})$ which has ghost number zero and Lorentz rank d+g+1 is s invariant, that is to say gauge invariant since it can only depend on classical fields $\omega^{i,0}_p$. From the previous analysis we then find that $d\Delta^0_{d+g}$ is a priori a group invariant form of Lorentz rank d+g+1 made from the field strenghts G^i and $*G^i$

$$d\, \Delta^0_{d+g} = Q^{inv}_{d+g+1}(G^i_{p+1}, (*G^i)_{d-p-1}) \,. \qquad (17)$$

Eq. (17) shows that $Q^{inv}_{d+g+1}(G^i, *G^i)$ is annihilated by d. From eq. (16-c) it is gauge invariant by construction. Therefore, as far as local properties are concerned, Q^{inv}_{d+g+1} has the requisite properties for being a topological invariant in d+g+1 dimension space. Conversely, if such an invariant exists, one has identically

$$\begin{array}{ccc} Q^{inv}_{d+g+1}(\tilde{G}^i, *\tilde{G}^i) & = & Q^{inv}_{d+g+1}(G^i, *G^i) \\ \| & & \| \\ \tilde{d}\, \Delta^0_{g+d}(\tilde{\omega}) & = & d\, \Delta^0_{g+d}(\omega) \,, \end{array} \qquad (18\text{-a})$$

since by definition $\tilde{G}^i = G^i$ and $*\tilde{G}^i = *G^i$. It implies

$$(d+s)(\Delta^0_{g+d}(\omega) + \Delta^1_{g+d-1}(\tilde{\omega}) + \ldots + \Delta^{g+d}_0(\tilde{\omega})) = d\, \Delta^0_{g+d}(\omega), \qquad (18\text{-b})$$

where we have defined

$$\Delta^0_{g+d}(\tilde{\omega}) = \Delta^0_{g+d}(\omega^0_p + \omega^1_{p-1} + \ldots + \omega^g_0) \equiv \Delta^0_{g+d}(\omega) + \Delta^1_{g+d-1}(\tilde{\omega}) + \ldots$$

$$\ldots + \Delta_0^{g+d}(\tilde{\omega}) .$$

Eqs. (18-b) gives directly the solution to the anomaly equation (13) by projection over the forms with ghost number g + 1.

In order to make contact with the usual anomaly problem, one sets g = 1. Then one must look for the existence of "topological" invariants in d + 1 + 1 = d + 2 dimension space. Whenever such an invariant exists, it implies the existence of a form $\Delta_d^1(\omega_p^{i,0}, \omega_{p-1}^{i,1})$ which satisfies the Wess and Zumino consistency condition (12). Up to a model dependent coefficient X, Δ_d^1 realizes an obstruction for imposing the Ward identities associated with the gauge symmetry. The broken Ward identity reads as follows in d dimensional space time

$$\int d^dx \left(s\, \omega_p^{i,0} \frac{\delta}{\delta\omega_p^{i,0}} + \ldots\right) \Gamma_d(\omega_p^{i,0}, \ldots, \omega_0^{i,p})$$

$$= h\, X \int \Delta_d^1\, (\omega_p^{i,0}, \omega_{p-1}^{i,1}) . \qquad (19)$$

Here Γ_d is the generating functional of 1PI vertices. It is function of all field sources, classical and ghost. The dots stand for terms of the Slavnov operator which depend upon the s-variations of ghosts. Now, since Δ_d^1 has ghost number 1, it depends linearly on the ghosts $\omega_{p-1}^{i,1}$. On the other hand, according to eq. (9), the term linear in the ghost field in $s\,\omega_p^{i,0}$ reads as follows

$$s\,\omega_p^{i,0} = -\, d\, \omega_{p-1}^{i,1} + \ldots \qquad (20)$$

Therefore, in order to determine the potentially anomalous vertices and the value of the model dependent coefficient X one can differentiate both sides of eq. (19) once with respect to the ghosts $\omega_{p-1}^{i,1}$, and a certain number of times with respect to the classical field $\omega_p^{i,0}$ in such a way that one obtains in the r.h.s. X times some constants depending upon the explicit form of Δ_d^1. In that way one gets

$$\partial_\mu \frac{\delta}{\delta\omega^{i,0}_{[\mu\nu\ldots]}} \frac{\delta}{\delta\omega^{i,0}_{[\mu_1,\mu_2\ldots]}} \cdots \frac{\delta}{\delta\omega^{i',0}_{[\mu_1',\mu_2'\ldots]}} \Gamma_d \propto X , \qquad (21)$$

where the tensorial structure which has been left implicit in the r.h.s. depends on the polynomial structure of Δ_d^1. Eq. (21) allows one to identify the potentially anomalous diagrams of the theory.

Examples

The pure Yang Mills case. Let $\mathcal{G}$ be the Lie Algebra of a compact Lie group G. The Yang Mills 1-form $\mathcal{A} = A_\mu dx^\mu + A_\alpha dy^\alpha$ is $\mathcal{G}$-valued. The matter field φ is either a spin 0 or a spin 1/2 field which is valued in some representation R of G.

$$\tilde{A} = A + c , \qquad \tilde{\phi} = \phi . \tag{22}$$

$A = A_\mu dx^\mu$ and $c = A_\alpha dy^\alpha$ are respectively identified with the classical Yang Mills field and the Faddeev Popov ghost. The covariant derivative (5) can only be of the following form

$$\tilde{D} = \tilde{d} + \tilde{A} . \tag{23}$$

Thus, the only possible field strenghts which can satisfy eq. (4) and (6) are as follows

$$\tilde{F} \equiv \tilde{d}\tilde{A} + 1/2\,[\tilde{A},\tilde{A}] = F \quad , \tag{24-a}$$

$$\tilde{G}_\phi \equiv \tilde{D}\,\phi = D\,\phi ,$$

which means

$$s\,A = -\,D\,c ,$$

$$s\,\phi = -\,c\,\phi , \tag{24-b}$$

$$s\,c = -\,1/2\,[c,c] .$$

Obviously the factor 1/2 in $\tilde{F}$ is uniquely defined once $\tilde{D}$ has been defined as in eq. (23).

One can easily check equations (4) and (6). Indeed one has $\tilde{D}\tilde{F} \equiv 0 \equiv DF$ (or equivalently $SF = -[c,F]$) on the one hand and $\tilde{D}\tilde{D}\phi = \tilde{F}\phi = F\phi = DD\phi$ on the other hand. Eq. (7) is clearly satisfied.

The Hodge transforms $*F$ and $*D\phi$ of F and $D\phi$ are respectively d-2 and d-1 forms

$$*F = \epsilon_{\mu_1 \ldots \mu_d}\, F^{\mu_1 \mu_2}\, dx^{\mu_3} \wedge \ldots \wedge dx^{\mu_d} ,$$

$$*D\phi = \epsilon_{\mu_1 \ldots \mu_d}\, D^{\mu_1}\phi\, dx^{\mu_2} \wedge \ldots \wedge dx^{\mu_d} . \tag{25}$$

The invariant d-form which can be made from F, $*F$, $D\phi$, $*D\phi$ and which is not a pure differential reads as follows

$$\mathcal{L}_d^{inv} = *F \wedge F + *D\phi \wedge D\phi . \tag{26}$$

All topological invariants $Q^{inv}_{g+d+1}(A,\phi)$ have been already classified. They can only depend on $\tilde{F}$ or $*F$. Thus Q^{inv}_{g+d+1} is non zero only if $g + d + 1$ is even. It is an invariant polynomial of F (or $*F$) and satisfies the Chern identity[11,12]

$$Q^{inv}_{d+g+1}\ (F,\ldots F) = d \int_0^1 dt\ Q^{inv}_{d+g+1}\ (A,F_t,\ldots F_t), \tag{27}$$

with $F_t = t\ dA + t^2/2\ [A,A]$. A detailed analysis for using the Chern formula (27) in the study of consequences of anomalies in renormalizable gauge theories can be found in refs. 13. Here we shall only explain briefly how one can determine the potentially anomalous diagrams in even d dimension Yang Mills theories. This will illustrates the general method described above. Consider the case when G is a simple group. Then the topological invariants in d+2 dimensions are (they don't exist in odd dimensions)

$$\mathrm{Tr}\ (\underbrace{F \wedge\ldots\wedge F}_{\frac{d}{2}+1}) = \frac{1}{\frac{d}{2}+1}\ d \int_0^1 dt\ \mathrm{Tr}(A \wedge \underbrace{F_t \wedge\ldots\wedge F_t}_{\frac{d}{2}}), \tag{28}$$

with $F_t = tdA + t^2/2\ [A,A]$. Thus, when one goes from 0 to 1 loop in d dimensional field theory, the broken Ward identity can only have the following form

$$\int d^dx\ (D_\mu c \frac{\delta}{\delta A_\mu} +\ldots\ )\ \Gamma_d(A,c,\phi) = \tag{29}$$

$$\mathrm{Tr}\ X \int_d c^a \frac{\delta}{\delta A^a}\Big|_F \int_0^1 dt\ \mathrm{Tr}(A \wedge \underbrace{F_t \wedge\ldots\wedge F_t}_{\frac{d}{2}}).$$

The r.h.s. can be expanded as a polynomial in A. It contains monomials with well defined coefficients whose degrees in the gauge field A are d/2, d/2 + 1, ..., d − 1. It implies that one obtains the coefficient X (times some group factors) by differentiating both sides of eq. (29) once with respect to c and d/2, d/2 + 1, ... or d − 1 times with respect to A. Thus X can be computed from the one-loop evaluation of any of the following d/2 vertices

$$\partial_\mu \frac{\delta}{\delta A^a_\mu(x)} \frac{\delta}{\delta A^{a_1}_{\mu_1}(x_1)} \cdots \frac{\delta}{\delta A^{a_k}_{\mu_k}(x_k)}\ \Gamma^{1-\mathrm{loop}}_d\ , \tag{30-a}$$

that is to say of diagrams

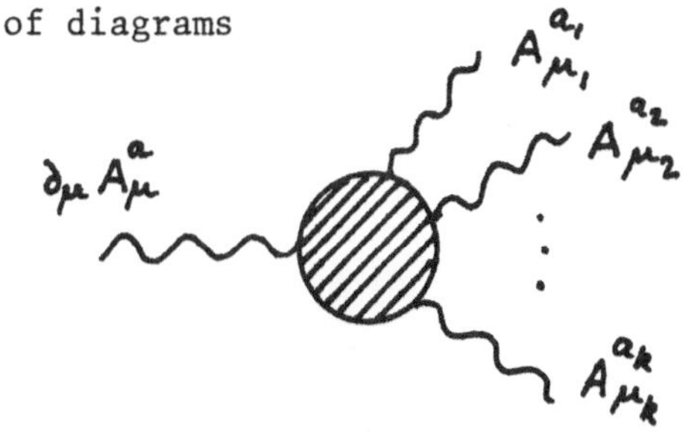

(30-b)

with $d/2 \leqq K \leqq d - 1$. Obviously the value of these diagrams depends on the couplings of gauge fields A to matter fields which can circulate within the loop.

An exotic example. Let us consider the following theory in $d = 6$ dimensions. It contains two Abelian skew tensors of rank 2, $B^i = B^i_{\mu\nu}\, dx^\mu \wedge dx^\nu$, with $i = 1, 2$ and a neutral skew tensor of rank 4, $D_4 = D_{\mu\nu\rho\sigma} dx^\mu \wedge dx^\nu \wedge dx^\rho \wedge dx^\sigma$.

The following field strenghts satisfy the criteria (i), (ii), (iii)

$$G^i_3 = d\, B^i_2 \qquad\qquad i, j = 1, 2. \tag{31}$$

$$G_5 = d\, D_4 + a_{ij}\, G^i_3\, B^j_2$$

where a_{ij} is an anti-symmetric 2x2 matrix $a_{ij} = -a_{ji}$. Indeed these field strenghts can be set as horizontal in the enlarged space $\{x^\mu, y^\alpha\}$ in a consistent way :

$$\tilde{G}^i_3 \equiv (d+s)\, \tilde{B}^i_2 = d\, B^i_2 \equiv G^i_3 \quad ,$$

$$\tilde{G}^i_5 \equiv (d+s)\, \tilde{D}_4 + a_{ij}\, \tilde{G}^i_3\, \tilde{B}^j_2$$

$$= d\, D_4 + a_{ij}\, G^i_3\, B^j_2 \equiv G^i_5 \quad , \tag{32}$$

where $\tilde{B}^i_2 = B^i_2 + B^{i,1}_1 + B^{i,2}_0$ and $\tilde{D}_4 = D_4 + D^1_3 + D^2_2 + D^3_1 + D^4_0$. The essential point is that eqs. (32) are compatible with their integrability conditions. Indeed, one has

$$\tilde{d}\, \tilde{G}^i_3 = 0 = d\, G^i_3 \quad ,$$

$$\tilde{d}\, \tilde{G}_5 = a_{ij}\, \tilde{G}^i_3\, \tilde{G}^j_3 = a_{ij}\, G^i_3\, G^j_3 = d\, G_5 \, . \tag{33}$$

Eq. (32) gives the BRS equation of the theory

$$s\, B^i_2 = -\, d\, B^{i,1}_1 \, ,$$

$$s\, B^{i,1}_1 = -\, d\, B^{i,2}_0 \, ,$$

$$s\, B^i_0 = 0 \quad ,$$

$$s\, D_4 = -\, d\, D^1_3 - a_{ij}\, d\, B^i_2\, B^{j,1}_1 \, ,$$

$$s\, D^1_3 = -\, d\, D^2_2 = a_{ij}\, d\, B^i_2\, B^{j,2}_0 \, ,$$

$$s\, D^2_2 = -\, d\, D^3_1 \, ,$$

$$s\, D^3_1 = -\, d\, D^4_0 \, ,$$

$$s\, D_0^4 = 0 \quad . \tag{34}$$

The infinitesimal gauge transformations derive from the substitution

$$B_1^{i,1} \rightarrow \epsilon_1^i = \epsilon_\mu^i \, dx^\mu \quad ,$$

and

$$D_3^1 \rightarrow \epsilon_3 = \epsilon_{[\mu\nu\rho]} \, dx^\mu \wedge dx^\nu \wedge dx^\rho \quad ,$$

where $\epsilon_\mu^i(x)$ and $\epsilon_{[\mu\nu\rho]}(x)$ are local real infinitesimal parameters

$$\begin{aligned} \delta\, B_2^i &= -\, d\, \epsilon_1^i \quad , \qquad\qquad i, j = 1, 2. \\ \delta\, D_4 &= -\, d\, \epsilon_3 - a_{ij}\, d\, B_2^i\, \epsilon_1^j \quad . \end{aligned} \tag{35}$$

One can check easily that $s^2 = 0$ and equivalently that the gauge transformation (35) build up a closed algebra.

The consistency equation (6) is clearly satisfied. It implies that the field strenghts are left invariant under s transformation (eq. 6-b)

$$\begin{aligned} s\, G_3^i &= 0 \Leftrightarrow \delta G_3^i = 0 \;, \qquad\qquad i = 1, 2 \quad \cdot \\ s\, G_5 &= 0 \Leftrightarrow \delta G_5 = 0 \;. \end{aligned} \tag{36}$$

It follows that the gauge invariant Lagrangian reads as follows

$$\begin{aligned} \mathcal{L}_6^{inv} &= \sum_i *(dB_2^i)\, dB_2^i + *(dD_4 + a_{ij} dB_2^i\, B_2^j)(dD_4 + a_{ij} dB_2^i\, B_2^j) \\ &= d^6x\, \{\sum_i (\delta_{[\mu} B^i_{\nu\rho]})^2 + (\delta_{[\mu} D_{\nu\rho\sigma\tau]} + a_{ij}\, B^j_{[\mu\nu} \delta_\rho\, B^i_{\sigma\tau]})^2\}. \end{aligned} \tag{37}$$

In order to get the anomaly we have to look for the topological invariants in 8 dimensions. They are any linear combination of both following 8-forms made from G_5 and G_3^i :

$$Q_8^{i,inv} = G_3^i\, G_5 \;, \qquad\qquad i = 1, 2 \;. \tag{38}$$

Indeed, on the one hand one has

$$d\, Q_8^{i,inv} = G_3^i\, d\, G_5 = G_3^i\, a_{kl}\, G_3^k\, G_3^l = 0 \;, \tag{39}$$

since $G_3^i\, G_3^i \equiv 0$. On the other hand $Q_8^{i,inv}$ is trivially gauge invariant (i.e. s invariant).

$$s\, Q_8^{i,inv} = 0 \Longleftrightarrow \delta\, Q_8^{i,inv} = 0 \;. \tag{40}$$

Since $d\, Q_8^{i,inv} = 0$, $Q_8^{i,inv}$ is the differential of a 7-form. One has explicitely

$$Q^{i,\mathrm{inv}}(G_3,G_5) = d\{D_4 G_3^i + \sum_{k=1,2} a_{ki}\, B_2^i\, B_2^i\, G_3^k\}$$
$$= Q^{i,\mathrm{inv}}(\tilde{G}_3,\tilde{G}_5)$$
$$= \tilde{d}\{(D_4 + D_3^1 + D_2^2 + D_1^3 + D_0^4)G_3^i + \sum_{k=1,2} a_{ki}(B_2^i + B_1^{i,1} + B_0^{i,2})^2 G_3^k\}. \tag{41}$$

The generating functional for the anomaly is therefore any linear combination of the following forms with ghost number 1 and Lorentz rank 6

$$\Delta_6^{i,1} = X^i(D_3^1\, G_3^i + 2 \sum_{k=1,2} a_{ki}\, B_1^{i,1}\, B_2^i\, G_3^k) \ . \tag{42}$$

Then the broken Ward identity reads as follows

$$\int d^6x\, (dB_1^{i,1} \frac{\delta}{\delta B_2^i} + dD_3^1 \frac{\delta}{\delta D_4} + \ldots)\, \Gamma_6(B_2^i, D_4, \ldots)$$
$$= X_i \int (D_3^1\, dB_2^i + 2 \sum_{k=1,2} a_{ki}\, B_1^{i,1}\, B_2^i\, d\, B_2^k) \ . \tag{43}$$

It implies that the potentially anomalous diagrams are the following ones

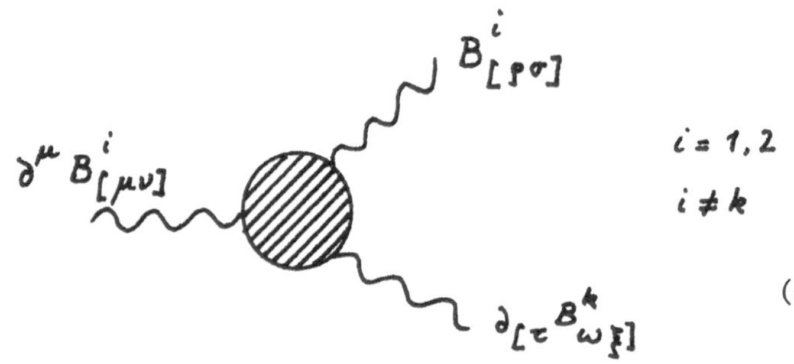

(44-a)

and

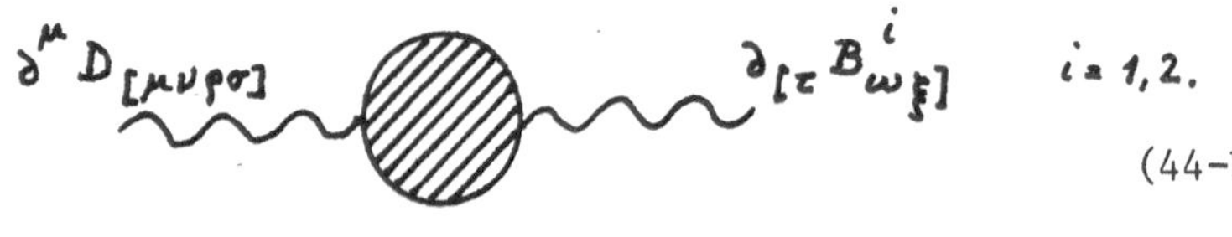

(44-b)

In order to evaluate the value of these diagrams one should precise the possible interactions of B_2^i and D_4 gauge fields with matter and/or with gravitation.

It is straightforward to build this theory for d > 6, even or odd. It gives a neat example of a theory which can have anomalies in odd dimensions. Indeed the following 11 form

$$Q_{11} = a_{ij}\, G_3^i\, G_3^j\, G_5 \tag{45}$$

is clearly such that

$$dQ_{11} = 0 \ . \tag{46}$$

It implies the existence of an anomaly in d = 9 dimensions. The corresponding anomalous diagrams are the same as in eq. (44) with one extra leg $\partial_{[\tau'} B^k_{\omega'\xi']}\, a_{ki}$.

Gravity. The same general method can be used to reconstruct the gravity in the vielbein formalism and determine the anomalies streaming from gravitational effects[5].

Interestingly enough the d = 11 dimension supergravity contains potential anomalies. They are associated with topological invariants in 13 dimensions. A simplest example is

$$\begin{aligned} Q_{13} &= (dC) \wedge \mathrm{Tr}(RR)\ d(T^a T^a) \\ &= d\ \{\mathrm{Tr}(\omega d\omega + \tfrac{2}{3}\ \omega\omega\omega)(dC)\ T^a\ dT^a\}\ , \end{aligned} \tag{47}$$

where $C = C_{[\mu\nu\rho]}\, dx^\mu \wedge dx^\nu \wedge dx^\rho$ is the abelian 3-tensor field of Cremmer and Julia theory, $T^a = de^a + \omega e^a$ is the torsion built from the vielbein 1-form e^a and the connection 1-form ω^{ab}, and

$$R^{ab} = d\omega^{ab} + \tfrac{1}{2}[\omega,\omega]^{ab}$$

is the curvature.

The potentially anomalous diagrams associated with the invariant defined in eq. (47) are the following ones.

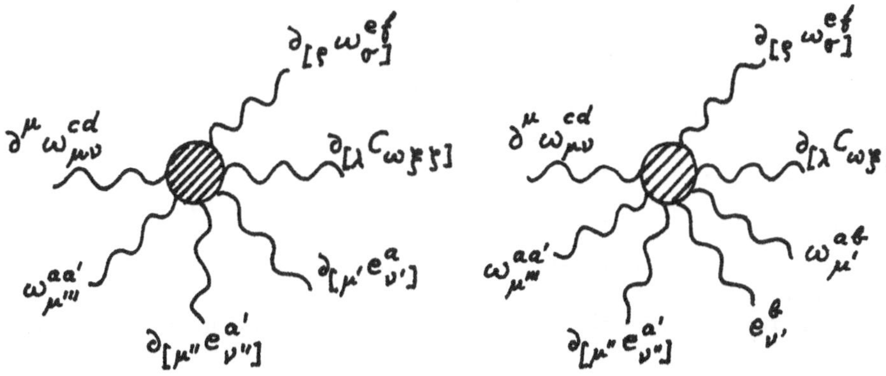

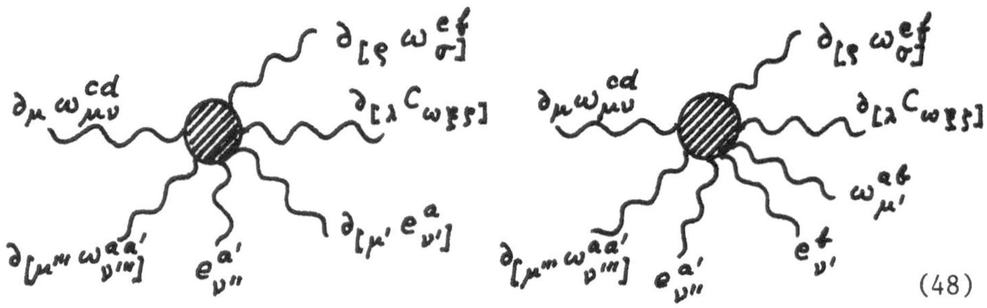

(48)

The latin indices are Lorentz indices and the greek ones are world indices.

AKNOWLEDGMENTS

It is a pleasure to thank my collaborator J. Thierry Mieg with whom I have elaborated many ideas contained here, and I. Sezgin with whom I have worked out the details of the example (2).

REFERENCES

1. J. Thierry-Mieg, J. Math. Phys. 21 (1980) 2834, Nuovo Cim. 56A (1980) 396.
   L. Bonora and M. Tonin, Phys. Lett. 98B (1981) 48.
2. L. Baulieu, J. Thierry-Mieg, Nucl. Phys. B228 (1983) 259.
   L. Baulieu, Phys. Lett. 126B (1983) 455 and Nucl. Phys. B227 (1983) 157.
3. M. Quirros, F.J. de Urries, J. Aoyos, M.J. Mazon, E. Rodriguez, J. Math. Phys. 22 (1981) 767.
4. L. Baulieu, J. Thierry-Mieg, Nucl. Phys. B197 (1982) 477.
5. L. Baulieu and J. Thierry-Mieg, in preparation.
6. L. Baulieu and J. Thierry-Mieg, On a new class of gauge transformations, LPTHE preprint 83/25, submitted to Physics Lett. B.
7. L. Alvarez Gaumé and L. Baulieu, Nucl. Phys. B212 (1982) 255.
8. F.R. Ore and P. Van Nieuwenhuizen, Nucl. Phys. B204 (1982) 317.
9. J. Wess, B. Zumino, Phys. Lett. 37B (1971) 95.
10. R. Stora, Cargèse lectures, 1976.
    C. Becchi, A. Rouet, R. Stora, Ann. of Phys. 98 (1976) 287.
11. S.S. Chern, Complex manifold without potential theory, 2nd (Springer, Berlin, 1979) Universitext.
12. R. Stora, Cargèse lectures (1983), LAPP preprint.
13. L. Baulieu, Anomalies and Gauge Symmetry, preprint LPTHE 83/26.
    Wu Yong Shi, A. Zee, B. Zumino, Seattle preprint, august 1983.
    L. Bonara, P. Pasti, Padova preprint IFPD 20/83.
    B. Zumino, Les Houches lectures 1983.
    L. Baulieu, Perturbative Gauge Theories, LPTHE preprint 83/31.

AN INTRODUCTION TO LATTICE GAUGE THEORIES

J. Zinn - Justin
(D. Ph. T)
CEN - SACLAY

F - 91191 Gif-sur-Yvette Cedex

WHY A LATTICE : AN EXAMPLE IN QUANTUM MECHANICS

By studying the renormalized perturbation theory of ordinary continuous gauge theories which corresponds to the classical lagrangian L :

$$L = \frac{1}{4g_0^2} \vec{F}_{\mu\nu} \vec{F}_{\mu\nu} ,$$
$$\vec{F}_{\mu\nu} = \partial_\mu \vec{A}_\nu - \partial_\nu \vec{A}_\mu + \vec{A}_\mu \times \vec{A}_\nu , \tag{1}$$

we learn from renormalized group arguments that the theory is simple at short distance, and complicated at long distance where the effective coupling constant becomes large. Therefore the perturbative spectrum of the theory which consists of massless vector mesons may not be the true spectrum of the theory. If a mass is generated by non perturbative effects, it will be a renormalization group invariant, and thus have the form for small coupling :

$$\beta(g) = - \beta_3 \, g^3 + O(g^5) , \tag{2}$$

$$m \sim \mu \, e^{-\frac{1}{2\beta_3 g^2}} , \quad g \to 0. \tag{3}$$

Here μ is the renormalization scale and g the renormalized coupling constant. Equation (3) shows that the mass has a behaviour which is consistent with the fact that it cannot be seen in perturbation theory for g small.

To answer this fundamental question of mass generation, perturbation theory, at least in its simplest form, is useless and we have to invent some new procedure to study the spectrum of the field theory. To understand what kind of method can be used to handle the problem we shall first study a few simpler examples which are similar in some respect.

DOUBLE-WELL POTENTIAL

Let us consider first a standard example in quantum mechanics corresponding to the following hamiltonian :

$$H = -\frac{1}{2}\left(\frac{d}{dx}\right)^2 + \frac{1}{2}x^2 (1 - xg)^2 . \tag{4}$$

The potential has the symmetry :

$$x \leftrightarrow \frac{1}{g} - x ,$$

therefore we can expand perturbation theory around each of the minima of the potential and to all orders in perturbation theory. We find two degenerate ground states corresponding to wave functions concentrated around these minima. In fact it is easy to prove that the ground state is not degenerate. Using familiar WKB or instanton methods one can calculate the energy difference between the two low lying states $\Delta E(g)$:

$$\Delta E(g) = \frac{2}{\sqrt{\pi g^2}} e^{-1/6g^2} (1 + O(g^2)). \tag{5}$$

This form is quite similar to the one predicted for masses in gauge theories. Since gauge theories contain also instantons (finite action solution of euclidean field equations) this has induced some authors into trying similar methods to explain confinement. Unfortunately the evaluation of instanton contribution is plagued by incontrolable I.R. problems and this method has up to now failed. So we shall not discuss it here. Rather we shall have to imagine another method to convince ourselves about the existence of $\Delta E(g)$ and to calculate it. It is clear that if for g small, $\Delta E(g)$ is very small; on the other hand for g large this is no longer the case. Thus if we could solve the theory for g large, we would have at least a qualitative understanding of the model. For g large the hamiltonian becomes :

$$H = -\frac{1}{2}\left(\frac{d}{dx}\right)^2 + \frac{1}{2} gx^4. \tag{5}$$

In some sense, for g large, we would like to be able to neglect the kinetic term with respect to the potential term. Unfortunately this is impossible, since the eigen functions of the potential are δ functions :

$$\frac{1}{2} g\, x^4\, \delta(x-x_\circ) = \frac{1}{2} g\, x_\circ^4\, \delta(x-x_\circ), \tag{6}$$

and a δ function has always an infinite kinetic energy, so that the kinetic term cannot be considered as a small perturbation. Indeed, it tells us that the wave function should be sufficiently regular. By omiting it we have changed the space in which the wave function lives. To circumvent this difficulty, one can discretize space, by replacing the Schrödinger equation by a finite difference equation :

$$-\frac{1}{2a^2}(\Psi_{n+1} + \Psi_{n-1} - 2\Psi_n) + \frac{1}{2} g\, n^4 a^4 \Psi_n = E\Psi_n\ , \tag{7}$$

a being the discrete space interval.

For g large the unique ground state is now :

$$\begin{aligned}\Psi_n &= \delta_{n0} + \frac{1}{ga^6}(\delta_{n,1} + \delta_{n,-1}) + O(\frac{1}{g^2})\ ,\\ E &= \frac{1}{a^2} - \frac{1}{2ga^8} + O(\frac{1}{g^2})\ .\end{aligned} \tag{8}$$

The energy of the lowest antisymmetric state can also be easily calculated in this limit :

$$\begin{aligned}\Psi_n &= \delta_{n,1} - \delta_{n,-1} + O(\frac{1}{g})\ ,\\ E &= \frac{1}{2} g\, a^4 + O(1)\ .\end{aligned} \tag{9}$$

Note that from eq.(7) that we have :

$$E = \frac{1}{a^2}\, \varepsilon\, (ga^6)\ . \tag{10}$$

If we wish to take the small lattice limit, this is equivalent to take the small coupling limit.

Now it is clear that if nothing dramatic happens when we go from large to small coupling, the ground state will never be degenerate. Thus going to a lattice allows us here to understand to spectrum structure and even to calculate energies provided there is no critical point in the coupling constant between the origin and infinity.

THE DISCRETIZED PATH INTEGRAL

It is not always convenient to discuss the spectrum problems in terms of wave functions if we want to apply similar arguments to field theory.

Another standard way for obtaining information about the ground state of a quantum mechanical vacuum is to calculate the density operator exp-βH in the large β limit. For a hamiltonian bounded from below one has :

$$\exp -\beta H = e^{-\beta E_\circ} |0\rangle\langle 0| [1 + O(e^{-\beta(E_1 - E_\circ)})] \; . \tag{11}$$

To obtain the equivalent of the discretized form of the Schrödinger equation, one can use the method which leads formally to the Feynman-Kac formula, the representation of exp(-βH) in terms of path integrals. For a hamiltonian of the form :

$$H = \frac{1}{2} p^2 + V(q) \; , \tag{12}$$

one can use Trotter's formula :

$$e^{-\beta[\frac{1}{2} p^2 + V(q)]} = \lim_{n\to\infty} [e^{-\frac{\beta}{2n} V(q)} e^{-\frac{\beta}{2n} p^2} e^{-\frac{\beta}{2n} V(q)}]^n . \tag{13}$$

In the configuration representation each factor can be calculated :

$$\langle q_1 | e^{-\frac{\beta}{2n} V(q)} e^{-\frac{\beta}{2n} p^2} e^{-\frac{\beta}{2n} V(q)} | q_2 \rangle$$

$$= \frac{1}{\sqrt{2\pi\beta}} e^{-\frac{\beta}{2n} V(q_1)} e^{-\frac{2n}{\beta}(q_1 - q_2)^2} e^{-\frac{\beta}{2n} V(q_2)} . \tag{14}$$

Let us set :

$$a = \beta/n . \tag{15}$$

The matrix element of $e^{-\beta H}$ can then be written :

$$\langle q' | e^{-\beta H} | q'' \rangle = \lim_{n\to\infty} \frac{1}{(2\pi\beta)^{n/2}} e^{-\frac{a}{2}[V(q')+V(q'')]} \int dq_1 \ldots dq_{n-1} \tag{16}$$

$$\exp -\frac{1}{2} \{\frac{1}{a} [(q'-q_1)^2 + (q_1-q_2)^2 + \ldots + (q_i - q_{i+1})^2 + \ldots + (q_{n-1} - q'')^2]$$

$$+ a[V(q_1) + \dots + V(q_i) + \dots + V(q_{n-1})]\}.$$

The formal limit of this expression is just the path integral representation of exp(-βH) and can be expressed in terms of the euclidean action. Conversely this expression can be obtained by discretizing time in the path integral, $q_1 \dots q_{n-1}$ representing successive positions at time a, 2a,... (n-1)a, and a being the discretized (imaginary) time interval.

The important point is that this discretized form is suitable for other calculations than perturbation theory. For example it can be used to generate a strong coupling expansion.

If we expand systematically all contributions coming from the kinetic term, factors of the form

$$e^{-\frac{1}{2a}(q_i-q_{i+1})^2} = \Sigma \frac{(-1)^k}{(2a)^k} \frac{1}{k!} (q_i-q_{i+1})^{2k} , \qquad (17)$$

then the multidimensional integral reduces to a sum of products of simple integrals of the form :

$$\int dq\, e^{-a\, V(q)}\, q^s . \qquad (18)$$

If we take again the example of the asymptotic potential

$$\frac{1}{2} g\, q^4$$

and perform the change of variables

$$agq_i^{\,4} = x_i^{\,4} , \qquad (19)$$

we observe that the expansion parameter is just $(a^3g)^{-1}$. We recognize that it is the same expansion as for the Schrödinger equation, the lattice spacing a here being a time is homogeneous to the square of the space lattice spacing there.

The strategy used here can be repeated in field theory examples. At finite volume it leads to a finite dimensional integral with some invariant measure, the integrand being the exponent of minus a space time discretized form of the euclidean action (i.e. in imaginary time).

A last remark : at β finite, when we approximate for instance the ground state energy by

$$E_o \sim \frac{1}{\beta} \ln \operatorname{tr} e^{-\beta H} ,$$

we make an error of the form

$$e^{-\beta(E_1 - E_0)} .$$

The field theory interpretation of this expression is that E_1-E_0 will be the smallest mass m in the spectrum, β the linear size L of the system, so that finite size effects vanish as a function of mL. The inverse of m is a length ξ called the correlation length.

On the other hand lattice effects are of order a/ξ to some power. Properties of the continuous theory will be seen in the domain :

$$a \ll \xi \ll L. \tag{20}$$

THE NON-LINEAR σ-MODEL

We shall now briefly discuss a field theory example closer to the model we are finally interested in, which shares with it many properties, and which is the source of many of the ideas which have been used to study gauge theories.

The euclidean action of this model is

$$A(\phi) = \frac{1}{2g^2} \int d^2x \, [\partial_\mu \vec{\phi}(x)]^2 , \tag{21}$$

in which $\vec{\phi}(x)$ is a n component vector of unit length :

$$\vec{\phi}^2(x) = 1 , \tag{22}$$

so that the model is O(n) symmetric.

This model can be shown to be renormalizable in two dimensions, and asymptotically free as gauge theories are in four dimensions.

In perturbation theory its spectrum has the following structure the minimum of the action corresponds to $\vec{\phi}(x) = \vec{n}$ in which $\vec{n}$ is a unit vector. Let us call then $\sigma(x)$ the component of $\vec{\phi}(x)$ along $\vec{n}$, which is thus close to $\vec{n}$, and solve the constraint equation by expressing $\sigma(x)$ in terms of the (n-1) other components $\vec{\pi}(x)$:

$$\sigma^2(x) + \vec{\pi}^2(x) = 1 , \tag{23}$$

$$\sigma(x) = \sqrt{1- \vec{\pi}^2(x)} . \tag{24}$$

Then ,

$$\partial_\mu\sigma(x) = - \vec{\pi}(x).\partial_\mu\vec{\pi}(x)(1-\vec{\pi}^2)^{-1/2} , \tag{25}$$

and the action becomes :

$$A(\pi,\sigma) = \frac{1}{2g^2} \int [(\partial_\mu \vec{\pi})^2 + (\partial_\mu \sigma)^2]\, d^2x$$

$$= \frac{1}{2g^2} \int d^2x\, [(\partial_\mu \vec{\pi})^2 + \frac{(\vec{\pi}.\partial_\mu \vec{\pi})}{1 - \vec{\pi}^2}]$$

The part of the action quadratic in π (in a small π expansion) is $\int (\partial_\mu \vec{\pi})^2 d^2x$, which shows that the perturbative spectrum consists in n-1 massless particles which are the goldstone bosons of the spontaneously broken O(n) symmetry. But a general theorem tells us that a continuous symmetry cannot be spontaneously broken in two dimensions (Mermin-Wagner theorem). The true theory must be symmetric with multiplets of n massive degenerate states.

From the point of view of renormalization group a few remarks can be made :

i) Since the theory is asymptotically free, the large distance behavior of the theory is governed by large effective renormalized coupling.

ii) If a mass m is spontaneously generated, it is a R.G. invariant and thus satisfies :

$$\mu \frac{\partial}{\partial \mu} m(\mu,g) \Big|_{g_0,\Lambda \text{ fixed}} = 0 \tag{27}$$

in which μ is a renormalization scale, g_0 the bare coupling constant, Λ a U.V. cut-off and g the renormalized coupling constant. But :

$$m = \mu\, f(g), \tag{28}$$

which transforms (27) into :

$$f(g) + \beta(g) f'(g) = 0\ , \tag{29}$$

$$f(g) \sim \exp -\int g dg'/\beta(g'). \tag{30}$$

If we expand $\beta(g)$ for g small :

$$\beta(g) = \beta_3 g^3 + \beta_5 g^5 + O(g^7) \beta_3 < 0\ , \tag{31}$$

we get :

$$m = \mu\, g^{\beta_5/\beta_3^2}\, e^{1/2\beta_3 g^2} [1 + O(g^2)]\ , \tag{32}$$

which is expontially small for small g.

iii) The bare coupling constant goes to zero at g fixed when Λ becomes large.

Indeed again :

$$\mu \frac{\partial}{\partial \mu}\bigg|_{g_0 \Lambda} g_0(g,\mu) = 0 \tag{33}$$

which translates in :

$$\mu \frac{\partial}{\partial \mu} g_0(g,\Lambda/\mu) + \beta(g) \frac{\partial}{\partial g} g_0(g,\Lambda/\mu) = 0 \,. \tag{34}$$

We see that in this equation the large Λ behaviour is an U.V. behavior so that g_0 goes to zero at g and μ fixed (g_0 is the effective coupling constant at the scale Λ) :

$$g_0^2 \sim \frac{1}{\ln \Lambda/\mu} \quad . \tag{35}$$

In terms of the bare coupling constant the physics is entirely concentrated near $g_0^2 = 0$.

iv) Finally note that for g_0 small, at cut-off fixed, g is small so that the mass m has to behave like :

$$g_0 \sim g \quad \text{for } g \text{ small} \;,$$

$$m = \Lambda g_0^{\beta_5/\beta_3^2} \, e^{1/2\beta_3 g_0^2} \quad [1 + O(g_0^2)] \;. \tag{36}$$

The logical conclusion of this set of arguments is that in terms of the renormalized coupling constant, the spectrum is a strong coupling problem. On the other hand in terms of the bare coupling constant the whole physics comes from the behaviour of the theory at g_0 small but non zero. To study the apparent strong coupling problem we shall introduce a space time lattice in complete analogy with what we have done in Q.M. It will then be easy for example to perform strong coupling expansions. Unfortunately to take the continuum or small lattice spacing limit we shall have to extrapolate back to small bare coupling.

To construct our lattice approximation we remember that, for imaginary times, space and time are equivalent, and that we have noticed in Q.M. that the lattice approximation was formally obtained by replacing the euclidean action by a lattice approximation. Thus on each lattice site i of the 2-dimensional lattice we have a vector $\vec{\phi}_i$ of unit length :

$$\vec{\phi}_i^2 = 1 \;, \tag{37}$$

and we have to calculate a partition function Z :

$$Z = \int \prod_i d\vec{\phi}_i \; \delta(\vec{\phi}_i^2 - 1) \exp - \frac{1}{2g_0^2} \sum_{i,\mu} (\vec{\phi}(i+\hat{n}_\mu) - \vec{\phi}(i))^2 , \tag{38}$$

in which $\hat{n}_\mu$ is the unit vector along the direction μ.

To get a qualitative understanding of the spectrum we can study the behavior of the two-point function at large euclidean distances. In a massive teory it behaves like :

$$G_2^{ab}(\vec{r}) = \langle \phi^a(\vec{r})\ \phi^b(0)\rangle \sim \delta_{ab} e^{-m|\vec{r}|} \quad \text{for } |\vec{r}| \to \infty . \tag{39}$$

It is given by :

$$G_2^{ab}(\vec{r}) = Z^{-1} \int \prod_i d\vec{\phi}_i\ \delta(\vec{\phi}_i^2 - 1)\ \phi^a(0)\ \phi^b(\vec{r}) \exp - A(\vec{\phi}_i). \tag{40}$$

As in QM a strong coupling expansion, which yields an integrand factorized in each variable, can be performed. By O(n) invariance the integral vanishes unless a = b and the first non vanishing contribution corresponds to a term in the expansion of exp $- A(\phi_i)$ which connects the origin to $\vec{r}$ such that all integration variables appear always twice. Consequently

$$G_2^{ab}(r) \sim \delta_{ab}(g_0^2)^{-|\vec{r}|} \sim e^{-|\vec{r}|\ln g_0^2}\ \delta_{ab} , \qquad g_0^2 \to \infty . \tag{40}$$

This is the propagator of n massive bosons of mass ln g_0^2 in unit of inverse lattice spacing. Of course this mass is therefore very large and we are very far from the continuum limit.

Notice that in a lattice theory in which all coupling constants g_i are dimensionless (the lattice spacing giving the dimension), a calculation of masses will yield a result of the form :

$$m = 1/a\ f(g_1, g_2, \ldots g_k) . \tag{41}$$

Therefore for generic values of the coupling constants m is very large and no particle really propagates. Looking for the continuum limit $a \to 0$, is then equivalent to look for a set of coupling constants for which at fixed lattice spacing (the lattice spacing being taken as unit scale), the mass vanishes or becomes very small.

In the non-linear σ-model, in order for the strong coupling result to be qualitatively the same as in the continuum limit, we must be able to extrapolate smoothly from large bare coupling to small lattice coupling, for which the mass becomes small in terms of the inverse lattice spacing. As such it is a numerical question based on series analysis.

Strong coupling expansion (also called high temperature series expansion by analogy with classical statistical mechanics in which g_0^2 would be the temperature) is not the only method available to study lattice model. Among the more analytic methods, also available for gauge theories, we can mention mean field theory. The idea is to construct an approximation which, as in strong coupling expansion, involves only calculations for decoupled sites. Instead of just expanding the kinetic part, method which is only suited to strong coupling, one tries to replace it at any coupling by an effective one-body potential which is optimally close to it.

Mean field theory for lattice models

Technically one proceeds in the following way : let us write a set of identies :

$$1 = \int \prod_i d\vec{S}_i \; \delta(\vec{S}_i - \vec{\phi}_i). \tag{43}$$

The δ functions are now replaced by an integral representation

$$1 = \int \prod_i d\vec{S}_i \; d\vec{H}_i \; e^{i \sum_i \vec{H}_i (\vec{S}_i - \vec{\phi}_i)}. \tag{44}$$

We then introduce this identity in the partition function, replacing in the integrand $\vec{\phi}_i$ by $\vec{S}_i$. Since we have the corresponding δ functions, except when we encounter $(\vec{\phi})^2$ which we replace by its value 1 :

$$Z = \int \prod_i d\vec{S}_i \; d\vec{H}_i \; \exp \{-\frac{1}{2g_0{}^2} \sum_{i,\mu} [2-2\, \vec{S}(i).\vec{S}(i+\hat{n}_\mu)] + i\sum_i \vec{H}_i \vec{S}_i\} \int \prod_i d\vec{\phi}_i \; \delta(\vec{\phi}_i^2 - 1) \exp -i \sum_i \vec{H}_i \vec{\phi}_i \; . \tag{45}$$

The integrals on $\vec{\phi}_i$ now decouple site by site and we have to calculate the Fourier transform of the measure of integration

$$\int d\vec{\phi}_i \; \delta(\vec{\phi}^2 - 1) \exp - i\vec{H}\vec{\phi} = \exp I(H). \tag{46}$$

The partition function becomes :

$$Z = \int \prod_i d\vec{S}_i \; d\vec{H}_i \; \exp \{-\frac{1}{2g_0{}^2} \sum_{i,\mu} [2-2\, \vec{S}(i).\vec{S}(i+\hat{n}_\mu)] + \sum_i (i\vec{H}_i.\vec{S}_i + I(H)). \tag{47}$$

The integral over the variables $\vec{S}(i)$ is now a pure gaussian integral which here can be performed exactly. Since the corresponding integral cannot be performed in the case of gauge theories we shall not perform it. Instead we shall look for a saddle point, in order to evaluate the integral by steepest descent. The advantage of having replaced $\vec{\phi}$ by $\vec{S}$ is that $\vec{S}$ is an unconstrained vector whose length can be arbitrarily small, in contrast with $\vec{\phi}$. The solution of the saddle point equation for $\vec{S}$ will be the expectation value of $\vec{\phi}$ in this approximation. Now it is clear that in general the expectation value of a unit vector is not a unit vector.

The calculation of the original integral by steepest descent gave

$\vec{\phi}$ a maximal expectation value, and this could only make sense a priori for very small coupling. In the mean field approximation the saddle point value for $\vec{S}$ can be arbitrary and therefore one can hope that the approximation makes sense for arbitrary couplings.

We are looking for translation invariant solutions, i.e. solutions for $\vec{H}$ and $\vec{S}$ independent of the lattice site :

$$\frac{2}{g_0{}^2} \vec{S} + i \vec{H} = 0 \quad , \tag{48}$$

$$i \vec{S} + \frac{\partial I}{\partial \vec{H}} = 0 \quad . \tag{49}$$

It is possible to eliminate the mean field $i\vec{H}$ between the two equations :

$$\vec{S} - i \frac{\partial I}{\partial \vec{H}} \left(i \frac{2}{g_0^2} \vec{S}\right) = 0 \quad . \tag{50}$$

For g_0^2 large this equation has only one solution $\vec{S} = 0$, the field has no expectation value, and we are in the symmetric phase. At a critical coupling g_{0c}^2 defined by

$$1 + 2\, I''(0)/g_{0c}^2 = 0 \quad , \tag{51}$$

a new non trivial solution appears so that $\vec{\phi}$ has now a non trivial expectation value, and the symmetry is spontaneously broken. The expectation value grows then until it reaches 1 for very small coupling. Therefore meanfield theory allows us to describe both phases of the theory. In this model for any dimension langer than two, meanfield theory gives a correct qualitative description of the phase diagram. For large dimensions d, the approximation becomes even exact. But for dimension two g_{0c}^2 in reality vanishes identically, so that the spontaneously broken phase never exists. From the equation one sees that the expectation value of $\vec{\phi}$ vanishes continuously when g_0^2 goes to g_{0c}^2. This is called a continuous phase transition, and a closer study shows that in such a situation, when g_{0c}^2 is approached from above, the mass of the corresponding states goes to zero with a power law :

$$m \sim (g_0^2 - g_{0c}^2)^{1/2} \quad . \tag{52}$$

Therefore coupling constants close to g_{0c}^2 correspond to the continuous field theory, and a study in the mean field approximation of the neighborhood of g_{0c}^2, allows us to examine some properties of theory which are valid even in dimension two (when $g_0^2 > g_{0c}^2$), and to construct an effective field theory, the O(n) invariant theory ϕ^4. We shall see that this unfortunately does not happen in gauge theories.

A last remark

We can write lattice approximations for a continuous theory in many different ways. The claim, called universality, and substantiated by a more detailed analysis, is that provided we respect the symmetry of the model, and the locality (we connect for example only sites at a finite distance) for large classes we get the same answer.

LATTICE GAUGE THEORIES

We shall mostly concentrate on pure lattice gauge theories (without fermions). Physically this means that we cannot study many properties of a realistic theory, but we can still try to answer a few questions: does the theory generate confinement, i.e. a force increasing at large distances, so that heavy quarks in the fundamental representation cannot be separated ? Are there massless vector parcticles in the theory ? Otherwise are there massive group singlet bound states in the spectrum (gluonium) ?

We shall consider the following theory :

$$A(\vec{A}_\mu) = \frac{1}{4g_0^2} \int d^4x \, \vec{F}_{\mu\nu}(x) \, \vec{F}_{\mu\nu}(x) \ , \tag{53}$$

in which A is the euclidean action, $\vec{A}_\mu$ a gauge vector field belonging to the adjoint representation of a simple group G and $\vec{F}_{\mu\nu}$ is the field strength (We shall assume that G is a unitary group)

$$\vec{F}_{\mu\nu}(x) = \partial_\mu \vec{A}_\nu - \partial_\nu \vec{A}_\mu + \vec{A}_\mu \times \vec{A}_\nu \ . \tag{54}$$

Infinitesimal gauge transformations of the form

$$\delta \vec{A}_\mu = \partial_\mu \vec{\Lambda} + \vec{A}_\mu \times \vec{\Lambda} \tag{55}$$

leave the action invariant.

We want to construct a lattice approximation of the theory. The main concepts we want to preserve are gauge invariance and locality.

The presence of a gradient term in the transformation law shows that, unlike the case of global symmetries, the transformation law itself will be affected by the discretization.

In QED if we set a charged field $\phi(x)$ at point x and $\phi^*(y)$ at point y and want to build a gauge invariant combination we have to introduce phase factors :

$$\phi(x) \, e^{ie \int_x^y A_\mu(s) \, ds_\mu} \, \phi^*(y) \ . \tag{56}$$

In a gauge transformation the various factors become :

$$\begin{aligned} \phi(x) &\to e^{ie\Lambda(x)}\, \phi(x) , \\ \phi^*(y) &\to e^{-ie\Lambda(y)} \phi^*(y) , \end{aligned} \tag{57}$$

and

$$\exp ie \int_x^y A_\mu(S)\, ds_\mu \to e^{ie\,[\Lambda(y) - \Lambda(x)]} \exp ie \int_x^y A_\mu(s) ds_\mu \tag{58}$$

so that the product is invariant.

Therefore the natural concept on a lattice on which points are separated is no longer gauge vector fields associated with Lie algebras, but rather group elements associated to links. Here the phase factor was belonging to U(1) and associated to a string going from x to y.

Thus on the lattice let us associate to each link going from site $\vec{i}$ to $\vec{i} + \hat{n}_\mu$ a matrix $U(\vec{i},\hat{n}_\mu)$ belonging to some representation of the group G. A gauge transformation will then read :

$$U(\vec{i},\hat{n}_\mu) \to S^{-1}(\vec{i})\ U\ S(\vec{i}+\hat{n}_\mu) , \tag{59}$$

in which $S(\vec{i})$ is an arbitrary group element associated to site $\vec{i}$. Note that it is natural and consistent to take :

$$U(\vec{i},\hat{n}_\mu) = U^{-1}(\vec{i}+\hat{n}_\mu,- \hat{n}_\mu).$$

To understand that we have not lost the connection with our initial problem we have to argue in the following way : we are going to construct an action which for small coupling favors group elements which are, up to gauge transformations, close to the identity in the same way as our original action was minimal for pure gauges, i.e. gauge transform of the trivial gauge field $\vec{A}_\mu = 0$.

Choosing the special gauge in which all U are close to the identity we can write

$$U(\vec{i},\hat{n}_\mu) = 1 + ia\ \vec{A}_\mu(\vec{x})\ \vec{t}\ \hat{n}_\mu + 0\,(a^2) , \tag{60}$$

in which $\vec{t}$ are the generators of the Lie algebra, $\vec{x}$ is $(a\ \vec{i})$. To the different group elements starting from a given site and going in the different directions μ we have associated the components of a vector field (it is just a parametrization). Then a gauge transformation on U, in which the group elements at site $\vec{i}$ and $\vec{i} + \hat{n}_\mu$ are close, becomes

$$1+ia\ \vec{A}_\mu(\vec{x})\vec{t}\vec{n}_\mu \to S^+(x)\ [1+ia\ \vec{A}_\mu(x)\vec{t}\hat{n}_\mu]\ [S(x)+a\hat{n}_\mu \partial_\mu\ S(x)] \tag{61}$$

At leading order in the lattice spacing a we get :

$$\vec{A}_\mu \vec{t} \to S^+(x)\ \partial\ S(x) + S^+(x)\ \vec{A}_\mu(x)\ \vec{t}\ S(x), \tag{62}$$

which is the usual gauge transformation.

The lattice action

We want now to construct a gauge invariant action, function of the group elements U. If U_{ij} links site i to site j, to construct a quantity invariant under gauge transformations involving the point j, we have to multiply it by a matrix U_{jk}. We use here the more convenient notation :

$$U_{\vec{I},\vec{I}+\hat{n}_\mu} = U(\vec{I},\hat{n}_\mu). \tag{63}$$

The same argument can be repeated with site i and k. Thus the only local gauge invariant quantities are associated to closed loops on the lattice and of the form

$$\mathrm{tr}\ [U_{i_1 i_2}\ U_{i_2 i_3} \cdots U_{i_n i_1}]\ . \tag{64}$$

On a hypercubic lattice the simplest possible gauge invariant interaction is then :

$$\mathrm{tr}\ U_{12}U_{23}U_{34}U_{41} \tag{65}$$

and associated to the square 1234 called plaquette. The partition function can then be chosen to be given by :

$$Z = \int_{\text{all links } ij} \pi dU_{ij}\ \exp 1/g_0^2 \sum_{\text{all plaquettes } ijkl} (\mathrm{tr}\ U_{ij}U_{jk}U_{kl}U_{li} + \text{c.c.}). \tag{66}$$

The measure of integration on the U's is the group measure. Since we are considering only compact groups the integral is perfectly finite at least as long as the number of sites is finite. In addition it is possible to normalize the volume of the group to 1.

In constrast to gauge theories in the continuum it is not necessary to fix the gauge on the lattice to be able to calculate.

Let us now show that the minimum of the action corresponds to U's gauge transform of the identity. Let us start from a first plaquette 1234. Without loss of generality we can set :

$$U_{12} = S_1^{-1}\ S_2 \tag{67}$$

by letting S_1 arbitrary and calculating S_2 from U_{12} and S_1. Then we can also set :

$$U_{23} = S_2^{-1}\, S_3 \;,$$
$$U_{34} = S_3^{-1}\, S_4 \;. \tag{68}$$

These relations define first S_3, then S_4. The minimum of the action is obtained when the real part of all traces are maximum, i.e. when the products of the group elements on a plaquette are 1. (The trace of unitary matrix U is maximum when all its eigenvalues are 1). In particular :

$$U_{12}U_{23}U_{34}U_{41} = 1 \;, \tag{69}$$

which yields :

$$U_{41} = S_4^{-1}\, S_1 \;. \tag{70}$$

If we take now an adjacent plaquette the argument can be repeated for all links but one which has already been fixed. In this way we can show that the minimum of the action is a pure gauge. Thus when "the coupling constant" g_0^2 becomes very small, all group elements are constrained to stay up a to gauge transformation, close to the identity. To make a small coupling expansion we have thus to fix the gauge, to separate the gauge degrees of freedom and integrate over them exactly. If we expand the U's around the identity, introducing a gauge vector field, as we have done before, and collect the leading terms in the lattice spacing, it is not difficult to verify that up numerical factors

$$\sum_{\text{Plaquettes } ijkl} [\operatorname{tr} U_{ij}U_{jk}U_{kl}U_{li} - 2 \operatorname{tr} 1 + \text{c.c.}] \sim - \int (\vec{F}_{\mu\nu})^2 d^4x \;. \tag{71}$$

Wilson's loop and a criterion of confinement

It has been proposed as a criterion of confinement to consider the behavior of expectation values of a loop when the loop becomes very large. Intuitively the behavior of a rectangular loop of time size T and space size R is related with the probability of separating by a distance R for a time T a pair of very heavy particles which initially formed a group singlet. For R and T large this probability is proportional to :

$$W(c) = \Big\langle \prod_{\text{loop } C(R,T)} U \Big\rangle \sim \exp - T\, E(R) \;, \tag{72}$$

where E(R) is the cost in energy due to the separation.

Let us calculate the average of Wilson's loop in a continuous abelian gauge theory with a short distance cut-off a. Since it is a free field theory it is easy to get the result :

$$W(C) = < \exp i \int_C ds_\mu \, A_\mu(s) > . \tag{73}$$

The effect of the Wilson's loop term is to add to the action an external current term

$$\int J_\mu(x) \, A_\mu(x) \, dx$$

with :

$$J_\mu(x) = i \int_C ds_\mu \, \delta(s - x) .$$

Then since we are in a free field theory, the field $A_\mu(x)$ can be replaced by the solution of the classical equation of motion. The result is then :

$$W(C) = \exp \left[- \frac{g_0^2}{2} \int_{CxC} ds_\mu \, ds'_\mu \, \Delta_{\mu\nu} \, (s - s')\right] , \tag{74}$$

in which $\Delta_{\mu\nu}(s)$ is the gauge field propagator in an arbitrary gauge. The gauge is irrelevant since Wilson's loop is gauge invariant, and therefore the result is gauge independent.

Since the propagator $\Delta_{\mu\nu}$ is singular at $s=s'$, the dominant contribution comes from s close to s'. The integral is not infinite since we have introduced a short distance cut-off. Keeping $|s-s'|$ fixed we can integrate along the curve and obtain a factor proportional to the perimeter of the loop. The integration in the direction transverse to the loop gives just a factor $1/a$

$$< W(C) > \sim \exp - g_0^2 \; P(C)/a . \tag{75}$$

In particular if we take a rectangle RxT, we get :

$$< W(C) > \sim \exp - g_0^2 \; (R+T)/a . \tag{76}$$

The coefficient of T is just a constant in R. It is a contribution to the charged state mass in the static limit. In a non abelian gauge theory, the result would be the same in the weak coupling limit since the interactions coming from the non abelian character vanish. This result reflects the fact that the potential between charged particles is $1/R$ and vanishes at large distances. We shall see also that it can reflect a phenomenon of charge screening as opposed to charge confinenement.

We shall now use our lattice approximation to perform a strong coupling calculation.

Strong coupling expansion for Wilson's loop

We shall assume here that the group we consider has a non trivial center for example SO(2) rather than SO(3)... This means that the heavy quarks creating the loop cannot be simply screened by the gauge field, in which case we would get again a perimeter law. For example we can assume that the gauge field on the lattice belongs to the fundamental representation of SU(n).

To calculate W(C) we shall expand the integrand in powers of $1/g_0^2$. We can perform the integrations which are just factorized group integrations. The only way to get a non zero result when we integrate on the gauge fields belonging to the loop is if they are canceled by plaquette terms. But now when the integration over the corresponding link is performed, we remain with a modified loop in which plaquettes have been added, so we can repeat the argument until the loop has been shrinked to a point. In this way we get a contribution to W(C) proportional to $(g_0^2)^{-A}$, in which A is the area of the surface bounded by the loop we have generated by adding plaquettes. The largest contribution corresponds to minimal area surfaces. For a rectangular loop RxT we just get :

$$\exp - RT \ln g_0^2 \quad . \tag{77}$$

If instead the center is trivial, then integrals of an odd power of a link variable can be now zero, so that it is possible to form a tube along the curve and one gets a perimeter law. For example if the group is SO(3), in the decomposition of a product of two spin 1 representations, one finds again a spin 1 which can be coupled to a third spin to form a scalar.

The interpretation of expression (76) is that we have found a linear rising potential at large distance between the heavy quarks. The coefficient in front the area is called the string tension σ

$$\begin{aligned} &\sigma(g_0) \sim \ln g_0^2 \\ &\qquad g_0 \to \infty \ . \end{aligned} \tag{78}$$

It has the dimension of a mass squared so that if no phase transition occurs when g_0 varies from zero to infinity, for g_0 small, it should behave according to the predictions of R.G. arguments as :

$$\sigma(g_0) \sim (g_0^2)^{-\beta_5/\beta_3^2} \exp - \frac{1}{\beta_3 g_0^2} (1 + O(g_0^2)) \ . \tag{79}$$

A physical quantity relevant to the continuous limit can then be obtained by dividing $\sqrt{\sigma}$ by its asymptotic behavior. Let us define Λ_L as :

$$\Lambda_L = a^{-1} (\beta_3 \, g_0^2)^{-\beta_5/2\beta_3^2} \exp - \frac{1}{2\beta_3 g_0^2} \ , \tag{80}$$

then $\Lambda_L/\sqrt{\sigma}$ has a continuous limit.

It is possible to systematically expand σ in powers of $1/g_0^2$. The possibility of verifying that confinement is realized in the continuous limit, depends on the possibility of analytically continuing the strong coupling expansion up to the origin. Unfortunately theoretical arguments lead to believe that, independently of the group, the strong tension is affected by a singularity associated to the roughening transition, which does not affect the bulk properties. At strong coupling the contributions to the string tension come from smooth surfaces. When g_0^2 decreases, one passes through a critical point g_{0R}^2, after which the relevant surfaces become rough. At the singular coupling g_{0R}^2, the string tension does not vanish but has a weak singularity. Still at this point the strong coupling expansion fails. Therefore it is impossible to extrapolate to arbitrarily small coupling. The usefulness of the strong coupling expansion depends then on the position of the roughening transition with respect to the onset of the weak coupling behavior. Notice that numerically in the neighborhood of the roughening transition rotational symmetry is approximatively restored (at least at large enough distance).

One can also calculate other quantities which are associated to bulk properties, and are therefore not affected by roughening singularities, such as the free energy (the connected vacuum-vacuum amplitude) or the plaquette-plaquette propagator. Even here extrapolations are not easy because many of these quantities are affected by a sharp transition between strong and weak coupling behavior. This is confirmed by results coming from Monte-Carlo experiments and is interpreted as the existence of complex singularities close to the real axis. From the numerical point of view it seems the plaquette-plaquette is the most promising case for strong coupling expansion.

Mean field theory

The calculations are completely similar to those we performed when we applied M.F.T. to the non-linear σ-model. We shall introduce in the partition function a δ function under the form :

$$1 = \int \prod_{l} ds_{l} \quad \delta(U_{l} - S_{l}) \,, \tag{81}$$

in which the S_l are general complex matrices. As before, after having replaced the U_l by S_l in the partition function we replace the δ function by an integral representation, introducing the new complex matrix H_l. The introduction of the unconstrained variables S_l obeys the same logic as before, the average value of a unitary matrix is not unitary in general. On the other hand the H_l will yield at leading order the mean field which approximates the plaquette terms which link the sites together. We then obtain the expression :

$$Z = \int \prod_l dS_l \, dH_l \, e^{\frac{1}{g_0^2} \sum_{\text{plaquettes}} [\mathrm{tr}\, S_{ph.} + c.c.] + \sum_{\text{links}} (\mathrm{tr}\, S_l H_l^+ + c.c.)}$$

$$\int \prod_l dU_l \, e^{-\sum_{\text{links}} \mathrm{tr}\, (H_l U_l^+ + U_l H_l^+)} . \qquad (82)$$

The integral over the group decouples and becomes a product of integrals for each link which itself yields in the unitary case an invariant function of $H_l H_l^+$. One can then look for saddle points in the variables H and S. Since H and S are general complex matrices, one may find a large number of saddle points. Both for simplicity and symmetry reasons one looks for solutions multiple of the identity (up to a gauge equivalence) independent of the link :

$$S_l = s \mathbb{1} , \qquad H_l = h \mathbb{1} . \qquad (83)$$

The action then (for h and s real) reduces to :

$$A(h,s) = V \, [1/g_0^2 \; s^4 + sh + I(h)], \qquad (84)$$

in which I(h) is the logarithm of the integral over the group when H is proportional to the identity :

$$I(h) = \ln \int dU \exp h \, [\mathrm{tr}\, U + \mathrm{tr}\, U^+] . \qquad (85)$$

V is the volume of the system, and some inessential numerical factors have been omitted. The saddle point equations are :

$$\frac{4}{g_0^3} s^3 + h = 0 , \qquad s + I'(h) = 0 . \qquad (86)$$

Eliminating s yields :

$$h = \frac{4}{g_0^3} I'(h)^3 . \qquad (87)$$

For h small I'(h) is linear in h. The essential difference with the non-linear σ-model case is that the right hand side of the equation is now proportional to h^3 for h small instead of h. Thus the equation has never a non trivial solution arbitrarily close to zero. For g_0 large the only solution is the trivial one, which according to the strong coupling analysis corresponds to a confined phase i.e. Wilson's loop has an area law. For a critical value g_{oc}^2, h jumps from zero to a finite value, sign of a first order phase transition.

Unfortunately at a first order phase transition, the correlation length in the system (the inverse mass) stays finite so that in the neighborhood of g_{oc}^2 there is no continuous field theory, in contrast to the nonlinear σ-model. This transition is deconfining. Since s is the leading order expectation value of U the average of Wilson's loop W(C) is just :

$$W(C) \sim \mathrm{tr} \prod_C s \mathbb{1} \sim s^{P(C)} . \tag{88}$$

Below g_{oc}^2, Wilson's loop has a perimeter law. Obviously this is not exactly the situation we expect in four dimensions, since we would like to find $g_{oc}^2 = 0$. Mean field theory is supposed to be good in high dimensions. The dimension four is, in the language of phase transitions, the lower critical dimension (as dimension 2 is for global continuous symmetries) and mean field theory cannot give a complete description of the physical picture there. On the other hand a more systematic analysis in particular in the case of more general interactions seems to indicate that useful quantitative results can be extracted from mean field theory calculations, in particular when used in connection with results coming from other methods like Monte Carlo calculations.

Monte Carlo method

A thorough discussion of Monte Carlo methods can be found in Parisi's lectures. We shall describe here only a few properties of these methods to contrast them with those of more analytic methods like weak or strong coupling expansion or mean field theory. Due to the limitations of the analytic methods when applied to four dimensional gauge theories, one tries to integrate explicity on the lattice degrees of freedom. These are really brute force calculations. A first obvious consequence is that one is restricted to finite and even small lattices (in four dimensions presently $\sim 10^4$). This introduces in addition to lattice effects finite size effects. Even so the number of degrees of freedom is much too big for normal integration methods, so that one uses statistical methods which sample the configuration space.

A few qualitative differences between series method can be observed :

- Series methods work with infinite volume systems while M.C. type methods are limited to finite systems.
- The accuracy of the results varies much more rapidly for series methods than for M.C. type methods. As a consequence M.C. methods are very well suited to explorations of the phase diagram, and the discovery of critical coupling constants and qualitative properties.
- The accuracy as a function of computing time can be roughly described in the following way : the number of terms in a series increases at most as a logarithm of the computing time, while the accuracy of results in generaly improves like a power of the number

of terms. If we call $\varepsilon(T)$ the error in the estimates we have typically :

$$\varepsilon(T) \sim \frac{1}{(\ln T)^{\alpha}} \quad . \tag{89}$$

For M.C. methods the error is statistical and thus in general :

$$\varepsilon(T) \sim \frac{1}{\sqrt{T}} \; . \tag{90}$$

Actually there is here some subtlety here. To extract quantitative results of physical interest, one has to work close to the critical coupling constant, where correlations both in space and Monte Carlo time become large, and the $1/\sqrt{T}$ law is not necessarily true if the configurations are correlated in time. This is the phenomenon of critical slowing down. If the correlations decrease not fast enough the power will be modified and could even become a logarithm (stricly speaking this is only true for the infinite system but will be more and more true when the size of the system increases).

Finally let us note that series once derived are there for ever and can be reanalysed if more ingenious methods of analysis are found, but adding terms to them demands a fast increasing amount of ingeniosity.

For these various reasons M.C. methods have been extremely useful to study the qualitative properties of many lattice theories. For what concerns quantitative results, series methods, even with their strong limitations, should not be abandonned. Best results may come from combinations of various approaches.

Improved lattice action

Since in Monte Carlo simulations one works with finite lattices, the correlation length ξ or inverse mass is bounded by the lattice size L and this introduces finite size effects function of the ratio ξ/L. On the other hand it is then impossible to get rid of lattice effects which at lowest order in weak coupling are in general of order $(a/\xi)^2$. Since many different lattice actions are supposed from universality arguments to give the same continuous limit, a natural question arises : is it possible to choose a special action which decreases the lattice effects ? The answer, at least in perturbation theory is yes. If we return to our discretized Schrödinger equation, we see that instead of using the discrete form of the second derivative :

$$\frac{1}{a^2}\,[\psi(x+a) + \psi(x-a) - 2\psi(x)] = \psi''(x) + \frac{a^2}{12}\,\psi^{(4)}(x) + O(a^4), \tag{91}$$

we could have used, at the price of making the discrete Schrödinger equation more complicated:

$$\frac{1}{3a^2}[-\psi(x+2a)-\psi(x-2a)+3\psi(x+a)+3\psi(x-a)-4\psi(x)]=\psi''(x)+O(a^4) \quad (92)$$

A more sophisticated analysis in the same spirit can be made for lattice gauge theories.

One can add to the one plaquette action, two plaquette-terms + etc ... with suitable coefficients to reduce in the same way the lattice effects. To calculate numerically these coefficients, one should impose that the Green's functions satisfy more accurately the bare renormalization group equations. This is impossible in practice because it involves too many calculations and the best one can do at present is to calculate these coefficients in perturbation theory at the tree level and one-loop order. Other practical drawbacks of the method are that the effective size of the lattice is decreased because the interaction terms have a bigger linear size, and the computing time is increased because there are more interaction terms. The effectiveness of the improvement becomes then a practical question which can only be answered by direct numerical experiments. First results seem to be encouraging.

MEAN FIELD THEORY : GENERAL FORMALISM

We shall explain here how to obtain a mean field theory result and how to expand systematically around it for the most general boson lattice theory. A brief account of the method has been included in a review article[13]. Let us first expose the method and make then various comments which hopefully clarify its significance.

We shall assume that we have some lattice action A function of some lattice variably S_i in which i is the lattice site. It is necessary for what follows to consider differents powers of the same lattice variable S_i as independent variables so that we shall write A as $A[S_i,S_i^2,\ldots S_i^k]$. We shall introduce two new sets of lattice variables $\sigma_i^{(\ell)}$ and $H_i^{(\ell)}$, the $H_i^{(k)}$'s being Lagrange multipliers, to express the conditions :

$$\sigma_i^{(\ell)} = S_i^\ell \;, \qquad \ell = 1, \ldots, k \,. \quad (a1)$$

The partition function, or vacuum amplitude, can then be formally written :

$$Z = \int \prod_i \rho(S_i)dS_i \, e^{-A(S_i,\ldots,S_i^k)} \quad (a2)$$

$$\sim \int \prod_i \rho(S_i)dS_i \prod_{i,\ell} d\sigma_1^{(\ell)} dH_i^{(\ell)} \exp - A(\sigma_i^{(1)},\sigma_i^{(2)},\ldots,\sigma_i^{(k)}) + \sum_{i,\ell} H_i^{(\ell)}[\ {}_i^{(\ell)} - S_i^\ell]. \quad (a3)$$

Note that the action A is a linear function in all variables $\sigma_i^{(\ell)}$. Now the integrations over the variables S_i are decoupled. Let us introduce the function $F_\circ$:

$$e^{-F_\circ(H^{(1)},\ldots,H^{(k)})} = \int dS\, \rho(S)\, e^{-\sum_{\ell=1}^{k} S^\ell H^{(\ell)}} . \tag{a4}$$

We assume here that the integration measure $\rho(S)$ is either of compact support or decreasing fast enough so that the integral exists. The partition function can then be rewritten :

$$Z = \int \prod_{i,\ell} d\sigma_i^{(\ell)} dH_i^{(\ell)} \exp\{-A[\sigma_i^{(1)},\ldots,\sigma_i^{(k)}] + \sum_{i,\ell} H_i^{(\ell)}\sigma_i^{(\ell)} - \sum_i F_\circ(H_i^{(1)},\ldots,H_i^{(k)})\}. \tag{a5}$$

Now the mean field approximation is just the leading order in the evaluation of this expression by steepest descent. The saddle point equations are :

$$\frac{\partial A}{\partial \sigma_i^{(\ell)}} = H_i^{(\ell)} ,$$

$$\frac{\partial F_\circ}{\partial H_i^{(\ell)}} = \sigma_i^{(\ell)} . \tag{a6}$$

It is clear from these equations that if A does not depend on a given $\sigma^{(\ell)}$, the corresponding field $H^{(\ell)}$ vanishes, and both disappear from the equations. Therefore one finds the same result as if we had omitted them right at the beginning. Notice also that if A is a quadratic function of the variables $\sigma^{(\ell)}$, the $\sigma^{(\ell)}$ integration can be performed explicitly and only integrals on the corresponding $H^{(\ell)}$ remain. This is in particular the case in many models with global symmetry and two body interactions of the form

$$\sum_{\text{nearest neighbours }(i,j)} S_i \cdot S_j .$$

Let us now explain why we have introduced two sets of variables $\sigma^{(\ell)}$ and $H^{(\ell)}$ instead of one as in the examples given in the lectures.

In the mean field approximation the variables S_i are replaced by some average value so that the average of a product is replaced by the product of averages :

$$\langle S_i,\ldots,S_{in}\rangle \to \langle S_{i_1}\rangle \langle S_{i_2}\rangle \ldots \langle S_{i_n}\rangle .$$

If it is conceivable that in some limit this can be a good approximation if the different variables are independent, this cannot be possibly true if we have a power of the same variable. For example in the Ising model

$$S_i^2 = 1 \neq (\langle S_i \rangle)^2 .$$

A variable is always correlated with itself, and the new Lagrange parameters allow to take into account these correlations. To give more specific examples, imagine that the action A is only function of S_i^2, then it is obvious that we should take S_i^2 as a variable and perform a mean field approximation on it. The procedure explained above does it automatically since only the parameters $\sigma^{(2)}$ and $H^{(2)}$ will be introduced, which are coupled to S^2. This procedure also solves a very simple problem. Terms in the action which, in the field theory language, are pure potential i.e. only sum of functions depending on the variable at one site as :

$$\Sigma \frac{a}{2} S_i^2 + \frac{b}{4!} S_i^4 \ldots, \tag{a7}$$

could be considered as well as part of the measure $\rho(S_i)$ or as part of the action. From this it would seem that the results of mean field theory depend on the formulation. The introduction of additional variables insures that this is not the case.

Let us for example make the transformation :

$$\begin{aligned} &A(S_i, S_i^2, \ldots S_i^k) \to A(S_i, \ldots S_i^k) + \frac{a}{2} \sum_i S_i^2 \quad , \\ &\rho(S) \to e^{a/2\, S^2} \rho(S) \ . \end{aligned} \tag{a8}$$

Obviously the lattice theory is independent of a. Consider now the modifications so induced in the equations (a4), (a5) and (a6)

$$e^{-\bar{F}_\circ(H^{(1)}, \ldots H^{(k)})} = \int e^{\frac{a}{2} S^2} \rho(S) dS \, e^{-\sum_{\ell=1}^{k} S^\ell H^{(\ell)}} \ . \tag{a9}$$

Comparing equations (a4) and (a9) it follows :

$$\bar{F}_\circ(H^{(1)}, \ldots H^{(k)}) = F_\circ(H^{(1)}, H^{(2)} - \frac{a}{2}, H^{(3)}, \ldots, H^{(k)}), \tag{a10}$$

and the saddle point equations become :

$$\begin{aligned} &\frac{\partial A}{\partial \sigma_i^{(\ell)}} + \frac{a}{2} \delta_{\ell 2} = H_i^{(\ell)} \quad , \\ &\frac{\partial F_\circ}{\partial H_i^{(\ell)}} (H^{(1)}, H^{(2)} - \frac{a}{2}, H^{(3)}, \ldots H^{(k)}) = \sigma_i^{(\ell)} \ . \end{aligned} \tag{a11}$$

We see that the equations are identical up to the change

$$H_i^{(2)} - \frac{a}{2} \to H_i^{(2)}. \tag{a12}$$

Therefore mean field theory results are insensitive to such changes as it should be.

It is now possible to correct the mean field approximation, by systematically expanding the effective action in (a3) around the saddle points (a6) and integrating term by term as one does in usual perturbation theory. We shall call this expansion the mean field expansion.

The natural question is then what is the expansion parameter in this steepest descent calculation of the partition function (a5). If we introduce a parameter L in front of the effective action (a5), then the mean field expansion is just a formal expansion in powers of 1/L.

$$Z_L = \int \prod_{i,\ell} d\sigma_i^{(\ell)} dH_i^{(\ell)} \exp L \{- A(\sigma_i^{(1)},\sigma_i^{(2)},\ldots,\sigma_i^{(k)})$$

$$+ \sum_{i,\ell} H_i^{(\ell)}\sigma_i^{(\ell)} - \sum_i F_\circ(H_i^{(1)},\ldots,H_i^{(k)})\} . \tag{a13}$$

The interpretation of the factor L in front of $F_\circ$ is, for L integer, that we have replaced each variable S by a sum of L independent variables having the same distribution $\rho(S)$.

$$S_i = \sum_{\alpha=1}^{L} S_i^\alpha \quad , \tag{a14}$$

$$\int \rho(S_i)dS_i \to \int \prod_\alpha \rho(S_i^\alpha)\, dS_i^\alpha \; . \tag{a15}$$

Let us introduce the distribution $\rho_L(S)$ of S defined by :

$$e^{- LF_\circ(H^{(1)},\ldots,H^{(k)})} = \int \rho_L(S)dS\, e^{-\sum_{\ell=1}^{k} H^{(\ell)}S^\ell} \; . \tag{a16}$$

For L large, $\rho_L(S)$ becomes a gaussian distribution of dispersion $\sqrt{L}$. Using definition (a16) we can rewrite Z_L :

$$Z_L = \int \prod_i \rho_L(S_i)dS_i \exp - LA(\frac{S_i}{L},\frac{S_i^2}{L},\ldots,\frac{S_i^k}{L}). \tag{a17}$$

We see that each interaction term in A is multiplied by a power of L which is equal to the number of sites which are connected by the interaction, minus one.

The parameter L is just a formal parameter since really we want to calculate for L=1, but its introduction leads us to some simple considerations. It shows that mean field approximation is exact when we have a large number of identical independent variables on each site provided the coefficients in front of the interaction terms are scaled down by some natural coefficient. Such a situation is for instance approximatively realized when the dimension d of space becomes large. Then in a cell of given linear dimension one finds an increasing number of independent variables. By reexpanding terms in the mean field expansion one can generate a systematic 1/d expansion. Expression (a17) can also be used to show some interesting connections between mean field expansion and strong or weak coupling expansion. Let us introduce a factor β in front of the action A and perform an expansion in powers of β, which is a strong coupling expansion. At order k in β we find a sum of contributions of the form

$$\beta^k L^k L^{-\sum_{\ell=1}^{k} c_\ell} L^n . \tag{a18}$$

The factor L^k comes from the L factor in front of the action. The integers c_ℓ count the number of sites linked by the corresponding interaction terms. Finally equation (a16) shows that the average of any power of the lattice variable with the weight $\rho_L(S)$ is proportional to L since it is equal to :

$$\frac{\int \rho_L(S)\, S^t dS}{\int \rho_L(S)\, dS} = \frac{\partial}{\partial H^{(t)}} \{- L\, F_\circ(H^{(1)},\ldots, H^{(k)})\}\Big|_{H=0} . \tag{a19}$$

Therefore in eq. (a18) n is the number of different lattice variables present in the strong coupling diagram.

For a connected tree diagram one has :

$$n = \sum_{\ell=1}^{k} c_\ell - (k-1) . \tag{a20}$$

Indeed each interaction term brings in c_ℓ independent variables. But to construct a connected diagram each interaction term must have a variable in common with another interaction term. This suppresses exactly k-1 independent variables.

Now each one adds one loop to the diagram, one suppresses one additional independent variable. Calling B the number of loops of the diagram we find then :

$$n = \sum_{\ell=1}^{k} c_\ell - (k-1) - B . \tag{a21}$$

The corresponding power of L for the diagram is then :

$$k - \sum_{\ell=1}^{k} c_\ell + n = 1 - B .$$

The mean field expansion is an expansion in powers of L. It resums therefore at a given order all high temperature or strong coupling diagrams with the same number of loops.

On the other hand the mean field expansion also contains the weak coupling expansion. Indeed let us rewrite equations (a6) with a parameter β in front of the action :

$$\beta \frac{\partial A}{\partial \sigma_i^{(\ell)}} = H_i^{(\ell)} ,$$

$$\frac{\partial F_\circ}{\partial H_i^{(\ell)}} = \sigma_i^{(\ell)} . \tag{a22}$$

For β large (low temperature or weak coupling) the $H^{(\ell)}$ variables become large. To study the limit we have thus to evaluate $F_\circ$ for H large. This will in general select for S some classical value S_c. As a direct consequence we have :

$$\frac{\partial F}{\partial H_i^{(\ell)}} = (S_{i,c})^\ell = \sigma_i^{(\ell)} . \tag{a23}$$

In this limit the variables σ and H play no role any more and one expands around the configuration which dominates the weak coupling (large β) expansion. To show more explicitly the relation let us write the measure as a Fourier transform :

$$\rho(S) = \frac{1}{(2i\pi)^n} \int_{-i\infty}^{i\infty} e^{\mu S} \tilde{\rho}(\mu) d\mu . \tag{a24}$$

For this problem ρ(S) has to decrease fast enough for |S| large, so that $\tilde{\rho}(\mu)$ is an entire function. Then for $H^{(\ell)}$ large $F_\circ$ is given by the saddle point :

$$\mu = \sum_\ell \ell H^{(\ell)} S^{\ell-1} , \tag{a25}$$

$$S + \frac{\partial}{\partial \mu} \ln \tilde{\rho}(\mu) = 0 . \tag{a26}$$

which yields :

$$F_\circ(H^{(\ell)}) = \sum_\ell H^{(\ell)} S^\ell - \mu S + \ln \tilde{\rho}(\mu) . \tag{a27}$$

The mean field saddle point equations are then :

$$\sigma_i^{(\ell)} = \frac{\partial F_\circ}{\partial H_i^{(\ell)}} = S_i^\ell , \tag{a28}$$

$$\beta \frac{\partial A}{\partial \sigma_i^{(\ell)}} = H_i^{(\ell)} . \qquad \text{(a29)}$$

In the same notations the relevant configuration for weak coupling is given by the equations :

$$\beta \frac{dA}{dS_i} = \sum_\ell \ell \; S_i^{\ell-1} \frac{\partial A}{\partial \sigma_i^{(\ell)}} = \mu_i \qquad \text{(a30)}$$

and

$$S_i + \frac{d}{d\mu_i} \ln \tilde{\rho}(\mu) = 0 . \qquad \text{(a31)}$$

But summing equation (a29) on ℓ after multiplication by $\ell S_i^{\ell-1}$ and using equation (a25) reproduces equation (a30), while equations (a26) and (a31) are identical.

The mean field expansion is also a partial resummation of the weak coupling expansion.

A final remark : to generalize mean field theory to fermion systems, one has to identify the variable S_i with the product $\bar{\psi}_i\psi_i$, in which ψ_i and $\bar{\psi}_i$ are the fermion fields and to use a measure appropriate to Grassmann variables.

REFERENCES

Lattice gauge theories are first described in the article of Wilson :

1. K. Wilson, Phys. Rev. D10 (1974) 2445 and then studied with the use of standard methods of statistical mechanics in
2. R. Balian, J.M. Drouffe, C. Itzykson, Phys. Rev. D10 (1974) 3376, D11 (1975) 2098, D11 (1975) 2104.
3. J.B. Kogut, L. Susskind,Phys. Rev. D11 (1975) 395.
   Since they have been the subject of a very abundant literature, and thoroughly discussed in various lectures and review articles to which the interested reader will be refered for more details and bibliography.
4. Cargèse lectures 1976, M. Lévy et al. ed. (Plenum Press, 1977)
   Cargèse lectures 1979, G. 't Hooft et al. ed. (Plenum Press, 1980).
5. L.P. Kadanoff, Rev. Mod. Phys. 49 (1977) 267.
6. J.B. Kogut, Rev. Mod. Phys. 51 (1979) 659.
7. J.B. Kogut, Les Houches lectures 1982, J.B. Zuber and R. Stora eds, North Holland to appear.
8. M. Creutz, L. Jacobs, C. Rebbi, Phys. Reports 95 (1983) 20.
9. J.M. Drouffe and J.B. Zuber, Phys. Reports 102 (1983) 1.
10. C. Rebbi "Lattice gauge theories and Monte Carlo simulations" World Scientific, Singapore 1983.

Additional references concerning specific examples given in these lectures may be useful.

For lecture I see for example :

11. J. Zinn-Justin, Les Houches lecture notes 1982, R. Stora and J.B. Zuber eds., North Holland.

For lecture II see for example

12. E. Brezin and J. Zinn-Justin, Phys. Rev. Lett. 36 (1976) 691 and Phys. Rev. B14 (1976) 3110.

Concerning lecture III and the problem of improved lattice action : the bare renormalization group equations have been first described in :

13. J. Zinn-Justin, Cargèse lectures 1973, Saclay preprint.

The idea of the improved action is due to :

14. K. Symanzik, Mathematical problems in theoretical physics, Lecture notes in Physics, R. Schroeder et al. eds. (Springer, Berlin, 1982); Improved action in lattice gauge theories, in non-perturbative field theory and QCD, R. Iengo et al. eds (World Scientific, Singapore) 1983 DESY preprints 83/016, 83/026.

and has been applied in :

15. G. Martinelli, G. Parisi and R. Petronzio, Phys. Lett. 100B (1981) 485.
B. Berg, S. Meyer, I. Montvay and K. Symanzik, Phys. Lett. 126B (1983) 467.
R. Musto, F. Nicodemi and R. Pettorino, Phys. Lett. 129B (1983) 95.
S. Belforte, G. Curci, P. Menotti and G. Paffuti, Phys. Lett. 131B (1983) 423 and 136B (1984) 399.
B. Berg, A. Billoire, S. Meyer and C. Panagiotakopoulos, Phys. Lett. 133B (1983) 359.
P. Weisz, Nucl. Phys. B212 (1983) 1.
G. Curci, P. Menotti and G. Paffuti, Phys. Lett. 130B (1983) 205.
F. Gutbrod and I. Montvay, Phys. Lett. 136B (1984) 411.

Finally the appendix is an expanded version of

16. E. Brezin, J.C. Le Guillou and J. Zinn-Justin "Phase transitions and Critical Phenomena", vol. 6, C. Domb and M.S. Green eds. (Academic, 1976) pp. 169-170.

AN ELEMENTARY INTRODUCTION TO COMPUTER SIMULATIONS OF HADRONIC PHYSICS

G. Parisi

Dipartimento di Fisica, Università di Roma II
"Tor Vergata", Roma, Italy and
Laboratori Nationali INFN, Frascati, Italy

ABSTRACT

We present a review of the recent progresss done in simulating the low energy hadronic physics an a computer in the framework of QCD.

INTRODUCTION

At the present days QCD seems to be the only viable theory of strong interactions;due to asymptotic freedom, the large momentum behaviour is computable in perturbation theory, while a non-perturbative treatment is needed in the low energy domain. Computer based numerical simulations overcome this difficulty and allow us to compute the low energy spectrum : indeed the only free parameters in QCD are the masses of the quarks and the constant Λ which controlls the asymptotic behaviour of the running coupling constant $\alpha_s(q^2)$: we can trade these parameters for the masses of a few hadrons, e.g. m_p, m_π, m_K, (m_D, m_B, m_F, etc...), and all others masses should be fixed by the theory.

Up to now only exploration computations have been done[1]; the second generation[2,3] computations (total CPU time equivalent to $0(10^3)$ hours CDC 7600 CPU time) have confirmed the results of the first generation computations ($0(10^2)$ hours CDC 7600 CPU time) : an accurate evaluation of the low energy mass spectrum(5-10% of well controlled error, statistical and systematic) can be done with a relatively modest effort (10^5-10^6 CDC 7600 CPU time) ; to decrease the error at the 1% level would be much more difficult : the total CPU time needed would increase by a factor 10^3-10^4 : such a task can likely be done by a seventh generation computer.

In these notes I will explain which are the basic principles of such computations : in section II in order to establish the notations, I will recall the formalism of Euclidean field theory; in section III, I will explain the σ-called Monte Carlo method ; in section IV,I will make some general remarks on numerical simulations; section V will be devoted to some special techniques that make the computation of the masses possible; in the last section I will present some of the recent results obtained for SU(3) in the σ-called quenched approximation in which fermionic loops are neglected.

EUCLIDEAN FIELD THEORY

While in Minkowski field theory a quantum field ϕ is an operator which, acting on the vacuum, creates one or many particles states, in Euclidean field theory $\phi(x)$ is just a function of x, whose probability distribution is given by

$$d\,P[\phi] \propto \exp\{-S[\phi]\}\,d[\phi] \;;\; \int d\,P[\phi] = 1 \;, \tag{1}$$

where $S[\phi]$ is a simple functional of ϕ (e.g.

$$S[\phi] = \int d^D x \,[\tfrac{1}{2}\,(\partial_\mu\phi)^2 + \tfrac{1}{2}\,m^2\,\phi^2],$$

free theory).

If $A[\phi]$ is a functional of the field ϕ, (e.g. $\phi(x)$, $\phi(x)\phi(y)$, $\phi^{10}(z)$,etc.) we can define its expectation value according to the probability distribution $d\,P[\phi]$:

$$\langle A[\phi]\rangle = \frac{\int d[\phi]\,A[\phi]\,\exp\{-S[\phi]\}}{Z} = \int d\,P[\phi]\,A[\phi] \;,$$

$$Z = \int d[\phi]\,\exp\{-S[\phi]\} \;. \tag{2}$$

In the same language in an high energy experiment the probability distribution of the events is proportional to the differential cross section formulae similar to (2) hold for the expectation value of a given quantity, i.e. $p_\perp^2$, where σ_T plays the role of Z.

The connected two point correlation functions have a very important status : they are defined by

$$\langle\phi(x)\;\phi(0)\rangle_c = \langle\phi(x)\;\phi(0)\rangle - \langle\phi\rangle^2, \tag{3}$$

or more generally,

$$\langle A(x)\;B(0)\rangle_c = \langle A(x)\;B(0)\rangle - \langle A\rangle\langle B\rangle \;, \tag{4}$$

where $A(x)$ depends only on the field ϕ and its derivatives at the point x; we have also assumed translational invariance i.e.

$$<\phi(x)> = <\phi(0)> \equiv <\phi> .$$

If the connected correlation is small , the two objects (A and B) are statistically uncorrelated : it is intuitive that the connected correlation function must go to zero at infinity; in the generic case they go exponentially to zero like $\exp(-|x|/\xi)$, the quantity ξ (the so called correlation length) depends in general on the operators A and B; for example in the three dimensional free theory

$$<\phi(x)\ \phi(0)>_c = \exp(-m|x|)/|x| \ ; \ \xi_\phi = 1/m , \tag{5}$$

$$<\phi^2(x)\phi^2(0)>_c = 2(<\phi(x)\phi(0)>_c)^2 = 2\exp(-2m|x|)/|x|^2 \ ; \ \xi_{\phi^2} = \frac{1}{2m} .$$

Similar formulae hold in other dimensions.

Now there is a very simple relation between the correlation functions of Euclidean field theory and the Minkowski theory : the simplest way to state the connection consists in picking an Euclidean direction and to call it time (t), the transverse directions will be called space ($\vec{x}$) : we can thus define a time dependent correlation function

$$C_{A,B}(t) = \int d^3\vec{x}\ <A(\vec{x},t)\ B(\vec{0},0)>_c . \tag{6}$$

Now a general theorem states that there is a Minkowski field theory, whose lagrangian is essentially given by $S[\phi]$, such that we have

$$C_{A,B}(t) = \Sigma_n \frac{1}{2m_n} <0|A|n><n|B|0> \exp(-m_n|t|), \tag{7}$$

where n labels the states $|n>$ at rest of energy (mass) m_n, different from the vacuum (which is denoted by $|0>$) such that the r.h.s. of (7) is different from zero. In presence of a continuous spectrum the sum over n is replaced by an integral.

Equation (7) thus allow us to reconstruct the mass spectrum of the behaviour of the Euclidean correlation functions. In particular when (t) goes to infinity

$$C_{AB}(t) \propto \exp(-m_{AB}|t|) , \tag{8}$$

where m_{AB} is the mass of the lowest energy state contributing to (7): in other words the lowest lying mass is the inverse of the correlation length.

Up to now we have considered the case of a scalar theory; similar considerations can be done for QCD (apart from some technical difficulties due to the Fermi statistics of the quarks). We thus see that the masses of the lower lying hadrons are connected to the rate of the exponential decay of the appropriate correlation functions. To be more explicite, if J_a is the electromagnetic current in the a direction we have (in the narrow resonance approximation) :

$$C_{J_a,J_b}(t) \alpha \delta_{ab} \sum_{1n}^{\infty} \Gamma_n m_n^2 \exp(-m_n|t|) , \qquad (9)$$

where n runs over all the resonances which are produced in the process $e^+e^- \to$ hadrons, m_n and Γ_n being the mass and their electromagnetic width respectively.

By studying the correlation functions in presence of a small background field we can get additional informations on relevant quantities e.g. in presence of a small background magnetic field H in the z direction, the difference in energy of the spin up proton and the spin down proton is proportional to H^4 times the proton magnetic moment; in the same way by changing the background field we can measure the electric dipole transition elements, the axial coupling like g_A^5 the electromagnetic form factor, the deep inelastic structure functions, the matrix elements of the high twist operators, some key matrix elements relevant for the weak decays, etc.

It is clear now that the two point correlation functions contain plenty of information on the low energy hadronic world. We have neglected however how to compute them. Eq.(2) is a functional integral i.e. an integral on all possible shapes of a function; also if we consider only a limited set of functions, we will see that in typical applications the number of integration variables is in the range 10^4-10^7, so that we face a very high dimensional integral : it must be computed using the appropriate method. One of the most popular techniques is the Monte Carlo method which we will expose in the next section.

THE MONTE CARLO METHOD

The Monte Carlo method is well suited for computing integrals like those appearing in eq.(2) i.e. in the finite dimensional case

$$\langle f(y)\rangle = \frac{\int_{-\infty}^{+\infty} dy_1 \ldots dy_n \, P(y) f(y)}{\int_{-\infty}^{+\infty} dy_1 \ldots dy_n \, P(y)} \qquad (10)$$

with P(y) being a non-negative function. Now the Monte Carlo method of statistical mechanics is different from the method used by the experimental physicists for their Monte Carlos. So it is better to discuss the relative advantages in a simple case :

$$P(y) = \exp\{-\tfrac{1}{2} y^2\}, \qquad f(y) = y^2,$$
$$y^2 \equiv \sum_{1i}^{n} y_i^2 . \qquad (11)$$

It is evident that $\langle y^2 \rangle = n$, more generally

$$\langle f(y^2) \rangle = \frac{\int_0^\infty dr\, \tilde{p}(r)\, f(r^2)}{\int dr\, \tilde{p}(r)} , \tag{12}$$

$$\tilde{p}(r) = r^{n-1} \exp(- \tfrac{1}{2} r^2) ,$$

as can be seen by using spherical coordinates. It is also evident that when $n \to \infty$ $\tilde{p}(r)$ becomes concentrated around $r^2 = n$; indeed $\langle (y^2)^2 \rangle = n(n+2)$ and consequently

$$\frac{\langle (y^2)^2 \rangle}{\langle y^2 \rangle^2} \to 1$$

when n goes to infinity.

The simplest method consists in first restricting the integration domain from $-\infty$ — $+\infty$ to $-R$, R, with R sufficiently large and to evaluate the integrals at the points

$$y_i = K_i \varepsilon , \tag{13}$$

where the K_i are integers which run from $-R/\varepsilon$ to R/ε. The correct result is obtained in the limit $\varepsilon \to 0$. This method is very good for obtaining an high precision result (the error is proportional to $\exp(- \pi^2/\varepsilon^2)$) and vanishes extrimely fast with ε : about 20 points for n=1 are enough for obtaining the results with 10 decimal places; it is unfortunately not suited for high dimensions : if N is the number of trials in each direction ($N = 2R/\varepsilon + 1$): we need N^n trials for all directions; unless we consider the case N=1, it is an exponentially large number for large n. We certainly cannot dream to use this method for $n = O(10^4)$.

The typical high energy physics Monte Carlo procedure is the following : using a random number generator, we construct a sequence of random numbers $y^{(\ell)}$ ($\ell=1,\ldots N$) (let us consider the case n=1, the generatization to high values of n is trivial) uniformly distributed in the internal $-R$, R, and we weight each number (i.e., a configuration of the system or an event) with its probability $p(y)$ (its cross section) : we finally get :

$$\langle f(y) \rangle \cong \frac{\sum_1^N{}_\ell\, p(y^{(\ell)})\, f(y^{(\ell)})}{\sum_1^N{}_\ell\, p(y^{(\ell)})} + O(1/N^{1/2}) . \tag{13}$$

In the limit $N \to \infty$ we find the correct result : here the error decreases quite slowly $O(1/N^{1/2})$ but the method works relatively well in not too high dimensions. However the method fails when $n \to \infty$: we know that the integral (12) will be dominated by configurations with $y^2 \sim n$, but these configurations are a tiny fraction of those contained in the cube of side 2R (an exponentially small fraction) : we must wait up to a very large ℓ before we extract randomly a typical configuration.

Something better can be done if we use the fact that p(y) is a non-negative function which may be interpreted as a probability. The first idea is the following: starting from the previous list of randomly chosen $y^{(\ell)}$, for each value of ℓ we extract a random number r_ℓ uniformly distributed between 0 and 1 and we say that $y^{(\ell)}$ is accepted if $0 \leqq r_\ell \leqq p(y^{(\ell)})$, otherwise is rejected; in other words we accept each configuration $y^{(\ell)}$ with probability $p(y^{(\ell)})$. We can now do the list of the accepted configurations $y^{(K)}$, $1 \leqq K \leqq M$ whose number is M (obviously $M \leqq N$). It is evident that

$$\langle f(y)\rangle = \frac{1}{M} \sum_{1K}^{M} f(y_K) + O(1/M^{1/2}). \tag{14}$$

Indeed the weight p(y) has been absorbed by the probability of accepting the configuration. Now the algorithm we have just described is not very efficient when n is large (indeed it is particularly inefficient), the important achievement is eq.(14), the configuration (the events) are generated with authomatically the correct weight (in the same way that real events are generated in a real experimental apparatus) : this method is called the importance sampling method because the configuration space is sampled according to its importance.

A much more efficient algorithm used to generate a sequence of $y^{(K)}$ satisfying eq.(14) is based on the following observation:it is normally true that if a configuration is important also a nearby configuration is important; while in the previous algorithm configurations where generated independently one from the other, it is useful to choose the next configuration not too far from the last one; in practice one can construct a random algorithm just by assigning the transition probability $P(y^{(K)},y^{(K+1)})$ from the configuration $y^{(K)}$ to $y^{(K+1)}$; it can be proven that equation (14) holds (apart from technical hypotheses) if the detailed balance condition is satisfied i.e. :

$$P(y_1,y_2)/P(y_2,y_1) = P(y_2)/P(y_1). \tag{15}$$

The simplest way to implement (15) is the following : we first construct an algorithm (the suggestor) that its probability transition S is symmetric

$$S(y_1,y_2) = S(y_2,y_1) . \tag{16}$$

We use the suggestor to generate from $y^{(K)}$ a suggestion $S^{(K+1)}$; we now extract a random number $r^{(K+1)}$ and we set

$$y^{(K+1)} = s^{(K+1)}, \quad \text{if } P[s^{(K+1)}] > P[y^{(K)}].r^{(K+1)},$$
$$y^{(K+1)} = y^{(K)}, \quad \text{otherwise.} \tag{17}$$

In the first case the suggestion is accepted, in the second it is rejected. This last procedure is normally called the Monte Carlo method[6] : its efficiency strongly depends on a good choice of the suggestor algorithm; very often if n is large the suggestor moves only one coordinate at a time; in the particular case of (11) the n dimensional algorithm will be computed as efficiently as the one dimensional algorithm. It is clear that an acceptance rate of the suggestion too near to one or to zero will be not efficient in most of the cases. The information content of the sequence acceptance-rejection would be too low. It is often stated that an acceptance rate of 50% is likely to be optimal although in some cases smaller acceptance rates (30 %) are more efficient. A pratical hint:monitor the acceptance rate to avoid 10% or 90% (unless you have special reasons) and only if you have to do a long run make some comparison in order to see which is the best algorithm i.e. the one which produces the smallest error.

One must be a little careful to estimate the error : the usual formula for the error

$$\frac{1}{\sqrt{M-1}}\left\{\Sigma_{1}^{M}{}_{K} \frac{f^2(y_K)}{M} - \left(\Sigma_{1}^{M}{}_{K} \frac{f(y_K)}{M}\right)^2\right\} \tag{18}$$

does not hold (it is an underestimation) because y_K and y_{K+1} are correlated; it is usual to divide the sample of the y_K in L subsamples of M/L elements.

One defines

$$f_\ell = \frac{1}{(M/L)} \Sigma_{(\ell-1)M/L+1}^{\ell M/L}{}_{K} \; f(y_K), \quad \ell = 1...L. \tag{19}$$

If M/L is sufficiently large and L is not too small the error may be estimated by

$$\frac{1}{(L-1)^{1/2}}\left\{\Sigma_{1}^{L}{}_{\ell} \frac{f_\ell^2}{L} - \left(\Sigma_{1}^{L}{}_{\ell} \frac{f_\ell}{L}\right)^2\right\}, \tag{20}$$

because the f_ℓ are now independent quantities.

GENERAL CONSIDERATIONS

For many reasons it is convenient to introduce a lattice (a mesh) in the Euclidean space-time and to define the fields only on the points (or on the links) of the lattice. If a is the lattice spacing, the maximum momentum (on an hypercubic lattice) is π/a so that it is natural to suppose that

$$\alpha_R((\frac{\pi}{a})^2) = \alpha_B \ . \qquad (21)$$

If equation (21) holds and we use the one loop approximation for the β function (i.e.

$$\alpha_R(q^2) = \frac{12\pi}{33 \ln(q^2/\Lambda^2)} \qquad)$$

we obtain for a pure SU(3) gauge theory without fermions

$$a\Lambda = \pi \exp(-11/2\pi\alpha_B) \ . \qquad (22)$$

If we notice that eq.(21) contains corrections of order α_B^2, α_B^3... and we use the exact expression for the β function we get that[7]

$$a\Lambda_{\overline{MS}} = \pi(\frac{11}{2\pi\alpha_B})^{51/121} \exp\{-\frac{11}{2\pi}(\frac{1}{\alpha_B} + C + O(\alpha_B))\}, \qquad (23)$$

$$C = 23 \ ,$$

where the $O(\alpha_B)$ corrections have not been computed yet, but can affect the determination of $a\Lambda$ by a factor 2 in the relevant region.

It is much more convenient to consider ratios of physical quantities like $m_G/\sqrt{K}$. (m_G and K being the glueball mass and the string tension respectively). For these quantities the error due to non zero lattice spacing is of order a^2 so it goes fast to zero when a goes to zero. It is therefore convenient to use (23) only as a guideline and to fix the value of a by computing a known quantity (like $\sqrt{K}$ which is supposed to be 400-450 MeV or better the mass of the ρ). It is reasonable to assume that good results should be obtained for a lattice spacing in the .1 - .15 F region (which corresponds to a momentum cut-off of about 6 - 4 GeV).

A box in space time must also be included if we consider a region of size L^3x 2T with periodic boundary conditions. A careful analysis shows that it is sufficient to take L larger or equal to the hadronic diameter, i.e. about .9 F and 1.8 F for particles with only strange quarks (like ϕ and Ω) and for particles with only light quarks respectively : in the same way T^{-1} should be definitely larger than the gap between the two resonances with the same quantum numbers ($T = (200 \text{ MeV})^{-1}$ seems enough). Putting everything together, lattices of 6^3 and 12^3 in the space direction (and somewhat longer in the time direction) are enough to study the strange quark and the light quark respectively.

I stress that such a small lattice is not sufficient for a careful computation, but they should be enough, if a is well chosen to see something interesting. The estimate of the error can be obtained only by studying the dependence of the results on a and L separately and for this aim one needs a definitely larger lattice[8].

It should also be noticed that, as far as the mass of the quark (m_q) is a free parameter, we have the option of computing the results as function of m_q and to extrapolate to small m_q. In doing this extrapolation it would be convenient to separate the effects of the pion from the others and to treat them as far as possible in an analytic way.

It is evident that the most difficult thing to obtain is, as usual, a good controll of the systematic error induced by the finite box (L) and by the non zero lattice spacing. If we take L $\sim$ 5F and a $\cong$ .02F the systematic errors will be by far negligible; unfortunately the CPU time and the memory requirements are proportional to $(L/a)^4$ (I assume T $\sim$ L) so a certain effort must be done for extracting the results without going at too large values of L/a.

As it was said before, the arguments of sections II + III are strictly valid for a bosonic theory; some modifications are needed for studying a theory with fermions; at the present moment the best algorithm for QCD is the one originally proposed in ref. 9 : it has been tested in two dimensional models[10,11] and a few equilibrium configurations have been produced in 4 dimensions[12,13] but no systematic investigations with this algorithm have been started; at the present moment all the computations are done in the so called quenched approximation where closed quark loops are neglected[10]. The main advantage of the quenched approximation is that the CPU time needed is decreased by about a factor 10, so that carefull investigations are possible. The quenched approximation is not unrealistic from the point of view of physics[14]: it essentially corresponds to neglecting the sea quarks inside the hadron and to considering only the valence quarks : the other technical advantage is that the resonances will have essentially zero width (the only diagram contributing to a decay are Zweig violating, see fig. 1) and we can apply eq. (9).

In a theory with N colours, when N goes to infinity, the quenched approximation becomes exact : an educated guess is that the effect of the quark loops on the mass ratios may be of the order of 10 - 20%. If this guess is correct, it is clear that if we consider a typical mass ratio (e.g. $R = m_p/m_\rho$) we can write

$$R_{ex} = R_{que} + (R_{ex} - R_{que}) . \qquad (24)$$

If $(R_{ex} - R_{que})/R_{ex} = 0$ (20%), a 10 % error on R_{ex} can be obtained by computing R_{que} with a 5 % prevision and $R_{ex} - R_{que}$ with 25 % precision. The present day strategy : first compute R_{que} at its best, trying to understand all the sources of systematic errors and to postpone the computation of $R_{ex} - R_{que}$ seems to be well justified by these remarks.

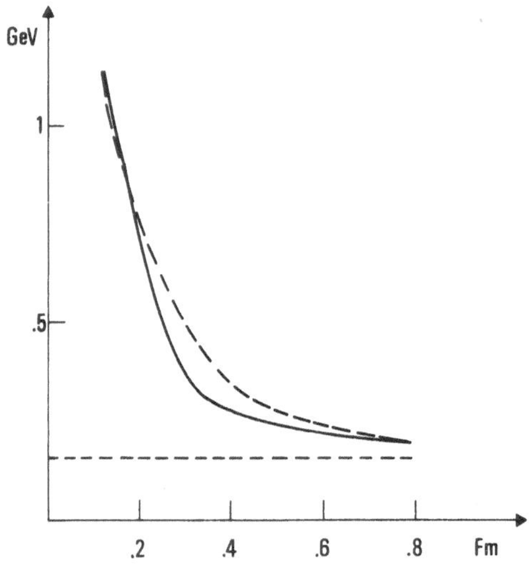

Fig. 1. The force (i.e. dV/dr) as function of the distance. The dashed line is the phenomenological fit of ref.26, while the full line is extracted from the computer simulations of ref.18, under the hypothesis of small finite volume corrections.

The interested reader can find a more careful discussion of the different sources of systematic errors in ref. 8 and 13.

SPECIAL TECHNIQUES

The most interesting information on the correlation functions is contained in their behaviour at large time t where they decay exponentially (at small t perturbation theory can be used) : if we use equation (8) for two equal operators we get :

$$C_{AA}(t) = \sum_{1}^{\infty} {}_n \frac{|\langle 0|A|n\rangle|^2}{2m_n} \exp(-m_n|t|) \to$$

$$\to \frac{|\langle 0|A|1\rangle|^2}{2m_1} \exp(-m_1|t|) \text{ for } t \to \infty . \tag{25}$$

We can thus define an effective t dependent mass

$$m_A(t) \equiv -\frac{d}{dt}\ln[C_{AA}(t)] \cong m_1 + O[\exp\{-(m_2 - m_1)/t\}] , \tag{26}$$

which is a monotonous decreasing function of t and is a reasonable approximation to the true mass only for $(m_2 - m_1)|t| \gg 1$; however in this region the correlation function is exponentially small : if a naive Monte Carlo method is used, the statistical error is essentially independent from the distance and, roughly speaking, proportional to $C_{AA}(0)/N^{1/2}$, N being the number of Monte Carlo steps. It is particularly difficult to observe an exponentially decreasing signal below a constant noise; moreover the situation is still worsened by the following phenomenon : if for example we take $A = F^2_{\mu\nu}$ in order to measure the mass of the glueball, dimensional analysis implies that in physical units $C_{AA}(0) \sim a^{-5}$ and consequently the noise goes up like a^{-5} when a goes to zero. Computations at small values of a are practically impossible by using the naive approach.

At the present moment essentially three different techniques are available :

A) We can change the operator A : the best should be to take an operator A such that $C_{AA}(0)$ is finite when a goes to zero. In gauge theories this requirement cannot be satisfied for gauge invariant local operators : apparently the only way consists in considering non-local operators in the Coulomb gauge; there are no difficulties in fixing the gauge in a Monte Carlo simulation (this has already been done).[16]

A complementary technique consists in considering more than one operator having the same quantum numbers (e.g. $A^{(K)}$, $K = 1,\ldots\ell$) and to measure their correlations; in this way we construct an $\ell \times \ell$ matrix

$$C_{KK'}(t) \equiv C_{A_K,A_{K'}}(t) . \tag{27}$$

The eigenvalues of the matrix

$$M(t) = \frac{d}{dt}[\ln(C(t))] \tag{28}$$

are asymptotically the first ℓ lower masses and the error on m_1 is now $O(\exp(-(m_{\ell+1} - m_1)|t|))$.

This method is particular useful for separating nearby states and for determining the first radial exitation.

B) We can measure the responce of the system with respect to an external perturbation at the place of the correlation function. Indeed if we consider the probability distribution

$$P_\varepsilon(\phi) \propto P_0(\phi)[1 + \varepsilon\phi(0)] , \tag{29}$$

and we call $< >_\varepsilon$ the expectation values with respect to $P_\varepsilon(\phi)$, we have

$$\frac{d}{d\varepsilon} <\phi(t)>_\varepsilon = <\phi(t)\phi(0)>_\varepsilon^c . \tag{30}$$

We can now use equation (30) for computing the connected correlation function. The best way would be to perform two different runs using two slightly different values of ε but the same random numbers; if the difference between the two runs remains of $O(\varepsilon)$, the derivative with respect to ε can be easily evaluated; this method works only if the computer code does contain no logical "if" in the upgrading procedure (e.g. it is so in the approach based on the Langevin[15,16] equation). An alternative method consists in writing[17]:

$$<\phi(t)>_{\varepsilon=1} - <\phi(t)>_{\varepsilon=0} \cong <\phi(t)\phi(0)>_\varepsilon^c , \tag{31}$$

where by "$\cong$" we mean that the exponential tail in t of the two quantities is the same.

C) The third method consists in mixing analytic and numerical computations : this is normally done for the fermions, however a technique similar in spirit has been used in ref. 18 for evaluating the string tension.

Let us consider an elementary example : we have only two variables (A and ϕ) which can be considered also as the values of two fields on an one point lattice, having the following probability distribution :

$$dP(\phi,A) \propto d\phi\, dA \exp[-\frac{1}{2}\phi^2(1 + A^2 g)]\, P_0(A). \tag{32}$$

If we need to compute $<\phi^2>$ it is useful to use the properties of the Gaussian integrals; we have that :

$$\int dP(\phi,A)\phi^2 = \int dP(\phi,A)(1 + gA^2)^{-1} = \langle(1 + gA^2)^{-1}\rangle . \quad (33)$$

This new expression for $\langle\phi^2\rangle$ is definitely less fluctuating than the original one, as can be seen at g=0; more generally we have :

$$\langle\phi^4\rangle - \langle\phi^2\rangle^2 = 2\langle(1 + gA^2)^{-2}\rangle + \langle(1 + gA^2)^{-2}\rangle - \langle(1 + gA^2)^{-1}\rangle^2 . \quad (34)$$

We can also introduce an induced probability distribution of the A field

$$dP_\nu(A) \propto dP_0(A)\ (1 + gA^2)^\nu , \qquad \nu = -1/2 , \quad (35)$$

such that

$$\langle\phi^2\rangle = \int dP_\nu(A)(1 + gA^2)^{-1} \equiv \langle(1 + gA^2)^{-1}\rangle_{\nu=-1/2} . \quad (36)$$

In this way the ϕ field can be completely eliminated.

If we substitute for ϕ an N component field or a fermionic field ψ we find respectively :

$$\langle\Sigma_i\phi_i^2\rangle = N\ \langle(1 + gA^2)^{-1}\rangle_\nu , \qquad \nu = -N/2 ,$$
$$\langle\bar{\psi}\ \psi\rangle = \langle(1 + gA^2)^{-1}\rangle_\nu , \qquad \nu = 1 . \quad (37)$$

If we set ν=0 in eq.37, we correspondingly neglect closed ϕ loops.

The same techniques can be used for QCD; we first discuss the case of continuum QCD and later we will mention the problems caused by the lattice. Neglecting various indices we find :

$$\langle\bar{q}(x)q(x)\ \bar{q}(0)q(0)\rangle_{I\neq 0} = \int dP_{n_f}[A]G(0,x|A)\ G(x,0|A)$$
$$\langle\bar{q}(x)q(x)\bar{q}(0)q(0)\rangle_{I=0} = \int dP_{n_f}[A]\{G(0,x|A)G(x,0|A) - n_f\ G(0,0|A)G(x,x|A)\} , \quad (38)$$

where by I=0 and I≠0 we indicate flavour singlet and flavour non singlet respectively; we also have :

$$(\not{D} + m)G(x,0|A) = \delta(x), \quad (39)$$
$$D_\mu = \partial_\mu + igA_\mu .$$

Here n_f denotes the number of different quark flavours and m is the quark mass (which for simplicity is supposedly flavour independent). If we set $n_f = 0$ in these equations we get the so called quenched approximation in which closed quark loops are neglected : within this approximation nonet symmetry is valid for the mesons : this is not a disaster for most of the mesonic nonets but for the pseudo-scalar one; the η' mass drops to zero while the η is approximatively 670 MeV, as follows from the Gell-Mann Okubo sum rule.

The procedure for computing fermionic correlation functions in the quenched approximation is rather clear : we use the same probabilistic procedure for constructing the A field configurations as in pure gauge theory and later we compute the Green function of the quark by an efficient analytic method. This combination of probabilistic and deterministic techniques is at the present moment the best for studying the spectrum of the fermionic states. A more precise extimate of the statistical error involved in a similar computation is described in ref.8, 14.

When we put fermions on the lattice peculiar problems arise : in the simplest method (Wilson fermions) chiral symmetry is explicitely broken and it is recovered only in the low energy region in the continuum limit : the quark mass is additively renormalized. In the usual parametrization the quark mass is traded for a parameter K, such that

$$m_q \cong 2\, a^{-1}\, (1/K_c - 1/K) \ , \tag{40}$$

K_c being the β-dependent value of K for which the mass of the quark is equal to zero; the mass of the quark can be evaluated only in perturbation theory; beyond perturbation theory K_c is fixed by the condition of having a zero mass pseudoscalar state. By computing the quark propagator (using a background gauge field configuration obtained via the Monte Carlo method) for different values of K we can study the dependence of various quantities on the quark mass.

In typical runs the computation of the quark propagator is done using the Gauss Seidel method and it takes about 90 % of the total CPU time (it is possible however that slightly faster methods could be found e.g. using Fast Fourier Transform in the Landau gauge). If quark loop effects are included this ratio is likely to be reversed.

It is evident that, if it is necessary, we can combine the various methods to decrease the statistical errors, The best method will change from case to case.

RESULTS

As it was explained in the introduction most of the results obtained up to now are affected by not well controlled systematic errors which can be removed by comparing the results taken for different values of the parameters.

Let us consider a pure gauge SU(3) theory; on the lattice it is usual to introduce a parameter β which is related to the bare coupling constant of the previous chapters by :

$$\beta = \frac{3}{2\pi\alpha_B} \quad . \tag{41}$$

The probability distribution of the gauge fields is proportional to :

$$\exp\{\beta \sum_p \mathrm{Tr}\, \Box_p\} \ , \tag{42}$$

where the sum runs over all the plaquettes of the lattice.

When we compute $\langle \mathrm{Tr}\,\Box \rangle$ as function of β, we see a very fast variation in the region $\beta = 5.5 - 5.7$; the low β and the high β expansion work reasonably well outside this cross-over region. If we are interested in the continuum physics, we must stay in the large β region; although it is possible that good results for dimensionless quantities are obtained also for not too high values of β, it is convenient to consider only the region $\beta > 5.7$.

The simplest quantity which can be obtained from simulations is the deconfinement temperature[19]; it is well known[20] (in certain cases it has also been proved rigorously[21]) that when the physical temperature is increased beyond a critical one (kT_c) pure gauge theories become deconfined. Now it is well known that at finite temperature Minkowski field theory corresponds to an Euclidean field theory defined on a strip of side $L = (kT)^{-1}$; in the confined phase the expectation value of the thermal loops (i.e. the Wilson loops winding around the lattice in the time direction) is strictly equal to zero, but it becomes different from zero in the deconfined region; by studying the expectation values of the plaquettes and of the thermal loops as function of β and L, we get precise information on the thermodynamics of the system.

A first order transition has been found at a critical temperature which is given by[22]:

$$\begin{aligned} aL_c &= 4 \ , \qquad \beta \sim 5.7 \ , \\ aL_c &= 6 \ , \qquad \beta \sim 6.1 \ . \end{aligned} \tag{43}$$

The value of the temperature is in good agreement with the hypothesis of constant $L_c\Lambda$, where Λ is computed from eq.(23).
This result confirms the belief that for $\beta > 5.7$ we should stay in the region where eq.(23) is approximatively valid. It is important that this is not true for $\beta < 5.7$[23], although the violations of the scaling law are reduced if one uses $\beta_{ef} \equiv \beta \langle \mathrm{Tr}\,\Box \rangle$ at the place of β[24]. Unfortunately the one loop corrections in eq.(23) have not yet been computed and no definite conclusion can be reached at this stage.

It is a wide spread belief that the string tension σ (as suggested by dual models) is 400 MeV, so that a measurement of σ would set the scale. The first estimates for the string tension where done by measuring the expectation value of a LxL Wilson loop; in the large L region it should decrease as :

$$\langle W(L)\rangle \sim \exp(-\sigma L^2) \, . \qquad (44)$$

It is clear that (44) is only an asymptotic behaviour; in the first computations the Wilson loops have been measured for rather small values of L and the string tension was extracted from the data by using eq.(44) beyond its range of validity; this wrong extimate[25] of the string tension (and consequently of the lattice spacing) was mainly responsable for the smallness of the lattice in physical units for the first generation runs with Fermions.

As far as the shape of the potential in the whole region .1-1 fm is of interest for the computation of the hadronic mass spectrum both for the non-relativistic quarks[26] (like the charm and the bottom quarks) and for the lightest quarks (where relativistic techniques must be used)[27], it would be very useful to extract the static potential from the lattice gauge theory in the least biased way ; the most safe measurement consists in taking a finite temperature lattice (i.e. a strip of size L much larger than kT_c) : in the limit of large L the correlation function between two thermal loops is given by

$$C_\ell(\vec{x}) \cong \exp[-LV(\vec{x})] \, , \qquad (45)$$

$\vec{x}$ being the space distance between two thermal loops and V(x) the quark static potential. The linear confinement hypothesis implies that at large distances

$$C_\ell(\vec{x}) \cong \exp[-L\sigma(x)] \, . \qquad (46)$$

In order to increase the ratio signal to noise it is convenient to measure (as usual) the correlation function integrated over the transverse directions :

$$\begin{aligned} \mathcal{C}(z) &= \int dxdy\, C(x,y,z) = \pi\int_0^\infty dr^2 C((r^2+z^2)^{1/2}) \, , \\ C(z) &= \frac{1}{2z}\frac{d}{dz}\mathcal{C}(z) \, . \end{aligned} \qquad (47)$$

In ref.18 a first attempt has been done for measuring the potential on a 10^3 x 20 lattice at $\beta = 6$ (it finally will be obtained that the dimensions of the lattice in fm are 1 fm^3 x 2 fm). The measurement is not easy because the correlation functions are rather small with respect to the noise; a special technique has been devised (based on the DLR equation[28]) which increases the signal to noise ratio by a factor 20. The final results give a value of the string tension of -.04 (in lattice spacing units) with a statistical error of about 20 % : the corresponding value of the lattice spacing is consequently $(2\ \mathrm{GeV})^{-1}$, i.e. 1 fm. The qualitative shape of the corresponding static potential which can be extracted from the data by using eqs.

45 - 47 is shown in fig. 1. Here the errors are much larger. Most of the computation has been done on the Cray. The CDC equivalent CPU time was about 80 hours; in reality the time needed was much less because the configurations of the gauge fields were produced for different aims (the direct computation of the meson spectrum); the extra time for doing the computation was about 10 hours. Before reaching definite conclusions on the shape of the potential in QCD a more careful computation on a large lattice should be done : by planning the computer simulation carefully not too much time should be needed.

Having settled the scale, we can come back to the thermodynamics : the deconfinement transition turns out to be at about 370 MeV, the latent heat[22] being about 10 GeV/fm^3; a unexpected large number.

In pure gauge theory the only existing particles are glueballs; unfortunately no evaluation exists in the scaling region ($\beta \geqq 5.7$); a simulation on a $7^3 \times 10$ lattice at $\beta = 5.8$ is nearly finished at Rome and the results will be available soon[15].

We can now discuss the spectrum of bound quarks. In the first generation experiments, in which the techniques discussed in the previous chapter have been introduced, in most of the cases the lattice was about $5^3 \times 10$ at $\beta = 6$; the size of the lattice would be adequate if the original estimates of the string tension were correct (about 1 fm); unfortunately using the correct estimate, the size is too small (about .5 fm) and the results suffer of relative large finite volume distorsions. It is interesting to note that not only the masses of the lower lying states have been computed but we have obtained also estimates of the magnetic moments of the proton and of the neutron (about 3 and -2 nuclear magnetons respectively[4]) and of the neutron axial coupling ($g_A/g_V \cong 1.35$) and of the relative D/F coupling ($\cong 1.80$)[5]. The statistical errors on these results are about 10 %, while the systematic error is not under controll. The extra computer time for extracting these numbers has been remarkably small (about 10 CDC hours) : it would be rather interesting to see the results on more realistic lattices.

The most extensive simulations with Fermions have been done on a Cray[3] on a $10^3 \times 20$ lattice. Thirty statistically independent gauge configurations have been generated by 4000 Monte Carlo sweeps of the whole lattice; for each configuration the Dirac equation has been solved for three different values of the quark mass. The output, consisting of about .6 10^9 numbers, has been stored in 30 high density tapes, the total CPU time on Cray is 70 hours; the same programs on a CDC would have taken about 1400 CPU hours plus input-output.

A first analysis of the data has been done (a more refined one is under way now). One proceeds in the following way; firstly one uses eq.(38) of the previous section to compute the correlation func-

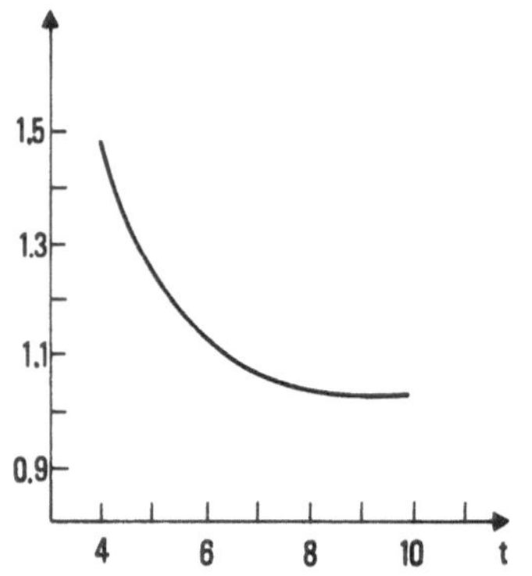

Fig. 2. The effective mass as function of the distance (from ref. 3).

tion of two pseudoscalar densities $\bar{q}\gamma_5 q$ and of two vector densities $\bar{q}\gamma_\mu q$ ($C_P(t)$ and $C_V(t)$ respectively). This is done for each configuration; by repeating the same computation for the whole set of configurations the final estimate for the correlation function is obtained (and also the statistical error).

The time t ranges from 0 to 10 lattice spacings : for small time t both $C_P(t)$ and $C_V(t)$ behave like t^{-3} as prediced by perturbation theory ($t \leqq 4a \cong (500 \text{ MeV})^{-1}$), in the large t region only the single particle state seems to give a relevant contribution and the central region $5a < t \leqq 10a$ is very well fitted by a single hyperbolic cosine

$$\text{ch}\,[(t - 10a)m]\ , \tag{48}$$

m being identified with the mass of the lowest particle (pseudo scalar or vector) in the corresponding channel. The same results are obtained if one considers a t-dependent mass :

$$m(t) = \frac{1}{a}\,\text{arch}[\frac{C(ta+a) + C(ta-a)}{2C(a)}]\ . \tag{49}$$

In the large t region m(t) is t independent and equal to the value obtained by the previous analysis, (see fig. 2).

Using three quark densities having the quantum numbers (spin and isospin) of the proton and of the delta, one obtains estimates of the respective masses; here the region where the asymptotic behaviour is seen is somewhat smaller and the estimated value may be somewhat higher than the correct value; this problem is under study at the present moment : a definite answer can be given using more than one operator having the same quantum numbers (the analysis will be finished soon). For the time being an analysis with the two operators[29] for the pseudo-scalars (i.e. $\bar{q}\gamma_5 q$ and $\bar{q}\gamma_5\gamma_0 q$) confirms the previous results for the lower mass and gives a splitting of about 1.2 GeV between the lowest particle and the first radial recurrence; it is likely that this number will decrease and hopefully stabilize when many operators will be included.

In this way we obtain the data in fig. 3. We observe that the pseudoscalar mass squared is well linear in K, so by extrapolating it to higher value of K we obtain that it vanishes at K about .1568; if for simplicity we stick to the previous discussion for the strange quark pseudoscalars in the presence of nonet symmetry (i.e. the mass of the η_λ should be 670 MeV) the value of the strange quark mass can be found by imposing that the pseudoscalar to vector mass ratio is the experimental one (.65); this operation is necessary to fix the scale. In this way one obtains that the value of K corresponding to the strange quark is .1548; using the value of K_c the strange quark mass turns out 80 MeV (the bare one!), the corresponding value of the renormalized mass, computed at the point where α_s is 1/3, is about 120 MeV, in agreement with the phenomenological estimates. By comparing the mass of the vector particle to the mass of the phi (1,019 MeV), the lattice spacing turns out to be about $(2.2\ \text{GeV})^{-1}$ in good agreement with the previously determined value. The masses of the proton in which the light quarks have the same mass as the strange one is (by using the Gell-Mann Okubo formula) about 1.51 GeV, the corresponding mass measured on the lattice is 1.55 GeV; the mass of the Δ turns out to be about 1.65 GeV against the experimental value of 1.67 GeV. The pseudoscalar decay constant has been estimated and it is about 120 MeV for strange quarks.

The above mentioned results have a typical statistical error of 10 %; the systematic error due to non zero lattice spacing should be large : it seems that a similar computation done by the Edinburgh group on a $8^3 \times 16$ lattice at $\beta = 5.7$ finds quite similar results;[30] the ratio of the lattice spacings ($\sim$1.4) is in agreement with the asymptotic freedom predictions, eq. (23). Finite volume corrections are probably small for strange quarks (they are proportional to $\exp(-m_p L)$ and $\exp(-m_p L)$ which are approximatively exp(-3) and exp(-4) at the largest value of K used). A direct computation of

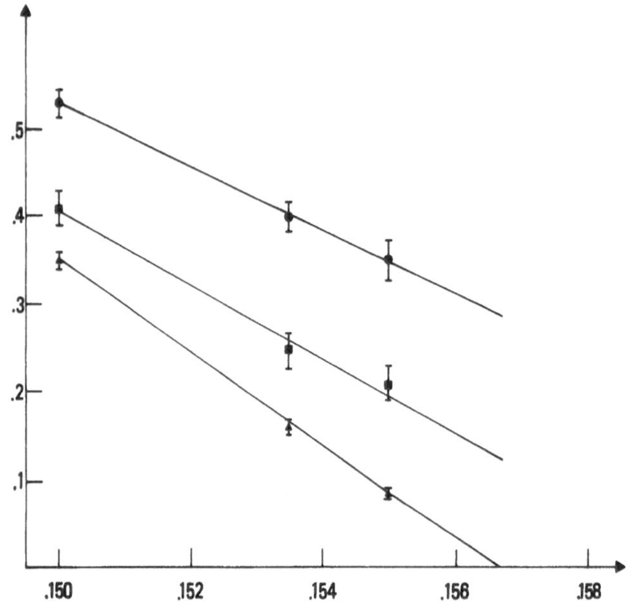

Fig. 3. The pion mass squared, the ρ mass squared, and the nucleon mass in lattice spacing units ($a^{-1} \cong 2.2$ GeV) as function of the hopping parameter K.

the light quark spectroscopy on this lattice would likely not be reasonable, because finite volume corrections may be too large; we can follow two solutions : extrapolate in K the data up to K very near to K_c, or wait for the results on a $16^3 \times 20$ at lower values of the quark mass, which should be ready for the end of the year. When these new data will be available, the effects due to finite volume will be much more under control. In these last two years there has been a large progress; many effects are much more understood and it is rather likely the for the next Cargèse school (i.e. two years from now) we should have results for the hadronic spectrum with well estimated systematic errors.

It seems that our understanding of QCD is not qualitatively different from QED; in QED the difficulty of the computation for going to the next loop increases of a factor 10 or more, while the precision of the result is increased of a factor about $10^{-3} \cong \alpha/2\pi$, in QCD we stay in the domain of strong interactions and it is not surprising that an order of magnitude gives you less than a factor 2 in precision. In both cases we have a formalism which allows us to compute measurable quantities with arbitrary small errors at the price of a sufficient long computation : the rate of convergence being obviously worse for strong interactions than for QED. It is surely worthwhile to push the approach here described to the point of producing sufficient evidence that QCD is the correct theory in the low energy domain.

REFERENCES

1. H. Hamber, G. Parisi, Phys. Rev. Lett. 47 (1982) 1792 ;
   E. Marinari, G. Parisi, C. Rebbi and Phys. Rev. Lett. 47 (1981) 1795;
   H. Hamber, E. Marinari, G. Parisi and C. Rebbi, D. Weingarten, Phys.Rev.Lett.109B (1988) 57;
   H. Hamber and G. Parisi, Phys. Rev. D27 (1983) 208;
   F. Fucito, G. Martinelli, C. Omero, G. Parisi, R. Petronzio and F. Rapuano, Nucl. Phys. B210 [FS6] (1982) 407;
   D. Weingarten, Nucl. Phys. B215 [FS7] (1983) 91.
2. K.C. Bower, E. Marinari, G.S. Pawley, F. Rapuano, D.J. Wallace, Edinburgh preprint 82/236, Nucl. Phys. B [FS] in press.
3. H. Lipps, G. Martinelli, R. Petronzio, F. Rapuano, CERN Preprint 3548 (1983).
4. G. Martinelli, G. Parisi, R. Petronzio, F. Rapuano, Phys. Lett. 116B (1982) 215.
   C. Bernard, T. Draper and K. Olynk, Phys. Rev. Lett. 49 (1982) 1796.
5. F. Fucito, G. Parisi, S. Petrarca, Phys. Lett. 115B (1982) 148.
6. See for example the book "Monte Carlo Methods", edited by Binder Springer Verlag, 1979.
7. A. Hasenfratz and P. Hasenfratz, Phys. Lett. 93B (1980) 165.
8. For a more careful discussion see G. Parisi,proceedings of the

"Workshop on non-perturbative QCD", Trieste 1982 and " Les Houches Winter Meeting", Physics Reports to be published.
9. F. Fucito, G. Parisi and C. Rebbi, Nucl. Phys. B190 [FS3] (1981) 234.
F. Fucito, E. Marinari, Nucl. Phys. B190 [FS3] (1981) 266.
10. F. Marinari, G. Parisi, C. Rebbi, Nucl. Phys. B190 [FS3] (1981) 734.
11. S. Otto and M. Randeira, Caltech preprint CALT 68-991 (1983).
12. H. Hamber, E. Marinari, G. Parisi and C. Rebbi, Phys. Lett. 124B (1983) 99.
13. H. Hamber, E. Marinari, G. Parisi and C. Rebbi, Saclay Preprint SPh. T/83/54.
14. G. Parisi, Phys. Lett. 61B (1976) 386.
15. M. Falcioni, M.L. Paciello, G. Parisi, B. Taglienti (work under progress).
15. G. Parisi and Wu Yong-shi, Scientia Sinica.
16. G. Parisi, Nucl. Phys. B180 [FS1] (1981) 378 and 190 [FS2] (1982) 337; M. Falcioni, E. Marinari, M.L. Paciello, G. Parisi, B. Taglienti and Zhang Yi Chen, Nucl. Phys. B215 (1983) 265.
17. K. H. Mutter and S. Chilling, Phys. Lett. 120B (1982) 251.
18. G. Parisi, R. Petronzio and F. Rapuano, CERN Preprint 3596.
19. N. Cabibbo and G. Parisi, Phys. Lett 59B (1975) 67.
20. L. Suskind, Phys. Rev. D20 (1978) 2610.
A.M. Polyakov, Phys. Lett. 72B (1978) 447.
21. C. Borgs and E. Seiler, Nucl. Phys. B215 (1983) 125.
22. J. Kogut, M. Stone, H.W. Wyld, W.R. Gibbs, J. Shigemitsu, S.H. Shenker and D.K. Sinclair, Phys. Rev. Lett. 50 (1983) 393.
23. T. Celik, J. Engels and H. Satz, Bielefield preprint (1980)
B. Svetisky, F. Fucito, Cornell preprint (1983).
24. G. Parisi, proceedings of the "XX Conference on high energy Physics ", Madison (1980).
25. M. Creutz, Phys. Rev. Lett. 45 (1980) 313.
26. See A. Martin contribution to this school.
27. J.-L. Basdevant, private communication.
28. R.L. Dobrushin, Theory Prob. Appl. 13 (1969) 197.
O.E. Lanford III, D. Ruelle Comm. Math. Phys. 13 (1969) 194.
29. G. Martinelli, G. Parisi, R. Petronzio and F. Rapuano (work in progress).

SUPERSYMMETRY AND SUPERGRAVITY : THE PHENOMENOLOGICAL VIEW

D.V. Nanopoulos

CERN

Geneva

ABSTRACT

In this review I discuss in some detail the structure and physical consequences of global and local supersymmetric (SUSY) gauge theories. Section 1 contains motivations for SUSY theories, whilst Sections 2 and 3 explain what supersymmetry is, and what are its physical properties. The observable consequences of SUSY at low energies and super-high energies are discussed in Sections 4 and 5. The physical structure of simple (N=1) local SUSY ($\equiv$ supergravity) is given in Section 6, whilst Section 7 contains the physics of simple supergravity both at super-high as well as at low energies. The experimental evidence (?) for supersymmetry is analyzed in Section 8, whilst Section 9 contain the conclusions. Amazingly enough, we find that gravitational effects, as contained in supergravity theories, may play a rather fundamental role at all energy scales. This strong interrelation between gravity and particle physics is unprecedented.

MOTIVATION(s) FOR SUPERSYMMETRY

Unification of all elementary particle forces has been the Holy Grail of theoretical physics. The first realistic step towards this end has been the highly successful unification between electromagnetic and weak interactions, now called electroweak interactions. The obvious (and natural) next step is then the amalgamation of electroweak and strong interactions, justifiably called Grand Unified Theories[1] (GUTs). The qualitative successes (e.g., charge quantization, equality of different coupling constants and equality of certain quark-lepton masses at super-high energies, natural understanding of quark and lepton quantum numbers, etc.), as well as the quantitative successes (e.g., disparity of coupling constants and of quark-lepton masses at low energies, determination of the electroweak mixing angle (θ_{e-w}), possible limit on the number of flavours, virtually massless neutrinos, etc.) are rather well-known[1]. It is also well known that the grand unification scale M_X is rather large

$$M_X \sim 10^{15} \text{ GeV}. \tag{1}$$

GUTs contain baryon and lepton number violating interactions, and the presently observed proton stability ($\tau_p > 10^{31-32}$ years) immediately puts a lower bound on M_X which more or less is saturated by (1). The existence of two scales, the electroweak scale ($M_W \sim 100$ GeV) and the GUT scale M_X, so different

$$\frac{M_W}{M_X} \leqq 10^{-13} \tag{2}$$

creates a fundamental problem for GUTs.

The gauge hierarchy problem

How is it possible to keep these two scales separate, incommunicado? In ordinary field theories, even if we fix the parameters at the tree level to satisfy (2), radiative corrections will undo it and we will have to adjust it at each order in perturbation theory. This does not sound right if we claim to have a natural theory. The heart of the problem is the existence of scalar fields. The only way to break a gauge theory and keep renormalizability is through Spontaneous Breaking (SB). The simplest way to achieve SB is to have certain scalar fields get a vacuum expectation value (v.e.v.). In GUTs we need two sets of scalars, one set to break the big group G which supposedly contains all interactions down to SU(3) x SU(2) x U(1) at M_X, and thus the v.e.v. of these fields $V \sim M_X$, and another set to break SU(3) x SU(2) x U(1) down to SU(3) x $U(1)_{E-M}$, and thus they should get v.e.v.'s $v \sim M_W$.

Since the masses of these Higgs fields are proportional to their v.e.v.'s, we end up with light Higgs of masses $O(M_W)$ and heavy Higgs of masses $O(M_X)$. Again this can be arranged at the tree level. But, since light and heavy Higgs are both coupled to gauge bosons, fermions and scalars, already at the 1-loop level there is a communication between the light and heavy Higgs, through the above mentioned fields running around the loop. That immediately creates corrections to light scalar masses $O(M_X)$, while only corrections at most $O(M_W)$ are allowed, i.e., end of the gauge hierarchy !

This happens because there is no symmetry able to keep scalars massless (or virtually massless : $M_W \lll M_X$), in contrast to gauge or chiral symmetries which keep gauge bosons or fermions massless.

One way out would be to abandon completely the use of scalar fields as means of SB. Then we have to attempt dynamical SB. This approach has been tried (technicolor, extended technicolour, ...) but led to fatal flaws, so it became evident that technicolour has created more heat than light[2]. So back to scalars again. One may try now a different way. Is there any possibility that when one adds up all the 1-loop corrections to light scalar masses, they practically vanish, becoming at most of order M_W? We may exploit the fact that thanks to the "spin-statistics theorem and all that", boson and fermion loops differ by an overall minus sign. Then if suitable relations exist between fermion and boson masses on the one hand and gauge, Yukawa and scalar self coupling constants on the other, the hope of cancellation between different 1-loop diagrams may be realized. Well, this is a very neat way to discover supersymmetry[3] (SUSY). Actually, one may rigorously prove that the only way to get the desired 1-loop cancellation is through supersymmetry[4]. As we will see later, a remarkable property of supersymmetry is that if the cancellation occurs at the 1-loop level, then it occurs automatically to all orders in perturbation theory. The technical aspect of the gauge hierarchy problems then has been solved.

Since exact supersymmetry implies equal fermion boson masses, which is not seen experimentally, supersymmetry has to be broken. The corrections to the light Higgs masses will be $O(M_{LESB})$ and they should better not exceed $O(M_W)$, otherwise the gauge hierarchy probem strikes back again, thus

$$M_{LESB} \leqq O(M_W), \tag{3}$$

where M_{LESB} refers to the Low Energy Supersymmetry Breaking, i.e., the SUSY breaking that the low energy world suffers. Discussing about hierarchies, another serious problem, this time concerning quantum chromodynamics (QCD), naturally comes to mind. Non-perturbative effects in QCD have the disturbing feature of adding a term :

$$\varepsilon_{\mu\nu\rho\sigma}\, F^a_{\mu\nu}\, F^a_{\rho\sigma}\, \theta \ , \tag{4}$$

where $F^a_{\mu\nu}$ is the gluon field strength and θ is a new parameter. There is a contribution from the non-perturbative term (4) to d_n, the Dipole Electric Moment of the Neutron (DEMON)

$$d_n \cong 10^{-16}\ \theta\ \text{e.cm}\ , \tag{5a}$$

which imposes a severe upper bound on θ

$$\theta \leqq O(10^{-9}) \tag{5b}$$

when the present experimental upper bound[5] $d_n < 0\ (10^{-25}$ e.cm) on DEMON is used. This is the strong CP-hierarchy problem. Again supersymmetry solves the technical aspect of this problem[6]. Starting with $\theta = 0$, one proves[6] that in spontaneously broken SUSY type theories, θ naturally lies below the limit posed by (5). The same type of miraculous cancellations, as mentioned above, occur again. Supersymmetric theories are very well behaved; they respect hierarchies.

We have by now enough physical motivation to have a close look at the structure of supersymmetric theories.

SUPERSYMMETRY (SUSY)

All kinds of symmetries that one normally uses in particle physics, global (like Isospin, Eightfold way, ...) or gauge (SU(3) x SU(2) x U(1), SU(5), O(10), E_6,...) do always transform fermions to fermions and bosons to bosons. Supersymmetry, or Fermion-Boson Symmetry[3], tries to bypass this prejudice and aims at a theory in which a fermion-boson transformation will also be possible. Indeed, such theories have been constructed[3] and are in full accord with all the standard laws of quantum field theory. In its simplest form, one has to extend the usual Poincaré Algebra of the generators of space-time rotations and translations, $M^{\mu\nu}$ and P^μ, to contain a self-conjugate (Majorana) spin 1/2 generator, Q_α, which turns boson fields to fermion fields and vice versa. Schematically,

$$Q_\alpha|\text{boson}\rangle = |\text{fermion}\rangle \quad ; \tag{6}$$

$\alpha = 1, 2, 3, 4$ is the spinor index.

They satisfy the following (anti) commutation rules :

$$[Q_\alpha, M^{\mu\nu}] = i(\sigma^{\mu\nu}Q)_\alpha \tag{7}$$

$$[Q_\alpha, P_\mu] = 0 \tag{8}$$

$$\{Q_\alpha, \bar{Q}_\beta\} = -\,2(\gamma_\mu)_{\alpha\beta}P^\mu \;, \tag{9}$$

in which

$$\sigma^{\mu\nu} = \frac{1}{4}\,[\gamma^\mu, \gamma^\nu]$$

and

$$\bar{Q} = Q^T\gamma^0 \;.$$

The Q_α are four hermitian operators and we use Majorana representation for Dirac Matrices. Because of the spinorial character of the generators Q_α, the extended Poincaré Algebra, called supersymmetry algebra, involves both commutation and anticommutation relations. It is not an ordinary Lie Algebra, but what the mathemeticians call a Graded Lie Algebra (GLA). The spinorial generator Q_α

is a grading representation of the Poincaré Lie Algebra. Supersymmetry extends this algebra in a rather non-trivial way; one can associate in irreducible representations a finite number of bosons and fermions. This fact is extremely important if one wants to construct conventional renormalizable quantum field theories invariant under supersymmetry and satisfying the usual Wightman axioms. Graded Lie Algebras play perhaps a unique role in particle physics, because they realize truly relativistic spin-containing symmetries in which particles of different spin belong to the same supermultiplet. It is remarkable that by making the spinorial generators transform as some representation of an internal symmetry group, the resulting albebra provides also a fusion between space-time and internal symmetry overcoming previous no-go theorems[7]. Irreducible multiplets combine in this case fermions and bosons with different internal quantum numbers.

The physical meaning of the (anti) commutation rules (7), (8) and (9) is rather apparent. Equation (7) simply states that Q transforms as a spinor, while Eq. (8) states that the spinor charges are conserved and are translation invariant. Presumably the most important is Eq. (9), and it suggests the terminology that the supersymmetry charges are the "square root of translations". Clearly, SUSY involves the structure of space-time and it is this connection that is fully developed in local supersymmetry or supergravity[8].

It is very easy to show that SUSY implies a relation between particles of different spin. Apply the spinor charge Q_α to a particle state $|P, S>$ of definite momentum and helicity :

$$Q_\alpha|P, S> = a|P, S + 1/2> + b|P, S - 1/2> \quad ; \tag{10}$$

because of (8), the RHS is a superposition of particles of the same momentum and energy - thus the same mass. Furthermore, addition of angular momentum implies that these particles have helicities $S \pm 1/2$. Supersymmetry transformations connect states which differ by 1/2 unit of spin. These states fill in the so-called supermultiplets. They therefore relate bosons to fermions. One may easily generalize the (anti) commutation rules (7), (8) and (9) to involve more than one spinorial charge, say Q^N_α, N = 1, 2,.., 8. Then we talk about N-extended supersymmetry in contrast to N = 1 or simple supersymmetry. Since here we are not going to be involved with N > 1 supersymmetries we stick to the above given (anti) commutation rules.

When trying to mix SUSY with gauge theories we better keep in mind certain general features and constraints that more or less come out from first principles :

1) In global supersymmetry the supercharges (Q_α) always commute with gauge symmetries; the commutator, if not zero, would be a supersymmetry transformation depending on the infinite number of para-

meters of the gauge symmetry, so would have to be a local SUSY. Thus,

$$[Q_\alpha^N, G] = 0 \quad . \tag{11}$$

There are two immediate consequences of this fact :

(i) all numbers of a supermultiplet have the same internal quantum numbers. This means that

(ii) only N=1 (global) SUSY makes phenomenological sense, since $N \geq 2$ (global) supersymmetric theories yield always fermions in real representations of SU(3) x SU(2) x U(1) in sharp contrast with what we observe experimentally at low energies (parity violation exists both in charged and neutral currents). For example in N = 2 global supersymmetry there are two types of supermultiplets containing spin 1/2 fermions : (1/2, 0, -1/2) and (1, 1/2, 0). Both of them provide vector-like theories. The (1/2, 0, -1/2) multiplet relates fermions of helicity 1/2 and -1/2 which would have to have the same quantum numbers. The (1, 1/2, 0) multiplet relates fermions of helicity 1/2 to massless bosons of helicity 1. But massless bosons of helicity 1 are always gauge bosons, transforming in the adjoint representation, which is real. So whether we consider the (1, 1/2, 0) or the (1/2, 0, -1/2) multiplet, the fermions in N = 2 (or N>2) global SUSY transforms in a real representation of the gauge group : helicity 1/2 and helicity -1/2 transform equivalently. This observation justifies our attitude of not considering very seriously any $N \geq 2$ global SUSY theories.

2) An immediate consequence of Eq. (9) is that in global SUSY theories with SB the Hamiltonian H is the sum of the squares of the supersymmetry charges

$$H = \sum_{\alpha=1}^{4} Q_\alpha^2 \quad . \tag{12}$$

Since H is the sum of squares of Hermitian operators, the energy of any state is positive or zero.

Clearly, if there exists a SUSY invariant state, that is a state annihilated by Q_α, then it is automatically the true vacuum state since it has zero energy, and any state that is not invariant under supersymmetry has positive energy. Thus, in contrast with ordinary gauge theories, if a SUSY state exists, it is the ground state and SUSY is not SB. Only if there does not exist a state invariant under SUSY, SUSY is SB. In this case the ground state energy is positive. Obviously, it is far more difficult to achieve SUSY SB than achieving gauge symmetry SB. The supersymmetric state would have to be ostracized from the physical Hilbert space.

In global SUSY theories, it is impossible to spontaneously break an N-extended SUSY to an N'(N > N' ≥ 1) SUSY, because all Q_α^N satisfy separately Eq. (12) and if one breaks down some N, then all of them break down as well. No step-wise extended SUSY SB in global supersymmetric theories is possible. Things are different though in supergravity.

We are ready now to move to the construction of N = 1 SUSY models.

PHYSICAL PROPERTIES OF SUPERSYMMETRY

From our previous discussion of the general features of supersymmetric theories, we recall the fact that only N = 1 global SUSY theories make phenomenological sense. Thus each supermultiplet contains only two kinds of particles, with identical internal quantum numbers but with a "spin-shift" of 1/2 unit.

Let us take an arbitrary gauge group, with gauge mesons A_μ^a (spin 1) and fermionic partners λ^a (spin 1/2) called gauginos, belonging to the adjoint representation of the gauge group. They consist of the vector multiplet. The gauginos have to have spin 1/2 and not 3/2 because of renormalizability. In addition, we may introduce left-handed fermions ψ_L^i in an arbitrary multiplet of the gauge group. They form supersymmetry multiplets

$$\phi_i \equiv \begin{pmatrix} \psi_i^L \\ s_i \end{pmatrix} \tag{13}$$

with complex scalar bosons s_i. They consist of the chiral multiplets. The right-handed fermion fields are the complex conjuguate of the left-handed fields, $\psi_{jR} = (\psi_L^j)^*$ and their SUSY partners are the complex conjugates s_j^* of the s^i. We have to use scalar fields and not spin-1 fields to complete the "fermion" multiplet because of renormalizability; the only allowed spin-1 bosons are gauge bosons, but then the fermions should belong to the adjoint representation of the gauge group, in contradiction to what we see experimentally. It is very interesting that renormalizability plus "observation" define the superpartners uniquely. The superpartners of the observed fermions (quarks, leptons) are called sfermions (squarks, sleptons) and have spin 0, while the superpartners of Higgs are called Higgsinos and have spin 1/2. We will see below why in N = 1 global SUSY, the usual Higgs fields cannot be identified with the superpartners of the observed quarks and leptons.

In addition, we introduce a function $f(\phi_i)$, which is known as the "superpotential". f must be an analytic function of the ϕ_i, i.e., a function of ϕ_i but not of their complex conjugates ϕ_j^*. For a renormalizable theory f should be at most cubic in the ϕ_i, otherwise f is restricted only by gauge invariance. The general form of f is $f(\phi_i) = a_i\phi^i + a_{ij}\phi^i\phi^j + b_{ijk}\phi^i\phi^j\phi^k$, where a_i, a_{ij}, b_{ijk} are gauge covariant tensors.

Notice that in SUSY theories the usual Higgs fields cannot be identified with the superpartners of the observed quarks and leptons (sfermions). In ordinary gauge theories we can use a Higgs doublet (H_2) to give masses to up quarks, while its charge conjugate (H_2^C) can provide masses to charged leptons and down quarks. In SUSY theories, since the superpotential f is function only of ϕ_i's and not of ϕ_j^*'s, H_2 and H_2^C should be chosen to be completely unrelated, different fields. Thus, even if we identified the (sν, se) doublet with H_2^C, we are missing H_2, i.e., we are left with massless up quarks (u, c, t)! Furthermore, in a GUT theory, quarks and leptons are sitting in the same multiplet; thus by identifying sleptons with "weak" Higgs fields, we have to interpret squarks as color Higgs fields which is catastrophic, since these color Higgs fields will make protons to decay instantly. For example, in SU(5), the Higgs doublet H_2 is sitting in the same multiplet, a 5-plet of SU(5), with a colored Higgs triplet, H_3. The color Higgs mediates proton decay and since it is coupled to quarks and leptons with the normal Yukawa couplings, it had better be superheavy ($\gtrsim 10^{10}$ GeV), otherwise, matter will disintegrate instantly! Thus, if we identified s-ups or s-downs with colored Higgs fields, we are in big trouble since s-ups and s-downs cannot weigh much more than M_W ($<< 10^{10}$ GeV) if we want to solve the gauge hierarchy problem. Finally, the existence of a pair of Higgs supermultiplets of opposite helicities is crucial in cancelling the Adler-Bell-Jackiw anomalies of the Higgsino sector.

It is an unfortunate fact of supersymmetric life that no known particle can be the spartner of any other particle. In N = 1 SUSY, all particles and their spartners must have identical $SU(3)_C \times SU(2)_L \times U(1)_Y \times$ global baryon and lepton quantum numbers. No known particles fit into the appropriate pairs, so we must invent a doubling of the elementary particle spectrum as seen in Table 1. Gauge theories including the above mentioned supermultiplets embody the following relations[3] between couplings :

$$g\bar{f}_{L \text{ or } R}\, \gamma_\mu\, f_{L \text{ or } R}\, G^\mu \rightarrow \sqrt{2}g\, (\bar{f}_{L \text{ or } R}\, \tilde{G}_{L \text{ or } R})\tilde{f}_{L \text{ or } R} + (\text{h.c.}) \tag{14}$$

$$P \equiv \lambda(f_L f_L')\, H \rightarrow \lambda\, (f_L \tilde{H}_L)\tilde{f}_L' + \tilde{f}_L(f_L'\tilde{H}_L)\ , \tag{15}$$

where the $f_{L,R}$ are left- (right-)handed fermions, G^μ are gauge bosons $\tilde{G}$ are gauginos, the products of two fermions are denoted by

$$(f_L f_L') \equiv \varepsilon^{\alpha\beta}\, f_\alpha f_\beta'\ , \tag{16}$$

and the $\tilde{f}$ ($\tilde{H}$) are "sfermion" ("shiggs") partners of conventional fermions and Higgs bosons.

Table 1. Spectrum of supersymmetric particles.

Particle	spin	Sparticle	spin
quark $q_{L,\ R}$	1/2	squark $\tilde{q}_{L,\ R}$	0, 0
lepton $\ell_{L,\ R}$	1/2	slepton $\tilde{\ell}_{L,\ R}$	0, 0
photon γ	1	photino $\tilde{\gamma}$	1/2
gluon g	1	gluino $\tilde{g}$	1/2
W	1	wino $\tilde{W}$	1/2
Z	1	zino $\tilde{Z}$	1/2
Higgs H	0	shiggs $\tilde{H}$	1/2

It is convenient to introduce the concept of superspace, which possesses anticummuting (Grassmannian) coordinates θ_α as well as the conventional space-time coordinates x_μ. Supermultiplets can then be represented by superfields which can be expressed as power series in the superspace coordinates θ_α, which are of finite order because they anticommute. Chiral superfields ϕ are spin-0 fields of the form

$$\phi(x) = s(x) + \sqrt{2}\theta\psi(x) + \theta\theta F(x) \ , \tag{17}$$

whereas vector superfields V are written as

$$V = -\ \theta\sigma^\mu\bar{\theta}V_\mu(x) + (-\ i\ \bar{\theta}\bar{\theta}\theta\lambda(x) + \text{h.c.}) + \frac{1}{2}\ \theta\theta\bar{\theta}\bar{\theta}[D(x) - i\partial_\mu V^\mu(x)] \ . \tag{18}$$

The so-called auxiliary fields F and D can be eliminated using the equations of motion, leaving us with physical degrees of freedom corresponding just to the supermultiplet structures. The interactions of the supermultiplets are derived from a supersymmetric action which can be written schematically as :

$$A = \int\ [\bar{\phi}\phi e^{2V} + f(\phi) + VV] \ , \tag{19}$$

where the integral over space and superspace denotes an $\int\ d^4x\ d^4\theta$ for the first term, which is a kinematic term for the chiral supermultiplet $\phi \equiv (s,\psi)$ and an $\int\ d^4x\ d^4\theta$ for the last term, which is a kinetic term for the gauge supermultiplets (V, $\tilde{V}$). The middle term in (19) is a $\int d^4x\ d^2\theta$ which gives Yukawa interactions, fermion masses and multiple scalar interactions. The object $f(\phi)$ is a cubic polynomial, introduced above

$$f(\phi) = a_{ij}\phi_i\phi_j + b_{ijk}\phi_i\phi_j\phi_k \tag{20}$$

called the superpotential. Fermion interactions are obtained from $f(\phi)$ (20) by removing two ϕ's and putting in their spin-1/2 ψ components, while taking the scalar components s of any remaining ϕ :

$$(\psi_i\psi_j) \frac{\partial^2 f}{\partial\phi_i\partial\phi_j}\bigg|_{\phi_k=s_k} = a_{ij}(\psi_i\psi_j) + b_{ijk}s_k(\psi_i\psi_j). \tag{21}$$

The first term on the right-hand side of (21) is a fermion mass term, while the second term is a conventional Yukawa interaction. The multiscalar interactions obtained from $f(\phi)$ are

$$\sum_i |F_i|^2 \equiv \sum_i \left|\frac{\partial f}{\partial\phi_i}\right|^2\bigg|_{\phi_i=s_i} = |a_{ij}s_j + b_{ijk}s_js_k|^2 . \tag{22}$$

We easily derive from (20) a $(\text{mass})^2$ matrix :

$$(m_s^2)_{ik} = a_{ij}a^*_{jk} = (m_\psi m^*_\psi)_{ik} , \tag{23}$$

which we see to be identical with the fermion $(\text{mass})^2$ matrix derived from (21). Thus, fermion and boson masses are identical, as we would expect from exact SUSY (6). Gauge interactions also give[3] quartic scalar interactions

$$\frac{1}{2}\sum_\alpha g_\alpha^2 \; |s^*T^\alpha s|^2 : \tag{24}$$

thus, the full scalar potential is given by

$$V(s_i, s_i^*) = \sum_i \left|\frac{\partial f}{\partial s_i}\right|^2 + \frac{1}{2}\sum_a g_a^2 |(s^*, T^a s)|^2 , \tag{25}$$

where the second sum runs over all generators a of the gauge group, the g_a are the gauge coupling constants and T^a are the generators of the gauge group acting on the representation of the group furnished by the s_i. If the gauge group is not semi-simple but contains U(1) generators, say Y_ℓ with charge $g_{Y\ell}$, then its contribution to (25) becomes

$$\frac{1}{2} g_{Y\ell}^2 |(s^*, Y_\ell s) + \xi_\ell|^2 , \tag{26}$$

where ξ_ℓ are arbitrary constants of mass dimension 2. Usually one defines

$$F_i \equiv \frac{\partial f}{\partial s_i} \tag{27}$$

and

$$D_a \equiv (s^*, T^a s) + \delta a Y_\ell \cdot \xi_\ell \ (T^{Y_\ell} \equiv Y_\ell) \quad , \tag{28}$$

and

$$V(s_i, s_i^*) = \sum_i |F_i|^2 + \frac{1}{2} \sum_a g_a^2 |D_a|^2 \quad . \tag{29}$$

In the classical approximation, the zero point energy of the fields may be neglected, and the energy of the ground state just equals the minimum of the potential $V(s_i, s_i^*)$. SUSY is unbroken if $V = 0$ for $s_i = \langle s_i \rangle$. As V is a sum of squares, $V = 0$ if each term separately vanishes. Thus the condition for unbroken SUSY is

$$F_i \equiv \frac{\partial f}{\partial s_i} = 0 \tag{30}$$

for each field s_i, and that

$$D_a \equiv (s^*, T^a s) + \delta a Y_\ell \cdot \xi_\ell = 0 \tag{31}$$

for every generator T^a of the gauge group.

If (30) and (31) have a simultaneous solution, SUSY is unbroken at the tree level. Otherwise, SUSY is spontaneously broken. It is not difficult to show, that a necessary condition to have spontaneous SUSY breaking at the tree level in a supersymmetric gauge theory, is to have one of the following two conditions satisfied :

(i) The group G should contain at least a neutral field X with linear terms in the superpotential f. This is called F-type breaking because (30) cannot be satisfied.

(ii) The group G should contain at least an Abelian factor U(1) with a non-vanishing ξ in (28). This is called D-type breacking because (31) cannot be satisfied.

Similarly to the SB of global symmetry where there are Goldstone bosons, in the case of SB of global SUSY a Goldstone fermion, called goldstino, should be present. In general, up to a normalization factor the goldstino is given by :

$$\psi_g = \frac{1}{2} g_a D_a \lambda^a + \frac{\partial f}{\partial \phi_i} \psi_i \quad . \tag{32}$$

Let us define the coupling M_S^2 of the supercurrent $S_{\mu\alpha}$ ($Q_\alpha = \int d^3x \ S_{0\alpha}$) to the goldstino by :

$$< 0|S_{\mu\alpha}|\psi_\beta > = M_S^2(\gamma_\mu)_{\alpha\beta} \quad . \tag{33}$$

There is a simple and fundamental relation between the value of the potential at the minimum (vacuum energy) V_0 and M_S^2 :

$$V_0 = (M_S^2)^2 = M_S^4 \ . \tag{34}$$

The physical meaning of M_S is apparent : M_S^2 is the "order parameter" of supersymmetry. A value of M_S different from zero implies that supersymmetry is spontaneously broken. In particular, if we denote by ε the coupling of the goldstino to a supermultiplet, then there is a mass splitting between the boson and the fermion inside the supermultiplet :

$$m_B^2 - m_F^2 = M_S^2\varepsilon \ . \tag{35}$$

The form of the mass splitting (35) is very suggestive. If one wishes, one may arrange things in such a way that, by making the coupling of the goldstino to certain supermultiplets small (ε<<1), these supermultiplets suffer mass splittings, much smaller than the primordial SUSY breaking scale M_S. This simple mechanism, the SUSY DECoupling mechanism (SUDEC), has been discovered only recently[9].

Its importance in constructing realistic SUSY models is difficult[9] to overestimate. The reason is very simple. As we have seen before (see (3)), because of the gauge hierarchy problem, M_{LESB}, the mass splitting that the low-energy world supermultiplets suffer has to be of order M_W. Then if we identify M_S with M_{LESB}, a la Fayet[10], no realistic model can be constructed[9,11]. All Fayet type models[10] ($M_S \sim M_{LESB}$) suffer either from Adler-Bell-Jackiw type anomalies or/and flavour changing neutral currents or/and other pathologies related to the standard established low energy phenomenology[9,11]. It seems that we definitely need

$$M_S >> M_{LESB} \tag{36}$$

which, in turn, means that necessarily the SUDEC mechanism[9] has to be employed (ε<<1). Actually, thanks to the "magic" properties of SUSY theories (non-renormalization theorems), one can prove[9,12] that (36) persists to all orders in perturbation theory, i.e., it is stable against large radiative corrections. Indeed, realistic models have been already constructed[9] where one usually finds that

$$M_S^2 \cong M_W \, M_{Planck} \tag{37}$$

implying that, by using (3) and (35),

$$\varepsilon_{LOW} \cong \frac{M_W}{M_{P\ell}} \ . \tag{38}$$

Things become very interesting because local SUSY or supergravity[8] cannot be neglected anymore[13]. In local SUSY the goldstino becomes the missing longitudinal components of the spin 3/2 gravitino, the gauge fermion of local supersymmetry (the superpartner of the spin 2 graviton) through the superhiggs effect[14,15]. In analogy with ordinary gauge theories in which the gauge boson mass is given by

$$M \sim g < \phi > \quad , \tag{39}$$

where g is the gauge coupling constant and $<\phi>$ is the v.e.v. of the scalar field causing the breaking, i.e., the order parameter of the gauge symmetry, the mass of the gravitino is given by[15]

$$m_{3/2} \sim \sqrt{G_{Newton}}\, M_S^2 = M_S^2 \;/\; M_{Pl} \tag{40}$$

in terms of the gravitational coupling constant $\sqrt{G_N}$ ($\equiv 1/M_{Pl}$) and M_S^2, the order parameter of supersymmetry. One then finds[16,17] extra contributions to the RHS of (35) proportional to $m_{3/2}^2$. By putting together (37) and (40) we find that

$$m_{3/2} \sim M_W \quad , \tag{41}$$

implying that the extra gravitational contributions to (35) proportional to $m_{3/2}^2$, cannot be neglected anymore[13], as they are of the same order of magnitude with the non-gravitational ones. One may even suspect that it is possible to create the whole M_{LESB} through gravitational effects. This possibility will be explored further. There is another way of breaking global supersymmetry : the soft way, i.e. interactions of dimensions less than 4 which do not lead to quadratic divergences[18]. For example, scalar and gaugino mass terms belong to the list[18] of allowed soft SUSY breaking terms. The ultraviolet properties of the theory, e.g., softening of divergences etc., hold true[18] in soft breaking as they do in SB, so soft breaking may be employed in physical applications as good as SB. Amazingly enough, recent progress has shown that it is almost impossible[9,11] to construct appealing SUSY models with spontaneous global SUSY breaking, while realistic SUSY models have appeared[19] satisfying all possible phenomenological constraints, with a very definite pattern of soft SUSY breaking emerging from the spontaneous breakdown of local supersymmetry[19]. Gravity seems to play a rather fundamental role here, as will be discussed later.

One of the central features of supersymmetric theories is that, thanks to their fermion-boson symmetry, there are a lot of cancellations between "badly" behaving graphs. This amounts to a much better behaved field theory at the ultraviolet, compared with the ordinary field theories, which make them very attractive. Actually, recently it has been proven that the N = 4 SUSY Yang-Mills (Y-M) field theory in real (4) space-time dimensions is FINITE[20]!

The remarkable property of SUSY gauge theories which has recently made them so interesting is their non-renormalization theorem[21]. Since all loop diagrams have an $\int d^4\theta$ form, there is no intrinsic renormalization of the superpotential $f(\phi)$, but only wave function renormalizations of the chiral supermultiplets :

$$\phi_i \rightarrow Z_{ij}\phi_j \tag{42}$$

and renormalizations of the gauge couplings g_α. There is no intrinsic renormalization of the Yukawa coupling parameters a_{ij}, b_{ijk} in $f(\phi)$ (20). This means that we can "set and forget" them : any small (or vanishing) superpotential term remains small (or vanishing) in all orders of perturbation theory.

"SET IT AND FORGET IT". It is this property that solves[22] the technical aspect of the gauge hierarchy and strong CP-hierarchy[6] problems that we discussed in the beginning as motivation for SUSY theories. With the same token, one can prove that if SUSY is unbroken at the tree level, it remains unbroken to all orders in perturbation theory[21]. It sounds like a magic world full of miracles. But still, it is true !

"LOW ENERGY" PHYSICS AND SUSY

Indeed, if SUSY provides the solution[22] to the gauge hierarchy problem, then the phenomenological implications are tremendous. Firmly, each one of the "standard" particles of the low energy world (quarks, leptons, gauge bosons, Higgs) should have their superpartners in a mass range at most not far above M_W. This fact makes the situation very exciting because the hope exists to discover these particles in the not very far future. Present experimental limits from PETRA and PEP put the mass of any new charged particles approximately above 20 GeV, which puts a lower bound on the mass of charged SUSY particles. I will not discuss here how to find SUSY particles since it has been discussed in lengthy detail elsewhere[23].

Here I will discuss the constraints that well-established phenomenological facts like absence of flavor changing neutral currents (FCNC), absence of strong CP-violation, g-2, etc., impose on SUSY models. Clearly, the introduction of new particles which are coupled to the low-energy world with ordinary gauge or Yukawa couplings and of mass not far above M_W, sounds like trouble. For example, analogously to the gauge boson-fermion-fermion coupling there is a gaugino-sfermion-fermion coupling of comparable strength (see Eq. (14)). Such kinds of couplings contribute to all kinds of rare processes, like Re and Im part of $K_L - K_S$, $K_L \rightarrow \mu\mu$, $\mu \rightarrow e\gamma$, etc.

We saw before that exact SUSY would require degenerate spin-0 and spin-1/2 particles :

$$m_0^2 = m_{1/2}^2 \quad . \tag{43}$$

Since all squarks and charged sleptons must have masses $\gtrsim$ O(20) GeV, it must be that SUSY is broken. A first approach to describing SUSY breaking might be to introduce soft SUSY breaking, i.e. interactions of dimension less than 4 which do not lead to quadratic divergences[18]. These could include the desirable

$$L_{SUSY} \ni - m_{\tilde{q}}^2 |\tilde{q}|^2 - m_{\tilde{\ell}}^2 |\tilde{\ell}|^2 - m_{\tilde{V}}(\tilde{V}\tilde{V}), \tag{44}$$

with $m_{\tilde{q}}^2$ and $m_{\tilde{\ell}}^2$ being general matrices in both flavour and helicity spaces. (Recall that since quarks and charged leptons have both left- and right-handed components q_L, q_R; ℓ_L, ℓ_R, there must exist all the corresponding spartners $\tilde{q}_L$, $\tilde{q}_R$; $\tilde{\ell}_L$, $\tilde{\ell}_R$). We must[24] be careful with our choice of $m_{\tilde{q}}^2$ and $m_{\tilde{\ell}}^2$ so as to avoid problems with flavour-changing neutral interactions. Recall first the usual formalism[25] for the Cabibbo-Kobayashi-Maskawa (CKM) mixing for quarks : a general mass matrix m_q is diagonalized to m_q^D by unitary transformations $U_{L,R}^q$ on the quark fields :

$$m_q \bar{q}_R q_L = m_q^D \bar{q}_R^D q_L^D \; ; \; m_q^D = U_R^q m_q U_L^{q\dagger} \; ; \; q_{L,R}^D = U_{L,R}^q \, q_{L,R} \quad . \tag{45}$$

When we make the transformations (45) the neutral gauge boson interactions remain flavour-diagonal :

$$\bar{q}_{L,R} \, \gamma^\mu \, q_{L,R} \, G_\mu^0 = \bar{q}_{L,R}^D (U_{L,R}^{q\dagger} \, U_{L,R}^q) \gamma^\mu \, q_{L,R}^D \, G_\mu^0 = \bar{q}_{L,R}^D \, \gamma^\mu \, q_{L,R}^D \, G_\mu^0 . \tag{46}$$

However, the left-handed charged currents acquire non-trivial CKM angles because the unitary rotations U_L^u, U_L^d are in general different :

$$\bar{u}_L \gamma^\mu d_L W_\mu^\dagger = \bar{u}_L^D (U_L^{u\dagger} U_L^d) \gamma^\mu d_L W_\mu^\dagger = \bar{u}_L^D U_{CKM} \gamma^\mu d_L W_\mu^\dagger \; ;$$

$$U_{CKM} = U_L^{u\dagger} U_L^d \; . \tag{47}$$

Consider now what could happen to the corresponding neutral gaugino interactions :

$$\bar{\tilde{q}}_{L,R} (q_{L,R} \, \tilde{G}_{L,R}^0) = \bar{\tilde{q}}_{L,R}^D (\tilde{U}_{L,R}^{q\dagger} \, U_{L,R}^q)(q_{L,R}^D \, \tilde{G}_{L,R}^0) \; , \tag{48}$$

where the new unitary rotations $\tilde{U}_{L,R}^q$ diagonalize $m_{\tilde{q}}$ to $m_{\tilde{q}}^D$:

$$m_{\tilde{q}}^{D^2} = U^q m_{\tilde{q}}^2 U^{q\dagger} \; ; \; q_{L,R}^D = U_{L,R}^q \, q_{L,R} \; . \tag{49}$$

The interaction (48) would contain flavour-non-diagonal neutral gaugino couplings unless we arrange that

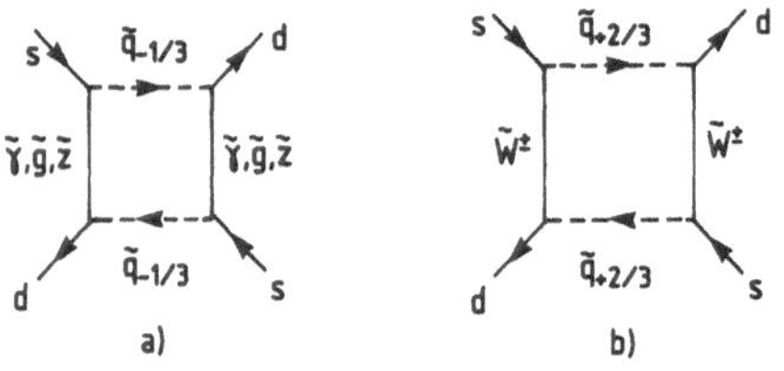

Fig. 1. (a) Neutral gaugino and (b) charged gaugino SUSY box diagrams contributing to the $K_1 - K_2$ mass difference[24,26]

$$\tilde{U}^q_{L,R} = U^q_{L,R} \quad , \tag{50}$$

which in turn implies that the SUSY CKM angles should be identical with the conventional CKM angles :

$$\tilde{U}_{CKM} \equiv \tilde{U}^{u\dagger}_L \tilde{U}^d_L = U^{u\dagger}_L U^d_L = U_{CKM} \quad . \tag{51}$$

You might wonder how important it is to ensure the absence of flavour-changing neutral gaugino interactions. If they were present, the $\tilde{\gamma}$ and $\tilde{g}$ SUSY box diagrams of Fig. 1 would make[24,26] disastrously large contributions to the $K_1 - K_2$ mass difference. Furthermore, if the photino $\tilde{\gamma}$ were light enough to be produced in $K^{\pm} \to \pi^{\pm}\tilde{\gamma}\tilde{\gamma}$ decay, the rate would greatly exceed the experimental upper limit on $K^{\pm} \to \pi^{\pm}+$ nothing observed, unless $\tilde{U}^q_{L,R} = U^q_{L,R}$[27]. Of course, it is not necessarily the case that $m_{\tilde{\gamma}} < (m_K - m_{\pi})/2$, so we may not need to worry about $K^{\pm} \to \pi^{\pm}\tilde{\gamma}\tilde{\gamma}$ decay, but the problem with $K_1 - K_2$ box diagrams exists for all $\tilde{\gamma}$ or $\tilde{g}$ with masses $O(100)$ GeV. To avoid this problem by ensuring the equality (50) of $\tilde{U}$ and U matrices, we must demand that the $m^2_{\tilde{q}}$ matrix be a simple function of the quark mass matrix m_q, presumably a quadratic function :

$$m^2_{\tilde{q}} = \tilde{m}^2 1 + C_1 m_q \tilde{m} + C_2 m^2_q \quad . \tag{52}$$

By making the Ansatz (52) we also avoid any problem[24,26,27] with $\tilde{W}^{\pm}$ box diagrams. Equation (52) guarantees that

$$m^2_{\tilde{c}} - m^2_{\tilde{u}} = O(1)\ (m^2_c - m^2_u) \tag{53}$$

if the $C_i = O(1)$, and therefore, the $\tilde{W}$ box diagrams will be suppressed by a super-GIM mechanism to

$$0\left(\frac{\alpha G_F \sin^2\theta_c}{4\pi's}\right)\left(\frac{m_c^2 - m_u^2}{m_{\tilde{w},\tilde{q}}^2}\right) = 0\left(\frac{\alpha G_F \sin^2\theta_c}{4\pi's}\right)\left(\frac{m_c^2 - m_u^2}{m_{\tilde{w},\tilde{q}}^2}\right), \quad (54)$$

which is of the same order as the $W^{\pm}$ box diagrams if $m_{\tilde{W}}$ and $m_{\tilde{q}}=0(m_W)$. Our problem is then to derive the desirable form (52) in a natural way. Lo and behold mass matrices of the form (52) do occur in most SB SUSY models of F- or D-type or in softly broken SUSY models where the soft breaking is provided from supergravity effects, as will be discussed later.

Similar results are obtained from the analysis of g-2[28] or of the absence of strong CP-violation[6]. Once again, we find that F- or D-type SB SUSY models, or supergravity induced softly broken SUSY models, easily satisfy very stringent types of constraints[28]. It is remarkable that realistic SUSY models satisfy automatically severe low energy phenomenological constraints[24,28], which have been the Nemesis of other alternatives like technicolor models[2].

Being very happy with the low energy front, let us move now to the GUT front.

SUPERSYMMETRIC GUTS

The main reason for supersymmetrizing Grand Unified Theories is of course the solution[22] of the cumbersome gauge hierarchy problem. We have seen that a proliferation of the "low energy" particle spectrum is then necessarily unavoidable. Every "known" particle, fermion, Higgs boson or gauge boson should have its corresponding superpartner with characteristic mass differences of order $0(M_W)$. Additional problems to the ones discussed in the previous section appear. The new "low energy" degrees of freedom will definitely modify the standard program of grand unification and, in general, there is the danger that the whole program will be mucked up. It is remarkable that in SUSY GUTs the standard success of ordinary GUTs remains more or less intact. So let us see how the unification program changes. Our SUSY GUT should contain at least the supersymmetrized SU(3) x SU(2) x U(1) model. This piece of information is enough to give a kind of general analysis. It is clear from the beginning that the unification point is going to be raised. The new "light" degrees of freedom involve fermions and scalars, thus their contribution to the various β functions has the effect of delaying the change of the various coupling constants with energy. Notably, the strong coupling constants fall down with energy much smoother than before and so it will take "longer" for the different coupling constants to "meet". At the same time, one expects a larger unification coupling constant. More precisely, in "minimal" type SUSY GUTs[29] one finds, for the coefficients of the SU(3), SU(2) and U(1) β-functions,

$$\beta_3 = 9 - f \ ,$$

$$\beta_2 = 6 - f - \frac{h}{2} \ ,$$

$$\beta_1 = - f - \frac{3h}{10} \ , \tag{55}$$

where f represents the number of flavors (f ≧ 6) and h stands for the number of "light" Higgs doublets (h ≧ 2).

Concerning the coupling constants we get, using Eq. (55),

$$\frac{1}{\alpha_3(m)} = \frac{1}{\alpha_{SG}} - \frac{1}{2\pi} [9 - f] \ln(\frac{M_{SX}}{m}) \ ,$$

$$\frac{1}{\alpha_2(m)} = \frac{1}{\alpha_{SG}} - \frac{1}{2\pi} [6 - f - \frac{h}{2}] \ln(\frac{M_{SX}}{m}) \ ,$$

$$\frac{1}{\alpha_1(m)} = \frac{1}{\alpha_{SG}} - \frac{1}{2\pi} [- f - \frac{3h}{10}] \ln(\frac{M_{SX}}{m}) \ , \tag{56}$$

where as usual $\alpha_i \equiv g_i^2/4\pi$ (i = 1,2,3), α_{SG} is the SUSY GUT unification fine structure constants and M_{SX} is the SUSY GUT unification mass and m a "low energy" mass scale larger than or equal to $\sim O(M_W)$. We can recast Eqs. (56) in a more useful form

$$\ln \frac{M_{SX}}{M_W} = \frac{2\pi}{18 + h} \left(\frac{1}{\alpha(M_W)} - \frac{8}{3} \frac{1}{\alpha_3(M_W)} \right) \ , \tag{57}$$

$$\sin^2\theta_{e-w}(M_W) = \frac{(3+h/2) + (10-h/3)\alpha(M_W)/\alpha_3(M_W)}{18 + h} \ , \tag{58}$$

$$\frac{1}{\alpha_{SG}} = \frac{(9-f)1/\alpha(M_W) - (6-(8f/3)-h)1/\alpha_3(M_W)}{18 + h} \ , \tag{59}$$

where, for simplicity, we have identified the supersymmetry breaking scale (M_{LESB}) with M_W.

Using Eqs. (57)-(59), and taking into account higher order corrections, we get[30,31]

$$M_{SX} \cong \begin{array}{ll} 6.10^{16}\ \Lambda_{\overline{MS}} & \text{for } h = 2 \\ 3.10^{15}\ \Lambda_{\overline{MS}} & \text{for } h = 4\ , \end{array} \tag{60}$$

where the present favorable value of $\Lambda_{\overline{MS}}$ (the QCD scale parameter evaluated in the modified minimal subtraction scheme with four flavors) is between 100 and 200 MeV. The electroweak angle is calculated to be[30,31]:

$$\sin^2\theta_{e-w}\,(M_W)\Big|_{\text{THEOR.}} = \begin{array}{ll} 0.236 \pm 0.003 & \text{for } h = 2 \\ 0.259 \pm 0.003 & \text{for } h = 4\ , \end{array} \tag{61}$$

while $\alpha_{SG} \cong 1/24$ to $1/25$ for six flavours and two light Higgs doublets. The present measured value of $\sin^2\theta_{e-w}\,(M_W)$ is[1]

$$\sin^2\theta_{e-w}\,(M_W)\Big|_{\text{EXPER.}} = 0.215 \pm 0.010 \pm 0.007\ . \tag{62}$$

We move next to the m_b/m_τ ratio in SUSY GUTs. Here we find[30,31]:

$$\frac{\left(\dfrac{m_b}{m_\tau}\right)_{SUSY}}{\left(\dfrac{m_b}{m_\tau}\right)_{ORD}} = \frac{\left[\dfrac{\alpha_3(M_W)}{\alpha_{SG}}\right]^{8/9}}{\left[\dfrac{\alpha_3(M_W)}{\alpha_G}\right]^{4/7}} \tag{63}$$

and substituting $\alpha_{SG} \cong 1/24$, $\alpha_G \cong 1/41$ and $\alpha_3(M_W) \cong 0.12$, we get

$$\frac{\left(\dfrac{m_b}{m_\tau}\right)_{SUSY}}{\left(\dfrac{m_b}{m_\tau}\right)_{ORD}} \cong 1 \tag{64}$$

and thus by recalling that $(m_b/m_\tau)_{ORD} \cong 2.8\text{-}2.9$, which literally coincides with its experimentally measured value, we declare that (64) is a very successful relation. We find this "coincidence" remarkable. The situation is rather clear. As was expected the unification scale moves upward and the unification coupling constant increases, as does the electroweak angle always compared to the ordinary GUTs results[1]. The m_b/m_τ remains unchanged, a surprise at least to me! Concernign the value of $\sin^2\theta_{e-w}$, it seems to be a bit high for the case of two light Higgs doublets compared with the experimental value (62). On the other hand, the increase of the grand

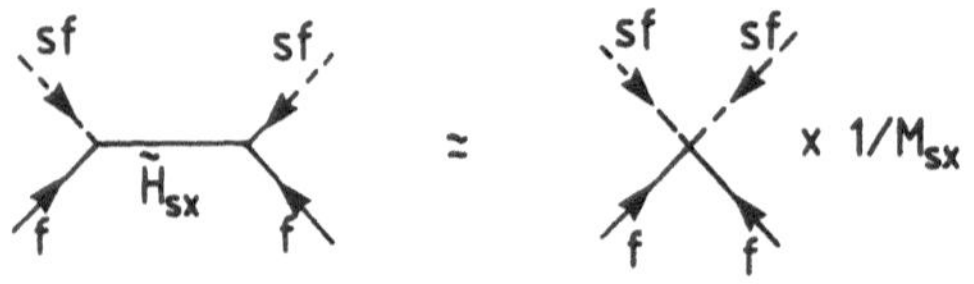

Fig. 2. Dimension five operators contributing to proton decay.

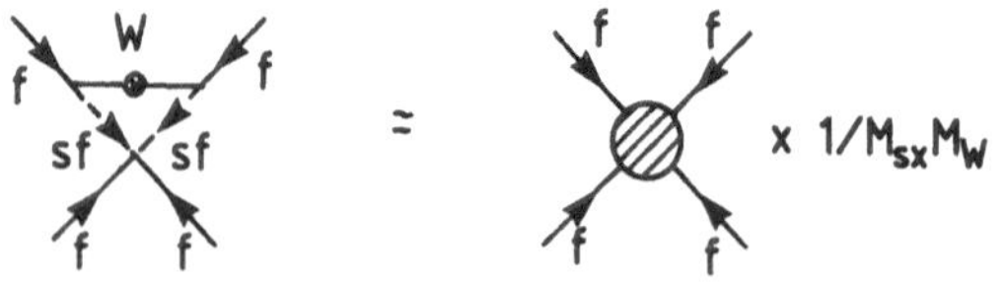

Fig. 3. "Looping" proton decay through dimension five operators.

unification scale by a factor of 0(10) with respect to ordinary GUTs, suppresses the conventional (gauge-boson mediated) proton decay mode, $p \to e^+\pi^\circ$, by a factor of (10^4) compared to the ordinary GUTs value, thus evading any conflict with the present experimental lower bound[1]. However, the show is not over! It has been remarked[32] that in a large class of SUSY grand unified theories, if there are no preventing symmetries, there are loop diagrams that may cause rapid proton decay. For example, by "dressing up" diagrams of the form of Fig. 2, where s_f and H_{SX} represent the SUSY partners of "light" fermions (f) and "superheavy" coloured Higgs triplets respectively, one may get "looping" proton decay where again $\tilde{W}$ stands for the SUSY partners of the charged weak bosons. The bizarre thing here is that $\tau_p \propto M_{SX}^2 M_W^2$ and not $\tau_p \propto M_{SX}^4$. One may then naively think that these kinds of SUSY theories are dead because they cause a too rapid proton decay[32]. A more careful analysis[31] showed though that things are different. Indeed, we have found[31] that in such theories the proton lifetime can easily be 10^{31} years or a bit longer (not much longer though), and with the very "peculiar" characteristic decay mode[31] $\bar{\nu}_\tau K^+$. The appearance of "strange" particles in the final decay products of the nucleon should not sound strange. As it is apparent from Figs. 2 and 3, "looping" nucleon decay involves Yukawa couplings, thus the nucleon will prefer to decay predominantly to the heaviest, energetically allowed quark, i.e. the strange quark! Thus in the so-called softly broken SUSY GUTs (without "preventing" symmetries) we find[31]

$$\tau_N \cong 0(10^{31\pm2}) \text{ years} ,$$

$$\tau(N\to\bar{\nu}_\tau K) >> \Gamma(N\to\bar{\nu}_\mu K) >> \Gamma(N\to\bar{\nu}\pi,\mu^+K) >> \Gamma(N\to\mu^+\pi)$$
$$>> \Gamma(N\to e^+K) >> \Gamma(N\to e^+\pi). \quad (65)$$

But the surprises are not over. Very recently we have found[33] that SUSY GUTs may solve naturally the monopole problem. In doing so though, we may upset the standard solution of the baryon asymmetry problem. One way to reconcile this puzzle and keep both solutions intact[33] is the existence of "light" superheavy triplets, i.e., $M_{H_3} \sim 10^{10}$ GeV. Further details on this interesting possibility will be given later. Actually we find[34] in this case $\sin^2\theta_{e-w} \cong 0.220$ much closer to the experimental value given by Eq. (62), than in other SUSY GUTs (see Eq. (61)). But it is well known[35] that such Higgsons mediate proton decay with lifetime $\sim 0(10^{31\pm 2})$ years, and we find that in SUSY GUTs the decay modes are given by[34]

$$\tau_N \cong 0(10^{31\pm 2}) \text{ years} ,$$
$$\Gamma(\bar{\nu}_\mu K^+,\mu^+K^0) : \Gamma(\bar{\nu}_e K^+,e^+K^0,\mu^+\text{"}\pi^0\text{"}) : \Gamma(e^+\pi^0,\bar{\nu}_e\pi^+)$$
$$\cong 1 : \sin^2\theta_c : \sin^4\theta_c$$
$$(\theta_c \cong \text{Cabibbo angle}) . \quad (66)$$

All these predictions have to be contrasted with the ordinary GUT predictions[1]:

$$\tau_N \cong 10^{29\pm 2} \text{ years} ,$$
$$B(N \to e^+ \text{ non-strange}, \bar{\nu}_e \text{ non-strange}, \mu^+ \text{ or } \bar{\nu}_\mu \text{ strange}) :$$
$$B(N \to e^+ \text{ strange}, \bar{\nu}_e \text{ strange}, \mu^+ \text{ or } \bar{\nu}_\mu \text{ non-strange})$$
$$= 1 : \sin^2\theta_c . \quad (67)$$

The contrast between Eqs. (65), (66) and (67) is rather dramatic. Apart from the case where the protons decay in the "conventional" way (Eq. (67)) but with $\tau_p \propto M_{SX}^4$ and M_{SX} as given by Eq. (60) (which will make life very, very difficult, if not impossible), all other possibilities are very interesting and hopefully not impossible to test experimentally. It should be emphasized once more that proton decay in SUSY GUTs at an observable rate always involves strange particles (K,...) in the final state. This striking difference in the proton decay modes between SUSY and ordinary GUTs is maintained also in supergravity models, as we will see later. Experiment will tell us!

One may wounder if there are at all realistic SUSY models encompassing all different phenomenological constraints previously mentioned. Indeed, realistic SUSY model building is not an easy task. However, any effort is worthwhile since SUSY models are left as the only candidates for a physical description of the world, at least up to energies of the Planck scale M_P. Any high standard(s) SUSY model should satisfy the following two SUSY golden rules :

1) It should provide naturally an acceptable form of SUSY breaking such that

$$\underset{\substack{\uparrow \\ \text{Experim.}}}{0((20\ \text{GeV})^2)} \leq m_B^2 - m_F^2 \equiv \tilde{m}^2 \leq \underset{\substack{\uparrow \\ \text{Gauge Hierar.}}}{0(M_W^2)} \ , \tag{68}$$

where $\tilde{m}^2$ is a typical boson-fermion mass splitting of a supermultiplet. The sparticle mass spectrum should be such that not only all types of low energy constraints are satisfied [e.g., Eq. (59)] but in addition some possible potential problems of the standard model should find a satisfactory resolution.

2) It should provide a complete solution to the three-fold gauge (scale) hierarchy problem : create, stabilize and dynamically explain the scale hierarchy. All (small) mass scales should be determined dynamically in terms of one fundamental one, the super-Planck scale $M \equiv (M_P/\sqrt{8\pi}) \cong 2.4.10^{19}$ GeV :

$$\frac{M_W}{M} \cong \frac{\tilde{m}}{M} \cong 0(10^{-16}) \ . \tag{69}$$

Surprisingly enough, such no-scale models[36-41] have recently been constructed. Since their construction involves a lot of very interesting physics, it is worth discussing the different steps that lead (uniquely?) to them. Furthermore, it should be recalled that we would like to understand in a natural, satisfactory way :

1) How to separate, at the tree-level the masses of the Higgs doublet and its GUT partner, the Coloured Higgs Triplet ?

2) How to incorporate gravitational interactions, etc.

3) The absence of the cosmological constant.

A possible answer to all these questions may be potentially found in the framework of local SUSY theories or supergravity[8], where we move next. It should be stressed that the move to supergravity theories is not only for aesthetical reasons but it is entailed by the structure of realistic SUSY models, as discussed above (see the remarks after Eq. (41)).

PHYSICAL STRUCTURE OF SIMPLE (N = 1) SUPERGRAVITY

We are then led to consider local SUSY gauge theories[16]. The effective theory below the Planck scale must be[42] N = 1 supergravity. The restriction to N = 1 follows from the apparent left-right asymmetry of the "known" gauge interactions. Since we are dealing with local SUSY, the breaking of SUSY must be spontaneous, not explicit, if Lorentz invariance or unitarity are not to be violated. It is remarkable that the effective theory below $M_{P\ell}$ has been determined[42] to be a spontaneously broken N = 1 local SUSY gauge theory[16].

We start with a reminder of the structure of N = 1 supergravity actions[16] containing gauge and matter fields (if not explicitly stated, we use natural units $k^2 \equiv 8\pi G_N = (8\pi/M_{P\ell}^2) \equiv 1/M^2 = 1$)

$$A = \int d^4x\, d^4\theta\, E(\Phi(\phi, \bar{\phi}e^2) + \mathrm{Re}[R^{-1}g(\phi)] + \mathrm{Re}[R^{-1}f_{\alpha\beta}(\phi)\, W_a^{\alpha}\varepsilon^{ab}\, W_b^{\beta}]) , \tag{70}$$

where E is the superspace determinant, Φ is an arbitrary real function of the chiral superfields ϕ and their complex conjugates $\bar{\phi}$, V is the gauge vector supermultiplet, R is the chiral scalar curvature superfield, g is the chiral superpotential, $f_{\alpha\beta}$ is another chiral function of the chiral superfields ϕ, and W_a^{α} is a gauge-covariant chiral superfield containing the gauge field strength. In addition, to all the obvious general co-ordinate transformations, local supersymmetry and gauge invariance, the action (70) is also invariant[16] under the transformations

$$J \equiv 3\ \ln(-\tfrac{1}{3}\Phi) \to J + K(\phi) + K^*(\bar{\phi})\ ;\ g(\phi) \to e^{K(\phi)}g(\phi) , \tag{71}$$

where $K(\phi)$ is any analytic function of ϕ. These transformations can be related to a description of the chiral superfields ϕ as co-ordinates on a Kähler manifold with Kähler potential G, defined by

$$G \equiv J - \ln(1/4\ |g|^2) , \tag{72}$$

and the transformations (71) are known as Kähler gauge transformations[43].

The general couplings of chiral and vector multiplets (13) to N = 1 supergravity is specified by two functions of the complex scalar fields ϕ_i contained in the chiral multiplets[16]. An analytic function $f_{ab}(\phi) = f_{ba}(\phi)$, related to the Y-M part of the Lagrangian, gives for the kinetic terms of the gauge fields,

$$\frac{-1}{4}\,(\mathrm{Re}\ f_{ab})\ F_{\mu\nu}^{\alpha}F^{\bar{b}\mu\nu} + \frac{i}{4}\,(\mathrm{Im}\ f_{a\bar{b}})\ F_{\mu\nu}^{a}\ F^{\bar{b}\mu\nu} , \tag{73}$$

(a, b are indices of the adjoint representation of the gauge group G). Then, a real gauge invariant function $G(\phi,\phi^*)$ (72) the Kähler potential, defines the scalar kinetic terms, given by

$$G^i_j(\partial_\mu\phi^j)(\partial^\mu\phi^*_i) \tag{74}$$

in the notation

$$G_i \equiv \frac{\partial G}{\partial\phi^i}, \quad G^i \equiv \frac{\partial G}{\partial\phi^*_i}, \quad G^i_j \equiv \frac{\partial^2 G}{\partial\phi^*_i\partial\phi^j} .$$

The kinetic terms have a form characteristic of supersymmetric non-linear σ models. The scalar fields ϕ_i in N = 1 SUGAR span a Kähler manifold with G^i_j (52) as its metric[43]. Clearly, the functions $G(\phi,\phi^*)$ and $f_{ab}(\phi)$ largely determine the physics of the N = 1 SUGAR Y-M theories[16]. Indeed, the scalar potential V has two terms[16,13,44]:

$$V = V_c + V_g . \tag{75}$$

The gauge potential V_g reads

$$V_g = \frac{1}{2}\,(\mathrm{Re}\, f_{ab}{}^{-1})\, D^a D^{\bar{b}} , \tag{76}$$

where the real functions D^a are

$$D^a = g_a\, G_j\, (T^a)^j_i\, \phi^i \tag{77}$$

(g_a is the gauge coupling constant associated to the normalized generator T^a). The "chiral" potential is

$$V_c = \exp G[G_i G^j (G^j_i)^{-1} - 3]. \tag{78}$$

It is apparent from the potential (78) that, unlike the global case, spontaneously broken SUSY does not imply $\langle V\rangle > 0$. This is fortunate since one can now obtain spontaneous breakdown of local SUSY (the super-Higgs effect) in Minkowski space ($\langle V\rangle = 0$). The theory contains also a gravitino mass term, $m_{3/2}\, \bar{\psi}_{\mu L}\, G^{\mu\nu}\psi_{\nu L}$, with

$$m_{3/2} = \langle\, \exp G/2\, \rangle . \tag{79}$$

The most naïve choice for the functions G and f_{ab} would be the ones that provide canonical scalar and vector kinetic terms,

$$G^j_i = \delta^j_i \quad ; \quad f_{ab} = \delta_{a\bar{b}} , \tag{80}$$

corresponding to a flat Kähler manifold. In such cases, one writes

$$G = \phi_i\phi^{i*} + \ln|f(\phi)|^2 , \tag{81}$$

where $f(\phi)$ stands for the gauge invariant superpotential. The "minimal" choice of G and f_{ab} in (80) has some rather unpleasant consequences. The cosmological constant $\langle V\rangle = \Lambda$ is zero due to some unbearable fine-tuning of the parameters in G. Furthermore, scalar boson masses are proportional[17] to the gravitino mass (79), as they are given by the curvature of the potential (78) at the minimum. In this case, the gravitino mass is essentially a free parameter and because of its relation[17] to scalar boson masses or equivalently to $\tilde{m}$ (68), it has to be chosen by hand $O(M_W)$. Dynamical determination of $\tilde{m}$ is excluded, no way to satisfy the second SUSY golden rule!

There is, however, a very elegant way to circumvent these two unsatisfactory points : there exist non-trivial Kähler potentials for which the chiral potential V_c is identically zero[131,132]. Supersymmetry is, however, broken. Vacuum expectation values are not determined by the classical theory, ditto for the gravitino mass. The cosmological constant is naturally zero[45,46]. As we will see later, radiative corrections are then used[36-41] to determine the various scales of gauge symmetry breaking, which in general will be closely related to the gravitino mass. Eureka! This is what we are aiming at to satisfy the second SUSY golden rule. In principle, it is sufficient to require zero chiral potential only in the direction of a gauge singlet complex scalar z field, the Polonyi field. In such a case, we may rewrite (79) as[45]

$$V_c = 9 \exp(\frac{4}{3} G)\, G^{-1}_{zz^*} \frac{\partial^2}{\partial z \partial z^*} \exp(-\frac{1}{3} G) , \tag{82}$$

and

$$V_c \equiv 0$$

implies

$$\frac{\partial^2}{\partial z \partial z^*} \exp(-\frac{1}{3} G) = 0$$

with solution[45]

$$G = -3 \ln(z + z^*) . \tag{83}$$

The scalar kinetic term $G_{zz^*}(\partial_\mu z)(\partial_\mu z^*)$ is never canonical, and the gravitino mass is[45]

$$m_{3/2} = \langle (z + z^*)^{-3/2} \rangle . \tag{84}$$

$m_{3/2}$ is undetermined but non-zero since

$$\langle G_{zz^*} \rangle = 3\, (m_{3/2})^{4/3} \neq 0 . \tag{85}$$

$V_c \equiv 0$ entails a very particular geometry of the Kähler manifold. The Kähler curvature

$$R_{zz^*} \left(\equiv \frac{\partial^2}{\partial z \partial z^*} \ln G_{zz^*} \right)$$

is given gy[45]

$$R_{zz^*} = \frac{2}{3} G_{zz^*} \tag{86}$$

or $R \equiv (R_{zz^*}/G_{zz^*}) = 2/3$. Equation (86) means that the Kähler manifold is an Einstein space (maximally symmetric space) i.e., that the scalar field z is a co-ordinate of the coset space SU(1,1)/U(1)[45,37]. The non-compact global SU(1,1) invariance can be checked explicitly in the whole Lagrangian apart from the gravitino mass term[37,47]. It is very interesting to notice[45] that the N = 1 Lagrangian for one chiral multiplet with vanishing potential corresponds, up to the gravitino mass term, to a particular truncation of N = 4 supergravity, which is known[48] to possess an SU(1,1) non-compact global symmetry. Vanishing chiral potentials for an arbitrary number n of chiral multiplets also exist[45,37-41,47]. In one case[38], the scalar fields ϕ_i, i = 1, 2, ..., n, are co-ordinates of an SU(n,1)/SU(n)x U(1) coset space, which is an Einstein space with curvature given by (86) but with (n+1) replacing 2 in the numerator.

Clearly, in both cases n = 1 or n > 1 the "flatness" of the potential implies massless scalar bosons, neglecting radiative corrections. There is some kind of "curvature-conservation" between the Kähler manifold spanned from the scalar fields of the chiral multiplets and the chiral potential V_c, as schematically represented by Fig. 4.

It should be stressed that higher than four ungauged extended supergravities also exhibit[49] invariances under non-compact global groups which contain SU(1,1) as a subgroup :

$$N = 5 : SU(5,1);\ N = 6 : SO^*(12);\ N = 7,8 : E_{7,7}\ . \tag{87}$$

In addition, extended N ≧ 4 and gauged N ≧ 2 supergravities do inevitably contain[48,49] non-minimal kinetic terms for the gauge smultiplets [$f_{ab} \neq \delta_{ab}$ in Eq. (73)]. As a result, tree level gaugino masses are introduced, of the form[16]:

$$m_{\tilde{V}} = m_{3/2} \frac{\langle 0|f'|0\rangle}{\langle 0|f|0\rangle} \frac{G'}{G} \tag{88}$$

for $f_{ab} = \delta_{ab} f(\phi_i, z)$, and with primes indicating z-differentiation. The usefulness of this remark will become apparent shortly. After this unavoidable disgression into the "esoterics" of N = 1 SUGAR, the road is now open to physical applications.

N = 1 supergravity Y-M field theories seem to be the only available framework for physics description below the Planck scale M.

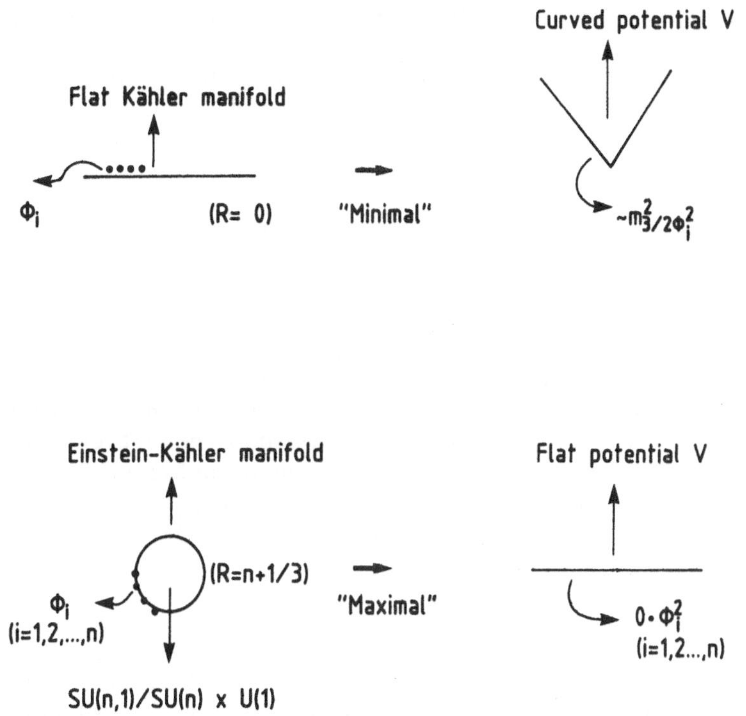

Fig. 4. Relation between the form of the Kähler manifold and the induced scalar (chiral) potential.

Alas, these theories are not renormalizable! This fact should not bother us, since anyway we are going to use the N = 1 SUGAR framework as an effective theory to describe physics for scales below M. Indeed, Ellis, Tamvakis and myself have suggested[16] interpreting Eq. (70) as an effective action suitable for describing particle interactions at energies << $M_{P\ell}$ just as chiral SU(N) x SU(N) Lagrangians were suitable for describing hadronic interactions at energies << 1 GeV. In much the same way as we know that physics gets complicated at E = 1 GeV, with many new hadronic degrees of freedom having masses of this order, we also expect many new "elementary particles" to exist with masses $O(M_{P\ell})$. It may well be that all the known light "elementary particles", as well as these heavy ones, are actually composite, and that at energies $\gg M_{P\ell}$ a simple preonic picture will emerge, analogously to the economical description of high-energy hadronic interactions in terms of quarks and gluons. It may even be that these preonic constituents are themselves ingredients in an extended supergravity theory[50]. But let us ignore these speculations for the moment and return to our pedestrian phenomenological interpretation of the action (70).

The well known rules of phenomenological Lagrangians[51] are that one should write down all possible interactions consistent with the conjectured symmetries (e.g., chiral SU(2) x SU(2)), and only place absolute belief in predictions which are independent of the general form of the Lagrangian (e.g., $\pi\pi$ scattering lengths). These are the reliable results which could also be obtained using current algebra arguments. It does not make sense to calculate strong interaction radiative corrections (read : supergravity loop corrections) to these unimpeachable predictions : these are ambiguous until we know what happens at the 1 GeV scale (read : $M_{P\ell}$), and our ignorance can be subsumed in the general form of the phenomenological Lagrangian, in which any and all possible terms are present a priori (read : non-trivial J, non-polynomial g and $f_{\alpha\beta}$). On the other hand, non-strong interaction radiative corrections can often be computed meaningfully (e.g., the $\pi^+ - \pi^0$ mass difference, large numbers of pseudo-Goldstone boson masses in extended technicolor theories). Similarly, it makes sense to compute matter interaction (gauge, Yukawa, Higgs) corrections to the tree level predictions of the effective action (70).

Since the supergravity action is non-renormalizable, and since both the ϕ and $f_{\alpha\beta}$ terms in the action (70) have a $\int d^4\theta$ form, we expect general variants of them to be generated by loop corrections. Presumably, radiative corrections maintain the essential geometry of the Kähler manifold[43]. Therefore, we expect loop corrections to fall into the class of Kähler gauge transformations (71). The only analogous transformation allowed in a conventional renormalizable theory is K = constant, corresponding to a wave function renormalization. In our case, more general gauge functions $K(\phi)$ might appear.

Thus it seems that we have enough justification to use the N = 1 SUGAR framework as an effective theory to describe physics below M, according to the general scheme

$$\mathcal{L}(N = 1\ \text{SUGAR}) \xrightarrow[E < M]{} \mathcal{L}(N = 1\ \text{SUSY}) + \mathcal{L}_{SOFT} , \qquad (89)$$

where $\mathcal{L}_{SOFT}$ stands for a highly constrained set of SUSY breaking terms. The passage described by (89) is carried out by making a choice for G (72) and f_{ab} (73) such that

1) Supersymmetry is spontaneously broken;
2) Certain fields associated with this breaking (z) decouple (hidden sector);
3) Certain fields become superheavy (X);
4) Remaining fields (ϕ_i) are to be observed in low energy theory (observable sector). After shifting all fields by their v.e.v.s, and discarding terms involving decoupled (z) or superheavy fields (X), $\mathcal{L}_{SOFT}$ is obtained[52,53,17]

$$\mathcal{L}_{SOFT} = m^2 \sum_i |\phi_i|^2 + (m \sum_n (A-3+n) f_n + \text{h.c.}) - \frac{1}{2} [m_{\tilde{g}} \tilde{g}\tilde{g} + m_{\tilde{W}} \tilde{W}\tilde{W} + m_{\tilde{B}} \tilde{B}\tilde{B} + \text{h.c.}] , \tag{90}$$

while more general forms[54] are certainly possible but irrelevant to our discussion here. A is a model-dependent parameter[53] of O(1) and f_n is the nth term in the superpotential $f(\phi_i) = \Sigma_n f_n = \Sigma_n C_n \phi_i^n$ with n = 1,2,3,..., not necessarily terminating at 3, because non-renormalizable terms are allowed in these effective theories. If we imagine that the low energy theory is embedded in a GUT model at some GUT scale below the Planck mass, then all gaugino masses are equal at the GUT scale, so that only one single parameter M_0 is needed (M_0 in principle may be of $O(M_W)$) :

$$m_{\tilde{V}} (M_X) = M_0 , \tag{91}$$

while at lower energies $m_{\tilde{V}}$ evolves in a manner identical for the gauge couplings :

$$\frac{m_{\tilde{V}}(\mu)}{M_0} = \frac{\alpha_\alpha(\mu)}{\alpha_G} , \quad \alpha = 1, 2, 3 , \tag{92}$$

with α_α, α_G the usual SU(3), SU(2), U(1) GUT fine structure constants. The mass parameters m and $m_{\tilde{V}}$ in (90) depend on the form of G and f_{ab}. We may distinguish three interesting cases (see also Fig. 4) :

i) "Minimal" (80)[52,53]
[All scalar fields z;ϕ_i satisfy (81)]:

$$\begin{cases} m = m_{3/2} \\ m_{\tilde{V}} = 0 \end{cases} \rightarrow \tilde{m} = O(m_{3/2}) .$$

ii) "Mini-Maxi"[36,37]
[All scalar fields ϕ_i satisfy (81), while the z (Polonyi) field satisfies (83)] :

$$\begin{cases} m = m_{3/2} \neq 0 \ (84) \\ \downarrow \\ \text{undetermined at} \rightarrow \tilde{m} = O(m_{3/2}) \\ \text{tree level} \\ m_{\tilde{V}} = \begin{cases} 0 \\ \text{or} \\ \text{given by (88)} \end{cases} \end{cases}$$

iii) "Maximal"[38-41]
[All scalar fields z, ϕ_i satisfy (83)-like G's] :

$$\begin{cases} m = 0 \ ; \ A = 0 \\ m_{\tilde{V}} \neq 0 \ (88) \\ \downarrow \qquad \rightarrow \quad \tilde{M} = O(m_{\tilde{V}}) \ . \\ \text{undetermined at} \\ \text{tree level} \end{cases}$$

It is interesting to notice that in the "Maximal" case iii), the emerging low energy theory[38] (observable sector) is globally supersymmetric (m = 0, A = 0) and so all the burden of the necessary global SUSY breaking shifts unavoidably to the gaugino mass $m_{\tilde{V}}$[38-41]. If the gaugino mass is non-zero (68), then radiative corrections will generate non-zero scalar masses. Thus, we expect the gaugino mass to be $O(M_W)$, but the gravitino mass could a priori be very different[39-41]. This possible decoupling[39-41] of the local SUSY breaking parameter ($m_{3/2}$) and the global SUSY breaking parameter ($\tilde{m}$) has some very interesting particle physics and cosmological implications to be discussed later. In order to cover all cases, $\tilde{M}$ will generically stand for either $m_{3/2}$ or $m_{\tilde{V}}$, both originating from supergravity and providing the "seed" for global SUSY breaking. Clearly, the form of $\mathcal{L}_{SOFT}$ as given by (90) is simple enough.

In the physics applications which follow, we shall make extensive use of two main characteristics of the general framework discussed above. First, since we are dealing with an effective theory (the N = 1 SUGAR action is non-renormalizable), the superpotential g is not anymore necessarily constrained by renormalizability to be at most cubic, but is may contain any higher powers, suitably scaled, by inverse powers of $M_{P\ell}$, the natural cut-off of the theory[42]. Secondly, because of the non-renormalization theorems[21] of SUSY (SET IT AND FORGET IT principle), we may set, as we wish, certain parameters equal to zero, even if no symmetry implies that - a very different situation from ordinary gauge theories. Here, no apologies are needed. As explained in detail before, most of the physics is contained in the "observable" sector superpotential $f(\chi_i)$. Here we shall assume that, in one way or another, the "hidden" sector has played its role, as discussed previously, and we shall concentrate on the form of $f(\chi_i)$. We follow the natural (cosmic) evolution of things starting at energies below $M_{P\ell}$ and "coming down" to M_W. So we distinguish physics around the GUT scale (M_X) and physcis around the electroweak (e-w) scale (M_W).

All physics from $M_{P\ell}$ down to (and including) low energies should emerge from such a program. We will show next that this is indeed possible.

PHYSICS WITH SIMPLE (N = 1) SUPERGRAVITY

Physics around the GUT scale (M_X)

The superhigh energy regime ($\sim 10^{16}$ GeV) is the theorists' paradise. There is a lot of freedom in building models, even though the constraints both from particle physics and cosmology become tighter and tighter. For definiteness, simplicity, and out of habit, we shall take as our prototype GUT an SU(5) type model[1]. All GUT physics information will be contained in f_{GUT}, the GUT part of the "observable sector" superpotential. There is no consensus about the definite form of this superpotential, but is should unavoidably contain a piece (f_I) that breaks SU(5) down to SU(3) x SU(2) x U(1) and if possible, a piece (f_{II}), providing some explanation about the tree level gauge hierarchy problem, so we write :

$$f_{GUT} = f_I + f_{II} \ . \tag{94}$$

For example, we may take[55]:

$$f_I = \frac{a_1}{M} X^4 + \frac{a_2}{M^2} X^2 \mathrm{Tr}(\Sigma^3) \tag{95}$$

and[56]

$$f_{II} = \bar{\theta} H(\lambda_1 \frac{\Sigma^2}{M} + \lambda_2 \frac{\Sigma^3}{M^2} + \ldots) + \bar{H}\theta(\lambda_1' \frac{\Sigma^2}{M} + \lambda_2' \frac{\Sigma^3}{M^2} + \ldots) + M_\theta \bar{\theta}\theta \tag{96}$$

where $X = \underline{1}$, $\Sigma = \underline{24}$, $\overset{(-)}{\theta} = \underline{\overset{(-)}{50}}$, $\overset{(-)}{H} = \underline{\overset{(-)}{5}}$ are chiral superfields of SU(5). The Higgs fields H and $\bar{H}$ couple to quark and lepton fields in the usual way. All components of θ and $\bar{\theta}$ have a bare mass M (which is taken to be of order M_X or larger), and so remain heavy after SU(5) breaks to SU(3) x SU(2) x U(1). After minimizing the potential, obtained by plunging into (81),(78) the sum of f_I and f_{II} as given by (95) and (96), we get zero v.e.v.'s for $\overset{(-)}{H}$ and $\overset{(-)}{\theta}$ but non-zero ones for :

$$\langle X \rangle = \left(\frac{m_{3/2}}{M}\right)^{3/8} M \ ,$$

$$< \Sigma > = (\frac{m_{3/2}}{M})^{1/4} M \quad . \tag{97}$$

Furthermore, we find[55] that the SU(3) x SU(2) x U(1) symmetric minimum is the lowest one for all values of a_1 and a_2, with a value :

$$V_{eff} \cong - (\frac{m_{3/2}}{M})^{5/2} M^4 \quad . \tag{98}$$

What do these results mean? First, since the v.e.v. of Σ sets the scale of SU(5) breaking, we find that the GUT scale M_X satisfies[55]:

$$M_X^4 \cong O(m_{3/2} M^3), \tag{99}$$

which is a highly successful relation. Using as an input the ratio $(M_X/M) \sim 10^{-2}- 10^{-4}$, we obtain that $m_{3/2} \sim O(100$ GeV)! More generally, relations of the form $M_X^{2p-2} \cong O(m_{3/2} M^{2p-3})$ with $p \geqq 3$, are also possible[55] by suitably modifying the exponents in (95). The supergravity hierarchy problem has been solved in a rather simple way.

Secondly, the SU(3) x SU(2) x U(1) symmetric minimum is lower in energy density than the SU(5) symmetric minimum $X = \Sigma = 0$ by an amount $(m_{3/2}/M)^{5/2} M^4$. Thirdly, the barrier between these two minima is never larger than $(m_{3/2}/M)^{5/2} M^4$, the same as the splitting between the states. Why this is so can be seen by noting that if we replace X by its v.e.v. (97) in (95), the effective renormalizable self-coupling of Σ is 10^{-12} $Tr(\Sigma^3)$. Thus we have generated[55] a small renormalizable coupling for Σ from our starting point of only non-renormalizable interactions among X and Σ. This small coupling suppresses the barrier the SU(5) and the SU(3) x SU(2) x U(1) phases. The consequences of this suppression for supercosmology[33] are difficult to overestimate. Simply, it now makes possible the transition from the SU(5) to the SU(3) x SU(2) x U(1) phase at temperatures $T \sim 10^{10}$ GeV, which was previously blocked, since the barrier between the two phases was of the order of $(M_X)^4$. Incidentally, in this picture, the number density of GUT monopoles is naturally suppressed[33] below its present experimental upper bound.

It should be clear that the basic result - small renormalizable couplings arising from non-renormalizable ones suppressed only by inverse powers of M - is quite general and does not depend on the detailed form of the superpotential (f_I)[55]. The main characteristics of these types of models[55,57] are that they provide relations of the type (99); they make possible "delayed" SU(5) to SU(3) x SU(2) x U(1) phase transitions at $T \sim 10^{10}$ GeV, and they contain[55,57] more "light" particles than the ones in the minimal SUSY SU(3) x SU(2) x U(1) model. This last fact may sound dangerous when calculating M_X, $\sin^2\theta_{e-w}$ and m_b/m_τ, since in general an arbitrary increase of "light" stuff gives an out-of-hand increase[30,31,58] and thus experimentally unacceptable values for the above-mentioned quantities. A more careful analysis[59] of these cosmological acceptable models (CAM)[59] shows that they make predictions as successful (for $\sin^2\theta_{e-w}$, m_b/m_τ,..) as at least the ones[30,31] of the phenomenologically acceptable minimal type models (MIM). For a detailed, thorough phenomenological analysis of CAMs, see Ref. 59.

Next, we discuss[56] physics related to f_{II} as given in (96). The v.e.v. of Σ does not only break SU(5) to SU(3) x SU(2) x U(1) but also provides a mass term which mixes the color triplets in H and $\bar{H}$ with those in θ and $\bar{\theta}$. However, there is no weak doublet in the $\underline{50}$, and so the weak doublets in H and $\bar{H}$ remain massless. The color triplets will have a mass matrix[56]:

$$\begin{pmatrix} 0 & M_X^2/M \\ M_X^2/M & M_\theta \end{pmatrix} , \tag{100}$$

where M_θ should be of order M_X or larger ($\leq M$), to avoid having particles from θ and $\bar{\theta}$ influencing the renormalization group equations at scales below M_X (or even M). The eigenvalues of this mass matrix are $O(M_\theta)$ and $O(M_X^4/M_\theta M^2)$; this latter eigenvalue is about 10^{10} GeV for $M_X \sim O(10^{16}$ GeV) and $M_\theta \sim O(M)$. In this case, the Higgs color triplet can be used to generate[33,55] the baryon number of the universe after the SU(5) to SU(3) x SU(2) x U(1) transition which, as discussed earlier, occurs at temperatures $T \sim 10^{10}$ GeV in CAMs[59]. It is remarkable that $O(10^{10}$ GeV) is the lower bound[35] allowed for color triplet Higgs masses from present limits[1] on proton decay ($\tau_p > 10^{31-32}$ years). If indeed there are 10^{10} GeV Higgs triplets, then protons should decay predominantly[34] to $\bar{\nu}_\mu K$, μK with a lifetime $\sim O(10^{31-32}$ years), as it was discussed before (see Eq. (66)). Of course in this case one has to control[34] the menace of dimension 5 operators mediating proton decay (see Figs. 2, 3), because they lead to a catastrophically short proton lifetime[34].

The role of supergravity in this natural explanation[56] of the Higgs triplet-doublet mass splitting ($\equiv$ tree level gauge hierarchy problem) is fundamental, in several aspects. The same kind of explanation had been suggested before[60] in the framework of renormalizable global SUSY GUTs, where Σ^2 in (96) was replaced by a $\underline{75}$ of SU(5) and higher than two powers of Σ were absent[60]. Unfortunately, the use of $\underline{75}$ drastically conflicts with cosmological scenarios[33,55,57] based on SUSY GUTs. The barrier between the SU(5) and SU(3)xSU(2)x U(1) phases is impossible to overcome unless most of the $\underline{75}$ is very light ($\sim M_W$). But then, all hell breaks loose. A light $\underline{75}$ makes the gauge coupling in the SU(5) phase decrease at lower energies so there is no phase transition at all[33,55]. Furthermore, the presence of these new light particles in the SU(3) x SU(2) x U(1) phase changes the renormalization group equations, and prevents perturbative unification. On the contrary, in SUGAR theories, since we may use non-renormalizable terms, we may replace the fundamental $\underline{75}$ by an "effective" $\underline{75}$ contained in Σ^2. Unlike a light $\underline{75}$, a light $\underline{24}$ neither makes the SU(5) gauge coupling decrease at energies below M_X, nor upsets perturbative unification[56]. The previously mentioned cosmological scenarios[33,55,57] can proceed without modification. In addition, SUSY non-renormalization theorems[21] ensure the stability of the triplet-doublet splitting to all orders in perturbation theory. Since the only modifications of the theory are at the GUT scale M_X, it seems that we have got[56] a harmless and elegant solution of the tree level, and for that matter, to all orders in perturbation theory, gauge hierarchy problem.

SUGAR models give good physics at the GUT scale - unique, cosmologically acceptable breaking of SU(5) to SU(3) x SU(2) x U(1), with an explanation of the smallness of the gravitino mass[55] ((99)-like relations), and a natural explanation[56] of the Higgs triplet-doublet splitting, cosmologically fitted and general enough. We believe that even if the very specific form of f_I in (95) may change, then f_{II} as given by (96) (or its obvious generalization to other GUT models) will be always a useful part of the f_{GUT}.

After finding plausible explanations for the SUGAR hierarchy problem (gravitino mass $\sim$ 0(100 GeV)), the tree level and higher orders gauge hierarchy problem (triplet-doublet Higgs splitting), it is time to explain the last gauge hierarchy problem (2), i.e., why does $M_W/M_X \leq 0(10^{-13})$? This problem brings us naturally to our next subject.

Physics around the electroweak (e-w) scale (M_W)

Although there is no consensus on the best way to incorporate grand unification in SUGAR models, a unique minimal low energy model has recently emerged[42,61-63]. In this model, the physics of the TeV scale is described by an effective SU(3) x SU(2) x U(1) gauge theory, in which the breaking of weak interaction gauge symmetry is induced by renormalization group scaling of the Higgs $(\text{mass})^2$ operators[42]. Much of the attractiveness of this model stems from the fact that no gauge symmetries or fields beyond those required in any low energy SUSY theory are included. Sometimes, it may happen, as is the case of Cosmological Acceptable Models[59](CAMs), that there are GUT relics which are light ($\sim M_W$), but they do not seem to play any fundamental role at low energies, so we may neglect them in our present discussion. Furthermore, adding random chiral superfields to the low energy theory may be problematic. For example, the presence of a gauge singlet superfield coupled to the Higgs doublets and added to trigger SU(2) x U(1) breaking[52-54], usually (but not always[64]) destroys[65] any hope of understanding the gauge hierarchy problem; the reason being[65] that in a GUT theory, the gauge singlet does not only couple to the Higgs doublets but also to their associate, superheavy color triplets. Then we have to try hard[64] to avoid $\sqrt{M_W M} \cong 10^{10}$ Higgs doublet masses, generated by[65] one-loop effects involving color triplets. Something smells fishy.

We focus then on the standard low energy SU(3) x SU(2) x U(1) gauge group, containing three generations of quarks and leptons, along with two Higgs doublets, as chiral superfields. The low energy effective superpotential (f_{LES}) of the model consists only of the usual Yukawa couplings of quark and lepton superfields to the Higgs superfields, along, in general, with a mass term coupling the two Higgs doublets, H_1 and H_2. Explicitly, in a standard notation :

$$f_{LES} = h_{ij} U_i^c Q_j H_2 + \tilde{h}_{ij} D_i^c Q_j H_1 + f_{ij} L_i E_j^c H_1 + m_4 H_1 H_2 \,, \qquad (101)$$

where a summation over generation indices (i, j) is understood and $Q(U^c)$ denote generically quark doublets (charge -2/3 antiquark singlet) superfields, while $L(E^c)$ refers to lepton doublets (charge +1 antilepton singlet) superfields. With the exceptions of the top quark (h_t) Yukawa coupling and the mass parameter m_4, which in principle may be of order $O(m_{3/2})$, all other parameters appearing in (101) contribute to the masses of the observed quarks and leptons and are known to be small. Neglecting these small couplings, the effective Low Energy Potential (V_{LEP}) can be written as (see (78) and (90)) :

$$\begin{aligned} V_{LEP} = \sum_{i=1}^{3} \{ & m_{L_i}^2 |L_i|^2 + m_{E_i}^2 |E_i^c|^2 + m_{Q_i}^2 |Q_i|^2 \\ & + m_{U_i}^2 |U_i^c|^2 + m_{D_i}^2 |D_i^c|^2 \} \\ & + m_1^2 |H_1|^2 + m_2^2 |H_2|^2 + A h_t m_{3/2} (U_3^c Q_3 H_2 + h.c.) \\ & + B m_{3/2} m_4 (H_1 H_2 + h.c.) + h_t m_4 (H_1^+ Q_3 U_3^c + h.c.) \\ & + h_t^2 (|Q_3|^2 |U_3^c|^2 + |Q_3|^2 |H_2|^2 + |U_3^c|^2 |H_2|^2) \\ & + \text{"D-terms"} \quad . \end{aligned} \qquad (102)$$

The effective parameters appearing in (102) take, at large scales ($\sim M_X$), the values :

$$m_1^2(M_X) = m_2^2(M_X) = m_{3/2}^2 + m_4^2(M_X)$$

$$m_{Q_i}^2(M_X) = m_{U_i}^2(M_X) = m_{D_i}^2(M_X) = m_{L_i}^2(M_X) = m_{E_i}^2(M_X) = m_{3/2}^2$$

$$A(M_X) = A; \qquad B(M_X) = A - 1; \qquad (i = 1, 2, 3) \qquad (103)$$

as dictated by (90). It should be stressed once more that the boundary conditions (103) are exact, if we only neglect corrections at the Planck scale, ignore the scaling of parameters from M to M_X, and pay no attention to corrections[54] at the GUT scale. All these effects are expected to be small and it is assumed that they do not seriously disturb (103) and the picture hereafter.

It is apparent from (102) that SUGAR models can easily succeed in giving weak interaction scale masses ($m_{3/2} \sim M_W$) to squarks, sleptons and gauginos (see (90)). Alas, SUGAR models also give large positive $(\text{mass})^2$ to the Higgs doublets, thus making the breaking of SU(2) x U(1) difficult. One way to overcome this difficulty is the introduction[52-54] of a gauge singlet coupled to H_1 and H_2, but, as mentioned above, with disastrous effects[65] for the gauge hierarchy. A particularly simple solution to the SU(2) x U(1) breaking relies upon the fact that the boundary conditions (103) need be satisfied only at M_X (or M), and that large renormalization group scaling effects can produce a negative value for m_H^2 at low energies[42]. The full set of renormalization group equations for the parameters in V_{LEP} (90) has been written eslewhere[66]. Here we concentrate on the most interesting equation, the one for the mass-squared of the Higgs (m_2^2), which gives mass to the top quark :

$$\mu\frac{\partial}{\partial\mu}\begin{pmatrix} m_2^2 \\ m_{U_3}^2 \\ m_{Q_3}^2 \end{pmatrix} = \frac{h_t^2}{8\pi^2}\begin{pmatrix} 3 & 3 & 3 \\ 2 & 2 & 2 \\ 1 & 1 & 1 \end{pmatrix}\begin{pmatrix} m_2^2 \\ m_{U_3}^2 \\ m_{Q_3}^2 \end{pmatrix} + \frac{|A|^2 h_t^2 m_{3/2}^2}{8\pi^2}\begin{pmatrix} 3 \\ 2 \\ 1 \end{pmatrix}$$

$$- \frac{h_t^2 m_4^2}{8\pi^2}\begin{pmatrix} 3 \\ 2 \\ 1 \end{pmatrix} - \frac{8\alpha_3}{3\pi} m_{\tilde{g}}^2 \begin{pmatrix} 0 \\ 1 \\ 1 \end{pmatrix} , \qquad (104)$$

where we have neglected gauge couplings other than the "colored" one, $\alpha_3 (\equiv g_3^2/4\pi)$, $m_{\tilde{g}}$ is the gluino mass (see (90) and (92)), and Yukawa couplings other than h_t, for the top, have been dropped. The physics content of (104) is apparent. Since μ is decreasing (we come from high energies down to low energies), the sign of the first two terms in (104) is such as to make all m_2^2, m_{U3}^2, m_{Q3}^2 smaller at low energy with the decrease of m_2^2 becoming more pronounced because of the 3:2:1 weighting. On the other hand, the sign of the last two terms in (104) is such as to make m_{U3}^2 and m_{Q3}^2 (the squark masses) larger at low energy, but have no direct effect on m_2^2 (notice the "zeros" in the corresponding matrices in (104)). Indirectly though, the net effect on m_2^2 of the last two terms in (104) is to enhance further its decrease at low energies, by decreasing $m_{U_3}^2$ and $m_{Q_3}^2$, which then drive down m_2^2 via the first two terms of (104). This is exactly what we are after! We want large ($\sim M_W^2$) and positive squarks and slepton (masses)^2, but negative Higgs (mass)^2 to trigger $SU(2) \times U(1)$ breaking. The ways of obtaining negative Higgs (mass)^2 now become clear (see (104)). We have to use either a large top Yukawa coupling (h_t), or large A, or large m_4, or a fourth generation to provide large Yukawa couplings, or some suitable, physically plausible combination of the above possibilities. There are pros and cons for every one of the above situations. In the case of large h_t, a lower bound on the mass of the top quark is set[42,61-63]:

$$m_t > 0(60 \text{ GeV}) , \tag{105}$$

which some people may find uncomfortable. We may avoid a large h_t by moving it into the large A (> 3) regime[61,67]. The price though is high. The phenomenological acceptable vacuum becomes unstable against tunnelling into a vacuum in which all gauge symmetries, including color and electromagnetism, are broken. We must[61,67] then arrange things in such a way that the lifetime for this vacuum decay process is greater than the age of the universe. Some people, not without reason, may find this possibility dreadful. We may avoid large h_t and/or large A by using[59] non-vanishing $m_4 (\sim m_{3/2})$ where a rather satisfactory picture then emerges[59]. Some people may object here to the basic assumption of large $m_4 (\sim m_{3/2})$, since in the case of natural triplet-doublet Higgs splitting-type models[56] (see (96)), m_4 has a tendency to be small, if not zero, even though other sources of m_4 may be available.

Finally, we come to the possibility of a fourth generation which, suitably weighted, may help us to avoid large h_t, A, or m_4. The problem here is that low energy phenomenology (evolution of coupling constants, m_b/m_τ, ...)[59] as well as firm cosmological results like nucleosynthesis (especially ^{4}He abundance)[68], may suffer almost unacceptable modifications. Furthermore, one has to watch out for the mass of the fourth generation charged lepton, since it is going to behave like m_2^2 in (104), and thus m_{L_4, E_4}^2 may easily go negative, breaking electromagnetic gauge invariance.

Whatever mechanism (if any) turns out to be correct, it is rather remarkable that in SUGAR-type models, there is a simple explanation of the breaking of $SU(2) \times U(1)$ and of the non-breaking of $SU(3) \times U(1)_{E-M}$. Furthermore, for the first time, we have a simple explanation of why $M_W \lll M_X$ (or M), i.e., a simple solution of the cumbersome gauge hierarchy problem. Starting with a positive Higgs (mass)^2, of order $m_{3/2}^2$ at M_X, and noticing that (see (104)) the evolution with μ^2 of the Higgs (mass)^2 is very slow (logarithmic), it is not surprising that we have to come down a long way in the energy scale, before the Higgs (mass)^2 turns negative and is thus able to trigger SU(2) x U(1) breaking. Another very amazing fact is that the values of the parameters of the low energy world seem to co-operate with us. Since quarks are feeling strong interactions,(104) tells us that quarks may enjoy large masses (Yukawa couplings) without making squark (masses)^2 negative, because of the last term ($\sim \alpha_3$), which easily balances off large Yukawa couplings, without any sweat. On the other hand, since leptons are not feeling strong interactions, the balance-off between the weak gauge couplings and large Yukawa couplings becomes extremely delicate and could be problematic. How nice that for all three generations, leptons and down quarks weigh less than 5 GeV and especially for the third generation that the top quark (t) is heavier than the bottom quark (b). An inverse situation would be disastrous, because in any reasonable GUT, a very heavy b quark would mean a very heavy τ lepton, thus making electromagnetic gauge invariance tremble in such SUGAR-type schemes. I will not go any further into the esoterics of this type of SU(2) x U(1) breaking models, since a rather thorough and detailed exposé of these types of theories and of their phenomenological consequences is now available[59]. It should be stressed that things are now very constrained, as we see from the Table taken from Ref. 59, where the whole low energy spectrum is worked out in terms of very few parameters, $m_{3/2}$, $A(M_X)$, $\xi \equiv M_0/m_{3/2}$ (see (91)) and $m_4(M_X)$. Eventually, with more theoretical insight, we hope to determine even these very few parameters, thus predicting uniquely the low energy spectrum. For example, we have already discussed ways of determining $m_{3/2}$ (see (99)), while some people may favor $A(M_X) = 3$ as a natural solution[45] to the absence of the cosmological constant problem, etc. Among other interesting things contained in the Table, the existence of a very light ($\sim$(3-6) GeV) neutral Higgs, with the usual Yukawa couplings to matter, should not escape our attention. Since such a particle is a common feature of a large class of models[59,61], a search in the $\Upsilon \to H^0 + \gamma$ channel, which is expected to be a few per cent of the $\Upsilon \to \mu^+\mu^-$ decay, may turn out to be very fruitful.

No-scale models

As already mentioned above, the complete solution of the gauge hierarchy problem demands both scales, M_W and $m_{3/2}$, to be explained. We saw before how by using non-renormalizable interactions, e.g. Eq. (95), we can relate the gravitino mass to the M_X e.g. Eq. (99), working only at the tree level. Some people may find this approach a bit ad-hoc and may be very much dependent on the specific form of the non-renormalizable terms, which deviates slightly from the standard philosophy of effective Lagrangians : put more weight to results emerging from the general symmetries and not from the very specific form of the effective Lagrangian. Well, here is a resumé or our new, more ambitious approach[36-41].

An interesting possibility in this framework is the determination of the weak interaction scale by dimensional transmutation[59,61].

Table 2. Particle Spectrum[59]*

		CAM	MIM	CAM	MIM	CAM	MIM
$A(M_X)$		3	3	2.8	2.8	2.0	1.6
$m_{3/2}$		15	15	15	15	15	15
ξ		2.8	2.2	3.2	3.1	3.5	1.9
$m_4(M_X)$		15	17	16	18	11	7
top		25	25	35	35	50	50
All families	$(\text{sleptons})_L$	29	27	32	36	35	25
All families	$(\text{sleptons})_R$	21	20	22	23	23	19
1st and 2nd families	$(\text{squark})_L$	58	77	66	108	72	67
1st and 2nd families	$(\text{squark})_R$	54	74	61	104	67	65
		54	74	60	103	66	65
3rd family	$(\text{sbottom})_L$	58	76	64	106	68	66
3rd family	$(\text{sbottom})_R$	54	74	60	104	66	65
3rd family	$(\text{stop})_L$	81	96	95	132	112	106
3rd family	$(\text{stop})_R$	26	54	23	78	21	37
	Charged Higgses	96	93	95	94	88	83
	Neutral Higgses	106	104	105	105	100	95
		3	3	4.4	5.3	6	5
	"Axion"	51	46	49	48	35	19
	Gluinos	42	84	47	118	52	72
	Photino	11	4.6	9.4	7.3	4	3
	WH, HW-inos	89	87	87	94	84	90
		78	82	79	79	82	75
	ZH, HZ-inos	99	108	99	116	101	106
		92	85	91	80	88	83
	Axino	26	23	24	24	17	9

*Physical mass spectrum of the cosmologically acceptable model (CAM) and the minimal model (MIM) corresponding to the same gravitino mass $m_{3/2} \cong 15$ GeV for top quark masses equal to 25 GeV, 35 GeV and 50 GeV respectively. ξ denotes the ratio of the gaugino to the gravitino mass at M_X. All masses are in GeV units. The light neutral Higgs gets its mass via radiative corrections.

The multiplicative renormalization (104) of the soft SUSY breaking mass parameters means that the renormaliation scale μ_0 at which m_H^2 goes negative is independent of the magnitude of $\tilde{m}^2$. The value of μ_0 is determined by the logarithmic rate of evolution specified by the renormalization group equations (104). Hence

$$\mu_0 = M \exp\left(- \frac{0(1)}{\alpha_t} \right) . \tag{106}$$

Once $m_H^2 < 0$, it is possible to have weak gauge symmetry breaking and $M_W = 0(\mu_0)$, implying through (106) the highly desirable relation

$$\frac{M_W}{M} \cong \exp\left(- \frac{0(1)}{\alpha_t} \right) \cong 10^{-16} . \tag{107}$$

The dynamical determination of M_W has been realized[59-61]. The first half (69) of the second SUSY golden rule has been satisfied. It should be emphasized that we have tacitly assumed $\tilde{m} < \mu_0$, otherwise the RGE will be frozen at some renormalization scale $\mu > \mu_0$, implying $m_H^2 > 0$ and thus no $SU(2)_L \times U(1)$ breaking. But who determines $\tilde{m}$? In the "minimal" case (93i), $\tilde{m}$ is put in by hand and that is no good. In the "Mini-Maxi" or "Maximal" cases (93), $\tilde{m}$ is undetermined at the tree level and we should use non-gravitational standard model radiative corrections to determine it, thus finally realizing the no-scale model dream[36-41].

To explore the basic mechanism[36] for this trick, we consider the usual low energy Higgs potential (102) of the SUSY standard model, in an idealized limit where the mixing between the two light Higgs doublets is neglected

$$V = \frac{g_2^2 + g'^2}{8} \left(|H_1|^2 - |H_2|^2\right)^2 + m_1^2|H_1|^2 + m_2^2|H_2|^2 . \tag{108}$$

We denote by H_2 the Higgs field coupled to the t quark. Radiative corrections (104) drive $m_2^2 < m_1^2$ and when $m_2^2 < 0$ weak $SU(2)_L \times U(1)_Y$ gauge symmetry is spontaneously broken and the vacuum energy can become negative :

$$V_{min} = \frac{-2|m_2^2|^2}{(g_2^2 + g'^2)} \propto (-)\, \tilde{m}^4 . \tag{109}$$

In writing Eq. (109), we have recalled that since the Higgs mass is multiplicatively renormalized (104) for $\mu > 0(\tilde{m})$, the Higgs mass is always $\propto \tilde{m}$, and hence $V_{min} \propto |m_2^2|^2 \propto \tilde{m}^4$. The negative coefficient in Eq. (109) means that increasing $|m_2^2|$ and hence $\tilde{m}^2$ is energetically preferred, at least for small values of $\tilde{m}$. However, if $\tilde{m}$ gets to be larger than the scale μ_0 at which m_2^2 falls through zero, then the evolution of m_2 with the renormalization scale μ will become truncated at $\mu = O(\tilde{m}_2^2) > \mu_0$, m_2^2 will never become negative, and the potential will always be positive semidefinite. The general form of $V_{min}(\tilde{m})$ is therefore as shown in Fig. 5. It decreases from zero to negative values as $\tilde{m}$ increases from zero, but then rises to zero again for some $\tilde{m} = O(\mu_0)$. (The precise value depends on the choice of renormalization scheme, but physical parameters such as particle masses do not depend on this choice.) It is apparent from Fig. 6 that there must[36,37] be a dynamically preferred value of $\tilde{m}$ in the range $(0, \mu_0)$. Indeed, one finds[36,37]

$$V_{min}(\tilde{m}) = - c^2 \tilde{m}^4 \ln^2 \frac{\tilde{m}^2}{d(\mu_0^2)} \quad , \tag{110}$$

with c and d calculable parameters. Minimizing the potential (110) with respect to $\tilde{m}$, one finds a minimum at

$$\tilde{m} \cong O(\mu_0) \ . \tag{111}$$

Remember that μ_0 is a dimensional transmutation scale (106), so that the preferred value of $\tilde{m}$ is also fixed by dimensional transmutation, leading through Eqs. (107) and (111) to the golden relation (69),

$$\frac{M_W}{M} \cong \frac{\tilde{m}}{M} \cong \exp\left(- \frac{O(1)}{\alpha_t} \right) \cong 10^{-16} \ . \tag{112}$$

This simultaneous and dynamical determination[36-41] of M_W and $\tilde{m}$ combined with the non-renormalization theorems[21] of global SUSY (stabilization) and with the classic "missing partner" mechanism[56,60] (90) (natural tree-level GUT Higgs-electroweak Higgs splitting), completely solves the gauge (scale) hierarchy problem.

No-scale SUSY standard models[36,37] contain three adjustable parameters : possible non-zero gaugino masses $\xi \equiv m_{\tilde{V}}/m_{3/2}$, non-zero H_1H_2 mixing characterized by a mixing parameter m_4 $(\hat{m}_4 \equiv m_4/m_{3/2})$ and the A parameter[53] (90). As seen in Fig. 6, we find[36,37] domains of ξ and $\hat{m}_4$ which give phenomenologically acceptable models for which all charged sparticles have masses above 20 GeV (denoted by P), and the cosmological density of the lightest stable neutral sparticle is less than 2 x 10^{-29} gm/cc (denoted by C). No-scale SUSY GUT models have been also constructed[38]. In most cases, it is obligatory[38] to use the "maximal" alternative (93iii) because global SUSY (m = 0, A = 0), at the GUT scale M_X is badly needed in order to get rid of the highly undesirable huge GUT vacuum energy. That is good news, since the decoupling entailed in principle between $m_{3/2}$ and $\tilde{m}$ is very welcome. There are two very interesting physical implications of this gravitino mass liberation movement[39-41], one cosmological and the other hierarchical. As is well known[69], standard cosmological constraints (critical energy density, primordial nucleo-

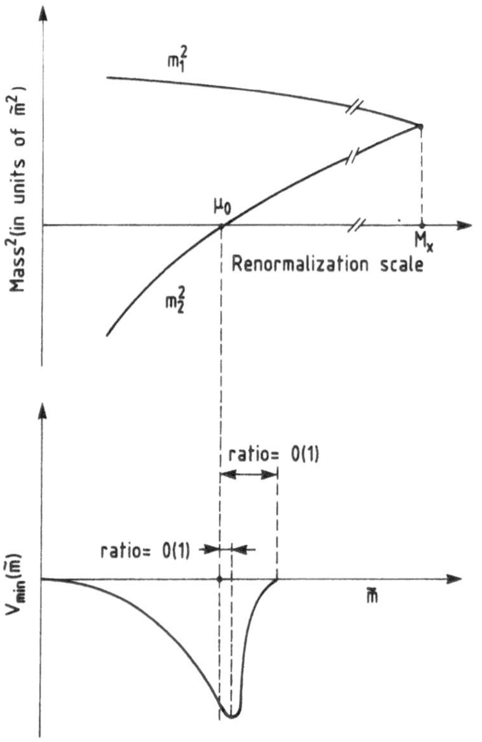

Fig. 5. Sketch of the variation of the SUSY breaking mass parameters m_1^2, m_2^2 with the renormalization scale μ. The Higgs (mass)^2 $m_2 2 = 0$ at a scale μ_0, which determines the dynamically preferred value of $\tilde{m}$, as seen in the bottom half of the figure.

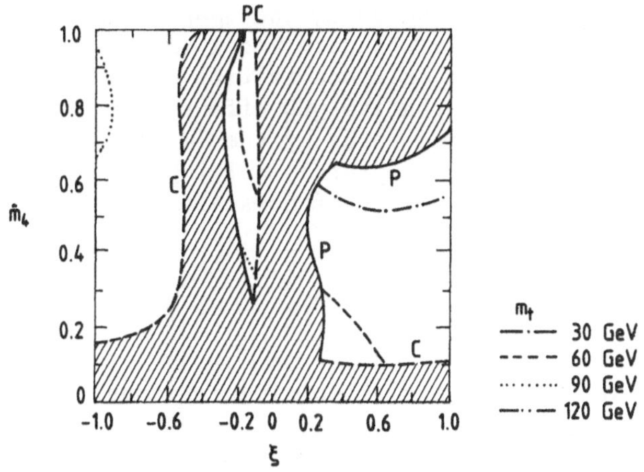

Fig. 6. Values of m_t (in GeV) for general values of the Higgs mixing parameter $\hat{m}_4$, $\xi = m_{\tilde{V}}/m_{3/2}$ and A = 3. Our present vacuum is unstable in the allowed region on the right of the figure. The shaded domain indicates values of $\hat{m}_4$ and ξ disallowed because of the absence of a charged sparticle with mass less than 20 GeV (P), and/or because of an excessive cosmological density of the lightest neutral sparticle (C).

synthesis, etc.) exclude gravitino masses in the region 1 keV to 10^4 GeV, thus excluding $m_{3/2} \cong O(M_W)$. Inflation can save[70] the gravitino by diluting its original number density, but gravitino regeneration processes[71] put a rather low upper bound[71-73] on the reheating temperature $T_R \leq O(10^{10}$ GeV). This is rather unfortunate because then, "standard" baryosynthesis[74] is in conflict with long-lived protons[75,76]. This is the gravitino problem. It seems that we need either superligt (< 1 keV) or (super)heavy (> 10^4 GeV) gravitinos. No-scale models encompassing both possibilities have been constructed[39-41], thus resolving the gravitino problem in a Gordian-Knot way by "cutting" the $m_{3/2} \sim O(M_W)$ relation. Actually, the superlight gravitino case[40,41] has a very interesting hierarchical implication[40,41]: it completely solves the strong CP problem (55). One simply notices from (73) that non-minimal gauge boson kinetic terms, necessarily needed for non-vanishing gaugino masses (98), give rise to $\theta_{QCD} \sim < \mathrm{Im}\, f_{ab}/\mathrm{Re}\, f_{ab} >$. Then, since there are two dynamical degrees of freedom (Rez, Imz), it is evident that these will determine dynamically two physical parameters. Since the gravitino is decoupled, we may naturally choose $m_{\tilde{V}}$ and θ_{QCD} as the two dynamically determined quantities. While standard model radiative corrections fix the gaugino mass $m_{\tilde{V}}$, the only non-trivial dynamical dependence on θ_{QCD} comes from non-perturbative QCD effects which favour $\theta_{QCD} = 0$. The dynamical freedom accorded to us by the Im f_{ab} field, enables us to "relax" to $\theta_{QCD} = 0$, thus completely solving[40,41] the strong CP problem. In this case, in a class of no-scale models[40,41], the gravitino mass is determined to be

$$m_{3/2} = 0\left(\frac{m_{\tilde{V}}^{p}}{M^{p-1}}\right) << m_{\tilde{V}} , \tag{113}$$

with $1 < p < 2$, in order to avoid too singular couplings to ordinary matter[10].

Turning now to baryon decay, an interaction of the form[42]

$$f \ni \frac{\lambda}{M} \bar{F}\ TTT , \tag{114}$$

where $\bar{F}$ is a $\underline{5}$ of matter (quark + lepton) chiral superfield in SU(5), T is a $\underline{10}$ of matter superfields and λ is some generic Yukawa coupling, could replace the Higgs exchange in the Weinberg-Sakai-Yanagida[32] loop diagram for baryon decay. The magnitude of the diagram with (114) relative to the conventional Higgs diagram (see Fig. 3) is :

$$\left(\frac{\lambda}{M}\right) / \left(\frac{\lambda^2}{M_{H_3}}\right) \cong \left(\frac{M_{H_3}}{\lambda M}\right) . \tag{115}$$

The ratio (115) could easily be > 1, making a non-renormalizable superpotential interaction the dominant contribution to proton decay. A careful analysis of SUGAR-induced baryon decay shows[77], surprisingly enough that the expected hierarchy of decay modes is similar to that[31-34] coming from conventional minimal SUSY GUTs as given by (65) and (66). One might have wrongly expected that no hard and fast predictions could be made about gravitationally-induced baryon decay modes. Anyway, this mechanism could give observable baryon decay even if the GUT mass $M_X \cong M$.

Incidentally, similar terms like (114) have been considered[78] in efforts to explain the "lightness" of the first two generations of quarks and leptons. One replaces[78] direct Yukawa couplings for the first two generations with (very schematically)

$$f \ni \frac{\tilde{\lambda}_2'}{M} \bar{H}\Sigma T_2 \bar{F}_2 + \frac{\tilde{\lambda}_2'}{M} H T_2 T_2 \Sigma + \frac{\tilde{\lambda}_1'}{M^2} \bar{H}\Sigma^2 T_1 \bar{F}_1 + \frac{\tilde{\lambda}_1'}{M^2} H T_1 T_1 \Sigma^2 + \ldots \tag{116}$$

which not only repairs[42,78] wrong relations like $m_d(M_X) \cong m_e(M_X)$, very difficult to correct[79] in conventional SUSY GUTs, but also provides reasonable masses for the first two generations. Indeed, it follows from (116) that the second generation is getting masses $(M_X/M)M_W \sim (0.1\text{-}1\ \text{GeV})$, while the first generation masses are $(M_X/M)^2 M_W \sim (1\text{-}10\ \text{MeV})$, exactly what was ordered. It is amazing that in SUGAR models, by increasing $M_X (\sim 10^6\ \text{GeV})$, relative to its ordinary GUT value (10^{14} GeV), and by decreasing $M_{P\ell}$, what is relevant is the superPlanck scale $M(\sim 10^{18}\ \text{GeV})$, the highly-desired ratio $(M_X/M) \sim 10^{-2}$, which appears naturally. It seems now, for the first time, that gravitational interactions may be responsible for the masses of at least the first two generations. Once more, non-renormalizable interactions contained in SUGAR models provide a simple solution[78] to another hierarchy problem, the fermion mass hierarchy problem.

I hope I have convinced you by now that (no-scale) phenomenological supergravity models are worth considering and I proceed next to discuss experimental evidence for (or against) them.

EXPERIMENTAL EVIDENCE (?) FOR SUPERSYMMETRY

Low energy supersymmetric models have a very rich structure that makes them experimentally vulnerable at accessible (present or very near future) energies, if SUSY indeed solves the gauge hierarchy problem [see Eq. (62)]. Since several reports exist[80], covering in rather lengthy detail, the numerous experimental consequences of SUSY, I shall limit my discussion here to the relation between SUSY and the new experimental results mentioned in the beginning.

Let me start with direct evidence, i.e., with SUSY particle production. As I emphasized before [see (14), (15)] , SUSY particles in general can only be produced in pairs. This means that among the decay products of every sparticle, there must be another sparticle, and hence the lightest sparticle must be stable. The lightest sparticle is probably[81] neutral and not strongly interacting and the most likely candidate[81] may be the photino $\tilde{\gamma}$. Thus, a characteristic signature for SUSY could be[80] missing energy-momentum carried away by weakly interacting photinos, for example, from gluino or squark-pair production[82]:

$$p + \bar{p} \rightarrow \tilde{g}\tilde{g} + X \quad ; \quad \tilde{q}\tilde{q} + X \qquad (117)$$
$$\tilde{g} \hookrightarrow q\bar{q}\tilde{\gamma} \qquad \tilde{q} \hookrightarrow q\tilde{\gamma},$$

or from supersymmetric W decay[83]:

$$p + \bar{p} \rightarrow W + X, \quad W \rightarrow \tilde{W} + \tilde{\gamma} \qquad (118)$$
$$\tilde{W} \rightarrow q\bar{q}\tilde{\gamma};\ \ell\nu\tilde{\gamma}\ .$$

It is highly exciting that the observed UA(1,2) "Zen"[84] or "Wen"[85] events, with their characteristic large missing energy-momentum, may be naturally due to production and decay of SUSY particles. Indeed, detailed studies[86-92] have shown that the UA(1) "Zen" (monojet) events[84] could be due to either squark or gluino production (and decay), according to (117), if either[86] $m_{\tilde{q}}$ or $m_{\tilde{g}} = O(40\ \text{GeV})$; lower masses down to O(30 GeV) cannot be yet firmly (?) excluded[87,90,92]. As is apparent from Fig. 7, for $m_{\tilde{g}} \leq O(40\ \text{GeV})$ and when all experimental cuts are taken into account, the total cross-section for $\tilde{g}\tilde{g}$ production is dominated by the one-jet final state[86,87]. Similar results hold true for the squark case[86,90]. So, despite the naîve expectations from (117) that $\tilde{g}\tilde{g}$ or $\tilde{q}\tilde{q}$ production should give four jet or two-jet final states respectively, surprisingly enough for $m_{\tilde{g}}$ or $m_{\tilde{q}} \leq O(40\ \text{GeV})$ one-jet final states dominate, in accordance with experiment[84]. Are then the observed events with large p_T^{miss} due to the production of either $\tilde{g}$ or $\tilde{q}$ with mass O(40 GeV)? Both are possible interpretations of the UA(1) data[84], but neither can be confirmed or refuted until more data are accumulated. Nevertheless, the squark interpretation has been favoured[26,90] on two grounds (i) the hardness of the observed missing $p_\perp$ spectrum, which is more naturally explained by two-body $\tilde{q} \to q + \tilde{\gamma}$ decays and (ii) the thinness of the observed monojets, which disfavours $\tilde{g} \to q + \bar{q} + \tilde{\gamma}$ decay which yields monojets with invariant masses up to O(20 GeV) and an average of O(10 GeV), while the squark decay yields monojet invariant masses O(2 GeV) consistent[86] with the observed ones. Other more contrived explanations are still possible. For example, if $\tilde{g}$ and $\tilde{\gamma}$ are both very light and approximately degenerate ($\sim$ 3 GeV) to ensure long enough $\tilde{g}$ lifetime, then $qg \to \tilde{q}\tilde{g} \to q(\tilde{\gamma}\tilde{g})$ with $m_{\tilde{q}} \sim 100$ GeV may also explain[89] the "monojet" events. Concerning the "photon" UA(1) event(s)[84], it may possibly be a monojet with a large collimated electromagnetic component containing one or more π^0's or η's whose charged multiplicity fluctuated down to zero. This event actually contains some soft charged tracks which are nearby in angle and could perhaps be associated in the "monojet". Another source of photons could be[91] $\tilde{q}q \to \tilde{\gamma}\gamma$, assuming $m_{\tilde{q}} \leq E_T^{trigger}$ such that the SUSY content of the proton is plausibly excited, but in this case[91], one should take care of the jet coming from the decay of the left-over spectator ($\bar{\tilde{q}}$) . Putting everything together, it does not look unreasonable to me to pursue the interpretation[86,90] of the UA(1) monojet data as squarks with $m_{\tilde{q}} \cong 40$ GeV, and infer a lower limit $m_{\tilde{g}} \gtrsim 40$ GeV on the gluino mass. Furthermore, such an assumption is not in contradiction with possible explanations of the UA(2) "Wen" events[85]. For example[88], $p\bar{p} \to \tilde{W}\tilde{g}$ or $\tilde{W}\tilde{q}$ production, followed by $\tilde{W} \to e\nu\tilde{\gamma}$ and $\tilde{g} \to q\bar{q}\tilde{\gamma}$ or $\tilde{q} \to q\tilde{\gamma}$, would produce "Wen" events (e + jet + large amounts of p_T^{miss}.). It has been argued[88] that with $m_{\tilde{q}} \cong (40\text{-}60)$ GeV, $m_{\tilde{g}} \cong (70\text{-}100)$ GeV and $m_{\tilde{W}} \cong (35\text{-}40)$ GeV, one may probably get suitable (?) rates. In addition, such an explanation[88] is consistent with the observation that in the three UA(2) "Wen" events[85], the missing "ν" vector is consistently larger than the p_T of the observed electron. They should on average be equal in $W \to e\nu$ decay, but could easily be different in

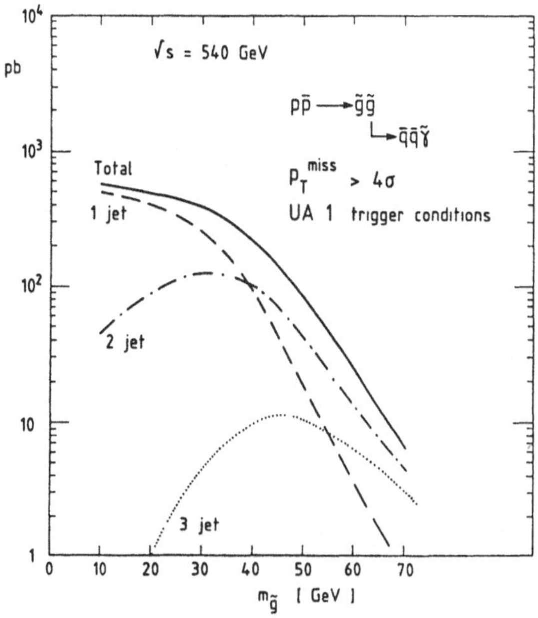

Fig. 7. The total and topological cross-sections for $\tilde{g}\tilde{g}$ production followed by $\tilde{q} \rightarrow q\bar{q}\tilde{\gamma}$ decay giving one- , two- and three-jet final states with $p_T^{miss.} > 4$, fulfilling the UA1 trigger requirements [from Ref. 87] .

SUSY, where the "ν" is actually a combination of one ν and two photinos $\tilde{\gamma}$.

If we indeed buy the squark explanation [$m_{\tilde{q}} = 40$ GeV; $m_{\tilde{g}} \geq O(40$ GeV)] of the Zen-Wen events, then its phenomenological implications are rather dramatic. In a large class of phenomenological supergravity models[52-61,36-41] discussed before, one may write down[59] convenient approximate formulae for physical squark and slepton masses at relevant renormalization scales $\mu = O(M_W)$:

$$m^2_{\tilde{q}} \cong m^2(1 + 7.6\ \xi^2)\ ,\quad \tilde{q} = \tilde{u},\tilde{d},\tilde{s},\tilde{c},\tilde{b} \tag{119}$$

and

$$\begin{aligned} m^2_{\tilde{\ell}_L} &\cong m^2(1 + 0.15\ \xi^2) \\ m^2_{\tilde{\ell}_R} &\cong m^2(1 + 0.5\ \xi^2)\ , \end{aligned} \tag{120}$$

where m and $m\xi(\equiv m_{\tilde{V}}(M))$ are respectively the scalar boson and gaugino masses at $\mu = M_X$ or M (m is commonly, but not always necessarily, identified with the gravitino mass). The corresponding formulae for gaugino masses take the simple form (see also (92))

$$m_{\tilde{g}} = \frac{\alpha_3}{\alpha_G}\, m_{\tilde{V}}(M) \quad ; \quad m_{\tilde{\gamma}} = \frac{\alpha_1}{\alpha_G}\, m_{\tilde{V}}(M)\ , \tag{121}$$

implying

$$\frac{m_{\tilde{\gamma}}}{m_{\tilde{g}}} = \frac{3}{8}\,\frac{\alpha}{\alpha_3}\,\frac{1}{\sin^2\theta_W}\ , \tag{122}$$

where it has been assumed that, thanks to grand unification, all gaugino masses are equal at M_X or M, not necessarily an unavoidable assumption[38]. Clearly enough, as it has been repeatedly emphasized in the literature[59,61,36-41] for some time now, the supergravity sparticle mass spectrum is rather tight. For example, assuming $m_{\tilde{q}} = 40$ GeV, $m_{\tilde{g}} \geq O(40$ GeV) and $m_{\tilde{\ell}_{L,R}} \geq 20$ GeV, it is trivial to show[94] that the set of equations (119)-(122) imply :

$$20\ \text{GeV} \leq m_{\tilde{\ell}_{L,R}} \leq 30\ \text{GeV}\ ,$$

$$5\ \text{GeV} \leq m_{\tilde{\gamma}} \leq 10\ \text{GeV}\ , \tag{123}$$

and

$$40\ \text{GeV} \leq m_{\tilde{g}} \leq 60\ \text{GeV}\ ,$$

a rather "light" and experimentally easily accessible spectrum. In addition, while SUSY in general entails the existence of at least two Higgs-doublets, dimensional transmutation[59] or no-scale type[36-41] supergravity models ask for the existence of a "light" [sometimes[59] $\leq$ 10 GeV] neutral Higgs boson. Actually, if indeed (123) holds, then the O(10 GeV) upper bound is certainly quite firm[59,94]. The suggestion was then made[59], some time ago, that $\Upsilon \to H^0 + \gamma$ is an excellent place[95] to look for such a "non-standard" light Higgs. If indeed (123) holds true, then news (good or bad) should come very soon from almost everywhere : $e^+e^- \to \tilde{\ell}^+\tilde{\ell}^-$; $e^+e^- \to (\tilde{\gamma}\tilde{\gamma})\gamma$[96], [one event has already been reported[97]], $p\bar{p} \to (\tilde{g}\tilde{g}$ or $\tilde{q}\tilde{g}) + X$, $p\bar{p} \to (W^+ \to \tilde{\ell} + \tilde{\nu}) + X$ or $p\bar{p} \to (Z^0 \to \tilde{\ell}\tilde{\ell}, \tilde{\nu}\tilde{\nu}) + X$ and $\Upsilon \to H^0 + \gamma$. Wait and see!

There are other, very interesting features concerning low energy phenomenology stemming out from the general form of V_{LEP} (see (102) and (103)) in SUGAR models. Very tight constraints coming from natural suppression of flavor changing neutral currents[24,26] (FCNC), absence of large corrections to $(g-2)$[28] and ρ[62,58] ($\equiv(M_W/M_Z\cos\theta_{e-w})^2$) as well as to θ_{QCD}[6], which have been the nemesis of SUSY models with arbitrary and explicit soft SUSY breaking, are satisfied in SUGAR models. The highly-constrained set of soft SUSY operators (80) in SUGAR models fits the bill[42,62]. Concerning FCNC, (103) guarantees the super-GIM mechanism, since the mass matrices for the quarks and leptons are diagonalized by the same transformation that renders the mass matrices for their scalar partners and gluino couplings generation diagonal. Despite the fact that this property does not survive, in general, after renormalization, it has been shown[99,100] that these effects are controllable. Furthermore, the Buras stringent upper bound[101] on the top quark mass ($<$O(40 GeV)), coming from kaon phenomenology ($K_L - K_S$ and $K_L \to \mu^+\mu^-$ systems), is avoided[99] in SUGAR models. There are a lot of cancellations between ordinary and SUSY contributions in K processes[102], such that the top quark mass may be stretched up to 100 GeV without problem . That sounds very satisfactory, especially for SUGAR models[61-63] that do need a large top quark mass for SU(2) x U(1) breaking. It looks like a self-service situation. Similar comments apply in the case of $(g-2)_\mu$ or ρ, where it has been shown that SUGAR model contributions are acceptable[59,62,98]. Typical values for SUGAR contributions are[59,103] $|\Delta(g-2)_\mu| \leq (3.10^{-9})$ and[62,98] $\Delta\rho \leq 0.01$, which compare favorably with the present experimental upper bounds of (4.10^{-8}) and (0.03) respectively, but are large enough to be interesting. Better experimental bounds, especially on $\Delta\rho$, could be revealing.

CONCLUSIONS

We have shown that gravitational effects, as contained in SUGAR theories, cannot be neglected anymore in the regime of particle physics. On the contrary, it may be that supergravitational effects are really responsible : for the SU(5) breaking at M_X with an automatic triplet-doublet Higgs splitting; for the SU(2) x U(1) breaking (and SU(3) x $U(1)_{e-m}$ non-breaking) at M_W, naturally exquisitely smaller than M_X; for the "constrained" soft SUSY breaking at $m_{3/2}$, hierarchically smaller than M in a natural way; for definite, at present experimentally acceptable departures from the "standard" low energy phenomenology (like the DEMON, or $\Delta\rho$, with values below, but not far from, their present experimental upper bounds, or the existence of very light (<0(10 GeV)) neutral Higgs bosons , as well as a rather well-defined low energy SUSY spectrum; for observable baryon decay even if $M_X \cong M$; and for the light fermion masses of the first two generations. Incidentally, it has been argued[104] recently that the observed[105] long b-quark lifetime and a "light" top quark[106] ($\sim$ 40 GeV) may not fit together in the conventional Kobayashi-Maskawa six-quark CP-violation model, as they give too small CP-violation. It has even been suggested[104] that a "light" top quark will be a signal for a fourth generation in the standard model. Well, this is not the case in a SUSY-K-M model[107]. We have found[107] that long τ_b and a "light" top quark provide enough CP-violation, if we supersymmetrize the standard K-M mode. For us[107] long lived b-quark and "light" top is maybe a signal for supersymmetry but not necessarily for a fourth generation[104]. Furthermore, supergravity theories may provide, for the first time, a problem-free cosmological scenario[108], from primordial inflation[108] through GUT phase transitions to baryon and nucleosynthesis, ostracizing troublesome particles such as GUT monopoles, gravitions[70-73], Polonyi fields[109] or other SUSY relics[81].

Putting the whole thing together, it becomes apparent that spontaneously broken N = 1 local SUSY gauge theories, with their prosperous and appropriate structure, may well serve as an effective theory describing all physics from $M_{P\ell}$ down to (and including) low energies, with well-defined and rich experimental consequences. What's next then? Well, we really have to understand where this highly successful theory comes from. There are reasons to believe[50] that N = 8 extended SUPERGRAVITY, suitably broken[110] down to N = 1 supergravity, may provide the fundamental theory. But this next move asks for a deep understanding of physics at Planck energies, which is as exciting as it is difficult, taking into account that even QUANTUM MECHANICS may need modification[111], if quantum gravitational effects have to be considered seriously.

ACKNOWLEDGEMENTS

I would like to express my deep appreciation to the Organizing Committee for setting up such an exciting Summer School both thematically and environmentally.

REFERENCES

1. For reviews on GUTs see :
J. Ellis, CERN preprint TH-3802 (1984);

D.V. Nanopoulos, Ecole d'Eté de Physique des Particules, Gif-sur-Yvette,(INEP3, Paris, 1980), p. 1 (1980);
P. Langacker, Phys. Rep. 72C (1981) 185.
2. For a review, see :
E. Farhi and L. Susskind, Phys. Rep. 74C (1981) 277.
3. Y.A. Gol'fand and E.P. Likhtman, Pis'ma Zh. Eksp. Teor. Fiz. 13 (1971) 323;
D. Volkov and V.P. Akulov, Phys. Lett. 46B (1973) 109;
J. Wess and B. Zumino, Nucl. Phys. B70 (1974) 39.
For reviews see :
P. Fayet and S. Ferrara, Phys. Rep. 32C (1977) 249;
J. Wess and J. Bagger, "Supersymmetry and Supergravity" (Princeton University Press, 1983).
4. M. Veltman,Acta Phys. Pol. B12 (1981) 437;
T. Inami, H. Nishino and S. Watamura, Phys. Lett. 117B (1983) 197;
N.G. Deshpande, R.J. Johnson, E. Ma, Phys. Lett. 130B (1981) 61.
5. I.S. Altarev et al., Phys. Lett. 102B (1981) 13;
W.B. Dress et al., Phys. Rev. D15 (1977) 9;
N.F. Ramsey, Phys. Rep. 43 (1978) 409.
6. J. Ellis, S. Ferrara and D.V. Nanopoulos, Phys. Lett. 114B (1982) 231.
7. S. Coleman and J. Mandula, Phys. Rev. 159 (1967) 1251.
8. For a review, see :
P. Van Nieuwenhuizen, Phys. Rep. 68C (1981) 189.
9. R. Barbieri, S. Ferrara and D.V. Nanopoulos, Zeit. für Phys. C13 (1982) 267; Phys. Lett. 116B (1982) 16.
10. P. Fayet, Proc. of Europhysics Study Conference on "Unification of Fundamental Interactions", ed. S. Ferrara, J. Ellis and P. Van Nieuwenhuizen (Plenum Press, N.Y., 1980), p. 587.
11. G.R. Farrar and S. Weinberg, Phys. Rev. D27 (1983) 2732.
12. J. Polchinsky and L. Susskind, Phys. Rev. D26 (1982) 3661.
13. R. Barbieri, S. Ferrara, D.V. Nanopoulos and K. Stelle, Phys. Lett. 113B (1982) 219.
14. D.V. Volkov and V.A. Soroka, JETP Lett. 18 (1973) 312.
15. S. Deser and B. Zumino, Phys. Rev. Lett. 38 (1977) 1433.
16. E. Cremmer et al., Phys. Lett. 79B (1978) 931; Nucl. Phys. B147 (1979) 1051; Phys. Lett. 116B (1982) 231; Nucl. Phys. B212 (1983) 413.
17. J. Ellis and D.V. Nanopoulos, Phys. Lett. 116B (1982) 133.
18. L. Girardello and M.T. Grisaru, Nucl. Phys. B194 (1982) 65.
19. For recent reviews, see :
J. Ellis, CERN preprint. TH-3802 (1984);
H.P. Nilles, Phys. Rep. C110 (1984) 2.
20. S. Mandelstam, Phys. Lett. 121B (1983) 30; Nucl. Phys. B213 (1983) 149; P.S. Howe, K.S. Stelle and P.K. Townsend, Nucl. Phys. B236 (1984) 125.
21. J. Wess and B. Zumino, Phys. Lett. 49B (1974) 52;
J. Iliopoulos and B. Zumino, Nucl. Phys. B76 (1974) 310;
S. Ferrara, J. Iliopoulos and B. Zumino, Nucl. Phys. B77 (1974) 413;

S. Ferrara and O. Piquet, Nucl. Phys. B93 (1975) 261;
M.T. Grisaru, W. Siegel and M. Rocek, Nucl. Phys. B159 (1979) 420.
22. L. Maiani, Proceedings of the Summer School of Gif-sur-Yvette, p. 3 (1979);
E. Witten, Nucl. Phys. B185 (1981) 153;
R.K. Kaul, Phys. Lett. 109B (1982) 19.
23. For an exhaustive review see :
"Supersymmetry Confronting Experiment" ed. D.V. Nanopoulos and A. Savoy-Navarro, Phys. Rep. 105, Nos. 1 and 2 (1984) 1.
24. J. Ellis and D.V. Nanopoulos, Phys. Lett. 110B (1982) 44.
25. J. Ellis, M.K. Gaillard and D.V. Nanopoulos, Nucl. Phys. B109 (1976) 213.
26. R. Barbieri and R. Gatto, Phys. Lett. 110B (1982) 211;
B.A. Campbell, Phys. Rev. D27 (1983) 1468;
T. Inami and C.S. Lim, Nucl. Phys. B207 (1982) 533.
27. M. Suzuki, UC-Berkeley preprint UCB-PTH/82/8 (1982).
28. J.A. Grifols and A. Mendez, Phys. Rev. D26 (1982) 1809;
J. Ellis, J. Hagelin and D.V. Nanopoulos, Phys. Lett. 116B (1982) 283;
R. Barbieri and L. Maiani, Phys. Lett. 117B (1982) 203.
29. S. Dimopoulos and H. Georgi, Nucl. Phys. B193 (1981) 150;
N. Sakai, Zeit. für Phys. C11 (1981) 153.
30. M.B. Einhorn and D.R.T. Jones, Nucl. Phys. B196 (1982) 475.
31. J. Ellis, D.V. Nanopoulos and S. Rudaz, Nucl. Phys. B202 (1982) 43.
32. S. Weinberg, Phys. Rev. D26 (1982) 187;
N. Sakai and T. Yanagida, Nucl. Phys. B197 (1982) 533.
33. D.V. Nanopoulos and K. Tamvakis, Phys. Lett. 110B (1982) 449;
M. Srednicki, Nucl. Phys. B202 (1982) 327; ibid. B206 (1982) 139;
D.V. Nanopoulos, K.A. Olive and K. Tamvakis, Phys. Lett. 115B (1982) 15.
34. D.V. Nanopoulos and K. Tamvakis, in Ref. 33; Phys. Lett. 113B (1982) 151; Phys. Lett. 114B (1982) 235;
J. Ellis, K. Enqvist, G. Gelmini, C. Kounnas, A. Masiero, D.V. Nanopoulos and A. Yu. Smirnov, Phys. Lett. 147B (1984) 27.
35. J. Ellis, M.K. Gaillard and D.V. Nanopoulos, Phys. Lett. 80B (1979) 360.
36. J. Ellis, A.B. Lahanas, D.V. Nanopoulos and K. Tamvakis, Phys. Lett. 134B (1984) 429.
37. J. Ellis, C. Kounnas and D.V. Nanopoulos, Nucl. Phys. B241 (1984) 406.
38. J. Ellis, C. Kounnas and D.V. Nanopoulos, CERN preprint TH.3826 (1984).
39. J. Ellis, C. Kounnas and D.V. Nanopoulos, Phys. Lett. 143B (1984) 410.
40. J. Ellis, K. Enqvist and D.V. Nanopoulos, Phys. Lett. 147B (1984) 99.
41. J. Ellis, K. Enqvist and D.V. Nanopoulos, CERN Preprint T.4036

(1984).
42. J. Ellis, D.V. Nanopoulos and K. Tamvakis, Phys. Lett. 121B (1983) 123.
43. B. Zumino, Phys. Lett. 87B (1979) 203.
44. J. Bagger and E. Witten, Phys. Lett. 115B (1982) 202; 118B (1982) 103;
J. Bagger, Nucl. Phys. B211 (1983) 302;
R. Arnowitt, A.H. Chamseddine and P. Nath, Phys. Rev. Lett. 49 (1982) 970; 50 (1983) 232 and Phys. Lett. 121B (1983) 33.
45. E. Cremmer, S. Ferrara, C. Kounnas and D.V. Nanopoulos, Phys. Lett. 133B (1983) 61.
46. N.P. Chang, S. Ouvry and X. Wu, Phys. Rev. Lett. 51 (1983) 327.
47. S. Ferrara and A. Van Proeyen, Phys. Lett. 138B (1984) 77;
J.-P. Derendinger and S. Ferrara, CERN preprint TH.3903 (1984).
48. E. Cremmer, J. Scherk and S. Ferrara, Phys. Lett. 74B (1978) 61.
49. E. Cremmer and B. Julia, Nucl. Phys. B159 (1979) 141.
50. E. Cremmer and B. Julia, Nucl. Phys. B159 (1979) 141;
J. Ellis, M.K. Gaillard and B. Zumino, Phys. Lett. 94B (1980) 343.
51. S. Weinberg, Phys. Rev. 166 (1968) 1568;
S. Coleman, J. Wess and B. Zumino, Phys. Rev. 177 (1968) 2239;
C. Callan, S. Coleman, J. Wess and B. Zumino, Phys. Rev. 177 (1968) 2247.
52. R. Barbieri, S. Ferrara and C.A. Savoy, Phys. Lett. 119B (1982) 343.
53. H.P. Nilles, M. Srednicki and D. Wyler, Phys. Lett. 120B (1982) 346.
54. L. Hall, J. Lykken and S. Weinberg, Phys. Rev. D27 (1983) 2359;
S.K. Soni and H.A. Weldon, Phys. Lett. 126B (1983) 215.
55. D.V. Nanopoulos, K.A. Olive, M. Srednicki and K. Tamvakis, Phys. Lett. 124B (1983) 171.
56. C. Kounnas, D.V. Nanopoulos, M. Srednicki and M. Quiros, Phys. Lett. 127B (1983) 82.
57. C. Kounnas, J. Leon and M. Quiros, Phys. Lett. 129B (1983) 67;
C. Kounnas, D.V. Nanopoulos and M. Quiros, Phys. Lett. 129B (1983) 223.
58. D.V. Nanopoulos and D.A. Ross, Phys. Lett. 118B (1982) 99.
59. C. Kounnas, A.B. Lahanas, D.V. Nanopoulos and M. Quiros, Phys. Lett. 132B (1983) 95; Nucl. Phys. B236 (1984) 438.
60. A. Masiero, D.V. Nanopoulos, K. Tamvakis and T. Yanagida, Phys. Lett. 115B (1982) 380;
B. Grinstein, Nucl. Phys. B206 (1982) 387.
61. J. Ellis, J. Hagelin, D.V. Nanopoulos and K. Tamvakis, Phys. Lett. 125B (1983) 275.
62. L. Alvarez-Gaumé, J. Polchinski and M. Wise, Nucl. Phys. B221 (1983) 495.
63. L. Ibanez and C. Lopez, Phys. Lett. 126B (1983) 54.
64. S. Ferrara, D.V. Nanopoulos and C.A. Savoy, Phys. Lett. 123B (1983) 214.
65. H.P. Nilles, M. Srednicki and D. Wyler, Phys. Lett. 124B (1983)

337;
A.B. Lahanas, Phys. Lett. 124B (1983) 341.
66. K. Inoue, A. Kakuto, H. Komatsu and S. Takeshita, Prog. Th. Phys. 68 (1982) 927.
67. M. Claudson, L.J. Hall and I. Hinchliffe, Nucl. Phys. B228 (1983) 501.
68. K.A. Olive, D.N. Schramm, G. Steigman and J. Yang, Ap. J. 246 (1981) 557.
69. H. Pagels and J.R. Primack, Phys. Rev. Lett. 48 (1982) 223;
S. Weinberg, Phys. Rev. Lett. 48 (1982) 1303.
70. J. Ellis, A.D. Linde and D.V. Nanopoulos, Phys. Lett. 118B (1982) 59.
71. D.V. Nanopoulos, K.A. Olive and M. Srednicki, Phys. Lett. 127B (1983) 30;
L. Krauss, Nucl. Phys. B227 (1983) 556.
72. M.Yu. Khlopov and A.D. Linde, Phys. Lett. 138B (1984) 265.
73. J. Ellis, J.E. Kim and D.V. Nanopoulos, Phys. Lett. 145B (1984) 181.
74. For a recent pedagogical review, see :
K.A. Olive, Fermilab Conf. 84/59-A Preprint (1984).
75. J.E. Kim, A. Masiero and D.V. Nanopoulos, Phys. Lett. 145B (1984) 187.
76. J. Ellis, K. Enqvist et al., Ref. 34.
77. J. Ellis, J. Hagelin, D.V. Nanopoulos and K. Tamvakis, Phys. Lett. 124B (1983) 484.
78. D.V. Nanopoulos and M. Srednicki, Phys. Lett. 124B (1983) 37.
79. L. Ibanez, Phys. Lett. 117B (1982) 403;
A. Masiero, D.V. Nanopoulos and K. Tamvakis, Phys. Lett. 126B (1983) 337.
80. "Supersymmetry Confronting Experiment", eds. D.V. Nanopoulos and A. Savoy-Navarro, Phys. Rep. 105 (1984) 1;
H.E. Haber and G.L. Kane, Univ. of Michigan preprint UM-HE TH-83-17 (1984), Physics Reports to be published.
81. J. Ellis, J.S. Hagelin, D.V. Nanopoulos, K.A. Olive and M. Srednicki, Nucl. Phys. B238 (1984) 453.
82. G.L. Kane and J.P. Leveille, Phys. Lett. 112B (1982) 227;
P. Harrison and C.H. Llewellyn Smith, Nucl. Phys. B213 (1982) 233 and B223 (1983) 542.
83. S. Weinberg, Phys. Rev. Lett. 50 (1983) 387;
R. Arnowitt, A. Chamseddine and P. Nath., Phys. Rev. Lett. 50 (1983) 232;
J. Ellis, J.S. Hagelin, D.V. Nanopoulos and M. Srednicki, Phys. Lett. 127B (1983) 233.
84. UA1 Collaboration : G. Arnison et al., Phys. Lett. 139B (1984) 115.
85. UA2 Collaboration : P. Bagnaia et al., Phys. Lett. 139B (1984) 105.
86. G.R. Blumenthal, S.M. Faber, J.R. Primack and M.J. Rees, Nature 311 (1984) 517.
87. J. Ellis and H. Kowalski, Phys. Lett. 142B (1984) 441; Nucl.

Phys. B246 (1984) 189.
88. E. Reya and D.P. Roy, Phys. Lett. 141B (1984) 442; Phys. Rev. Lett. 52 (1984) 881.
89. H.E. Haber and G.L. Kane, Phys. Lett. 142B (1984) 212.
90. V. Barger, K. Hagiwara, I. Woodside and W.Y. Keung, Phys. Rev. Lett. 53 (1984) 641.
91. A.R. Allan, E.W.N. Glover and A.D. Martin, Univ. of Durham preprint DPT/84/20 (1984);
E.N. Argyres, A.P. Contogouris and H. Tanaka, McGill Univ. preprints (1984).
92. N.D. Tracas and S.D.P. Vlassopoulos, National Technical Univ. (Athens) preprint (1984).
93. P.G. Ratcliffe, Cavendish Lab. preprint-HEP-84/4 (1984).
94. J. Ellis and M. Sher, CERN preprint TH. 3968 (1984).
95. F.A. Wilczek, Phys. Rev. Lett. 39 (1977) 1304;
J. Ellis, M.K. Gaillard, D.V. Nanopoulos and C.T. Sachrajda, Phys. Lett. 83B (1979) 339.
96. P. Fayet, Phys. Lett. 117B (1982) 460;
J. Ellis and J.S. Hagelin, Phys. Lett. 122B (1983) 303.
97. R. Prepost, Contribution to the Leipzig Conference 1984.
98. R. Barbieri and L. Maiani, Nucl. Phys. B224 (1983) 32;
C.S. Lim, T. Inami and N. Sakai, INS Rep., 480 (1983).
99. A.B. Lahanas and D.V. Nanopoulos, Phys. Lett. 129B (1983) 461.
100. J.F. Donoghue, H.P. Nilles and D. Wyler, Phys. Lett. 128B (1983) 55.
101. A.J. Buras, Phys. Rev. Lett. 46 (1981) 1354.
102. T. Inami and C.S. Lim, Nucl. Phys. B207 (1982) 533.
103. D.A. Kosower, L.M. Krauss and N. Sakai, Phys. Lett. 133B (1984) 305.
104. P.H. Ginsparg, S.L. Glashow and M.B. Wise, Phys. Rev. Lett. 50 (1983) 1415.
105. E. Fernandez et al., Phys. Rev. Lett. 51 (1983) 1022;
N.S. Lockyer et al., Phys. Rev. Lett. 51 (1983) 1316.
106. UA1 Collaboration : G. Arnison et al., CERN Preprint CERN-EP/84-134.
107. J.-M. Gérard, W. Grimus, A. Masiero, D.V. Nanopoulos and A. Raychaudhuri, Phys. Lett. 141B (1984) and CERN preprint TH.3920 (1984).
108. J. Ellis, D.V. Nanopoulos, K.A. Olive and K. Tamvakis, Nucl. Phys. B221 (1983) 524;
D.V. Nanopoulos, K.A. Olive, M. Srednicki and K. Tamvakis, Phys. Lett. 123B, 41 (1983);
D.V. Nanopoulos, K.A. Olive and M. Srednicki, Phys. Lett. 127B (1983) 30;
G. Gelmini, D.V. Nanopoulos and K.A. Olive, Phys. Lett. 131B (1983) 53;
G. Gelmini, C. Kounnas and D.V. Nanopoulos, CERN preprint TH-3777 (1983);
K. Enqvist and D.V. Nanopoulos, Phys. Lett. 142B (1984) 349 and CERN preprint TH-4027 (1984);

For reviews see :
K.A. Olive, Ref. 74;
D.V. Nanopoulos, CERN preprint TH-3778 (1983).
109. D.V. Nanopoulos and M. Srednicki, Phys. Lett. 133B (1983) 287.
110. R. Barbieri, S. Ferrara and D.V. Nanopoulos, Phys. Lett. 107B (1981) 275;
J. Ellis, M.K. Gaillard and B. Zumino, Acta Phys. Pol. B13 (1982) 153.
111. S. Hawking, Comm. Math. Phys. 87 (1982) 395;
J. Ellis, J.S. Hagelin, D.V. Nanopoulos and M. Srednicki, Nucl. Phys. B241 (1984) 381.

e^+e^- PHYSICS AT PETRA - THE FIRST FIVE YEARS

Sau Lan Wu

Department of Physics, University of Wisconsin, Madison,

WI, USA, and DESY, Hamburg, Germany

ABSTRACT

PETRA (Positron-Electron Tandem Ring Accelerator) is located at DESY in Hamburg, Germany, and is the highest energy electron-positron storage ring in the world. This report reviews a few selected topics of the experimental investigations carried out at PETRA for the first five years of its operation beginning in 1978 by the five Collaborations CELLO, JADE, MARK J, PLUTO and TASSO. The physics objectives in the original proposals have largely been fulfilled. The emphasis of this review, based mainly on journal publications before July 15, 1983 (the fifth anniversary of the first successful storage of an electron beam at PETRA), is on the physics results, ranging over strong, electromagnetic and weak interactions. The topics covered include quark and gluon physics, and electroweak interference.

Although cited for comparison, no attempt has been made for any systematic coverage of the corresponding results from PEP of SLAC.

INTRODUCTION

It has been just over five years since the electron beam was first successfully stored in PETRA. The purpose of this review is to summarize the experimental physics for several selected topics at

*These lectures already appeared as part of Phys. Rep. 107 (1984) 59, and are reproduced with kind permission of North-Holland Publishing Company, Amsterdam.

**Supported in part by the US Department of Energy Contract number DE-AC02-76ER00881.

PETRA during these five years. Throughout this period, PETRA (Positron-Electron Tandem Ring Accelerator) has been the highest energy electron-positron storage ring in the world. The maximum center-of-mass energy reached is above 43 GeV.

Another motivation for this review is the realization that the physics objectives in the original proposals have been largely fulfilled, although the results are not always positive. The most exciting results from PETRA include the following :

(A) Discovery of the gluon and the determination of the quark-gluon coupling constant,

(B) First observation of the interference between weak and electromagnetic interactions in $e^+e^- \to \mu^+\mu^-$.

Most of the hadronic events (i.e., $e^-e^+ \to$ hadrons) at PETRA are in the form of two back-to-back jets, naturally interpreted as the production of a quark-antiquark pair. The measurement of the total hadronic cross section and the properties of these two-jet events is presented in the first chapter. Further evidence for this quark interpretation is the experimental observation of the long-range charge correlation in opposite jets.

The second chapter reviews the discovery of the gluon in the form of a jet, rather similar to the quark jet. The resulting events therefore take the form of three jets, and hence provide a handle to the experimental measurement of the quark-gluon coupling constant α_s. From the distribution of these three-jet events, the spin of the gluon has been found to be 1, in agreement with gauge theory.

The last chapter gives the observation of the interference of weak and electromagnetic interactions. This is most successful in the pair production of muons $e^-e^+ \to \mu^-\mu^+$, the forward-backward asymmetry being almost ten standard deviations away from zero.

In the preparation of this review, the emphasis is mostly on the journal publications of the five Collaborations. At various places, however, efforts are made to update the results when more data become available after publication and the method of analysis is at most only slightly modified.

HADRONIC EVENTS IN e^+e^- ANNIHILATION

We begin by giving a summary of the numerous experimental results obtained so far. There are many possible ways to arrange the order of presentation; we choose here as the guiding philosophy to go from the general features to the specific channels. Since the various results are intimately related to each other, it is neither desirable nor possible to adhere to this philosophy too rigidly.

Orientation

When an electron and a positron interact with each other, many different phenomena may occur. For purposes of orientation, we show in Fig. 1 a rather coarse-grained classification of such events. This classification is not meant to be exhaustive. For ease of reference, the reactions listed are numbered. In the left two columns labelled as electron and photon, the well-known QED processes are given together with some of their Feynman diagrams. The amplitudes for such reactions are conveniently classified according to their orders in the electromagnetic coupling constant e. Thus to order e^2 there are only Bhabha scattering[1]

$$e^+e^- \to e^+e^- ,$$

labelled as 1e in Fig. 1 and annihilation into photons

$$e^+e^- \to \gamma\gamma ,$$

labelled as 1γ there. To order e^3, the only possibilities involve merely the addition of a photon to the final state. Thus 2e and 2γ refer respectively to the processes

$$e^+e^- \to e^+e^-\gamma ,$$

and

$$e^+e^- \to \gamma\gamma\gamma .$$

Some of the Feynman diagrams for these processes are shown in Fig. 1 and the others may be obtained by attaching the photon line to another segment of the electron line or by permuting the photon lines.

To order e^4, it is possible to add a further photon to the final state, leading to

$$e^+e^- \to e^+e^-\gamma\gamma$$

and

$$e^+e^- \to \gamma\gamma\gamma\gamma$$

respectively. These processes are not shown in Fig. 1. However, these are not the only possibilities. To this order, a new phenomenon occurs. The incoming electron and the incoming positron can each emit a virtual photon, and then the two virtual photons annihilate to give a new electron-positron pair. Although this process $e^+e^- \to e^+e^-e^+e^-$ is a fourth order in the coupling constant e, at the high energies available at PETRA, the two virtual photons can be very close to the mass shell so that the cross section is quite

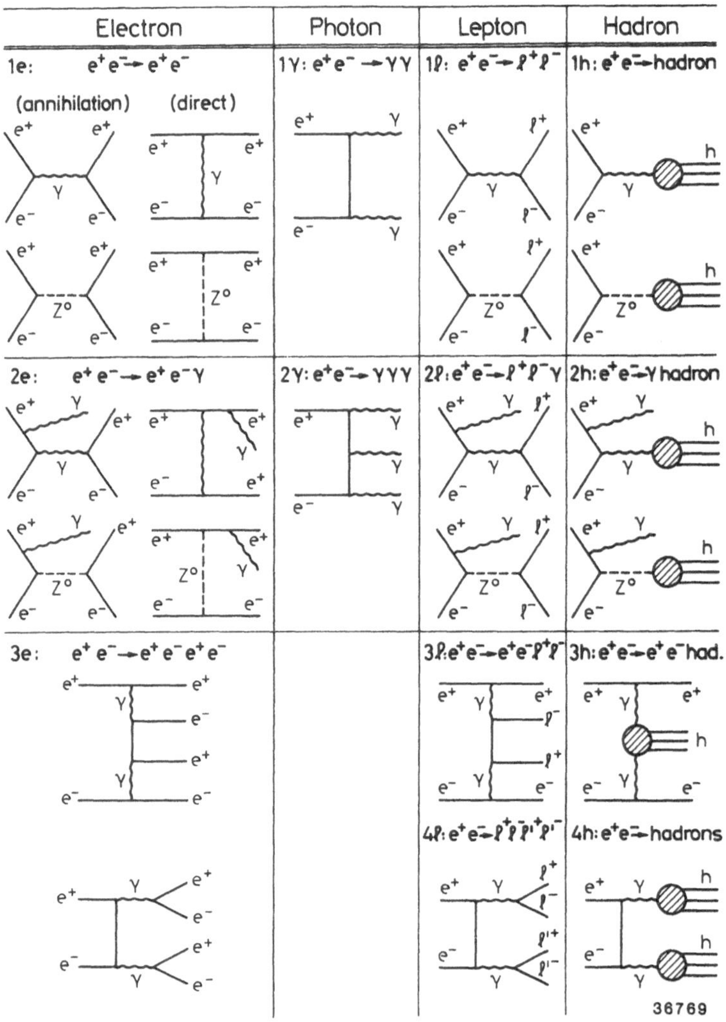

Fig. 1. Classification of e^+e^- reactions.

sizeable. We also show in the left column of Fig. 1 one of the numerous other Feynman diagrams for this process 3e; because of the absence of the enhancement factors due to the photons, such additional diagrams give negligible contributions to the integrated cross sections. The other diagrams in the left column involving $Z^{\circ 2}$ will be discussed later in this section.

With this understanding of the QED processes involving only electrons and photons, it is now straightforward to carry out an entirely similar classification of other e^+e^- reactions. First, the electron and the positron in the final state can be replaced by another lepton pair, either the muon or the τ[3] which was discovered at SPEAR several years ago, or perhaps some other charged lepton yet to be found. The resulting reactions are listed in the third column of Fig. 1 under the heading "lepton". This differs from the first column in two respects : the "direct" diagrams with a photon exchange in the cross channel are no longer allowed; and the two diagrams given under 3e lead to different reactions after the replacement of the electron by other leptons. For the upper diagram, only one e^+e^- pair can be replaced by $\ell^+\ell^-$ and hence the reaction $e^+e^- \to e^+e^-\ell^+\ell^-$ results. This is enhanced at high energies by the small virtual masses of the photons as discussed above. From the lower diagram, $e^+e^- \to \ell^+\ell^-\ell'^+\ell'^-$ is allowed but not enhanced, where ℓ and ℓ' may be μ and μ, τ and τ, or μ and τ.

Finally, the lepton pairs may be replaced by hadrons, and the resulting diagrams are shown in the last column of Fig. 1. While the electromagnetic interaction of leptons is theoretically understood and experimentally well verified, both before PETRA and at PETRA, that of hadrons is much more complicated and will be the central subject of study in the next several chapters. Accordingly, in Fig. 1, the electromagnetic interaction of hadrons is shown merely as a blob. Reaction 4ℓ evolves into one with only hadrons in the final state, designated as 4h.

So far we have discussed only reactions that are mediated by a virtual photon. Fifteen years ago, in unifying the electromagnetic and weak interactions[2], Weinberg and Salam proposed a neutral massive vector meson Z°. Recently the direct observation of this Z° has been accomplished at the CERN proton-antiproton collider[4]. Thus all these reactions can also proceed with the virtual photon replaced by a virtual Z°. Because of the large mass of Z°, measured to be about 95 GeV, diagrams with a virtual Z° are much less important, even at PETRA energies. For this reason, in Fig.1 these additional diagrams are shown only for the processes 1e, 1ℓ, 1h, 2e, 2ℓ, and 2h. For other processes, the replacement of a virtual photon by a virtual Z° leads to negligible correction, because of both the large Z° mass and the removal of the enhancement factor which depends crucially on the massless nature of the photon.

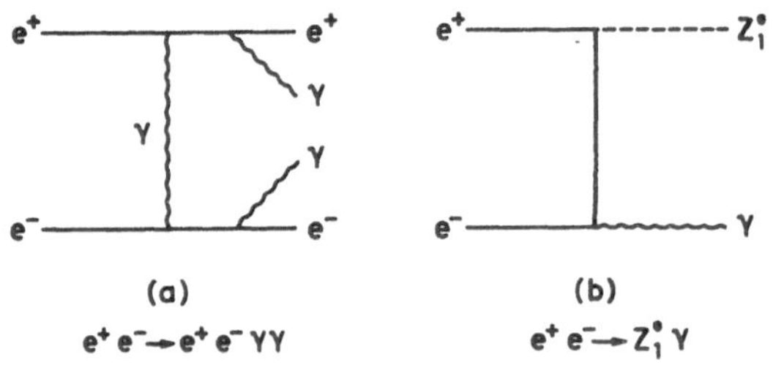

Fig. 2. Possible additional reactions not included in Fig. 1.(b) can occur at PETRA only if the neutral vector meson Z_1° exists and is of sufficiently low mass.

It is not the intention of Fig. 1 to give exhaustively all e^+e^- reactions with sizeable cross sections. Two examples are shown in Fig. 2 for additional reactions, one actual and one speculative. The reaction of Fig. 2.(a), $e^+e^- \to e^+e^-\gamma\gamma$, is expected to give a small but observable correction to luminosity measurements at PETRA. That of Fig. 2.(b) involves a neutral vector boson Z_1°, which is present in some theoretical models of electroweak interaction and is just like Z°, but with a lower mass.

Experimental Data on Total Hadronic Cross Section

The e^+e^- reactions shown in Fig. 1 can be broadly classified into two types : those where hadrons are produced, as shown in the right column, and those where only leptons and photons are in the final states, as shown in the other three columns. Of the second type, the cross sections are given by QED, modified by the presence of Z°. Radiative corrections should of course be taken into account. Processes of the first type, on the other hand, involve the blob, or more precisely, two kinds of blobs :

(A) γ-hadron coupling ,

and

(B) $\gamma\gamma$-hadron coupling .

Here (B) appears only in 3h, while (A) appears in all the rest. This process 3h, $e^+e^- \to e^+e^-$ hadrons, is called the two-photon process, while this chapter and the next one are devoted to 1h, $e^+e^- \rightarrow$ hadrons. Following the guidelines of going from general features to specific properties, we begin with the total cross section for $e^+e^- \rightarrow$ hadrons.

Because of the similarly of this process 1h to 1ℓ, $e^+e^- \rightarrow \ell^+\ell^-$, it is often convenient to use the ratio of these processes. Thus the R for the total hadronic cross section is defined as the ratio

$$R = \frac{\text{total cross section for } e^+e^- \rightarrow \text{hadrons}}{\text{total cross section for } e^+e^- \rightarrow \mu^+\mu^-} \tag{1}$$

at the same energy. More precisely, the denominator is not the actual total cross section for $e^+e^- \rightarrow \mu^+\mu^-$, but rather the theoretical total cross section in the lowest-order QED without weak interactions, i.e., without the Z° diagram shown under 1ℓ of Fig. 1. This theoretical total cross section is

$$\sigma(e^+e^- \rightarrow \mu^+\mu^-) = \frac{2\pi\alpha^2}{3s} \beta(3-\beta^2), \tag{2}$$

where β is the velocity of the produced muons, and the mass of the electron has been neglected. α is the fine-structure constant and s is the square of the center-of-mass energy.

In (2) the factor $\beta(3-\beta^2)$ is extremely close to 2; it is for example $(2-10^{-4})$ for a beam energy of 1 GeV. Replacing this factor by 2, the result is

$$\sigma(e^+e^- \rightarrow \mu^+\mu^-) = \frac{86.856}{s} \text{ nb} \tag{3}$$

with s in units of GeV^2. Therefore

$$R = \frac{s \text{ in GeV}^2}{86.856} (\text{total cross section for } e^+e^- \rightarrow \text{hadrons in nb}). \tag{4}$$

The precise determination therefore involves the careful identification of hadronic events, subtraction of backgrounds, calculation of acceptance, measurement of luminosity etc., but not any experimental determination of the total cross section for μ pair production.

We discuss briefly the experimental selection of hadronic events, not limited to the determination of the total cross section. This selection depends in detail on the characteristics of the detectors, and indeed is different for different purposes with the same detector. However, the general principle is as follows.

Since the processes 2h and 4h have relatively small cross sections, the purpose of the method of selection is to discriminate 1h against the following :

(i) Photon and lepton processes as shown in the left three columns of Fig. 1,
(ii) $\gamma\gamma$ process 3h : $e^+e^- \rightarrow e^+e^-$ hadrons; and
(iii) Interaction of the beams with either the residual gas in the beam pipe or the beam pipe itself.

With the notable exception of the production of $\tau^+\tau^-$ pairs, background (i) is characterized mostly by the presence of, at most, two charged particles entering the detector. Therefore a requirement is imposed to have more than two tracks. Since the enhancement of the cross section for background (ii) is due to the virtual photons being near their mass shells, most of the time the electron and positron in the final state are nearly in the direction of the incident beams and hence go down the beam pipes. Accordingly, both background (ii) and background (iii) can be suppressed by requiring sufficiently high total energy of the detected particles, called the total visible energy.

Fig. 3 gives an overall view of the measured values of R from low energies to the highest available energy with significant amount of data. This figure is based on the compilation by the TASSO Collaboration[5]. From the data[6] obtained before PETRA, it is seen that, at low energies, the behavior of R is rather complicated, showing peaks at the masses of the vector mesons ρ, ω and ϕ. With increased energy, there are two sets of high peaks, one between 3 and 4 GeV at the masses of J/ψ and ψ'[7] and the other set around 10 GeV at the masses of the upsilons Υ, Υ' and Υ''[8]. In both cases, immediately above the high peaks, there are other peaks attributed to the various resonances such as ψ (3770) and Υ (10570)[7,8]. In the PETRA energy range of 12 GeV and above , the data are from CELLO[9], JADE[10,11], MARK J[12], PLUTO[13], and TASSO[14,15]. It is seen that PETRA has reached the energy range above the known resonances and hence the R value is nearly constant. These PETRA data are also tabulated in Table 1. The theoretical lines will be discussed in the next section.

Since the JADE[11] value is the most accurate one in the energy range of PEP and PETRA, it is described in this section. The various contributions to the systematic error of JADE are shown in Table 2. When added in quadrature, the total is 3% for the center-of-mass energies W between 22 and 37 GeV. In Fig. 4(a), the measured distribution in the number of charged prongs is compared with the Monte Carlo simulation. In Fig. 4(b), the distribution of the vertex along the beam direction is given. In both cases, the cuts used are also shown and are quite far in the tails of the distributions.

As alsready mentioned, a cut is introduced in the total visible energy in order to remove the backgrounds (ii) and (iii). For the JADE detector, the total visible energy is the sum of energies carried by the charged tracks and photons. The observed distributions, the Monte Carlo simulation and the cuts used, are all shown in

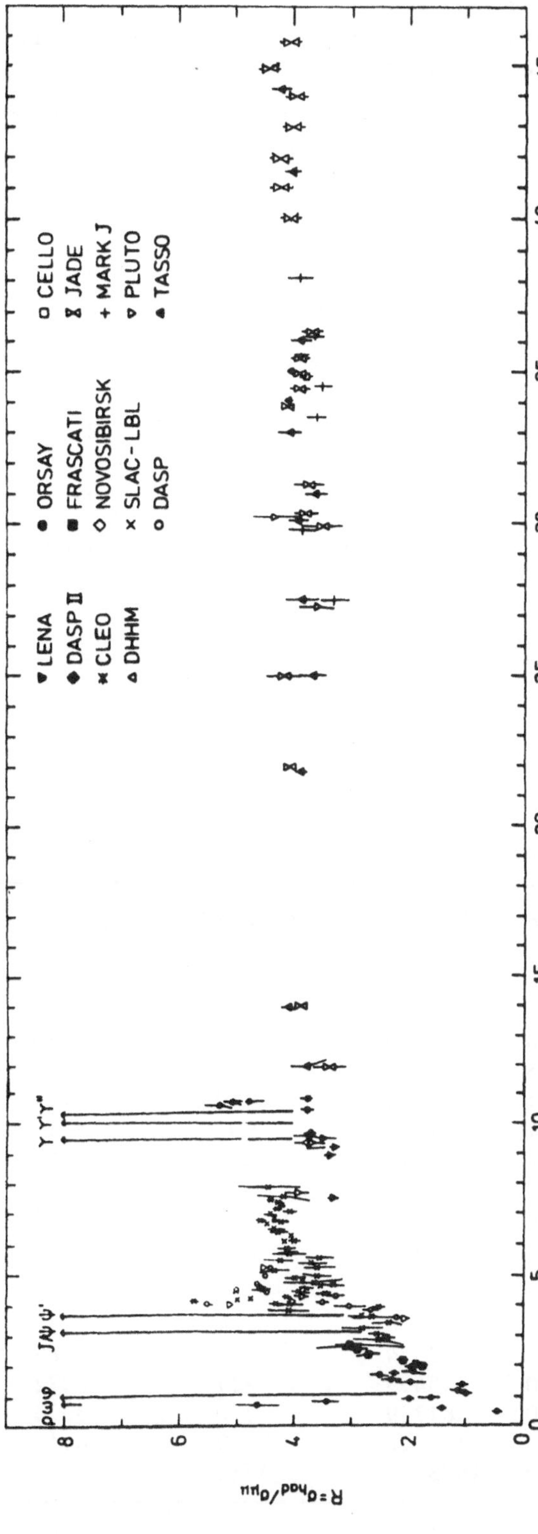

Fig. 3. The ratio of $R = \sigma(e^+e^- \to \text{hadrons})/\sigma_{\mu\mu}$ where $\sigma_{\mu\mu} = 4\pi\alpha^2/3s$. The data are taken from a compilation in Ref. 5.

Table 1(a). Values for R from JADE[11].
The errors quoted include the statistical and point-to-point systematic errors.

W (GeV)	number of events	Luminosity nb^{-1}	R ± ΔR	
12.00	219	106.39	3.45	0.27
14.04	2649	1462.62	3.94	0.14
22.00	1871	2405.87	4.11	0.13
25.01	290	470.81	4.24	0.29
27.66	84	181.90	3.85	0.48
29.93	101	276.24	3.55	0.40
30.38	642	1664.35	3.85	0.19
31.29	251	693.09	3.83	0.28
33.89	3785	11279.52	4.16	0.10
34.50	570	1880.32	3.93	0.20
35.01	4162	13951.49	3.93	0.10
35.45	679	2362.49	3.93	0.18
36.38	420	1623.35	3.71	0.21

Table 1(b). Values for R from TASSO[15] as a function of c.m. energy for runperiod I (1979,1980) and runperiod II (1981). The errors quoted include the statistical as well as the point-to-point systematic error. An overall uncertainty of ± 4.5% has to be added. The relative uncertainty between the two runperiods is ± 2 %.

W[GeV]	$\bar{W}$[GeV]	Luminosity [nb^{-1}]	number of events	R ± ΔR
Runperiod I				
12.0	12.0	96.3	186	3.80 ± 0.28
27.4 - 27.7	27.5	337.0	141	3.91 ± 0.32
29.9 - 30.5	30.1	1309.4	460	3.94 ± 0.18
30.5 - 31.5	31.1	1317.6	407	3.66 ± 0.18
32.5 - 33.5	33.2	762.6	262	4.48 ± 0.28
33.5 - 34.5	34.0	1422.2	410	4.09 ± 0.20
34.5 - 35.5	35.0	2224.4	592	4.03 ± 0.17
35.5 - 36.7	36.1	2213.3	543	3.93 ± 0.17
Runperiod II				
14.0	14.0	1631.0	2704	4.14 ± 0.30
22.0	22.0	2785.4	1889	3.89 ± 0.17
25.0	25.0	454.9	231	3.72 ± 0.38
33.0	33.0	817.9	220	3.73 ± 0.27
34.0	34.0	11143.0	3269	4.13 ± 0.13
35.0	35.0	15786.5	4532	4.22 ± 0.09

Table 1(c). Values for R from MARK J[12]

W(GeV)	R ± ΔR
13	4.6 ± 0.5 ± 0.7
17	4.9 ± 0.6 ± 0.7
27.4 to 27.7	3.8 ± 0.3 ± 0.6
31.57	4.0 ± 0.5 ± 0.6
33	2.9 ± 0.6 ± 0.3
35	3.8 ± 0.4 ± 0.4
35.8	4.4 ± 0.7 ± 0.4
37.94 to 38.63	3.91 ± 0.19 ± 0.3

Table 1(d). Values for R from PLUTO[13]

W(GeV)	R ± ΔR
13	5.0 ± 0.5
17	4.3 ± 0.5
22.0	3.41 ± 0.73
27.6	3.64 ± 0.31
30.0	4.38 ± 0.37
31.6	3.59 ± 0.52

Table 1(e). Values for R from CELLO[9]

W(GeV)	R ± ΔR
33.0 to 36.7	3.85 ± 0.12 ± 0.31

Table 2. Systematic errors of the JADE[11] measurement of R.

	Ecm ≦ 14 GeV	22 - 37 GeV
Background Subtraction	± 1.6 %	± 1.6 %
Radiative Corrections	1.1	0.8
Detection Efficiency	2.5	1.5
Luminosity Point-to-Point	1.0	1.0
Luminosity Overall normalization	1.5	1.5
Total	3.6 %	3.0 %
Point-to-Point	2.7 %	1.8 %
Overall normalization	2.4 %	2.4 %

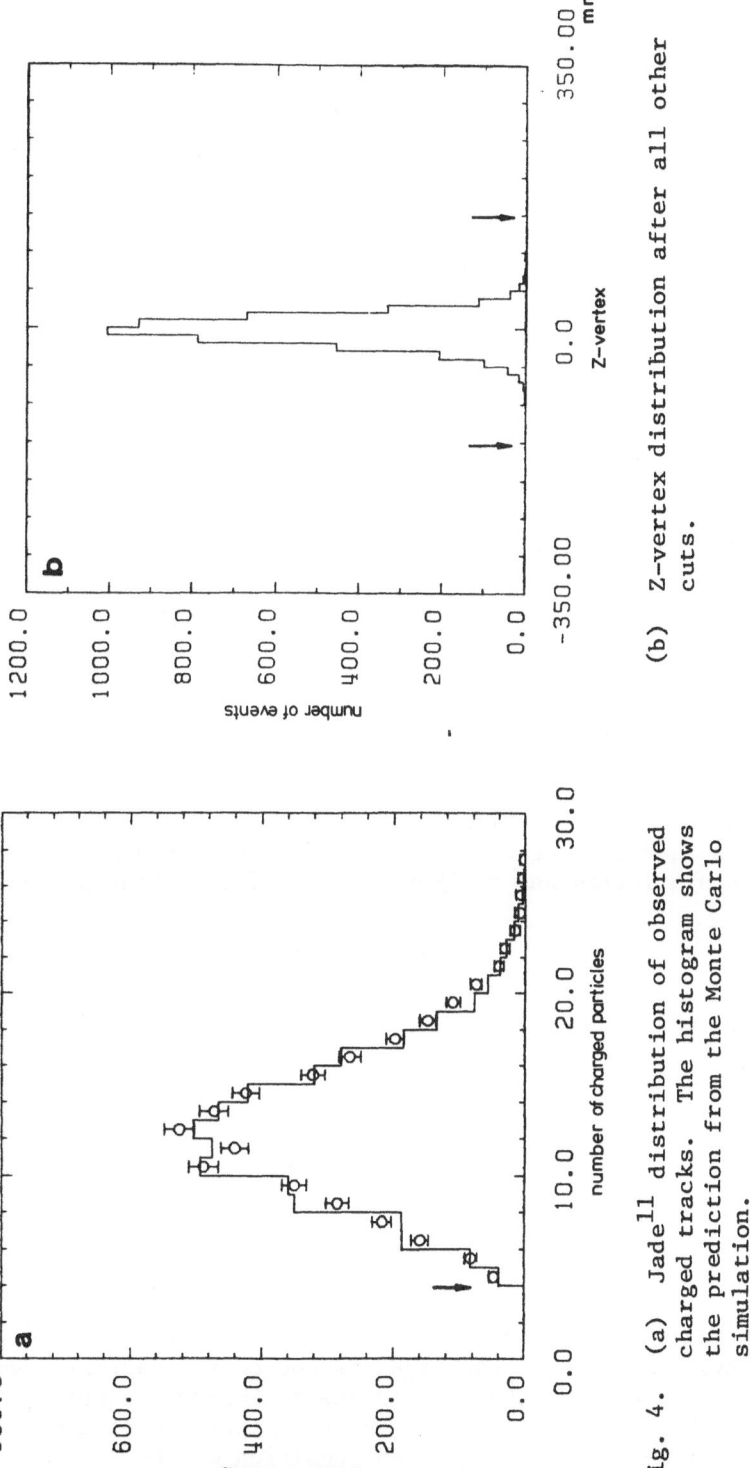

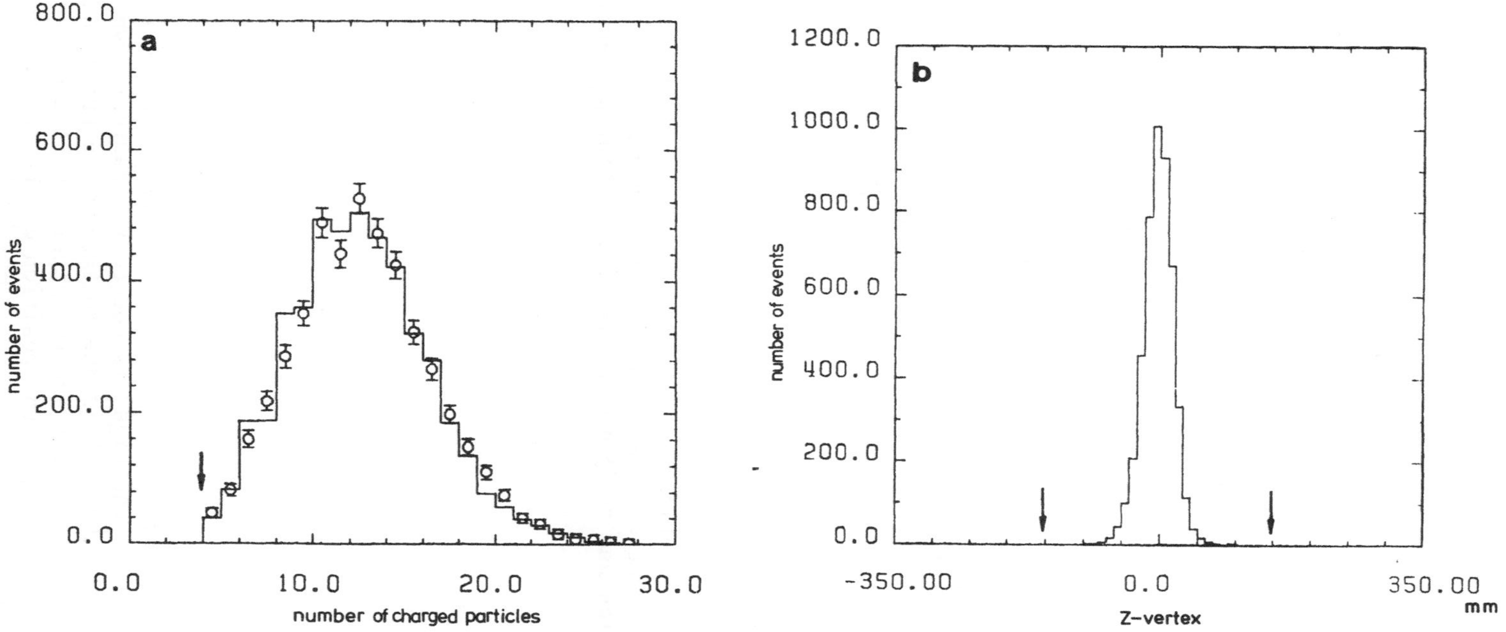

Fig. 4. (a) Jade[11] distribution of observed charged tracks. The histogram shows the prediction from the Monte Carlo simulation.

(b) Z-vertex distribution after all other cuts.

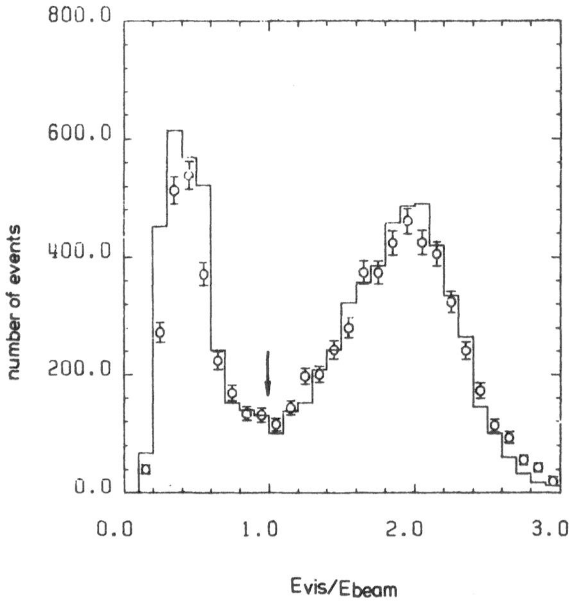

Fig. 5. JADE[11] distribution of the visible energy. The histogram shows the prediction from the Monte Carlo simulations which include the hadron production process via single photon annihilation and the VDM-like γ-γ interaction process.

Fig. 5 as functions of the ratio of the total visible energy to the beam energy. The peak on the right is due to the hadronic events of interest, while that on the left is due to two-photon processes. Because of the good agreement between the experimental data and the Monte Carlo simulation, changing the position of the cut has relati-

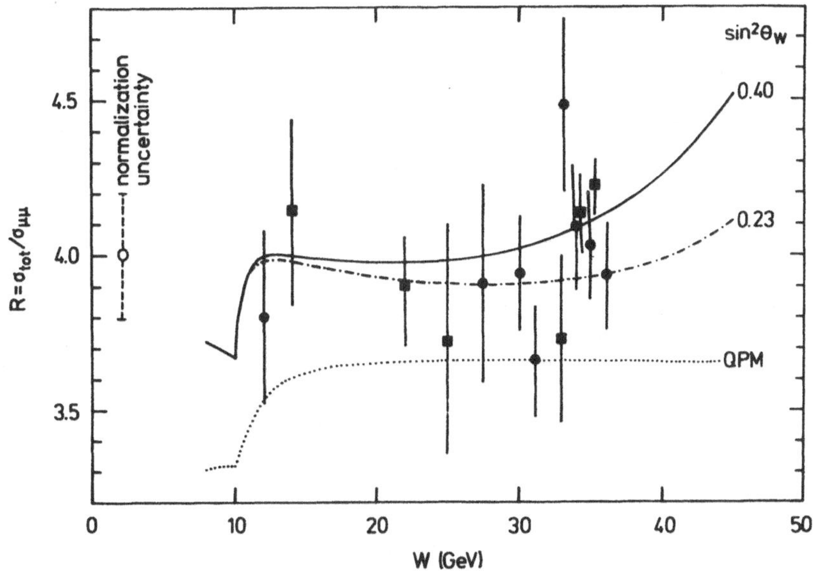

Fig. 6. TASSO[15] results on R for total hadronic cross section. The points marked by a circle are from the runs in 1979 and 1980, while those marked by a square are from 1981. The errors shown include the statistical and point-by-point systematic uncertainty, while the overall normalization uncertainty is indicated separately on the left. The dotted line shows the expectation from the quark parton model (QPM). The full line represents the best fit including weak contributions, while the dashed-dotted line was computed with α_s (s = 1000 GeV^2) = 0.18 and $\sin^2\theta_W$ = 0.23.

vely little effect on the value of R. For example, even with the rather extreme cuts at 1.0 and 1.6, R changes by less than one per cent.

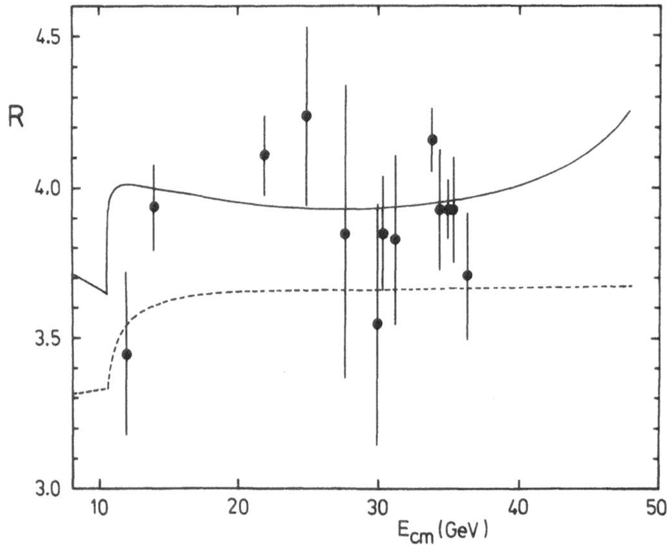

Fig. 7. JADE[11] results on R for total cross section. The error bars include the statistical and the point-to-point systematic errors. The solid curve represents the best fit with α_s (s = 1000 GeV^2) = 0.20 and $\sin^2\theta_w$ = 0.23. The prediction from the simple quark-parton model is also shown by the dashed curve.

The TASSO[15] and JADE[11] results on R are shown as functions of energy, in Fig. 6 and Fig. 7 respectively. The data between 12.0 and 36.4 GeV are consistent with a constant value of R with an average value $\langle R \rangle = 3.97 \pm 0.05 \pm 0.10$ (the second error is the overall normalization error) for JADE and $\langle R \rangle = 4.01 \pm 0.03 \pm 0.20$ for TASSO.

Hadronic Events at PETRA

It is seen from the preceding section that experimentally the total cross section for e^+e^- hadrons is approximately constant in the range of W from 12 GeV to 37 GeV. What kind of final states contributes to this total cross section?

At lower energies, it is customary to list the exclusive channels that are important. At the high energies of PETRA, it is neither possible nor even necessarily desirable to try to measure the cross sections of exclusive channels. Two of the reasons for this assertion are that the number of important final states is too large to be usefully enumerated and that the percentage of hadronic events where all the particles in the final state are detected is quite small. Both are, of course, closely related to the observed large multiplicity.

In order to give some idea about the hadronic events at PETRA, we show in this section some examples of such events. As already seen at SPEAR[16] but much more apparent at PETRA, the majority of the hadronic events consists of two back-to-back jets, where the term "jet" is used to describe a group of particles moving in nearly the same direction. An example of such two-jet events from CELLO is shown in Fig. 8. These two-jet events will be studied in detail in the remainder of this chapter.

In Fig. 9 a three-jet event is shown from TASSO. It is, in fact, the observation of three-jet events from electron-positron annihilation at PETRA that lead to the discovery of the gluon. The properties of the three-jet events and the gluon are described in the next chapter.

Since the majority of the hadronic events consists of two jets and a sizeable fraction consists of three jets, it is natural to expect to see some four-jet events. Such an example from JADE is shown in Fig. 10.

In Fig. 8 and Fig. 9, what has been shown are the views from the beam axis, and the bending of the tracks is, of course, due to the solenoidal magnetic field. There is no guarantee in general that a two-jet event does look like a two-jet event in this particular view. For example, if the axis of the back-to-back jets is close to the beam direction, then such an event does not look like a two-jet event from this view down the beam line, but of course, does look that way from a perpendicular direction. For these figures, the events have been chosen to show clearly their jet characteristics, and thus cannot be said to be typical.

In Fig. 11 we give an event of the type 3h, $e^+e^- \to e^+e^-$ hadrons from PLUTO. Since the e^+ and e^- in the final state go down the beam pipe and are hence not detected, the view in Fig. 11 perpendicular to the beam bears a great deal of resemblance to the same view in Fig. 8. Actually, these two-photon events are quite different : the two hadron jets are coplanar with the beam axis, but not colinear as in two-jet events, and the total visible energy is relatively small, as seen from Fig. 5.

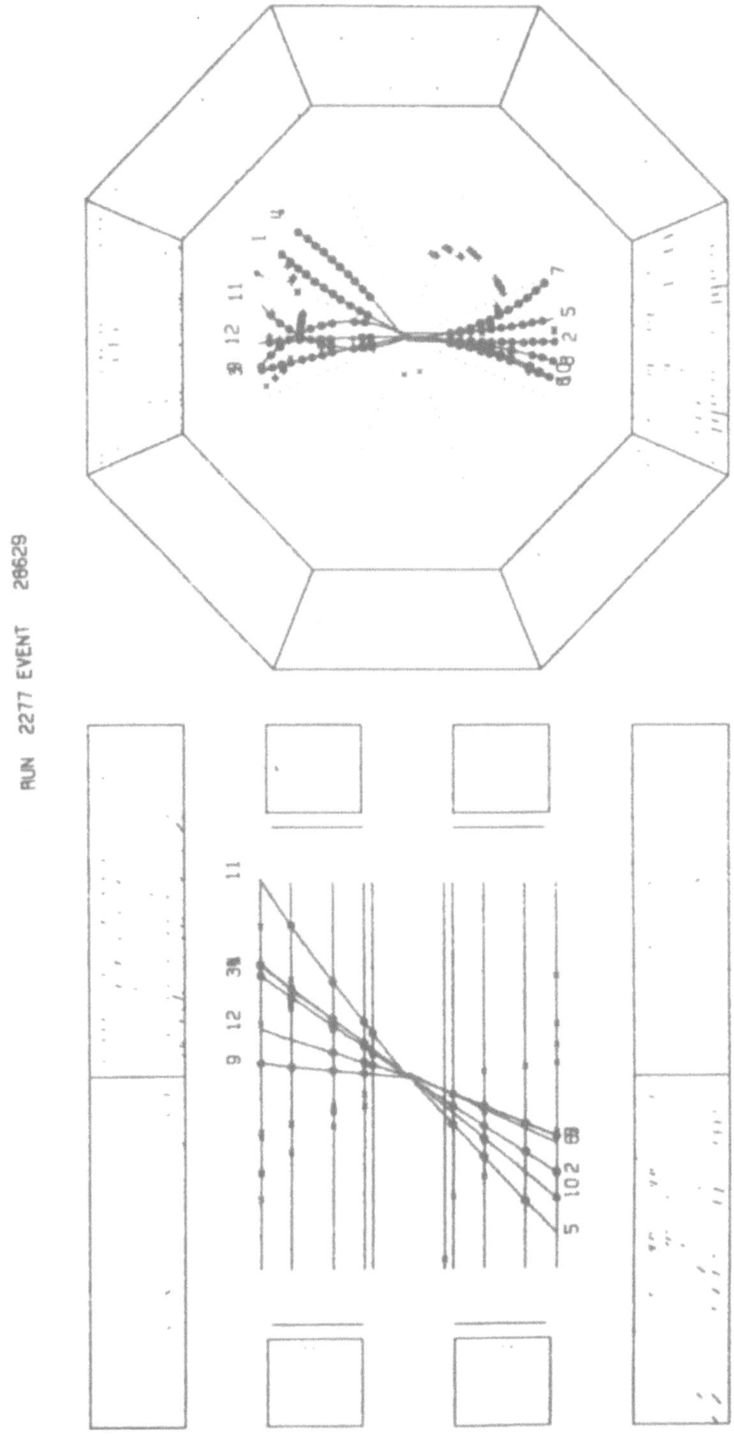

Fig. 8. Two-jet event from CELLO.

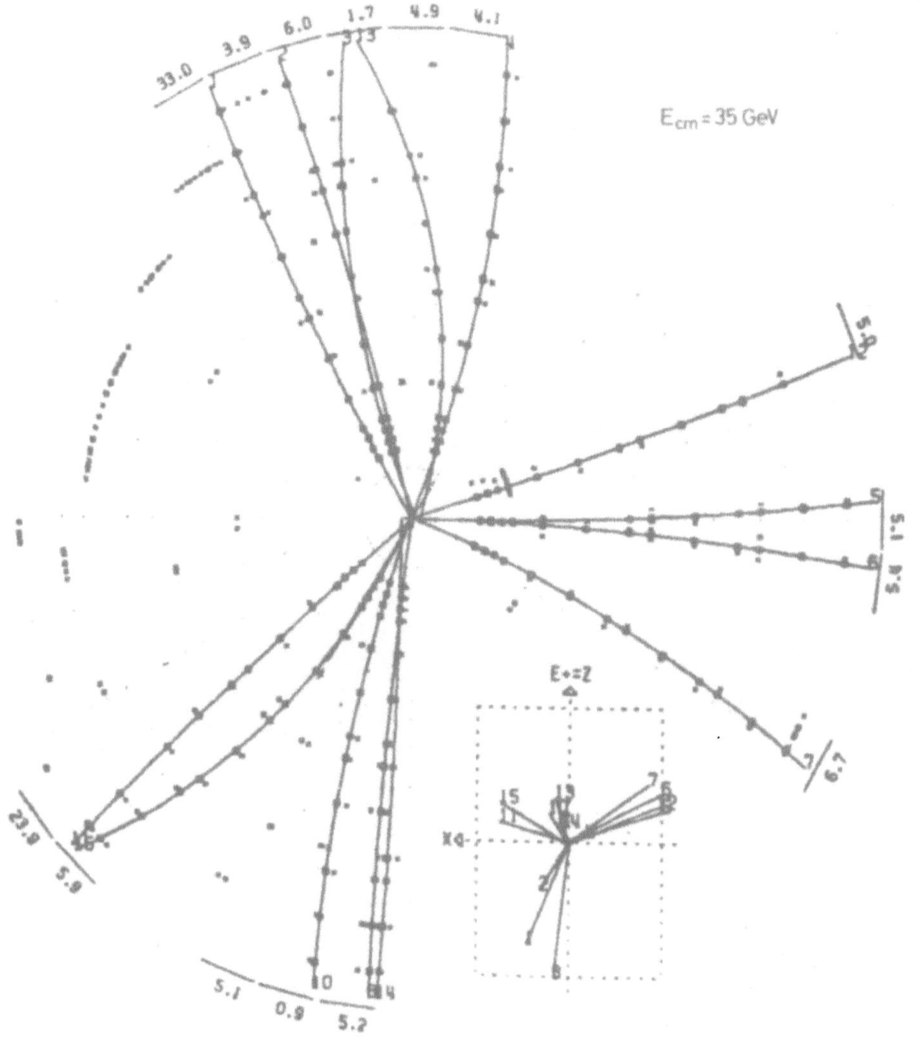

Fig. 9. Three-jet event from TASSO.

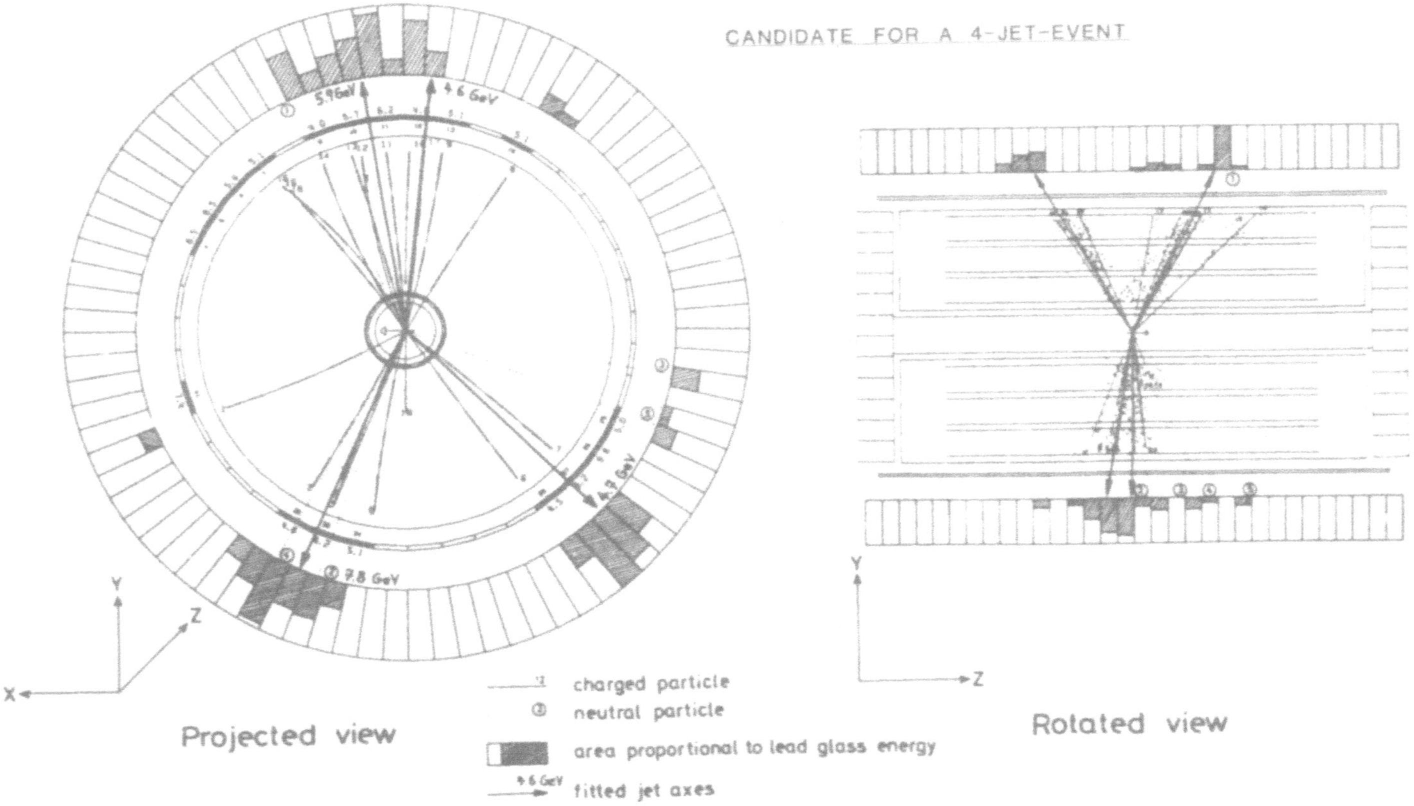

Fig. 10. Four-jet event from JADE.

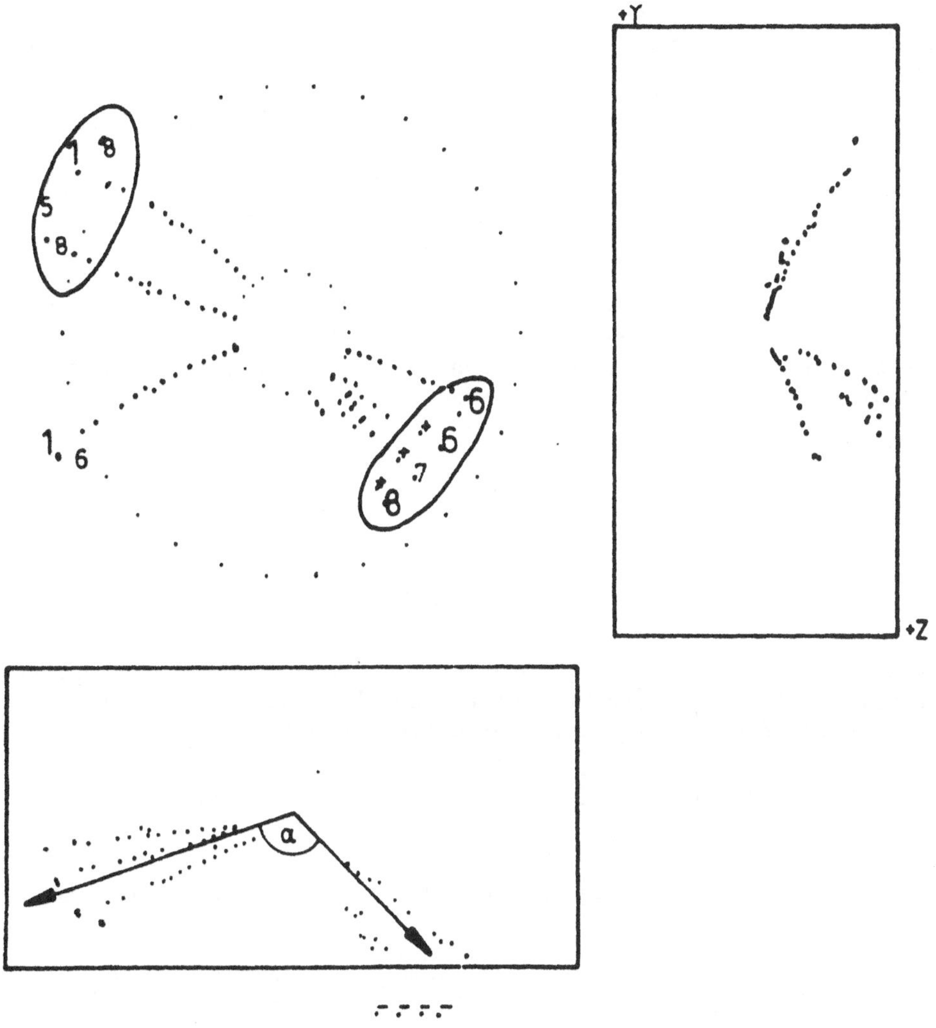

Fig. 11. Two-photon event $e^+e^- \rightarrow e^+e^-$ hadrons from PLUTO. The e^+ and e^- in the final state are not detected.

Quark Picture and Quantum Chromodynamics

The predominance of two-jet events indicates that something is pair produced, as represented by the Feynman diagrams of Fig. 12. If it is a point fermion, then the matrix element is proportional to its charge e_f, and hence

$$\frac{\sigma(e^+e^- \to f\bar{f})}{\sigma(e^+e^- \to \mu^+\mu^-)} = \frac{e_f^2}{e^2} , \tag{5}$$

when the center-of-mass energy is much larger than the fermion mass.

Quarks were invented in 1964 by Gell-Mann and Zweig[17] to explain the SU(3) multiplet structure of the observed hadrons. They are fermions with fractional charge and fractional baryon number. For reasons of statistics, the quark has a new internal quantum number called the color, which can take on three values. Since the cross section for the production of a quark pair with a given color is given by (5), the total cross section for the production of quark pairs of all colors is larger by a factor of 3 :

$$\frac{\sigma(e^+e^- \to q\bar{q})}{\sigma(e^+e^- \to \mu^+\mu^-)} = 3\,\frac{e_f^2}{e^2} , \tag{6}$$

again on the basis of the diagram of Fig. 12.

Since the discovery of the upsilon[8] , there are five known quarks, as listed in Table 3, together with some of their elementary properties. If (6) is used to get the total cross section for producing pairs of these five types of quarks, the result is

$$\frac{\sigma(e^+e^- \to u\bar{u},d\bar{d},c\bar{c},s\bar{s},b\bar{b})}{\sigma(e^+e^- \to \mu^+\mu^-)} = 3[(\tfrac{2}{3})^2 + (\tfrac{1}{3})^2 + (\tfrac{2}{3})^2 + (\tfrac{1}{3})^2 + (\tfrac{1}{3})^2] = \frac{11}{3} . \tag{7}$$

This compares favorably with the observed R as given in Table 1 and Figs. 3, 6 and 7.

Because of color symmetry, the quarks must interact through a Yang-Mills[18] non-Abelian gauge field called the gluon. Similar to QED, but much more complicated, the theory of the interaction between quarks and gluons is called quantum chromodynamics, or QCD for short. On the basis of QCD, first order radiative correction to $e^+e^- \to q\bar{q}$ gives an extra factor of $1 + \alpha_s/\pi$[19]:

$$\frac{\sigma(e^+e^- \to u\bar{u},\ d\bar{d},\ c\bar{c},\ s\bar{s},\ b\bar{b})}{\sigma(e^+e^- \to \mu^+\mu^-)} = \frac{11}{3}\left(1 + \frac{\alpha_s}{\pi}\right) , \tag{8}$$

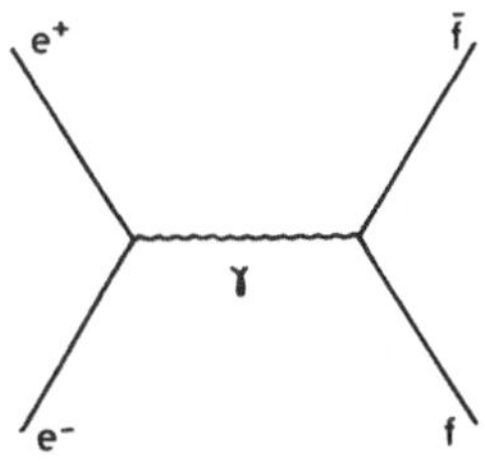

Fig. 12. Feynman diagram for the production of a pair of charged particles.

Table 3. Elementary properties of five quarks

quark	u	d	c	s	b
Baryon number	$\frac{1}{3}$	$\frac{1}{3}$	$\frac{1}{3}$	$\frac{1}{3}$	$\frac{1}{3}$
charge	$\frac{2}{3}$ e	$-\frac{1}{3}$ e	$\frac{2}{3}$ e	$-\frac{1}{3}$ e	$-\frac{1}{3}$ e
isospin I	$\frac{1}{2}$	$\frac{1}{2}$	0	0	0
I_z	$\frac{1}{2}$	$-\frac{1}{2}$	0	0	0
mass (GeV)	small	small	1.9	0.5	5.3

where α_s is the so-called running coupling constant for the strong interaction between quarks and gluons. The experimental determination of this fundamental coupling constant at PETRA energies will be discussed in the next chapter. Theoretically

$$\alpha_s(s) = \frac{12\pi}{(33 - 2N_f)\ \ln\ (s/\Lambda^2)}\ , \tag{9}$$

where N_f is the number of quarks with mass below the beam energy, and Λ is a characteristic strong interaction mass, believed to be about a few hundred MeV. Using the experimentally determined value of α_s, the PETRA data on the total hadronic cross section are in good agreement with the simple picture of pair producing five types of quarks, especially if the lowest-order QCD correction is included.

In table 3, there are three quarks of charge -1/3, but only two of charge 2/3. Therefore many physicists expect the existence of a sixth quark, called t, with baryon number 1/3, charge 2/3, and isospin $I = I_z = 0$.

Momentum Tensor

As shown in the preceeding section, the quark picture on the basis of the diagram of Fig. 12 leads to eq. (7) which is in good agreement with the experimental measured value of R for the total cross section at PETRA energy. Yet these produced quark pairs of fractional charge have never been observed directly. This dilemma leads to the working hypothesis that somehow the quarks turn into a group of hadrons through strong interactions. Another way of stating this hypothesis is that quarks are "confined" so that in the final state each quark must combine with other quarks or antiquarks to form hadrons. We shall accept this working hypothesis in this and the next chapters.

Independent of the mechanism of turning the quark into hadrons, the hadrons are expected to retain some memory of the quark momentum. In other words, if the quark is produced in the x-direction, the resulting hadrons are expected to have, on the average, larger momenta in the x-direction than the y- or z-directions, especially at high energies. This situation is illustrated in Fig. 13. From this point of view, the occurence of jets is natural. If the hadron momenta transverse to the quark direction of flight are limited and the number of produced hadrons grows only slowly with energy, the emitted hadrons will be more and more collimated around the primary quark direction as the total energy increases.

Motivated by such considerations, the jet structure in e^+e^- annihilation was first looked for and found at SPEAR[16]. Their method is based on the analog with the problem of the moment of inertia in classical mechanics. For a system of particles or a rigid body, the moment of inertia, as a function of the direction of axis of rotation, is an ellipsoid. If the rigid body is cigar shaped, then this

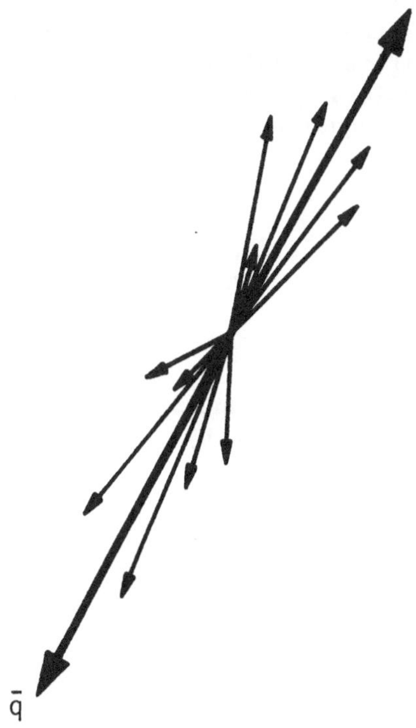

Fig. 13. e^+e^- annihilation in the quark model : production of a $q\bar{q}$ pair followed by hadronization.

ellipsoid of inertia is a pancake such that the momenta of inertia is small when rotated in the direction of the cigar. Thus the SPEAR group calculated for each event the momentum analog of the inertia tensor[20]

$$T_{\alpha\beta} = \Sigma_j(\delta_{\alpha\beta}\vec{p}_j^{\,2} - p_{j\alpha}\, p_{j\beta}) , \tag{10}$$

where the summation is over all detected particles and α and β refer to the three spatial components of each particle momentum $\vec{p}_j$. Since $T^{\alpha\beta}$ is a symmetrical tensor, it can be diagonalized to give the eigenvalues $\bar{\lambda}_1$, $\bar{\lambda}_2$ and $\bar{\lambda}_3$. If $\bar{\lambda}_3$ is the smallest of the three, then the sphericity S is defined as

$$S = \frac{3\bar{\lambda}_3}{\bar{\lambda}_1 + \bar{\lambda}_2 + \bar{\lambda}_3} = \frac{3(\Sigma_i p_{Ti}^2)_{min}}{2\Sigma_i\, \vec{p}_i^{\,2}} . \tag{11}$$

S approaches 0 for events with bounded transverse momenta and approaches 1 for events with large multiplicity and isotropic phase-space particle distributions. The jet structure is established by studying the energy dependence of sphericity.

Sphericity gives essentially the relative magnitude of $\bar{\lambda}_3$ compared with the other two. If we want to analyze the shape of the event in more detail, it is useful to study the relative magnitudes of all the three eigenvalues. These eigenvalues $\bar{\lambda}_1$, $\bar{\lambda}_2$ and $\bar{\lambda}_3$ satisfy the triangular inequalities, i.e.,

$$\bar{\lambda}_1 \leqq \bar{\lambda}_2 + \bar{\lambda}_3 \ , \tag{12}$$

etc. Since triangular inequalities are not easy to deal with, it is more convenient to use the momentum tensor[21,22]

$$M_{\alpha\beta} = \sum_j p_{j\alpha}\, p_{j\beta} . \tag{13}$$

Let λ_1, λ_2, λ_3 be the eigenvalues and $\hat{n}_1$, $\hat{n}_2$, $\hat{n}_3$ the corresponding eigenvectors of M which are ordered by

$$0 \leqq \lambda_1 \leqq \lambda_2 \leqq \lambda_3 , \tag{14}$$

and are related to those of T by

$$\bar{\lambda}_i = (\lambda_1 + \lambda_2 + \lambda_3) - \lambda_i \ , \tag{15}$$

for i = 1, 2, 3. Thus (15) guarantees the triangular inequalities (12). Since

$$\lambda_1 + \lambda_2 + \lambda_3 = \Sigma_j\, \vec{p}_j^{\,2} \ , \tag{16}$$

let

$$Q_k = \lambda_k / \sum_j \vec{p}_j^{\,2} . \tag{17}$$

These normalized eigenvalues Q_k satisfy

$$Q_1 + Q_2 + Q_3 = 1 \ , \tag{18}$$

and

$$0 \leqq Q_1 \leqq Q_2 \leqq Q_3 , \tag{19}$$

and their physical meanings are as follows :

$Q_1 = \min\limits_{\hat{n}} \Sigma_j\, (\vec{p}_j . \hat{n})^2 / \Sigma_j \vec{p}_j^{\,2}$ gives the flatness of the event ($\hat{n} = \hat{n}_1$),

$Q_2 = \min\limits_{\hat{n} \perp \hat{n}_1} \Sigma_j\, (\vec{p}_j . \hat{n})^2 / \Sigma_j \vec{p}_j^{\,2}$ gives the width of the event ($\hat{n} = \hat{n}_2$), and

$Q_3 = \max\limits_{\hat{n}} \Sigma_j\, (\vec{p}_j . \hat{n})^2 / \Sigma_j \vec{p}_j^{\,2}$ gives the length of the event ($\hat{n} = \hat{n}_3$).

Collinear events are characterized by $Q_2 \ll Q_3$, and similarly coplanar events by $Q_1 \ll Q_2$. In terms of the Q's, the sphericity S is

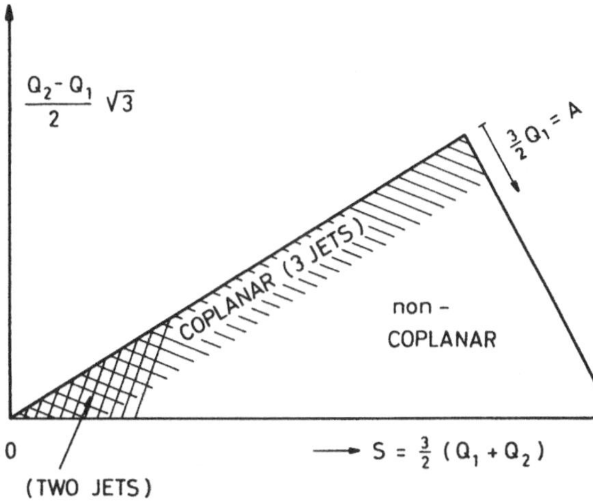

Fig. 14. Triangular plot to separate two-jet, three-jet and noncoplanar events.

$$S = \frac{3}{2} (Q_1 + Q_2). \tag{20}$$

A triangular plot, with the coordinate variables chosen to be S and $Y = \sqrt{3}\,(Q_2 - Q_1)/2$, can be used to separate two-jet, three-jet and non-planar events as shown in Fig. 14. The aplanarity A is naturally defined as

$$A = (3/2)\,Q_1 . \tag{21}$$

Fig. 15 shows the event distribution on the triangular plot for W = 14, 34 and 41.5 GeV. Fig. 16 shows the energy dependence of the average sphericity <S> as measured at PETRA[5]. As shown in this figure, <S> decreases rapidly with increasing W, i.e., the particles become more and more collimated in clear distinction to a phase space behavior. Fig. 17 shows the sphericity distribution measured at

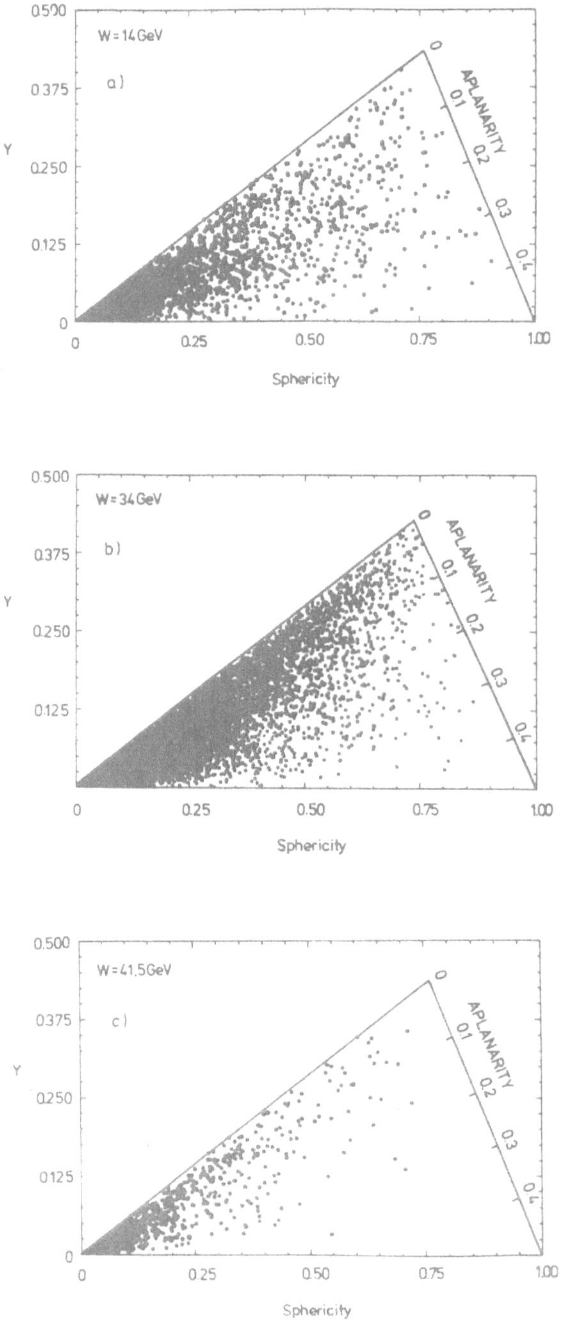

Fig. 15. The TASSO[5] distribution of sphericity versus aplanarity at W = 14, 34 and 41.5 GeV.

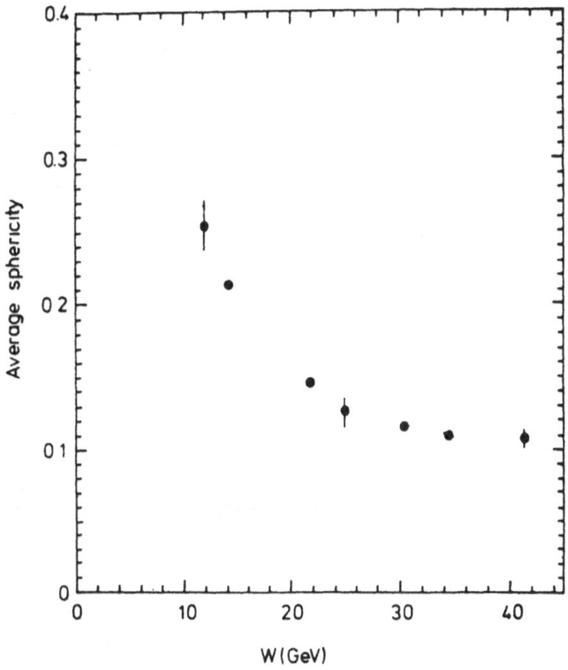

Fig. 16. The average sphericity as a function of the center-of-mass energy from TASSO[5].

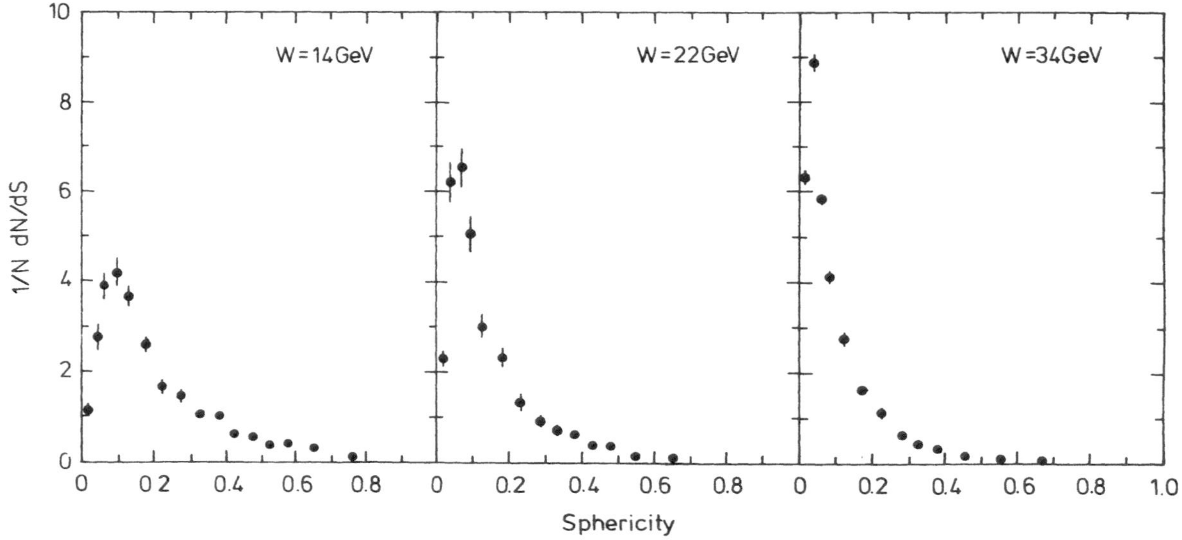

Fig. 17. The normalized sphericity distributions at W = 14, 22 and 34 GeV from TASSO[5].

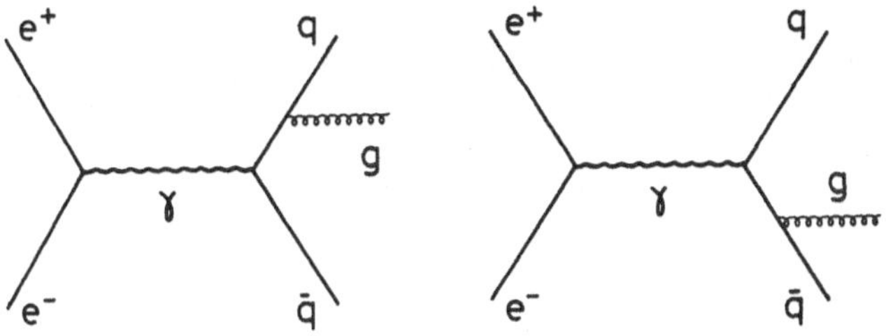

Fig. 18. Feynman diagrams for $e^+e^- \to q\bar{q}g$.

W = 14, 22 and 34 GeV[5]. The strong preference for small S values is clearly seen, 80 % of the events have S < 0.25. This shows clearly that two-jet events dominate at PETRA energies.

Thrust

In the context of electron-positron annihilation, the usefulness of thrust, defined by

$$T = \max_{\hat{n}} \frac{\sum_i |\vec{p}_i \cdot \hat{n}|}{\sum_i |\vec{p}_i|} , \tag{22}$$

where $\hat{n}$ is a unit vector, was first discussed by Farhi[23]. This same quantity was used previously by Brandt et al.[24] in connection with an attempt to analyze high energy hadronic collisions in order to test the "intermediate bodies" model.

As explained by Farhi, the reason for using T is entirely theoretical. The underlying question is : what quantities can be reliably calculated in quantum chromodynamics (QCD) by perturbation theory? Some theoreticians believe that the working hypothesis stated at the beginning of the preceding section can be derived from QCD, but no clear derivation has yet been given. The difficulty of applying perturbative QCD to e^+e^- annihilation is, however, of an entirely different nature, namely the uncertainty about quark masses.

For definiteness, consider the QCD correction to the two-body process $e^+e^- \to q\bar{q}$, as shown in Fig. 12. Thus it is necessary to study the three-body process $e^+e^- \to q\bar{q}g$, where g stands for the gluon. To the lowest order, the diagrams are those of Fig. 18 and the calculations are entirely similar to, but simpler than those of quantum electrodynamics (QED) for $e^+e^- \to \mu^+\mu^-\gamma$. Since in both cases gauge invariance obtains only for all fermions on their respective mass shells, it is necessary to know the masses of the quarks. Since quarks have never been observed directly, their masses are not known. More precisely, while good estimates of the masses of the s, c and b quarks can be obtained from the ϕ, the J/ψ and the Υ mesons, there is a great deal of uncertainty about the masses of the u and d quarks, which may be as light as 10 or 20 MeV, (see Table 3).

In view of this difficulty, there are at least two distinct approaches. The more straightforward approach is to introduce quantities that cannot be calculated within the framework of perturbative QCD, such as quark and gluon fragmentation functions into hadrons. An alternative approach is to restrict attention to quantities that do not depend critically on the masses of the u and d quarks. In particular, perturbative QCD is to be applied only to quantities which have the property of being finite order by order in the unphysical limit where the quark mass m approaches zero[25].

On a purely theoretical level, this second approach is perhaps to be preferred. Practically, however, the two approaches are not entirely different, because fragmentation functions are in any case needed for the analysis of experimental data.

It is on the basis of the second approach that Farhi introduced the thrust[23]. As he puts it, the basis is a physical assumption yet to be contradicted : quantities which in the massless case are physically sensible, i.e., measurable in principle, will have a perturbative expansion free of $m \to 0$ singularities. The main problem with massless particles is that they are kinematically allowed to split into several massless particles all moving in the original direction. Therefore only quantities that do not change under this splitting can be used in perturbative QCD. The thrust T of (22) indeed has this property. Roughly speaking, what is needed is linearity in the momenta.

The energy dependence of the average thrust $\langle T\rangle$ is shown in Fig. 19, and the normalized thrust distributions at W = 14, 22 and 34 GeV are presented in Fig. 20. A comparison with Fig. 16 and 17 shows that the sphericity and thrust behave in the same way. Similar data have been obtained by all other PETRA Collaborations. At PETRA, both sphericity and thrust are used extensively, and they are about equally useful.

Spin of the Quarks

On the basis of the angular distribution of the sphericity axis $\hat{n}_3$, Hanson et al.[16] found the spin of quarks to be 1/2 from the SPEAR data at W = 7.4 GeV. When W is increased by a factor of five to

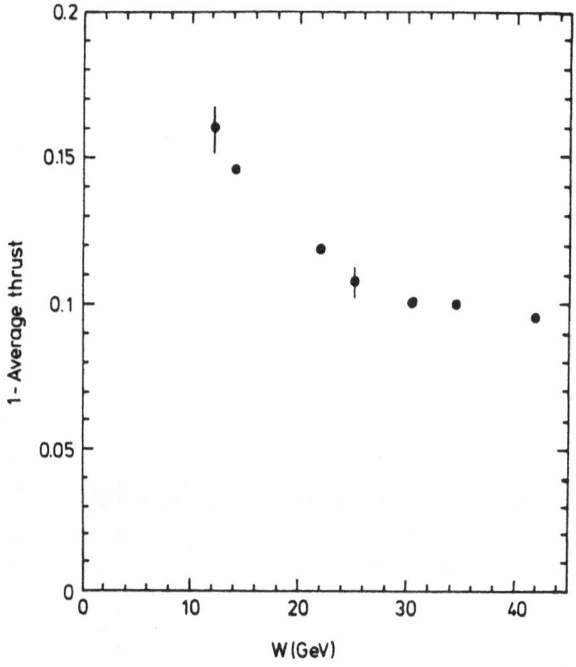

Fig. 19. The average value of 1-thrust, <1 - T>, as a function of the c.m. energy W from TASSO[5].

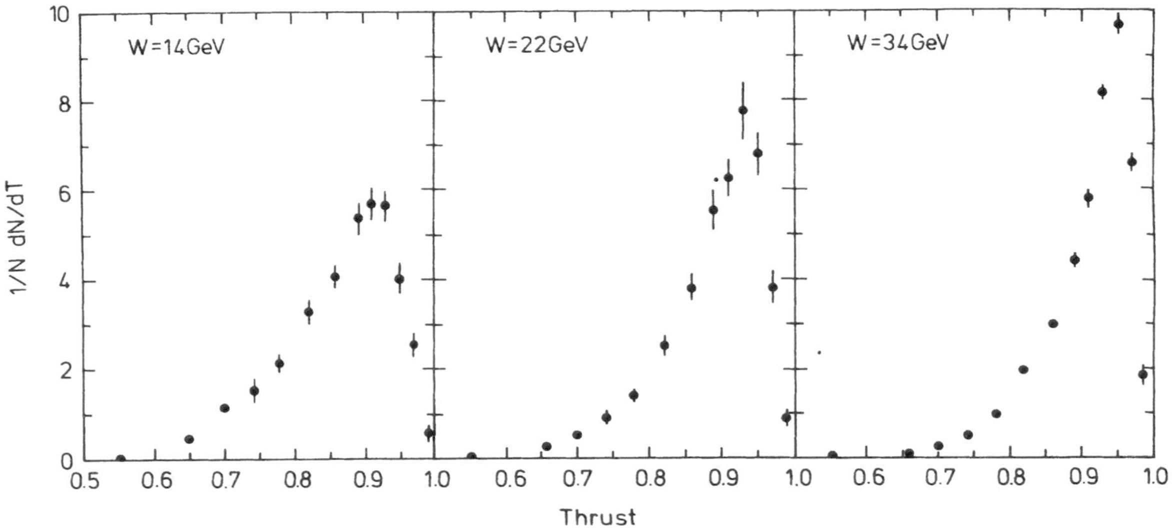

Fig. 20. The normalized thrust distributions at W = 14, 22 and 34 GeV from TASSO[5].

PETRA energies, this angular distribution, as shown in Fig. 21 remains $1 + \cos^2\theta$, where θ is the angle between the beam direction and the sphericity axis. When the thrust axis is used instead of the sphericity axis $\hat{n}_3$, there is no substantial change in the angular distribution. Since the angular distribution of the jet axis is

$$\begin{aligned} &1 + \cos^2\theta \qquad \text{for a quark of spin } 1/2 \\ &\sin^2\theta \qquad \text{for a quark of spin } 0 \end{aligned} \tag{23}$$

on the basis of the Feynman diagrams of Fig. 12, the spin of the quarks is verified to be 1/2 in agreement with the earlier SPEAR result.

Strictly speaking, since there are five quarks instead of one, this data by itself is not sufficient to give the spins of the quarks separately. However, there is by now so much information about the quarks that the spin is no longer much of an issue, since it is related to numerous quantities, such as the step in R and the properties of the $q\bar{q}$ bound states.

Field-Feynman Fragmentation

Previously, we introduced the working hypothesis that somehow the quarks turn into a group of hadrons through strong interactions. In the absence of a theory based on first principles, Field and Feynman[27] pioneered the development of a phenomenological description of this fragmentation process. This work of Field and Feynman is of fundamental importance to the analysis of the experimental data from PETRA.

Field and Feynman assume that quark jets can be analyzed on the basis of a recursive principle. The ansatz is based on the idea that quark of type "a" coming out at some momentum creates a color field in which new quark-antiquark pairs are produced. Quark "a" then combines with an antiquark, say "$\bar{b}$", from the new pair $\bar{b}b$ to form a meson "$a\bar{b}$" leaving the remaining quark "b" to combine with further antiquarks. The "meson" $a\bar{b}$, called the primordial or primary meson state, may be directly observed as a pseudoscalar meson, or it may be a vector or higher spin unstable resonance which subsequently decays into the observed mesons.

Fig. 22 of Field and Feynman gives a clear illustration of the phenomenological description.

In the original paper of Field and Feynman, only three quarks namely u, d and s are taken into account. The inclusion of the known c and b quarks and even the hypothetical top quark t has been carried out first by Meyer[28] of the TASSO Collaboration.

The description of the fragmentation process involves the

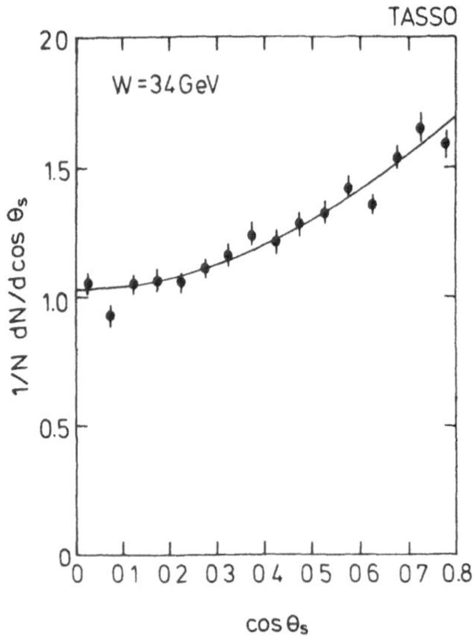

Fig. 21. The angular distribution of the jet axis determined by sphericity. The curves are proportional to $1 + \cos^2\theta_s$.

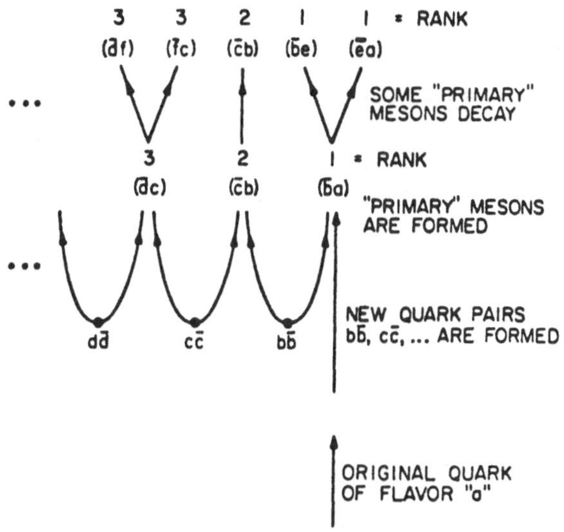

Fig. 22. Illustration of the "hierarchy" structure of the final mesons produced when a quark of type "a" fragments into hadrons in the Field-Feynman scheme[27]. New quark pairs $b\bar{b}$, $c\bar{c}$, etc., are produced and "primary" mesons are formed. The "primary" meson $\bar{b}a$ that contains the original quark is said to have "rank" one and primary meson $\bar{c}b$ rank two, etc. Finally, some of the primary mesons decay and all the decay products are assigned to have the rank of the parent. The order in "hierarchy" is not the same as order in momentum or rapidity.

following three parameters :

(i) σ_q. The distribution of the transverse momentum k_T of the quarks in the jet cascade is assumed to be $\sim\exp(-k_T^2/2\sigma_q^2)$.

(ii) P/(P+V). Here P/V is the ratio of primordial pseudoscalar mesons to vector mesons produced in the fragmentation process.

(iii) a_F. The primordial quark fragmentation function $f^h(z)$ of a quark into a hadron h in the Field-Feynman model is taken to be the same for u, d, and s quarks,

$$f^h(z) = 1 - a_F + 3a_F(1-z)^2, \tag{24}$$

where

$$z = (E + p_\parallel)_h/(E + p_\parallel)_q. \tag{25}$$

It is a peculiarity of the Field-Feynman fragmentation that only mesons are produced in the hadronic events of e^+e^- annihilation, but not baryons. The reason is that, in this type of models, it is quite difficult to bring three quarks or three antiquarks together. Indeed, it was one of the major surprises from PETRA that experimentally baryons are produced rather copiously.

Experimental Determination of the Field-Feynman Parameters

The TASSO Collaboration[29] has determined the Field-Feynman parameters a_F, σ_q, and P/(P+V) from their experimental data. They use only the data from the region sphericity S<0.25, where two-jet events dominate and gluon bremsstrahlung is of minor importance. It is noted that the average squared transverse momentum $\langle p_T^2\rangle_{out}$ $(=\langle(\vec{p}.\hat{n}_1)^2\rangle)$ perpendicular to the event plane is most sensitive to σ_q, the single particle fractional momentum distribution dN/dx_p (where x_p = momentum of particle/beam momentum) is most sensitive to a_F, and the distribution of charged multiplicity n_{ch} is most sensitive to P/(P+V). These distributions are well described by the parameters

$$a_F = 0.57 \pm 0.20, \quad \sigma_q = 0.32 \pm 0.04 \text{ GeV/c},$$

$$P/(P + V) = 0.56 \pm 0.15 \tag{26}$$

These parameters are used in the determination of the quark-gluon coupling constant α_s, as to be discussed later on. With (26) and the value of α_s, numerous comparisons have been carried out between the experimental data and QCD with Field-Feynman fragmentation. The agreement is generally excellent. Some examples of these comparisons are shown in Fig. 23.

Actually, in these and other comparisons, there is a fourth parameter related to the fragmentation of the gluon. As the gluon is assumed to split into a quark-antiquark pair, the distribution of the transverse momentum q_T of the quarks with respect to the gluon is taken to be $\exp(-q_T^2/2\,\sigma_g^2)$. However, the result of the analysis is insensitive to the value of this parameter, the reason being that in the gluon jets, the transverse momentum spread due to σ_g and that due to σ_q from the hadronization of quarks are hardly distinguishable Changing σ_g from 0 to 0.5 GeV/c makes almost no difference in the results.

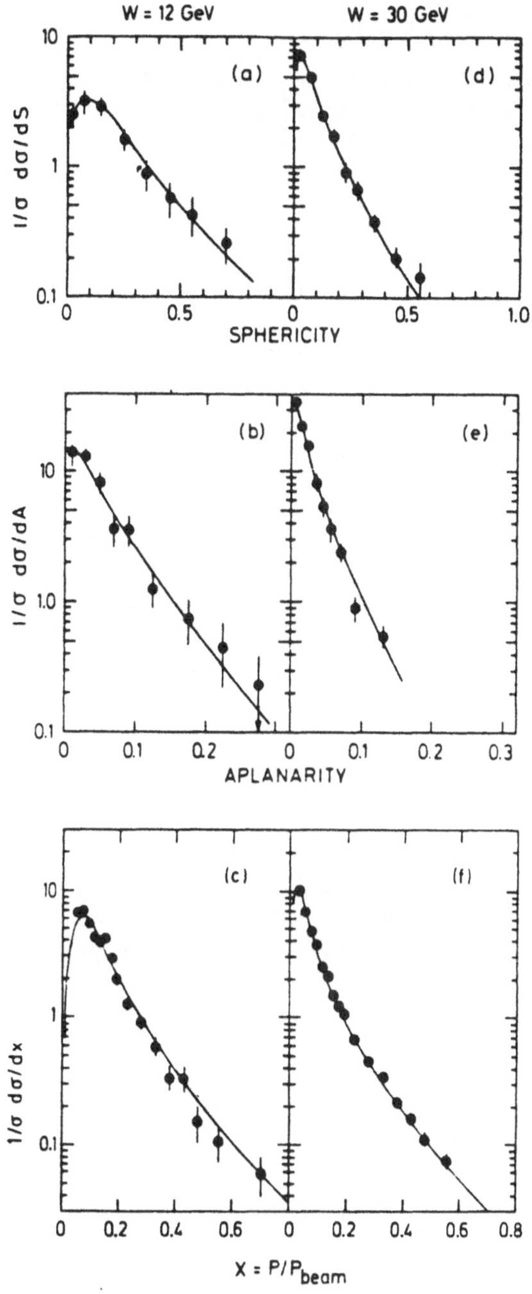

Fig. 23. Comparison of the TASSO[29] data with the QCD model (curves) at W = 12 GeV for (a) sphericity S, (b) aplanarity A and (c) the single particle inclusive x distribution for charged particles; and at W = 30 GeV for (d) S, (e) A, and (f) x. The experimental as well as the theoretical distributions are corrected for acceptance and radiative effects.

Long-Range Charge Correlation in Opposite Jets

As shown in Fig. 22, Field-Feynman fragmentation applies to each quark, or antiquark, separately. Actually quarks are never produced singly. In particular, in electron-positron annihilation, a $q\bar{q}$ pair is produced as shown in Fig. 12. Therefore a better representation of the fragmentation process is shown schematically in Fig. 24, where for simplicity the decays of the primordial mesons are omitted. Also the charged mesons are numbered as h_1, h_2,... but the neutral ones are not numbered. Since the quark charges are -1/3 and +2/3, we have with this ordering $e_1 = 1$, $e_2 = -1$, $e_3 = 1$, $e_4 = -1$, etc. Although this ordering cannot be determined experimentally, it is strongly correlated with the rapidity ordering.

Accordingly, the presence of a negative charge for large negative y increases the probability of the presence of a positive charge for large positive y', and vice versa, where the rapidity y is defined as usual by $y = (1/2)\ln[(E + p_\parallel)/(E - p_\parallel)]$, with $p_\parallel$ the component of the momentum parallel to the jet axis. In order words it is a prediction of this picture of jet formation via a $q\bar{q}$ pair as shown in Fig. 24 that there is a long-range charge correlation in opposite jets.

From the two-jet events, the TASSO Collaboration[30,31], has observed this long-range charge correlation. If n is the charged multiplicity of the event, then we define a charge correlation function, the compensating charge flow $\bar{\phi}$ by

$$\bar{\phi}(y,y') = -\frac{1}{\Delta y \Delta y'} < \frac{1}{n} \sum_{k=1}^{n} \sum_{i \neq k} e_i(y) e_k(y') >, \tag{27}$$

where, for example, for the ith particle with rapidity y_i, $e_i(y)=+1$ or -1 according to the charge of this particle if y_i is inside an interval Δy around y and $e_i(y) = 0$ otherwise. Here < > means averaging over all events.

This charge correlation function is to be compared with the corresponding particle density function defined by

$$\bar{\rho}(y,y') = \frac{1}{\Delta y \Delta y'} < \frac{1}{n(n-1)} \sum_{k=1}^{n} \sum_{i \neq k} |e_i(y)||e_k(y')| > , \tag{28}$$

where the sum is over all charged particles. The denominators n and n(n-1) in (27) and (28) are chosen such that

$$\int dy \int dy' \ \bar{\phi}(y,y') = \int dy \int dy' \ \bar{\rho}(y,y') = 1 , \tag{29}$$

because

$$\sum_{i \neq k} e_i e_k = -n \quad \text{and} \quad \sum_{i \neq k} |e_i||e_k| = n^2 - n. \tag{30}$$

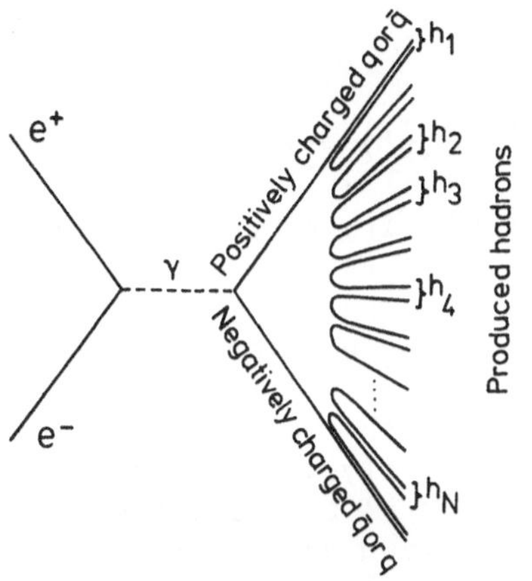

Fig. 24. A schematic diagram of hadron production in e^+e^- annihilation.

Let $P(y_+,y'_-)$, for example, be the probability of having a positively charged particle with rapidity y and a negatively charged particle with rapidity y'. In terms of this probability, $\bar{\rho}$ and $\bar{\phi}$ are

$$\bar{\rho}(y,y') \propto P(y_+,y'_-) + P(y_-,y'_+) + P(y_+,y'_+) + P(y_-,y'_-) \;, \qquad (31)$$

and

$$\bar{\phi}(y,y') \propto P(y_+,y'_-) + P(y_-,y'_+) - P(y_+,y'_+) - P(y_-,y'_-). \qquad (32)$$

For practical purposes, it is more convenient to plot instead

$$\rho(y,y') = \bar{\rho}(y,y') \;/\; \int dy\; \bar{\rho}(y,y') \;, \qquad (33)$$

and

$$\phi(y,y') = \bar{\phi}(y,y') \;/\; \int dy\; \bar{\phi}(y,y') \;, \qquad (34)$$

such that

$$\int dy\; \rho(y,y') = \int dy\; \phi(y,y') = 1. \qquad (35)$$

Thus $\rho(y,y')$ is the probability that a charged particle with rapidity y' finds another charged particle with rapidity y, while $\phi(y,y')$ is the probability that the charge of a particle with rapidity y' is compensated by another particle of opposite charge with rapidity y.

Using the TASSO data with the additional cut that the total observed charge is 0 or ±1, $\rho(y,y')$ is plotted in Fig. 25 for (a) $-0.75 \leq y' \leq 0$ and (b) $-5 \leq y' \leq -2.5$. Also $\phi(y,y')$ is plotted in Fig. 26 for (a) $-0.75 \leq y' \leq 0$ and (b) $-5 \leq y' \leq -2.5$. A comparison of these figures shows the following features.

(A) For small values of y', Fig. 25(a) shows a broad distribution while Fig. 26(a) shows a narrower distribution which peaks at $y \sim y'$. This is the evidence for a short range charge correlation. Moreover as y' becomes larger, there exhibits a peak adjacent to y' as shown in Fig. 26(b) but not so in Fig. 25(b). This provides further evidence for short range charge correlation.

(B) As shown in Fig. 26(b), there is a noticeable rise in the ϕ distribution near large positive y when y' is large and negative. The area beyond $y \geq 2.5$ is 0.101 ± 0.033 from Fig. 26(b) compared with the corresponding value of 0.011 ± 0.014 from Fig. 26(a) for small y'. This is the evidence of a long-range charge correlation in opposite jets from e^+e^- annihilation.

THREE-JET EVENTS AND PROPERTIES OF THE GLUON

Three-Jet Events

As discussed in the preceding chapter, at PETRA energies most

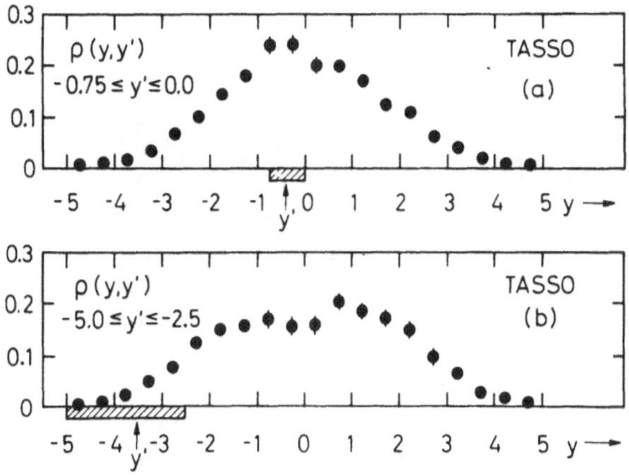

Fig. 25. The particle density function $\rho(y,y')$ as a function of y for (a) $-0.75 \leq y' \leq 0$ and (b) $-5 \leq y' \leq -2.5$. Here y and y' are the rapidity variables for charged particles. The data are from center-of-mass energies of 27 to 36 GeV by TASSO[30,31].

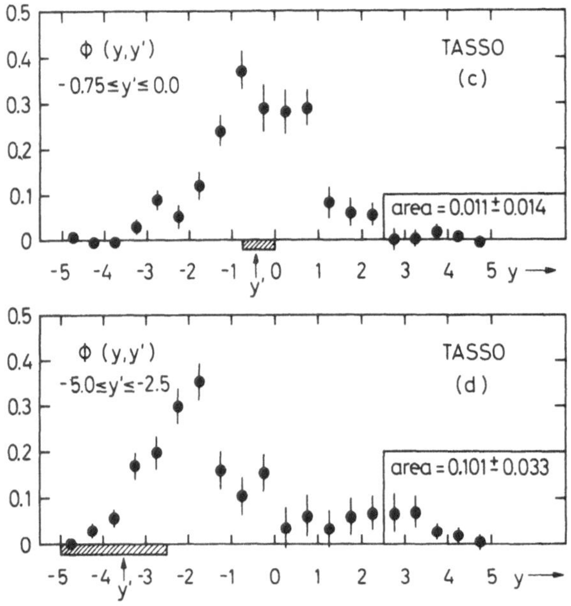

Fig. 26. The charge correlation function $\phi(y,y')$ as a function of y for (a) $-0.75 \leqq y' \leqq 0$ and (b) $-5 \leqq y' \leqq -2.5$. Here y and y' are the rapidity variables for charged particles. The data are from center-of-mass energies of 27 to 36 GeV by TASSO[30,31].

of the hadronic events consist of two back-to-back jets, and are interpreted as the production of a quark-antiquark pair : $e^+e^- \to q\bar{q}$. In May 1979, for the first time PETRA had a run at center-of-mass energy above 27 GeV. The TASSO Collaboration, using the three-jet analysis method of Wu and Zobernig[32], found from their data the first few events with distinctly different shapes. The first such event, with three jets instead of two, was reported[33] late that spring, and several more such events shortly thereafter[34].

By late summer, the number of observed three-jet events had increased rapidly[35-38]. By now, there are several thousand such events from each of the PETRA experiments; the precise number is not well defined and depends on the somewhat artificial formulation adopted, because a three-jet event gradually changes over to a two-jet event if the angle between two of the jets is reduced. In spite of this ambiguity, or perhaps because of it, these three-jet events are most naturally explained by hard non-colinear gluon bremsstrahlung $e^+e^- \to q\bar{q}g$[39], where the quark, the anti-quark, and the gluon each materialize as a jet of hadrons with limited transverse momentum.

The three-jet events therefore give the most direct way to study the properties of the gluon, including in particular the quark-gluon coupling constant α_s. The Feynman diagrams for $e^+e^- \to q\bar{q}g$ are already given in Fig. 18.

Methods of Three-Jet Analysis Used at PETRA

Wu-Zobernig Method. Hansen et al.[40] discovered the two-jet structure of hadrons produced in e^+e^- annihilation by studying sphericity. In view of this success, Wu and Zobernig devised the method[32] of generalized sphericity in order to discover three-jet events. This method has been widely used by the TASSO Collaboration[35,41].

The momentum tensor has been discussed in the preceding chapter and the unit vector $\hat{n}_3$ defined there is identified as the direction of the two back-to-back jets. This treatment of two-jet events can be reformulated as follows such that it can be generalized to three-jet events. Let $\hat{m}_1$ and $\hat{m}_2$ be two unit vectors for the directions of the two jets; since they are back-to-back, $\hat{m}_1$ and $\hat{m}_2$ satisfy

$$\hat{m}_1 = - \hat{m}_2. \tag{36}$$

The momenta $\vec{p}_j$ are partitioned into two non-overlapping sets S_1 and S_2 such that

$$\vec{p}_j \cdot \hat{m}_1 \geq 0 \qquad \text{if } \vec{p}_j \text{ is in } S_1,$$

and

$$\vec{p}_j \cdot \hat{m}_2 \geq 0 \qquad \text{if } \vec{p}_j \text{ is in } S_2. \tag{37}$$

Then $\hat{m}_1$ and $\hat{m}_2$ are determined, under the constraints (36) and (37), by maximizing the sum

$$\sum_{S_1} (\vec{p}_j \cdot \hat{m}_1)^2 + \sum_{S_2} (\vec{p}_j \cdot \hat{m}_2)^2. \qquad (38)$$

Because of (36), for the two-jet events the partition does not matter, and hence $\hat{m}_1$ or $\hat{m}_2$ can be identified with $\hat{n}_3$.

To generalize this formulation to three-jet events, let $\hat{m}_1$, $\hat{m}_2$, and $\hat{m}_3$ be three unit vectors for the directions of the three jets. By momentum conservation, $\hat{m}_1$, $\hat{m}_2$, and $\hat{m}_3$ must be coplanar; in other words, there is a unit vector $\hat{N}$ such that $\hat{N}$ is perpendicular to all three $\hat{m}$'s :

$$\hat{m}_1 \cdot \hat{N} = \hat{m}_2 \cdot \hat{N} = \hat{m}_3 \cdot \hat{N} = 0. \qquad (39)$$

This is the generalization of (36). The momenta $\vec{p}_j$ are partitioned into three non-overlapping sets S_1, S_2 and S_3 such that

$$\vec{p}_j \cdot \hat{m}_1 \geqq 0 \qquad \text{if } \vec{p}_j \text{ is in } S_1,$$

$$\vec{p}_j \cdot \hat{m}_2 \geqq 0 \qquad \text{if } \vec{p}_j \text{ is in } S_2,$$

and

$$\vec{p}_j \cdot \hat{m}_3 \geqq 0 \qquad \text{if } \vec{p}_j \text{ is in } S_3. \qquad (40)$$

This is the generalization of (37). The three-jet axes $\hat{m}_1$, $\hat{m}_2$, and $\hat{m}_3$ are determined, under the constraints (39) and (40), by maximizing the sum

$$\sum_{S_1} (\vec{p}_j \cdot \hat{m}_1)^2 + \sum_{S_2} (\vec{p}_j \cdot \hat{m}_2)^2 + \sum_{S_3} (\vec{p}_j \cdot \hat{m}_3)^2. \qquad (41)$$

This is the generalization of (38).

Unfortunately, this formulation is not useful practically, because the maximization of (41) is very difficult even on modern high-speed computers. The trouble stems entirely from the exceedingly large number of partitions : the number of ways to partition N observed tracks into three non-empty sets is

$$(3^{N-1} - 2^N + 1)/2. \qquad (42)$$

For example, this is 5.8×10^8 for N = 20, 1.4×10^{11} for N = 25, and 3.4×10^{13} for N = 30.

It is the basic idea of the procedure of Wu and Zobernig to use the following approximation

$$\hat{N} = \hat{n}_1 . \qquad (43)$$

In other words, the approximation is that the three jet axes $\hat{m}_1$, $\hat{m}_2$, and $\hat{m}_3$ are taken to lie in the plane defined by $\hat{n}_2$ and $\hat{n}_3$ of the preceding chapter called the event plane.

In general (43) is not exact; but it is exact in some limiting cases. Clearly it is exact in the case where all $\vec{p}_j$ are in the event plane. It is also exact if each $\vec{p}_j$ in the set $S_\alpha (\alpha = 1, 2, 3)$ is of the form

$$\vec{p}_j = a_j \hat{\ell}_\alpha + b_j \hat{n}_1 , \qquad (44)$$

where $\hat{\ell}_1$, $\hat{\ell}_2$ and $\hat{\ell}_3$ are three unit vectors in the event plane, and b_j are sufficiently small in the sense

$$\sum_{S_\alpha} b_j^2 < \sum_{S_\alpha} a_j^2 . \qquad (45)$$

In other words, (43) is exact when there is no spread out of the event plane <u>or</u> there is no spread in the event plane. Roughly speaking, the error of the approximation (43) is of the order of the product of the spreads of the jets in and perpendicular to the event plane. At PETRA energies, this error is quite small.

With this approximation (43), three-jet analysis becomes very much simpler because the components of $\vec{p}_j$ in the direction of $\hat{n}_1$ are no longer relevant. Let $\vec{q}_j$ be the projection of $\vec{p}_j$ into the event plane P. The important point here is that vectors in a plane, unlike those in a three-dimensional space, are naturally arranged in a cyclic order according to the polar angle. With this cyclic ordering, in the maximization of (41), or equivalently of

$$\sum_{S_1} (\vec{q}_j \cdot \hat{m}_1)^2 + \sum_{S_2} (\vec{q}_j \cdot \hat{m}_2)^2 + \sum_{S_3} (\vec{q}_j \cdot \hat{m}_3)^2 , \qquad (46)$$

it is sufficient to consider only partitions where each of the three sets S_1, S_2 and S_3 consists of contiguous momenta. This reduces the number of partitions from (42) to

$$\binom{N}{3} = N(N-1)(N-2)/6. \qquad (47)$$

This is 1140 for N = 20, 2300 for N = 25 and 4060 for N = 30; or reductions by factors of 5×10^5, 6×10^7 and 8×10^9 respectively.

Let the rectangular and polar coordinates of $\vec{q}_j$ be respectively (q_{j1}, q_{j2}) and (q_j, θ_j), and let the N momenta be labeled such that

$$0 \leqq \theta_1 \leqq \theta_2 \leqq \theta_3 \leqq \theta_N \leqq 2\pi . \qquad (48)$$

The three sets S_1, S_2 and S_3 are specified as

$$S_1 = \{N_1, N_1 + 1, \ldots\ldots, N_2 - 1\}$$

$$S_2 = \{N_2, N_2 + 1, \ldots\ldots, N_3 - 1\}$$

and

$$S_3 = \{N_3, N_3 + 1, \ldots\ldots N, 1, 2, \ldots\ldots, N_1 - 1\} , \quad (49)$$

where

$$1 \leqq N_1 < N_2 < N_3 \leqq N. \quad (50)$$

Furthermore, the conditions (40) imply the restrictions

$$\theta_{N_2-1} - \theta_{N_1} < \pi ,$$

$$\theta_{N_3-1} - \theta_{N_2} < \pi ,$$

and

$$2\pi + \theta_{N_1-1} - \theta_{N_3} < \pi . \quad (51)$$

In case $N_1 = 1$, $\theta_\circ$ is defined to be $\theta_N - 2\pi$.

The second step is to define three 2 x 2 matrices $M^{(1)}$, $M^{(2)}$, and $M^{(3)}$ by

$$M^{(\tau)}_{\alpha\beta} = \sum_{j \text{ in } S_\tau} q_{j\alpha}\, q_{j\beta} \quad (52)$$

for $\alpha, \beta = 1, 2$ and $\tau = 1, 2$ and 3. For each of these three 2 x 2 matrices, let $\Lambda^{(\tau)}$ be the larger eigenvalue and $\hat{m}_\tau$ the corresponding normalized eigenvector. Here $\Lambda^{(\tau)}$ is given explicitly by

$$\Lambda^{(\tau)} = \frac{1}{2} \{M^{(\tau)}_{11} + M^{(\tau)}_{22} + [(M^{(\tau)}_{11} - M^{(\tau)}_{22})^2 + 4(M^{(\tau)}_{12})^2]^{1/2}\}.$$

In terms of these $\Lambda^{(\tau)}$, we maximize

$$\Lambda(N_1, N_2, N_3) = \Lambda^{(1)} + \Lambda^{(2)} + \Lambda^{(3)} \quad (53)$$

over all partitions where (40) is satisfied. This maximizing partition gives the three jets and the corresponding $\hat{m}_1$, $\hat{m}_2$ and $\hat{m}_3$ the directions of the jet axes. From these directions, the total energies of the jets can be easily found if the approximation is made that the invariant masses of the jets are neglected. Let

$$\alpha_1 = \sin(\phi^{(3)} - \phi^{(2)}) ,$$

$$\alpha_2 = \sin(\phi^{(1)} - \phi^{(3)}) ,$$

$$\alpha_3 = \sin(\phi^{(2)} - \phi^{(1)}) ,$$

where the $\phi^{(\tau)}$s are the polar angular coordinates in the event plane for $\hat{m}_\tau$ for the particular partition N_1, N_2, N_3 that maximizes $\Lambda(N_1, N_2, N_3)$. From the directions of the three jet axes, one obtains the total energy $E^{(\tau)}$ of each jet by

$$E^{(\tau)} = W\alpha_\tau/(\alpha_1 + \alpha_2 + \alpha_3) , \tag{54}$$

where W is the center-of-mass energy.

As a measure of how three-jet like an event is, it is useful to define the quantity trijettiness J_3 by

$$J_3 = \frac{1}{N-3} \sum_{\tau=1,2,3} \{ \sum_{j \text{ in } S_\tau} \frac{(\vec{q}_j \times \hat{m}_\tau)^2}{\Delta_\tau^2} \} , \tag{55}$$

where $\Delta_\tau^2 = (1/2) \langle p_T^2 \rangle_\tau$. $\langle p_T^2 \rangle_\tau$ is the average transverse momentum squared of a jet with energy E_τ and may be assumed to be equal to that for two-jet events. An alternative possibility is to use

$$J_3' = \frac{1}{2N-5} \sum_{\tau=1,2,3} \{ \sum_{j \text{ in } S_\tau} \frac{(\vec{p}_j \times \hat{m}_\tau)^2}{\Delta_\tau^2} \}. \tag{56}$$

As emphasized in the original paper, this procedure has the following desirable features :

(1) All three jet axes are determined.

(2) It is not necessary to have the momenta of all produced particles. For examples, this procedure can be used where there is no detection of neutral particles. Of course the loss of information leads to a larger error.

(3) All measured momenta can be used; in other words, there is no need to introduce a cutoff for low momenta.

(4) Computer time is moderate.

Brandt-Dahmen Method. In the three-jet analysis of Brandt and Dahmen[42], first used by the PLUTO Collaboration[37], the quantity to be maximized is the generalized thrust

$$\sum_{S_1} \vec{p}_j \cdot \hat{m}_1 + \sum_{S_2} \vec{p}_j \cdot \hat{m}_2 + \sum_{S_3} \vec{p}_j \cdot \hat{m}_3 , \tag{57}$$

instead of (41). This maximization is to be taken over all choices of $\hat{m}_1$, $\hat{m}_2$, and $\hat{m}_3$, and also over all admissible partitions.

Brandt and Dahmen consider a case where the N measured momenta

$\vec{p}_j$ satisfy

$$\sum_j \vec{p}_j = 0 . \tag{58}$$

This is true when the momenta of all produced particles are measured. In this case, the $\hat{m}$'s can be found as follows. Given S_τ the maximum of

$$\sum_{S_\tau} \vec{p}_j \cdot \hat{m}_\tau$$

over $\hat{m}_\tau$ is

$$\max_{\hat{m}_\tau} \sum_{S_\tau} \vec{p}_j \cdot \hat{m}_\tau = \Big|\sum_{S_\tau} \vec{p}_j\Big| . \tag{59}$$

The choice of $\hat{m}_\tau$ is

$$\hat{m}_\tau = \Big(\sum_{S_\tau} \vec{p}_j\Big) \Big/ \Big|\sum_{S_\tau} \vec{p}_j\Big| . \tag{60}$$

The important point here is that the condition (58) guarantees that the $\hat{m}$'s satisfy (39), i.e. they are coplanar.

For any experimental data, (58) is almost never satisfied. It is therefore necessary to introduce a ficticious momentum $\vec{p}_{N+1}$ defined by

$$\vec{p}_{N+1} = - \sum_{j=1}^{N} \vec{p}_j . \tag{61}$$

Thus the N+1 momenta, taken together, do satisfy (58). If these N+1 momenta are partitioned into three non-empty sets S_1, S_2 and S_3 then (60) can be used, and the quantity to be maximized over all such partitions is

$$\Big|\sum_{S_1} \vec{p}_j\Big| + \Big|\sum_{S_2} \vec{p}_j\Big| + \Big|\sum_{S_3} \vec{p}_j\Big| . \tag{62}$$

A measure of the jettiness of a three-jet event is then the triplicity defined by

$$T_3 = \frac{\Big|\sum_{S_1} \vec{p}_j\Big| + \Big|\sum_{S_2} \vec{p}_j\Big| + \Big|\sum_{S_3} \vec{p}_j\Big|}{\sum_{j=1}^{N+1} |\vec{p}_j|} \tag{63}$$

for the maximizing partition of the N+1 momenta into the sets S_1, S_2, and S_3.

Because of the ficticious $\vec{p}_{N+1}$, the problem of the large number of partitions to be considered is further aggravated. This number is obtained by the replacement $N \to N+1$ in (42) : $(3^N - 2^{N+1} + 1)/2$. This is 261625 even for N = 12, 7×10^6 for N = 15 and 1.9×10^8 for N = 18. Accordingly, for multiplicities commonly seen at PETRA, it is practically not possible to search through all partitions. The procedure actually followed is to take about the 16 or 17 largest momenta, carry through the analysis by ignoring the rest, and then reintroduce the neglected momenta one by one. Even with this procedure, the computer time needed is quite extensive.

Combined Methods. It is clear that the methods described in the previous two subsections can be combined in a variety of ways. A bewildering array of such combinations has been tried at one time or another. Among these numerous possibilities, only two will be described and discussed in this section.

(A) Modified Brandt-Dahmen Method. The conservation of momentum, as expressed by (58), is of central importance in the Brandt-Dahmen method, but not in the Wu-Zobernig method. The ficticious momentum $\vec{p}_{N+1}$ is introduced for this reason in the Brandt-Dahmen method, and is treated on equal footing with the actually measured momenta. The question is therefore naturally raised whether this ficticious momentum $\vec{p}_{N+1}$ can be avoided.

Since the constraints (40) are of no importance, $\vec{p}_{N+1}$ can be avoided by reverting to the initial statement of the procedure, namely to maximize (57) under the constraint (39). Given a partition, let

$$\vec{P}_\tau = \sum_{S_\tau} \vec{p}_j \tag{64}$$

for $\tau = 1, 2, 3$, then $\hat{N}$ is determined by maximizing

$$|\vec{P}_1 \times \hat{N}| + |\vec{P}_2 \times \hat{N}| + |\vec{P}_3 \times \hat{N}| . \tag{65}$$

This expression (65) can also be written as

$$[\vec{P}_1^2-(\vec{P}_1 . \hat{N})^2]^{1/2} + [\vec{P}_2^2 - (\vec{P}_2 . \hat{N})^2]^{1/2} + [\vec{P}_3^2 - (\vec{P}_3 . \hat{N})^2]^{1/2} \tag{66}$$

and therefore $\hat{N}$ satisfies

$$\sum_\tau [\vec{P}_\tau^2 - (\vec{P}_\tau . \hat{N})^2]^{-1/2} (\vec{P}_\tau . \hat{N})\vec{P}_\tau = \text{constant } \hat{N}. \tag{67}$$

While (67) is not difficult to solve numerically, it is nevertheless too complicated because it needs to be solved for each choice of partition. Hence the approximation is introduced to expand each term of (66) by the binomial theorem and keep only the first non-trivial term. In this approximation, $\hat{N}$ is determined by minimizing

$$(\vec{P}_1 \cdot \hat{N})^2/|\vec{P}_1| + (\vec{P}_2 \cdot \hat{N})^2/|\vec{P}_2| + (\vec{P}_3 \cdot \hat{N})^2/|\vec{P}_3| \,. \tag{68}$$

In other words,

$$\hat{N} = \text{the } \hat{n}_1 \text{ for the three momenta}$$

$$|\vec{P}_1|^{-1/2}\,\vec{P}_1\,,\ |\vec{P}_2|^{-1/2}\,\vec{P}_2\,,\ \text{and}\ |\vec{P}_3|^{-1/2}\,\vec{P}_3\,, \tag{69}$$

when (58) is satisfied, $\vec{P}_1 + \vec{P}_2 + \vec{P}_3 = \vec{0}$, and clearly $\hat{N}$ is normal to the plane determined by $\vec{P}_1$, $\vec{P}_2$, and $\vec{P}_3$.

So far as computer time is concerned, this modified Brandt-Dahmen method is comparable to the original Brandt-Dahmen method of triplicity. The modification reduces the number of partitions by roughly a factor of 3, but the computation of $\hat{N}$ via (69) takes some additional time. The modified triplicity T_3' is

$$T_3' = \frac{|\sum_{S_1} \vec{p}_j| + |\sum_{S_2} \vec{p}_j| + |\sum_{S_3} \vec{p}_j| - \lambda_1}{\sum_{j=1}^{N} |\vec{p}_j|} \tag{70}$$

for the maximizing partition of the N observed momenta $\vec{p}_j$ into the sets S_1, S_2 and S_3, wher λ_1 is the λ_1 defined by (14) for the three momenta of (69) with (64).

(B) Modified Wu-Zobernig method. The basic idea of the Wu-Zobernig method is to reduce the three-dimensional momentum vectors to two-dimensional ones so that the cyclic ordering can be used to reduce drastically the number of partitions. Once this reduction is carried out, the method is not limited to maximizing the generalized sphericity (41). A natural modification is to replace the generalized sphericity by the generalized thrust (57). However, this modification is not as straightforward as it may seem.

The difficulty stems from the fact that the fundamental Wu-Zobernig approximation (43) $\hat{N} = \hat{n}_1$ is accurate only for the generalized sphericity, as discussed in the paragraph containing eqs. (44) and (45). For the generalized thrust, this approximation (43) is exact only when there is no spread out of the event plane, not when there is no spread in the event plane, as clearly demonstrated in connection with the modified Brandt-Dahmen method. Therefore, in the modified Wu-Zobernig method where generalized thrust is used, an iterative procedure is needed.

The steps are as follows.

(a) From the momentum tensor, calculate $\hat{n}_1$, $\hat{n}_2$ and $\hat{n}_3$ as usual.

(b) Let $\vec{q}_j$, j = 1, 2, ... N, be the projections of $\vec{p}_j$ into the $\hat{n}_2$ - $\hat{n}_3$ plane. Put these $\vec{q}_j$ in cyclic order and define the contiguous partitions S_1, S_2 and S_3 of (49) in accordance with the Wu-Zobernig procedure. Maximize, over these contiguous partitions, the quantity

$$\left|\sum_{S_1} \vec{q}_j\right| + \left|\sum_{S_2} \vec{q}_j\right| + \left|\sum_{S_3} \vec{q}_j\right| . \tag{71}$$

(c) Using the maximizing partition found in the preceding step, define the three vectors

$$\frac{\sum_{S_\tau} \vec{p}_j}{\sqrt{\left|\sum_{S_\tau} \vec{p}_j\right|}} \tag{72}$$

for τ = 1, 2, 3. Recalculate $\hat{n}_1$, $\hat{n}_2$, $\hat{n}_3$ using these three momentum vectors of (72).

(d) On the basis of these new $\hat{n}$'s, repeat step (b). Iterate if necessary.

(e) From the final maximizing partition S_1, S_2 and S_3 and the final $\hat{n}_2$ - $\hat{n}_3$ plane, a measure of the jettiness of the three-jet event is the modified triplicity

$$T_3'' = \frac{\left|\sum_{S_1} \vec{q}_j\right| + \left|\sum_{S_2} \vec{q}_j\right| + \left|\sum_{S_3} \vec{q}_j\right|}{\sum_{j=1}^{N} |\vec{p}_j|} , \tag{73}$$

and the three-jet axes are given by the unit vectors

$$\hat{m}_\tau = \frac{\sum_{S_\tau} \vec{q}_j}{\left|\sum_{S_\tau} \vec{q}_j\right|} \tag{74}$$

for τ = 1, 2, 3.

So far as computer time is concerned, this modified method is comparable to the original method. While the iteration increases the computer time, it is faster to calculate (71) than (53). Methods intermediate between the original and the modified versions have also been used at PETRA[43].

Ellis-Karliner Method. In the four methods of three-jet analysis discussed so far, the three jets are always treated on equal footing. While it is highly desirable to maintain this symmetry, it is not absolutely compelling. In the Ellis-Karliner method[44], this symmetry is not maintained. Indeed, the method was originally devised for the determination of the spin of the gluon, and will play an important role later on.

This method consists of the following steps.

(A) Determine the thrust axis $\hat{n}$ by (22). At this stage, the sign of $\hat{n}$ remains arbitrary.

(B) Designate one of the two jets as thin while the other one as fat according to the relative magnitudes of

$$\sum_{\vec{p}\cdot\hat{n} > 0} |\vec{p}_j \times \hat{n}|$$

and

$$\sum_{\vec{p}\cdot\hat{n} < 0} |\vec{p}_j \times \hat{n}| \ .$$

Choose the sign of $\hat{n}$ such that $\hat{n}$ is in the direction of the thin jet.

(C) Carry out a Lorentz transformation to the center of mass of the fat jet. In this new coordinate system and with only the tracks associated with the fat jet, determine again the thrust axis. In this way the fat jet is split into two jets.

(D) Go back to the laboratory system by reversing the Lorentz transformation.

The Lorentz transformation used in this procedure is shown schematically in Fig. 27.

Let us compare in more detail this Ellis-Karliner method[44] with those of the preceding sections. Aside from the asymmetrical treatment of the three jets as already mentioned, the main feature of this method is the application of Lorentz transformation. Therefore, at least in principle, not only the $\vec{p}_j$ but also the corresponding E_j are needed. Since experimentally E_j can be found only with particle identification, all particles are assumed to be pions. The Ellis-Karliner method is used by the JADE Collaboration[38], and for spin

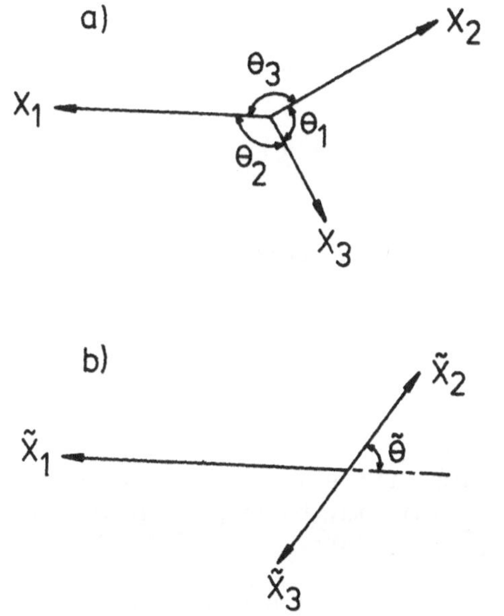

Fig. 27. a) Momenta and angles of a $q\bar{q}g$ three jet final state in the center-of-momentum frame. b) The $q\bar{q}g$ final state transformed to the rest frame of jets 2 and 3.

determination by the TASSO Collaboration[45].

By contrast, in applying the method of Wu-Zobernig, Brandt-Dahmen or their modifications, we are instructed to remain in the center-of-mass system. Indeed, it is only in this system that sphericity, thrust, etc. are defined. Nevertheless, it is reasonable to raise the question whether there is a natural measure of the jettiness of a three-jet event with Lorentz invariance maintained. With the question posed in this way, a possible answer is the jet mass. More precisely, in the spirit of (41) and (57), the optimal partition of the four-momenta $(E_j, \vec{p}_j)$ into three non-overlapping sets S_1, S_2 and S_3 is determined by minimizing

$$M_1 + M_2 + M_3 \tag{75}$$

where the jet masses M_τ are defined by

$$M_\tau^2 = (\sum_{S_\tau} E_j)^2 - (\sum_{S_\tau} \vec{p}_j)^2. \tag{76}$$

Since such methods have not been developed systematically, it is not known what the advantages and shortcomings are.

<u>Cluster Analysis</u>. In this and the next subsections, we shall describe two other methods of jet analysis, which do not use the full information of the events as those of the preceding four subsections. These methods are the cluster analysis and energy flow. While the method of energy flow has low discriminating power and uses excessive computer time, there is a distinct possibility that the method of cluster analysis may become the most suitable method for HERA, LEP, pp, and $p\bar{p}$ colliders.

In the present context, the basic idea of cluster analysis is to avoid as an input the number of jets, i.e., to let the procedure determine the number of jets. Hence cluster analysis is the application of pattern recognition to identify clusters. As such the method has wide applications in many areas, including library science, visual scene analysis, medicine, and pattern recognition in spark and bubble chambers. There is a great deal of literature on this subject, including a textbook[46]; more recently, there are also several papers[47-50] in the specific context of electron-positron annihilation. The fact that the details are quite different in these papers means that cluster analysis for e^+e^- annihilation is quite flexible. We begin with a general discussion.

As before, let $\vec{p}_j$ be the momenta of the observed particles. For each pair $i \neq j$, a number ρ_{ij} is assigned as a measure of the likelihood that the particles with momenta $\vec{p}_i$ and $\vec{p}_j$ are in the same jet. In general $\rho_{ij} = \rho_{ji}$ and the smaller the ρ_{ij}, the more likely for them to be in the same jet. The obvious first question is : how should ρ_{ij} be chosen? There are at least two distinct types of answers :

(A) The values ρ_{ij} should give only a relative ordering of the likelihood. More precisely, if $\rho' = f(\rho)$ is a strictly increasing function of ρ, then the cluster analysis should give the same answer independent of whether ρ_{ij} or $\rho'_{ij} = f(\rho_{ij})$ is used. This philosophy is advocated especially by Lanius, Roloff and Schiller[50].

(B) The values of ρ_{ij} are themselves meaningful. In the language of pattern recognition, these $\vec{p}_j$ are embedded in the so-called feature space, and ρ_{ij} is identified with the metric in this feature space[51]. However, it is interesting to note that, in some of the recent papers[47,48], the ρ_{ij} used do not satisfy the triangular inequality $\rho_{ij} + \rho_{jk} \geqq \rho_{ik}$.

In addition to the choice of ρ_{ij}, the proposed procedures to determine the jets are widely different. Indeed, even for each of the procedures, there are several parameters that need to be chosen. For example, the procedures of refs. 48 and 49 have each four parameters. Accordingly, some of the interesting questions concerning cluster analysis are the following. (i) How should the ρ_{ij} be chosen? In particular, are there desirable properties that they should have? (ii) Given ρ_{ij}, how can the quality of a proposed procedure be evaluated? And (iii) Given a proposed procedure, how can its parameters be chosen?

The procedure of Daum, Meyer, and Bürger[49] is the one actually used at PETRA, for example by the PLUTO and JADE Collaborations[52,53]. This procedure is based on a two-step algorithm in which
(1) all particles - charged and neutrals - are used. The particle is taken to have momentum $\vec{p}_i$, direction $\hat{n}_i = \vec{p}_i/|\vec{p}_i|$ and energy E_i.
In the first step preclusters are defined such that
(2) each particle is a member of only one precluster,
(3) any two particles i, k (i≠k) belong to the same precluster if

$$\rho_{ik} = \hat{n}_i \cdot \hat{n}_k > \cos\alpha \tag{77}$$

for a predefined value of α. Thus the choice of ρ_{ij} here is such that it depends only on the direction, but not the magnitudes, of the momenta involved. A typical value for α is 30°.

The direction of the precluster D_i is defined by

$$\hat{n}_{D_i} = \sum_{k\varepsilon D_i} \vec{p}_k \Big/ \Big| \sum_{k\varepsilon D_i} \vec{p}_k \Big| \,. \tag{78}$$

In the second step the preclusters are concatenated to clusters
(4) each precluster is a member of exactly one cluster,
(5) any two preclusters D_i, D_k (i≠k) belong to the same cluster if

$$\hat{n}_{D_i} \cdot \hat{n}_{D_k} > \cos\beta \tag{79}$$

for a predefined value of $\beta \geqq \alpha$. A typical value for β is 45°.

The energy and the direction of a cluster C_i are defined by

$$E_{C_i} = \sum_{k\varepsilon C_i} E_k \text{ and } \hat{n}_{C_i} = \sum_{k\varepsilon C_i} \vec{p}_k \Big/ \Big| \sum_{k\varepsilon C_i} \vec{p}_k \Big| . \tag{80}$$

Finally some cuts are made to define a jet :
(6) the multiplicity of clusters n_c is defined by the minimal number of clusters which fulfill

$$\sum_{i=1}^{n_c} E_{C_i} > E_{tot} \cdot (1-\varepsilon) \tag{81}$$

for a predefined value of ε, where E_{tot} is the energy sum of all particles. The clusters are accepted if they belong to the set of n_c most energetic clusters.
(7) Each accepted cluster is called a jet if $E_{C_i} > E_{th}$ for a predefined value of E_{th}.

While this procedure forms the basis of cluster analysis used at PETRA, various modifications are sometimes used in actual application. For example, neutral particles are not always included, see (1) above. Although theoretical questions can be readily raised, this cluster analysis works well in practice.

Method of Energy Flow. The method of energy flow to find three-jet events was originally proposed by De Rújula, Ellis, Floratos, and Gaillard[54]. However, the method of energy flow actually used by MARK J at PETRA is quite different in detail. In this section, only the version used by MARK J[36] will be discussed. So far as we know, the original version has never been employed to analyze e^+e^- data, and hence may not have the features of the MARK J version.

The momentum tensor discussed above gives not only the jet axis $\hat{n}_3$ but also an orthonormal system $\hat{n}_1$, $\hat{n}_2$ and $\hat{n}_3$. By contrast, the thrust as defined by (22) gives only the jet axis but not a coordinate system. In the method of energy flow, the first step is to construct an orthonormal system $\hat{e}_1$, $\hat{e}_2$, and $\hat{e}_3$, using the energies and directions of the observed particles instead of the momentum vectors. More precisely, instead of $\vec{p}_j$, define the energy flow

$$\vec{E}_j = (E_j/|\vec{p}_j|)\, \vec{p}_j . \tag{82}$$

The unit vector $\hat{e}_1$ is defined analogous to the thrust T :

$$F_{thrust} = \max \frac{\Sigma_i |\vec{E}_i \cdot \hat{e}_1|}{\Sigma_i E_i} \quad . \tag{83}$$

The orthonormal system is defined through $\hat{e}_2$ and $\hat{e}_3 = \hat{e}_1 \times \hat{e}_2$, such that

$$F_{major} = \max \frac{\Sigma_i |\vec{E}_i \cdot \hat{e}_2|}{\Sigma_i E_i} \qquad \hat{e}_2 \perp \hat{e}_1 \,, \tag{84}$$

and

$$F_{minor} = \frac{\Sigma_i |\vec{E}_i \cdot \hat{e}_3|}{\Sigma_i E_i} \quad . \tag{85}$$

The oblateness O is defined by

$$O = F_{major} - F_{minor}. \tag{86}$$

The method of energy flow used by MARK J[36] consists of the following. First make a cut in F_{thrust}. For these events with F_{thrust} not close to 1, project the $\vec{E}_i$ into the thrust-major plane. Orient these projections such that the longest jet is near the direction of $\hat{e}_1$ at 0° and the second longest jet is in a chosen half-plane. Make one energy-flow plot by superimposing all events, and look for three-lobe structure in this plot.

It was originally thought that the method of energy flow allowed a non-magnetic detector to have a quick look into jet physics. However this method is less effective for a number of reasons. For example, the jet axes are not determined event by event. But the most undesirable feature is that, as pointed out by the MARK J Collaboration[36] and to be discussed in what follows, this procedure introduces three-lobe structure even if none is actually present.

Four-Jet Analysis. Since there are three-jet events, there should also be four-jet events. With the three-jet events interpreted as gluon bremsstrahlung, double gluon bremsstrahlung leads to four-jet events :

$$e^+e^- \to q\bar{q}gg.$$

Alternatively, the bremsstrahlung gluon may materialize into a $q\bar{q}$ pair :

$$e^+e^- \to q\bar{q}q\bar{q} \ .$$

The lowest-order QCD diagrams for these two types of four-jet events are shown respectively in Fig. 28 (a) and 28 (b).

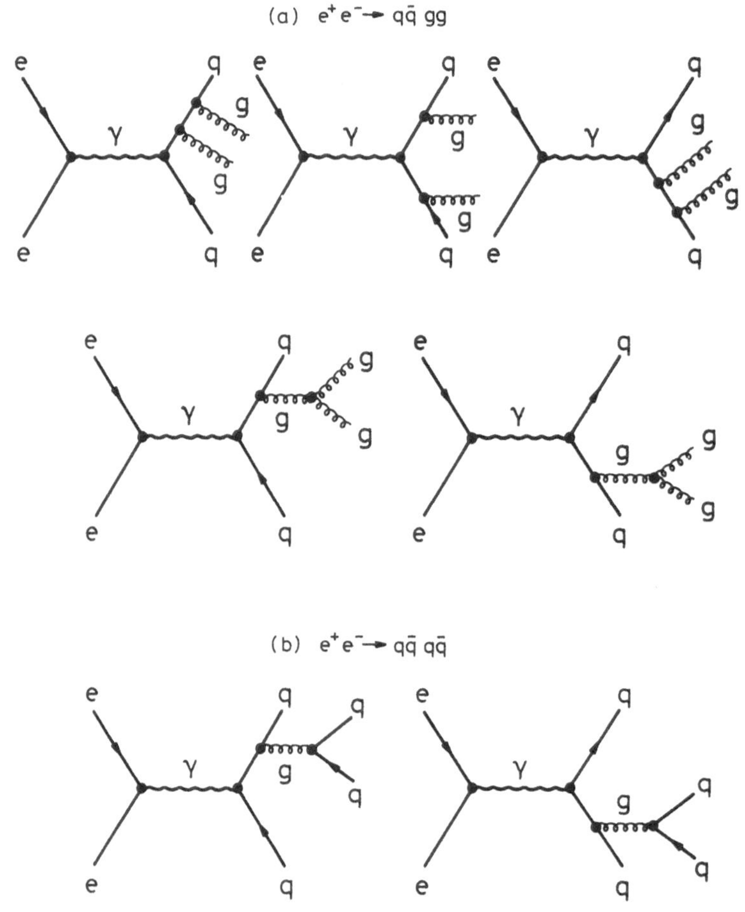

Fig. 28. Feynman diagrams for four-jet events.

From the point of view of physics, four-jet events are important for the following reason. The most fundamental difference between Abelian and non-Abelian gauge theories is the presence of self-interactions of the gauge particle in the non-Abelian case. For e^+e^- annihilation at high energies, this self-interaction is important not in three-jet events but in four-jet events, as shown in the last two diagrams of Fig. 28 (a).

The major difficulty of four-jet analysis is again the large number of possible partitions. The number of ways to partition N observed tracks into four non-empty sets is

$$\frac{1}{6}\,(4^{N-1} - 3^{N} + 3.2^{N-1} - 1)\ . \tag{87}$$

which is even much larger than (42). It is for example 4.5×10^{10} for N = 20 and 4.7×10^{13} for N = 25.

The basic idea of the four-jet analysis to be presented here is, taking the inspiration from Ellis and Karliner[44], to apply the three-jet analysis twice. The actual procedure followed at TASSO[56,57] is as follows :

(A) Use either the original or the modified Wu-Zobernig program for the three-jet analysis[32].

(B) Determine the three jet axes by adding the momenta of the tracks in the jet. In general these three jet axes are not coplanar.

(C) Find the average transverse momenta for the three jets with respect to their own jet axes as defined above. To avoid possible confusion with the average transverse momenta used in the three-jet analysis, let us call them <PT'>.

(D) Among the three jets being considered so far, define jet A as the one with the smallest value of <PT'>.

(E) Remove all the tracks of jet A, and study the remaining tracks, i.e. the tracks that are not in jet A.

(F) Add up the energies and momenta of these remaining tracks to find the motion of the center of mass. Perform a Lorentz transformation to the center-of-mass system of these remaining tracks.

(G) Apply three-jet analysis to the transformed momenta for these remaining tracks. This gives the three new partitions. In particular, determine whether this is itself a three-jet event.

(H) If these remaining tracks form an acceptable three-jet event, go back to the original laboratory coordinate system.

(I) In the laboratory coordinate system, determine the four jet axes by adding up the momenta of the tracks in each jet.

(J) Find the angles between all the momentum directions and all the four jet axes. Each momentum should make the smallest angle with respect to the jet axis that it belongs to. If this is indeed the case, then the jet axes have been found properly.

(K) If not, such a momentum should be reassigned to the jet that gives the smallest angle. After reassignment, repeat steps I and J.

(L) The energies of the four jets are determined under the assumption that the jet masses can be neglected. Let $\hat{n}_i$ with i = 1, 2, 3 and 4 be the unit vectors in the directions of the jet axes, then the energies E_i of the jets are determined by the energy-momentum conservation

$$\sum_{i=1}^{4} E_i \hat{n}_i = 0 \ ,$$

$$\sum_{i=1}^{4} E_i = E_{cm} \ . \qquad (88)$$

A variation of eq. (88) by introducing the observed jet mass is given in ref. 57.

This procedure of four-jet analysis has all the four advantages listed at the end of the preceding section for the Wu-Zobernig three-jet analysis. It has the further advantage that it requires very little additional programming on the computer. The computer time required is dominated by the two three-jet analyses of steps A and G. Since there are fewer momenta in step G than step A, the computer time of four-jet analysis is not much more than that of three-jet analysis for the same number of observed tracks.

Observation of Three-Jet Events

The first three-jet event, interpreted as $e^+e^- \to q\bar{q}g$, was discovered in the spring of 1979 at PETRA. As already mentioned, by now each of the PETRA and PEP experiments has thousands of three-jet events. It is however extremely interesting to retrace the historical excitements of the first observation of these events by each of the PETRA experiments.

TASSO. The TASSO Collaboration is the first group to observe three-jet events[33,34,35]. In the spring of 1979 when the center-of-mass energy of PETRA reached 27 GeV, the computer program based on the Wu-Zobernig method[32] was used and the first three-jet event observed is shown in Fig. 29. This first event was presented by Wiik of the TASSO Collaboration[33] at the Bergen Conference in late spring, and was much discussed there. As seen from Fig. 29, it had three clear, well separated jets, and was considered to be more convincing than a good deal of statistical analysis. Indeed, before the question of statistical fluctuation could be seriously raised, TASSO found a number of other three-jet events. Less than two weeks after the Bergen Conference, several further three-jet events from TASSO, shown in Fig. 30, were presented by Söding[34] at the EPS Conference in Geneva. Comparisons of event shape distributions with the QCD predictions were also included. These additional events further demonstrated unambiguously the existence of planar three-jet events.

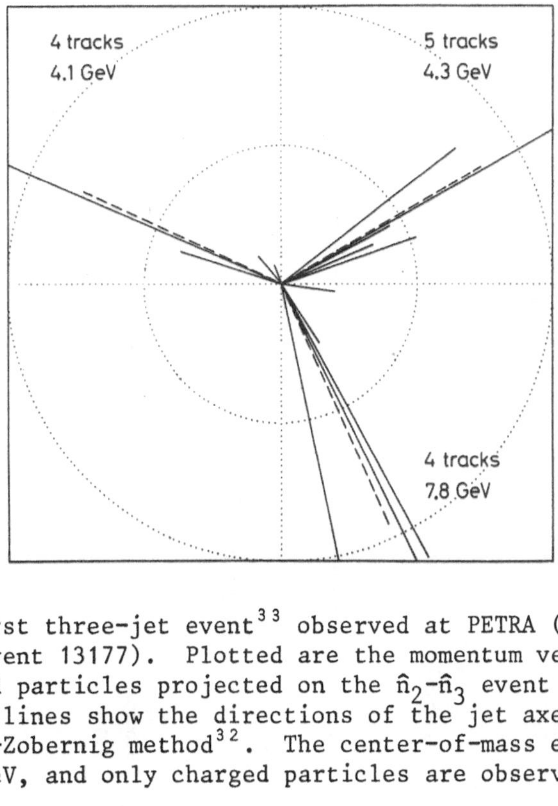

Fig. 29. The first three-jet event[33] observed at PETRA (TASSO run 447, event 13177). Plotted are the momentum vectors of the charged particles projected on the $\hat{n}_2$-$\hat{n}_3$ event plane. The dotted lines show the directions of the jet axes found by the Wu-Zobernig method[32]. The center-of-mass energy is 27.4 GeV, and only charged particles are observed.

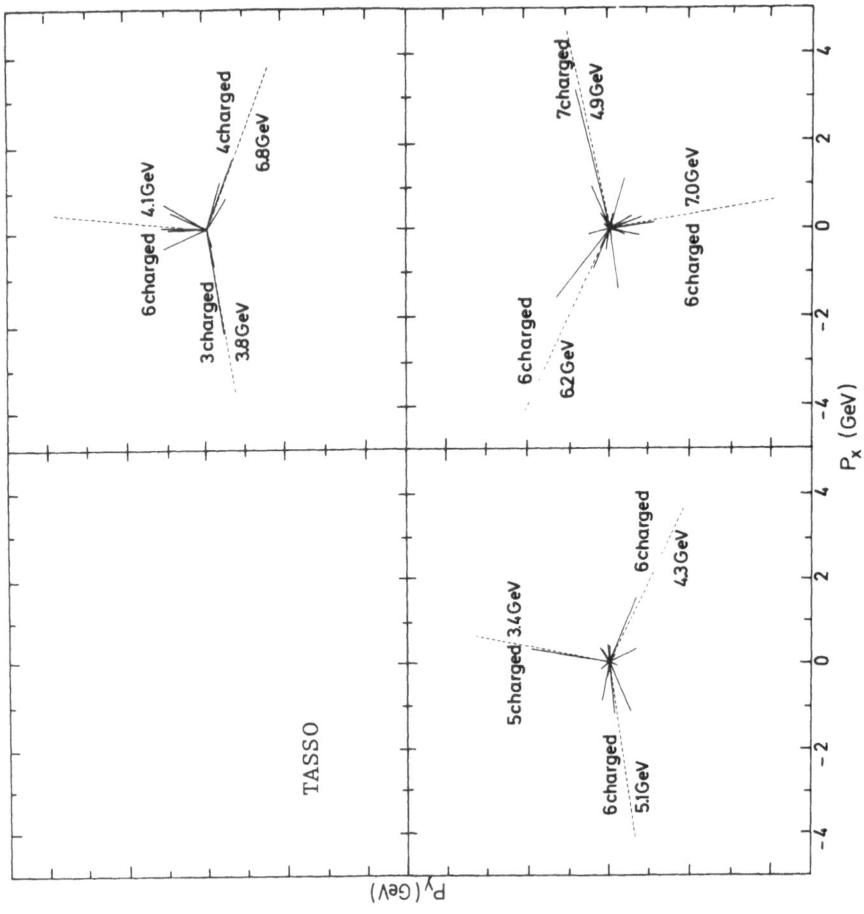

Fig. 30. Further three-jet events[34] observed at PETRA projected in the event plane. The center-of-mass energy is 27.4 GeV, and only charged particles are observed.

At Fermilab Conference[58] several months later, all four experiments at PETRA gave more extensive data, confirming the earlier observation of TASSO. Since these experiments were run simultaneously, the amounts of data were the same within about 10 - 20 %. In a period of three months, between August 29 and December 7, these more extensive data were submitted for publication by TASSO[35], MARK J[36], PLUTO[37], and JADE[38].

With the three-jet events understood as $e^+e^- \to q\bar{q}g$[32-39], no sharp separation between two-jet and three-jet events is expected : when the angle between two of the jets becomes smaller and smaller (or alternatively when the momentum of one of the jets becomes less and less) the three-jet event looks more and more like a two-jet event. One simple way to see this gradual transition is shown in Fig. 31[34], which compares, on the basis of the very early TASSO data, the distribution of

$$\langle p_T^2 \rangle_{out} = \frac{1}{N} \sum_j (\vec{p}_j \cdot \hat{n}_1)^2 = Q_1 \langle p^2 \rangle \tag{89}$$

(=square of momentum component normal to the event plane averaged over the charged particles in one event) with that of

$$\langle p_T^2 \rangle_{in} = \frac{1}{N} \sum_j (\vec{p}_j \cdot \hat{n}_2)^2 = Q_2 \langle p^2 \rangle \tag{90}$$

(=square of momentum component in the event plane and perpendicular to the jet axis averaged the same way). Also shown in Fig. 31 is $\langle p_T^2 \rangle_{in,\ 3\,jet\ axes}$, which is defined the same way as (90) but, for each jet, the jet axis found by the Wu-Zobernig method is used. For comparison, the corresponding results from Monte Carlo calculations on the basis of $q\bar{q}$ jets[59] are also given. Fig. 32[35] gives the energy dependence of the $\langle p_T^2 \rangle_{in}$ and $\langle p_T^2 \rangle_{out}$ distributions on the basis of a larger data sample. While there is little change in the $\langle p_T^2 \rangle_{out}$ distribution from low (W = 13, 17 GeV) to high energies (W = 27.4, 31.6 GeV), the distribution of $\langle p_T^2 \rangle_{in}$ becomes much wider at high energies; there is a long tail of events with large $\langle p_T^2 \rangle_{in}$.

Hadrons resulting from pure $q\bar{q}$ jets will on the average be distributed uniformly around the jet axis. However, some asymmetry between $\langle p_T^2 \rangle_{in}$ and $\langle p_T^2 \rangle_{out}$ is caused by the bias introduced in choosing the axes. Good agreement with the $q\bar{q}$ model using $\sigma_q = 0.3$ GeV/c is found at low center-of-mass energies for both distributions and also at high energies for the $\langle p_T^2 \rangle_{out}$ distribution, but the $q\bar{q}$ model fails to reproduce the long tail of the $\langle p_T^2 \rangle_{in}$ distribution at high energies, as shown in Fig. 32. Furthermore, as seen from the bottom histogram of Fig. 31, the planar events exhibit three axes, the average transverse momentum of the hadrons with respect to these

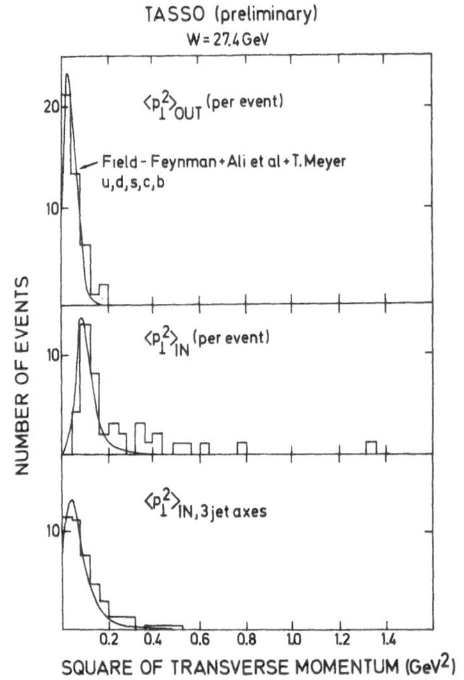

Fig. 31. TASSO distribution[34] of the average squared transverse momentum component out of the event plane (top), and in the event plane (center), for events at W = 27.4 GeV (averaging over charged hadrons only). The curves are for $q\bar{q}$ jets without gluon bremsstrahlung. The bottom figure shows $\langle p_\perp^2 \rangle$ per jet when 3 jet axes are fitted, again compared with the q jet model.

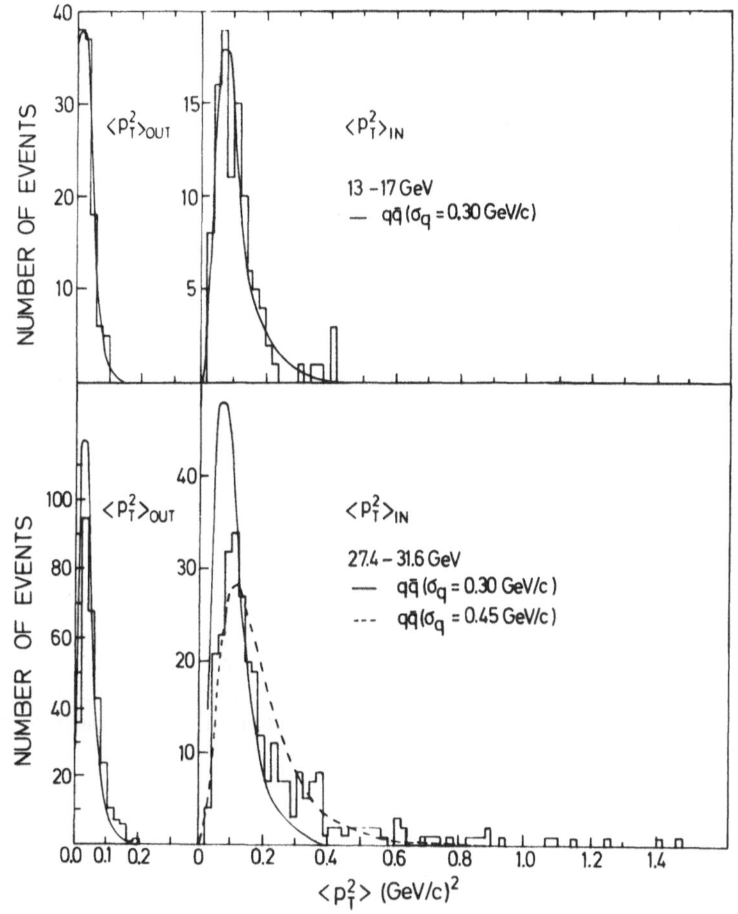

Fig. 32. The mean transverse momentum squared normal to the event plane $\langle p_T^2\rangle_{out}$ and in the event plane $\langle p_T^2\rangle_{in}$ per event for the low-energy and the high-energy TASSO data[35]. The predictions from the $q\bar{q}$ model are shown assuming σ_q = 0.30 GeV/c (solid curves) and σ_q = 0.45 GeV/c (dotted curves). The model includes u, d, s, c and b quarks.

axes being the same 0.3 GeV/c. Indeed, the data are in agreement with predictions based on perturbative QCD.

Fig. 33[41] compares the transverse momentum distribution, with respect to the individual jet axis, of three-jet events at the center-of-mass energy W of 30 GeV with that of two-jet events at W = 12 GeV. They are virtually identical. Also shown is the distribution of trijettiness J_3, defined by eq. (55). The peak at low J_3 values is a consequence of the strong collimation of the particles around the three jet axes. For Fig. 33, cuts have been made in sphericity and aplanarity : $S > 0.25$ and $A < 0.08$.

MARK J. The MARK J Collaboration[36] used the method of energy flow as described in the preceding section. The central quantity that the MARK J Collaboration uses is the oblateness O defined by (86). Fig. 34 (a) shows the event distribution as a function of this oblateness for their data at W = 17 GeV. The data agrees with both the $q\bar{q}$ and the $q\bar{q}g$ models. Fig. 34 (b) shows the event distribution as a function of oblateness at $27.4 \leqq W \leqq 31.6$ GeV as compared with the predictions of the $q\bar{q}$ and the $q\bar{q}g$ models. For the $q\bar{q}$ model, MARK J uses both $\langle p_T \rangle = 0.325$ GeV/c and $\langle p_T \rangle = 0.425$ GeV/c. The data have more oblate events than the $q\bar{q}$ model predicts, but agree with the $q\bar{q}g$ model very well. The situation is therefore similar to that of the $\langle p_T^2 \rangle_{in}$ distributions of Figs. 31 and 32.

More detailed analysis is carried out by MARK J by dividing the energy distribution of each event into two hemispheres using the plane defined by the major and minor axes $\hat{e}_2$ and $\hat{e}_3$. The forward hemisphere contains the narrow jet and the other contains the broader jet just as in the Ellis-Karliner method[44]. From the broader jet by itself, F^b_{major}, F^b_{minor}, and

$$O_b = 2(F^b_{major} - F^b_{minor}) \tag{91}$$

are calculated[60]. The events from W = 27.4, 30, and 31.6 GeV are used with the cuts

$$F_{thrust} < 0.8 \quad \text{and} \quad O_b > 0.1 \; . \tag{92}$$

From these events, the resulting energy-flow plot is shown in Fig. 35 (a). Although three-jet structure can be clearly seen in this plot, it was already emphasized in the original paper of MARK J[36] that "phase space distribution will show three nearly identical lobes due to the method of selection used."

This vital point is later discussed in great detail[55]. In this study with high statistics and higher energies W between 33 and 36.7 GeV, the cuts used are slightly different. The additional quantities used are F^n_{thrust} for narrow jet and θ_{minor}, which is the angle between

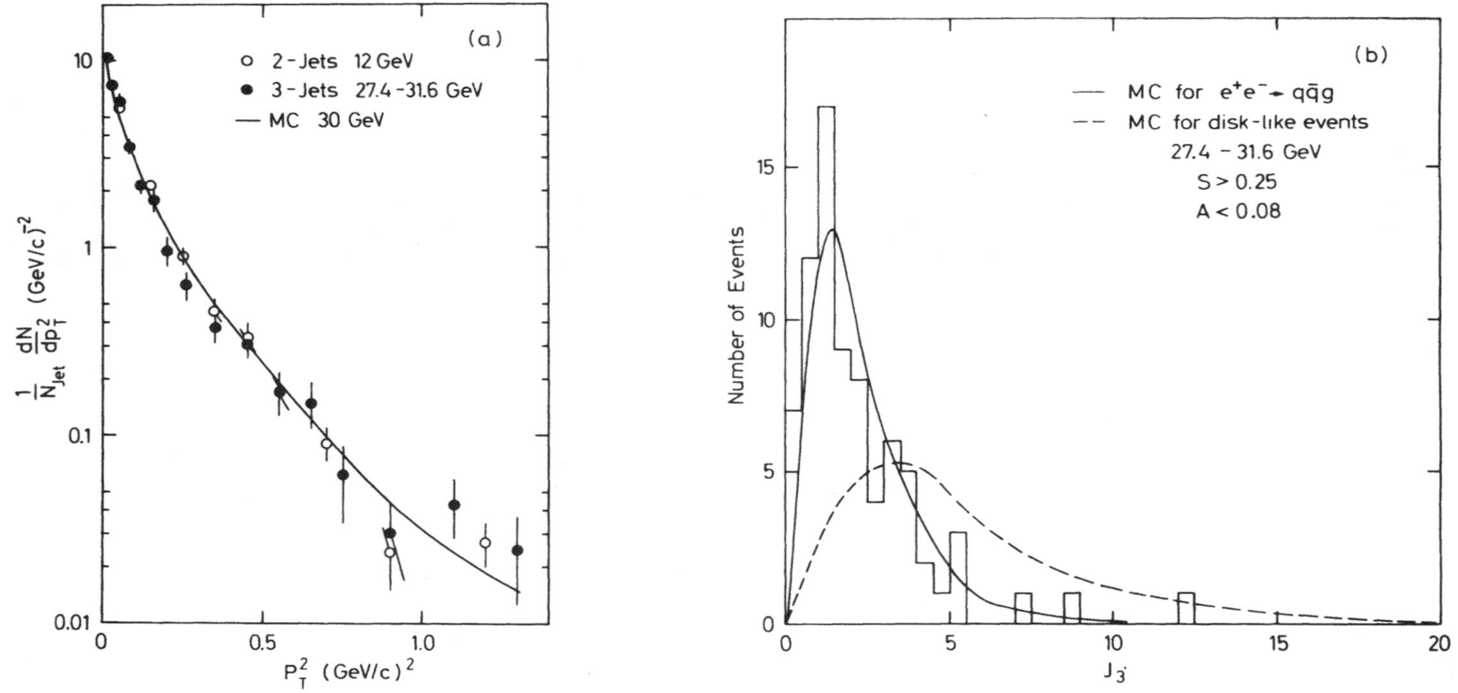

Fig. 33. (a) Observed transverse momentum distribution of the hadrons from the planar region ($S > 0.25$, $A < 0.08$) with respect to the three axes found with the Wu-Zobernig method[32], at $W = 30$ GeV (.). It is compared with the transverse momentum distribution relative to the sphericity axis for all events (no S or A cut) at $W = 12$ GeV, analyzed as two-jet events (O). It is also compared with the result from the QCD model at 30 GeV (curve). (b) Comparison of the trijettiness distribution for the planar event ($S>0.25, A<0.08$) sample at $W = 30$ GeV, with the distribution for disk-like events (dashed curve) and the QCD model (solid curve). The curves are normalized to the number of observed events. These data are from TASSO[41].

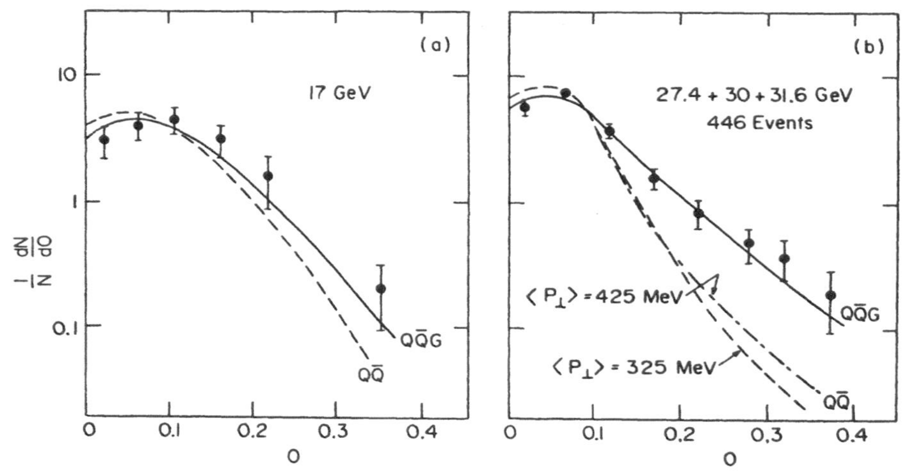

Fig. 34. (a) The MARK J[36] distribution N^{-1} dN/dO as function of oblateness at W = 17 GeV. (b) The distribution N^{-1} dN/dO as function of oblateness at W = 27.4 - 31.6 GeV. In both (a) and (b) the solid curves are the predictions based on the $q\bar{q}g$ model and the dashed curve is based on the standard $q\bar{q}$ model with $\langle p_\perp \rangle$ = 325 MeV/c. The dash-dotted curve in (b) is the $q\bar{q}$ model prediction with $\langle p_\perp \rangle$ = 425 MeV/c.

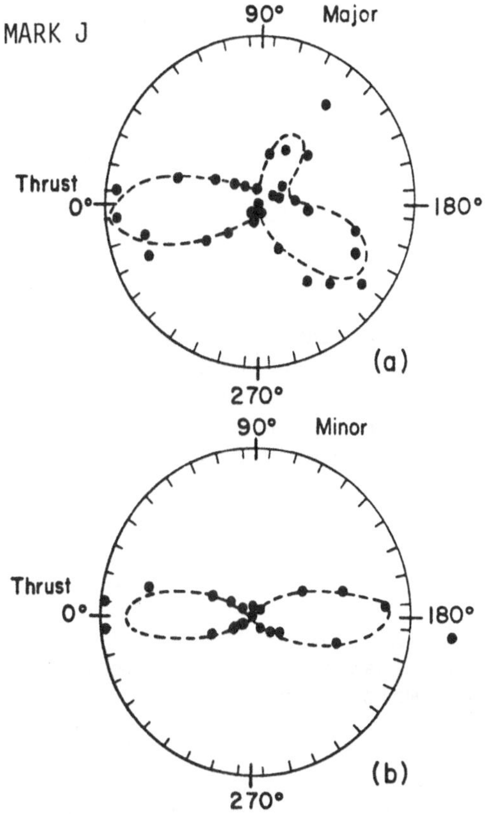

Fig. 35. (a) Energy distribution from MARK J[36] in the plane as defined by the thrust and the major axes for all the events with thrust < 0.8 and oblateness > 0.1 at W = 27.4, 30, and 31.6 GeV. The energy value is proportional to the radial distances. The superimposed dashed line represents the distribution calculated with the $q\bar{q}g$ model. (b) The measured and calculated energy distribution in the plane as defined by the thrust and the minor axes.

the minor axis and the beam direction. The results are shown in Fig. 36. It is seen clearly from Fig. 36 (d) that the phase space model also gives three jets due to the method of selection used. However as seen from the same figure the solid curve (QCD) fits the data best.

The question may be raised why the method of energy flow was employed in the first place. The original reason was that, since individual tracks were not necessarily measured separately in the MARK J detector, it was thought that the method of energy flow was less sensitive to the combination of tracks. Actually, none of the methods described in de preceding section is sensitive to this possible lack of resolution. Even in the Wu-Zobernig method, while the values of Q_1, Q_2 and Q_3 are sensitive to the combination of tracks (i.e. the replacement of say $\vec{p}_1$ and $\vec{p}_2$ by $\vec{p}_1 + \vec{p}_2$), the jet directions $\hat{m}_1$, $\hat{m}_2$ and $\hat{m}_3$ are much less sensitive.

It is instructive to compare the early data of TASSO and MARK J taken in 1979. Since moving into the beam in October 1978, both Collaborations had been struggling to get the various components to function properly. During the spring and summer of 1979 after PETRA energy reached 27 GeV, MARK J accumulated a somewhat higher integrated luminosity by about 10 - 20%, and hence a larger number of total hadronic events. However it is not meaningful to compare the total number of three-jet events. While the event of Fig. 29 consists clearly of three jets, reduction of the angle between two of the jets changes smoothly to a two-jet event. Thus the number of three-jet events depends sensitively on the cut used. The MARK J[60] advantage of having a slightly larger number of hadronic events compared with TASSO is, however, more than offset by the more effective method of three-jet analysis employed by TASSO.

PLUTO. The analysis of the PLUTO data[37] was carried out by the method of Brandt and Dahmen[42]. In this method, three-jet events are characterized by a high triplicity T_3 but a relatively low thrust T, say < 0.8. The original result of PLUTO are shown in Fig. 37, where for $\tau = 1, 2, 3$ C_τ^* denotes the partition S_τ that gives the triplicity. An example of a three-jet event from PLUTO is shown in Fig. 38, where the triplicity plane denotes of course the plane determined by the three $\hat{m}_\tau$ of (60).

As an independent approach, the PLUTO group has also analysed[52] all hadronic events in terms of separate hadronic jets as obtained by a cluster analysis method. The specific procedure followed is the one by Daum, Meyer, and Bürger[49]. The result is shown in Fig. 39. It is seen from this figure that the number of three-jet events is about 45% of that of two-jet events. This ratio gives a clear demonstration of the fact that the number of three-jet events depends sensitively on the cut used, as discussed before. This cluster analysis also gives a fair number of four-jet events, but virtually no five-jet or six-jet events.

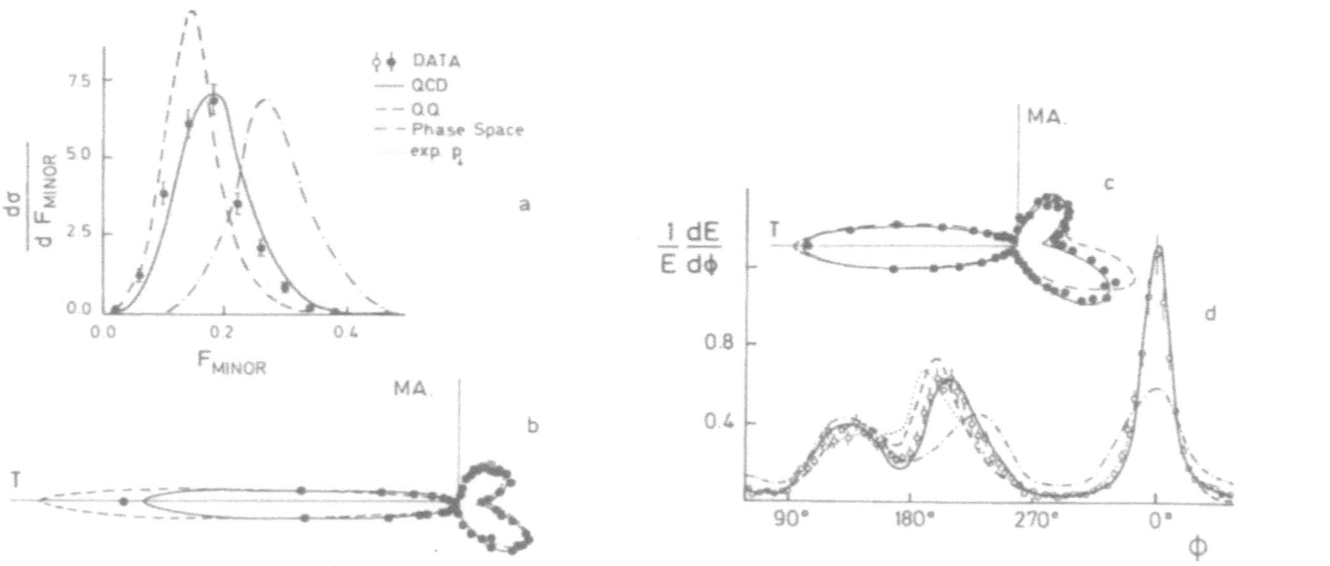

Fig. 36. (a) The MARK J[55] distribution $N^{-1}\ dN/dF_{minor}$ in the fraction of the visible energy flow of the entire event which is projected along the minor axis (perpendicular to the event plane). (b) Comparing the data with QCD, and $q\bar{q}$ models, using energy flow diagrams in the thrust major event plane for events with $O_B > 0.3$, $F^n_{thrust} > 0.98$ or $\theta_{minor} > 60°$. (c) Same as Fig. 36 (b) but with $F^n_{thrust} < 0.98$ and $\theta_{minor} < 60°$. (d) The unfolded energy flow diagram of 36 (c) compared with the models of QCD, $q\bar{q}$, phase space, and a $q\bar{q}$ model with a exp $(-P_T/650\ MeV/c)$ fragmentation distribution.

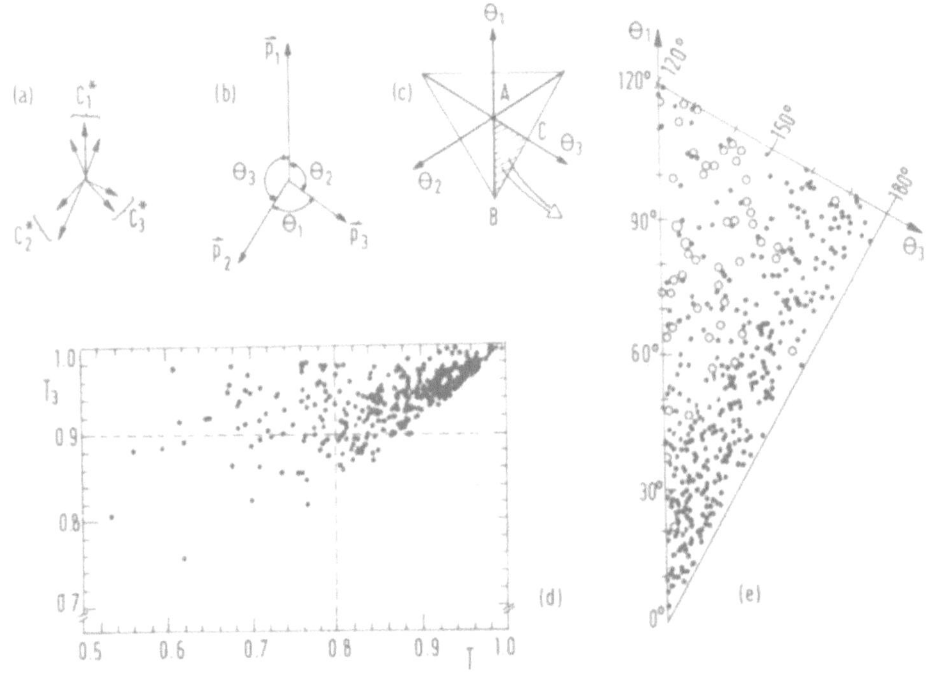

Fig. 37. Momentum configuration of hadrons (a) and jets (b) obtained by grouping hadrons into three classes according to the Brandt-Dahmen method[42]. A Dalitz plot (c) can be spanned by the angles between the jets whose shaded area only is populated. Nearly symmetrical three-jet events will be situated near point A. The data at W = 27.6, 30 and 31.6 GeV are shown in a scatter diagram of triplicity T_3 versus thrust T (d) and in the angular Dalitz plot (e). In (d) planar events will be in the upper left of the plot characterized by low thrust and high triplicity, e.g., $T < 0.8$ and $T_3 > 0.9$. Events falling in this category show up as large circles in (e). The data are from PLUTO[37].

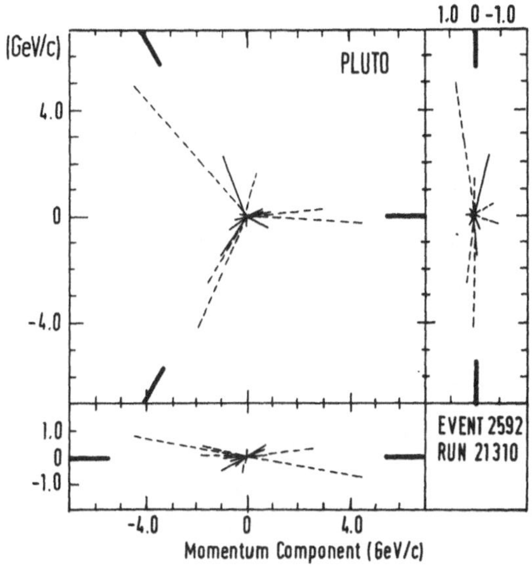

Fig. 38. Momentum vectors of a PLUTO[37] event (W = 31.6 GeV) with high triplicity and low thrust projected onto the triplicity plane (top left), onto a perpendicular plane normal to the fastest jet (top right) and onto a plane containing the direction of the fastest jet (bottom). Solid and dotted lines correspond to charged and neutral particles, respectively. The directions of the jet axes are indicated as fat bars near the margins of the figures.

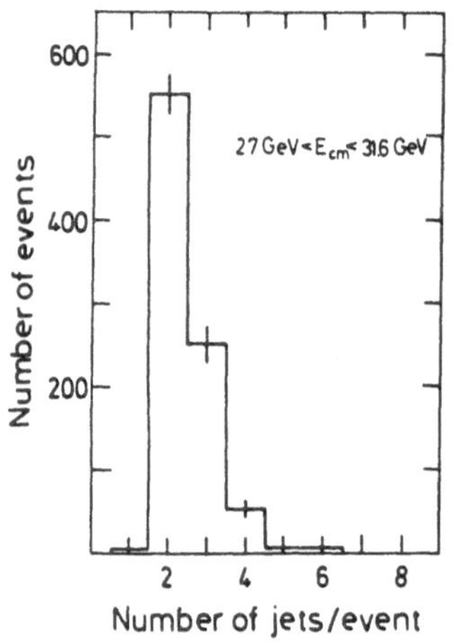

Fig. 39. PLUTO[52] distribution of number of jets as reconstructed by the cluster algorithm from the combined W = 27 - 32 GeV data.

JADE. An example of the JADE three-jet events is shown in Fig. 40[61]. The data on the planarity $Q_2 - Q_1$ distribution is given in Fig. 41[38]. JADE has also used the Ellis-Karliner method[44] described earlier. In this method, the thrust of the jets 2 and 3 in their own center-of-mass system is called T^* (see Fig. 27). Assuming that all observed particles are massless, JADE made the interesting compa-

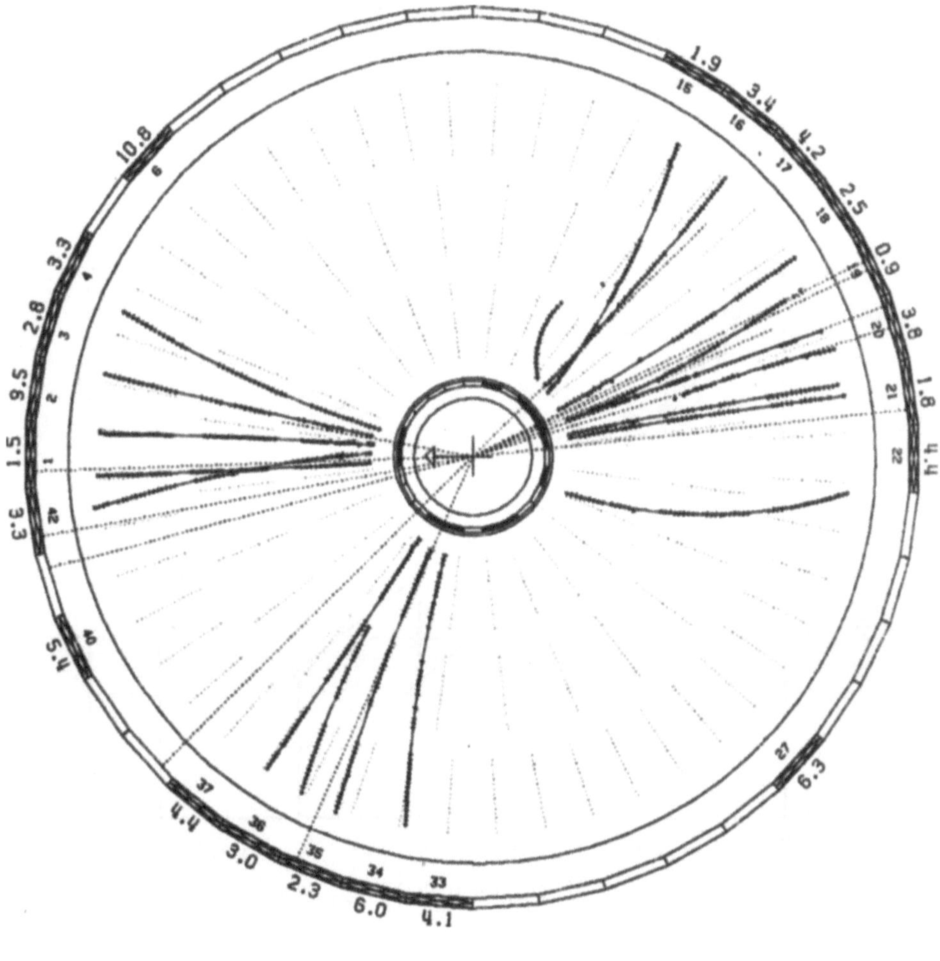

Fig. 40. A JADE three-jet event viewed along the beam axis.

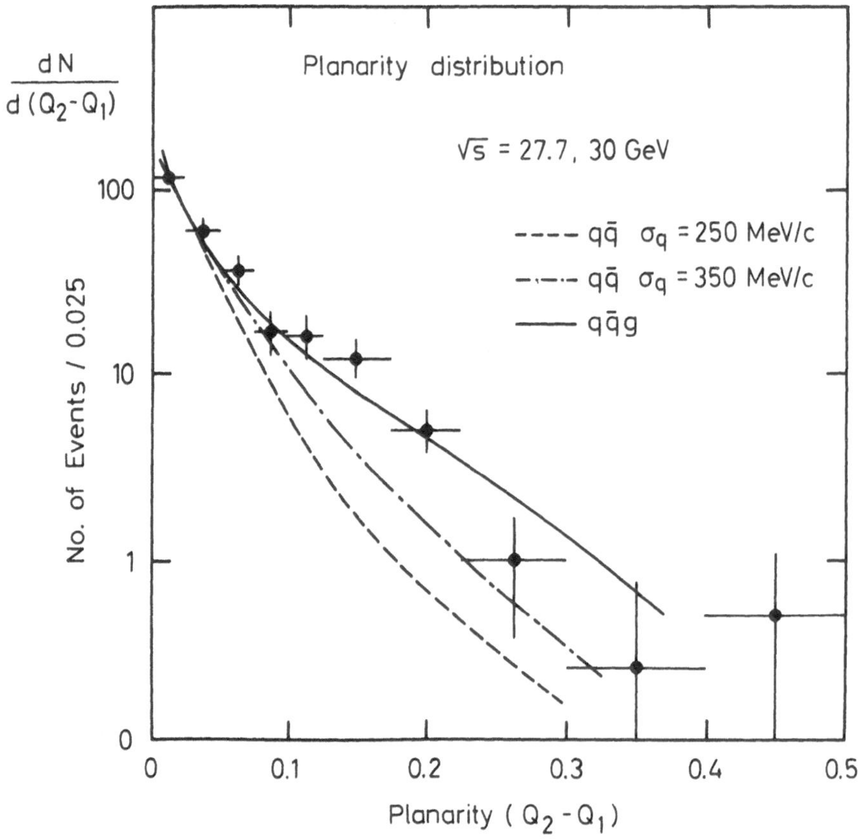

Fig. 41. The JADE[38] planarity distribution compared with model predictions.

rison between the T^* at W = 30 GeV for planar non-colinear events defined by $Q_2 - Q_1 > 0.07$ with the thrust T of all events at W = 12 GeV. This comparison is shown in Fig. 42. The two distributions fall on top of each other, leading to the same conclusion as drawn from Fig. 31, namely, for high energy planar events the particles are as collimated around three axes as particles are collimated around the common jet axis at lower energies.

CELLO. By the time the CELLO detector was moved into the beam in March 1980 each of the other Collaborations had already collected a large number of three-jet events. We present here only one figure based on their data[62]. In Fig. 43, the p_T and thrust distributions are compared with $q\bar{q}$ models with two different fragmentation processes. That both models fail to describe the data implies again the necessity of including $e^+e^- \rightarrow q\bar{q}g$.

Although the data presented in this Section are taken from the early publications and hence the statistics are relatively low, all the conclusions discussed here are supported by later data with much larger numbers of events.

Determination of the Quark-Gluon Coupling Constant α_s

Results. The MARK J Collaboration is the first group to determine the quark-gluon strong coupling constant α_s, defined in a way analogous to the fine-structure constant α of QED. By now, each of the five Collaborations at PETRA has published at least one determination of α_s, mostly at center-of-mass energy W around 30 GeV. The results are listed in Table 4 arranged in the order of publication. Only journal publications with the entire Collaboration as the author have been included, but not the numerous conference reports and summaries.

Theorists tell us that it is very easy to determine α_s : it is essentially the ratio of the numbers of three-jet to two-jet events. Actually it is not so simple due to the fragmentation of quarks and gluons and the resulting ambiguity between two- and three-jet events already mentioned.

Consequently, the actual procedure of obtaining α_s, to be explained later, is quite complicated. This is to be contrasted with the observations of the three-jet events discussed in the preceding section. There, once a three-jet analysis is used to determine the direction along which the event should be viewed, the three jets are clearly visible to the naked eye, as illustrated for example in Fig. 29. Different track fitting programs may give slightly different momenta, and different three-jet analyses may give slightly different directions, but that the event has a qualitatively different topology from the two-jet events remains unchanged. Here, to find the value of α_s, we must at least ascertain a suitable cut, fit for the

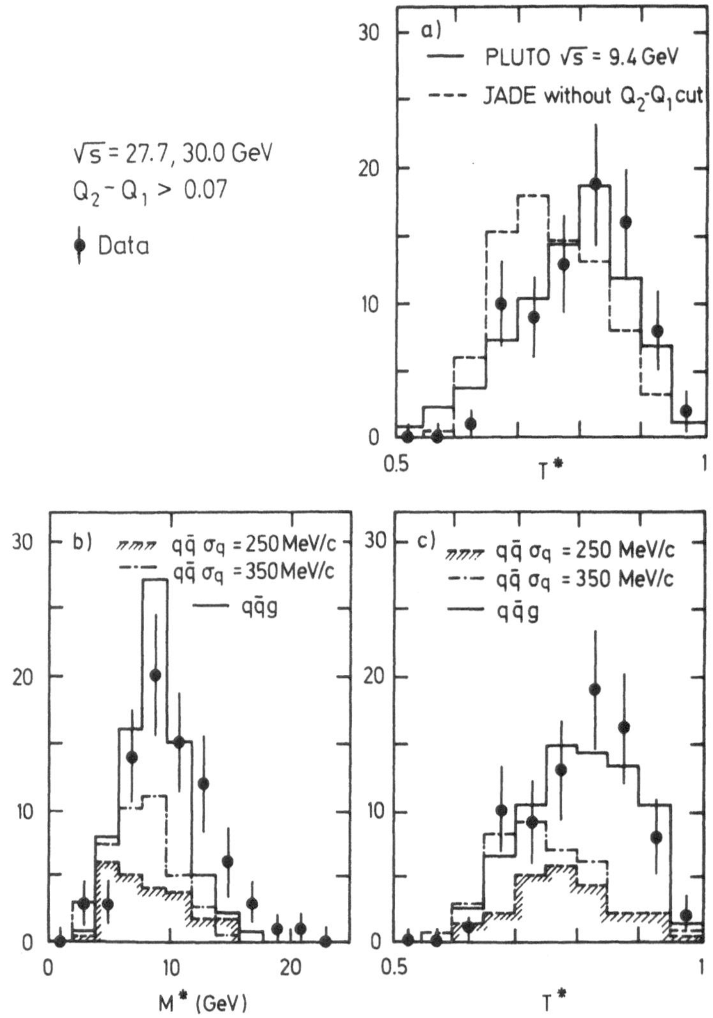

Fig. 42. The three-jet nature of the "planar" ($Q_2 - Q_1 > 0.07$) events of JADE[38]. (a) The observed distribution of T^* (the thrust of the fat jet in its rest system) for the planar events compared with the two-jet thrust distribution obtained by PLUTO at W = 9.4 GeV (full-line histogram). The broken-line histogram shows the normalized T^* distribution for all events without the planarity cut. (b) The observed invariant mass (M^*) distribution of the fat jet system for the planar events compared with the distributions expected from the $q\bar{q}$ model with σ_q = 250 MeV/c (shaded, broken-line histogram) and with σ_q = 350 MeV/c (dot-dashed histogram). The full-line histogram represents the M^* distribution predicted by the $q\bar{q}g$ model. (c) The same observed T^* distribution as shown in (a) compared with the predictions of the $q\bar{q}$ and $q\bar{q}g$ models.

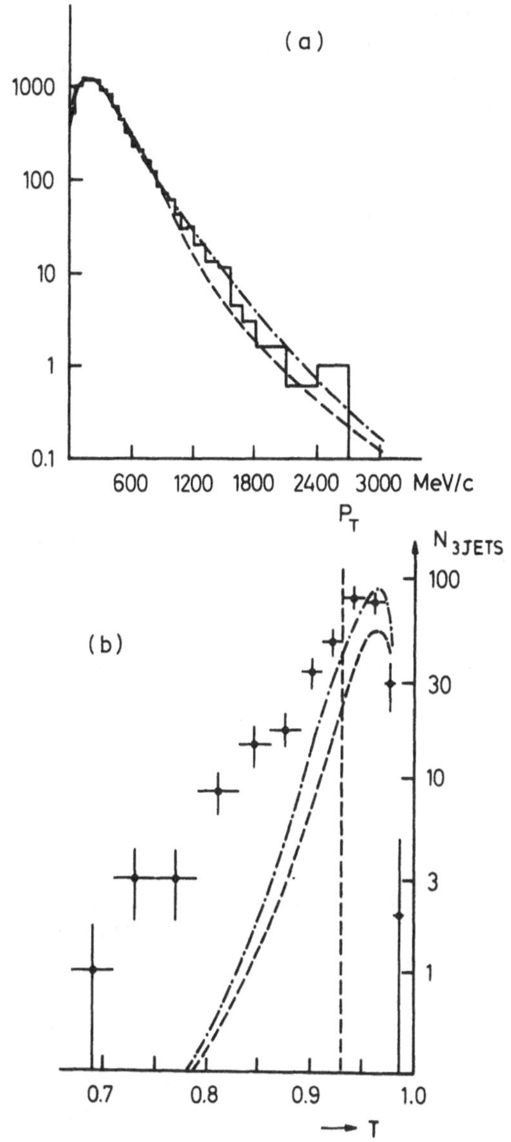

Fig. 43. Comparison of CELLO[62] data with different $q\bar{q}$ fragmentation models. (a) Distribution of p_T in a jet for charged particles in the 2 jet sample, (i)(histogram) experimental distribution, (ii)(dashed curve) Gaussian dependence in the fragmentation process with σ_q = 300 MeV/c. (iii) (dashed-dotted) exponential dependence with σ_q = 420 MeV/c. High p_T values are overestimated. (b) Distribution of thrust T calculated from the jet momenta in 3 jet events. Experimental distribution and $q\bar{q}$ model calculations (dashed and dashed-dotted curves) using the p_T distribution (ii) and (iii) described in (a). Both models fail to describe the data.

Table 4. List of experimental results on the quark-gluon coupling constant α_s published by the Collaborations at PETRA. The symbol * means that the systematic error is not given in the original publication.

Coll.	Values of α_s	Order	Method of analysis	Date of receipt by journal	Ref.
MARK J	0.23±0.02±0.04	First order	Oblateness distribution	November 1979	63
JADE	0.17±0.04± *	First order	Planarity distribution	December 1979	38
TASSO	0.17±0.02±0.03	Partial Second order	Sphericity and aplanarity distribution	May 1980	41
PLUTO	0.15±0.03±0.02	First order	Cluster analysis and parton thrust (x_1) distribution	October 1980	52
PLUTO	0.20±0.02± *	First order	Energy-energy correlation	November 1980	64
CELLO	0.16±0.01±0.03	First order	Thrust and oblateness distribution	April 1982	65
CELLO	0.21±0.01± *	First order	Energy-energy correlation	May 1982	66
CELLO	0.15±0.02± *	First order	Energy-energy correlation asymmetry	May 1982	66
JADE	0.16±0.015±0.03	Second order	Cluster analysis and parton thrust (x_1 and $x_\perp$) distributions	September 1982	67
CELLO	See text and reference	First order	Various methods of analysis	October 1982	68
MARK J	0.13±0.01±0.02	Second order	Energy-energy correlation asymmetry	May 1983	69

fragmentation parameters, and select a QCD model for comparison. Consequently, intimate acquaintance with the details of the experimental apparatus and data is essential.

QCD Diagrams. Before describing the QCD models useful in the experimental determination of α_s, we describe briefly the Feynman diagrams involved.

The lowest-order diagram for $e^+e^- \to q\bar{q}$ is shown once again in Fig. 44 (a). For radiative correction to the order α_s, there are, as shown in Fig. 44 (b), three diagrams, two of which (the second and third ones) have the radiative correction on the external legs. To avoid drawing too large a number of diagrams, those with radiative corrections on the external legs are omitted for radiative corrections to order α_s^2; the rest are shown in Fig. 44 (c).

Similarly, for the process $e^+e^- \to q\bar{q}g$, the lowest-order QCD diagrams are shown once again in Fig. 45 (a) while those for radiative correction to order α_s are given in Fig. 45 (b). The QCD diagrams for the four-jet processes $e^+e^- \to q\bar{q}gg$ and $e^+e^- \to q\bar{q}q'\bar{q}'$ are shown in Fig. 28.

Any differential cross section is calculated from the square of the corresponding matrix element, obtained by adding together the contributions from various Feynman diagrams. We list here the various interferences that contribute orders of α_s^0, α_s, and α_s^2 in the cross sections. For brevity, we omit the words "the diagrams of"; thus for example "44 (a) and 44 (b)" means the contributions to the cross section from the interference of the diagrams of Fig. 44 (a) with those of 44 (b).

To order α_s^0 : $e^+e^- \to q\bar{q}$ only from the square of 44 (a).

To order α_s : (i) two jets - 44 (a) and 44 (b).

(ii) three jets - square of 45 (a).

To order α_s^2 : (i) two jets - square of 44 (b), 44 (a) and 44 (c).

(ii) three jets - 45 (a) and 45 (b).

(iii) four jets - square of 28 (a); square of 28 (b).

QCD Models. The value of α_s is determined by comparing the experimental data with a suitable QCD model. By a QCD model, we mean a perturbative QCD calculation followed by a fragmentation scheme. Unavoidably, some parameters are needed to describe quark fragmentation, and further ones for gluon fragmentation. These parameters are also determined by the experimental data.

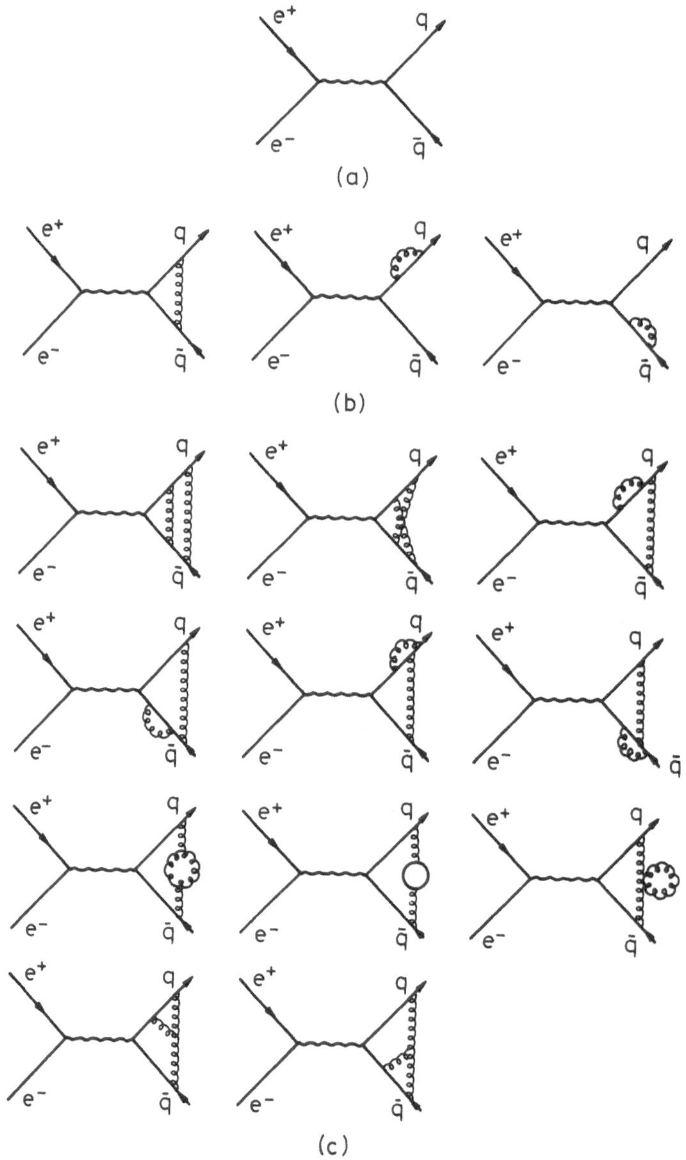

Fig. 44. Diagrams for $e^+e^- \rightarrow q\bar{q}$: (a) lowest order; (b) radiative correction to order α_s ; and (c) radiative correction to order α_s^2.

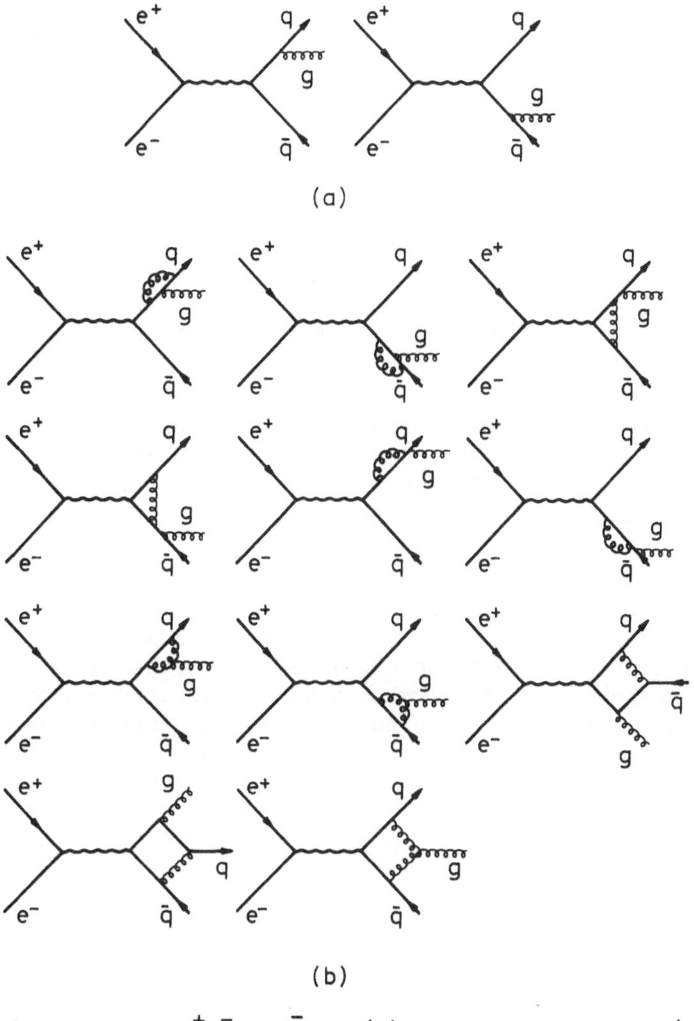

Fig. 45. Diagrams for $e^+e^- \to q\bar{q}g$: (a) lowest order; and (b) radiative correction due to the gluon to order α_s.

There are two general classes of QCD models used at PETRA : the Hoyer et al. or independent fragmentation models, and the Lund or string models. We describe both classes.

(A) Hoyer et al. or independent fragmentation models. The pioneering work on QCD model is due to Hoyer, Osland, Sander, Walsh, and Zerwas[70]. They use the complete first-order theory[39], and apply a Field-Feynman jet fragmentation[59] independently to the quark, the anti-quark, and the gluon. Their original program has been improved by Meyer[59] of the TASSO Collaboration with the inclusion of heavy quarks. At both PETRA and PEP, this improvement is used in all applications.

The basic idea of the models of this type is shown schematically in Fig. 46 (a). Later improvements of the original model include taking more QCD perturbative diagrams into account. Three general versions are currently in use at PETRA.

(a) First order. This is the original version of Hoyer et al., with the Meyer improvement.

(b) Partial second order. Ali, Pietarinen, Kramer, and Willrodt[71] have modified the above first-order model by including the effects of producing four jets. In other words, with reference to the list of the preceding section for the order α_s^2, (iii) is included but not (ii). Such a partial inclusion of α_s^2 effects require a more detailed discussion.

In QED, the masslessness of the photon leads to infrared divergences. Similarly, in QCD the masslessness of the gluon leads to infrared divergences of a somewhat more complicated nature. Physically meaningful results are obtained through cancellations in cross sections, such as between (ii) and (iii) or order α_s^2. Because of this necessary cancellation, in principle it is not acceptable to include (iii) but not (ii). Ali, Pietarinen, Kramers, and Willrodt get around this problem by introducing an acoplanarity cut of

$$A > 0.05$$

for four jets. Thus this partial second-order model contains this additional parameter.

(c) Second order. It is clearly desirable to have a complete second-order model. The computation of the large number of Feynman diagrams shown in Fig. 45 (b), which contribute to (ii) of order α_s^2, requires a heroic effort. Such a complete second-order QCD calculation was first accomplished by Ellis, Ross, and Terrano[72], and also later by Vermaseren, Gaemers, and Oldham[73] and by Fabricius, Schmitt, Kramer, and Schierholz[74]. With reference to Table 4, the version of Fabricius et al. is used by the JADE Collaboration[67],

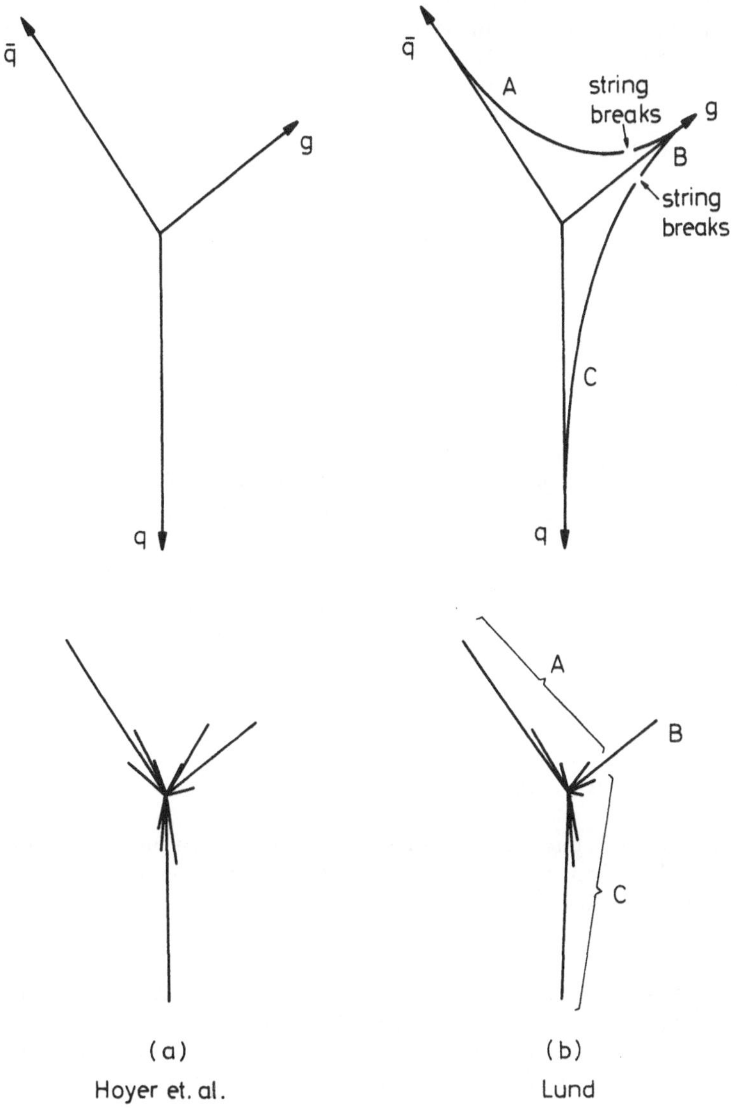

Fig. 46. Schematic representation of fragmentation models.

while that of Ellis et al. is used by the MARK J Collaboration[69] in their respective second-order determination of α_s. Because the jets fragment independently, the case of four jets is essentially the same as that of three jets.

(B) Lund or string models. The Lund group of Andersson, Gustafson, and Sjoestrand[75] had the idea of combining fragmentation with the concept of the color string, which is stretched from the quark via the gluon to the anti-quark. This color string is envisaged to break first to produce a hadron carrying a sizeable fraction of the gluon energy. The remainder of the gluon energy is then shared between the two pieces of the broken string, which are fragmented independently in their respective center-of-mass systems[75]. This Lund fragmentation process is shown schematically in Fig. 46 (b).

(a) First order. This is the original version, and is used extensively at PETRA.

(b) Second order. This is obtained by combining the Lund idea with the complete second-order QCD calculation mentioned above.[72,73,74] However, unlike the Hoyer et al. or independent fragmentation model, the extension of the Lund or string model to the case of four jets is not entirely straightforward. The basic problem is the following. For the process $e^+e^- \rightarrow q\bar{q}gg$, the Lund model requires the specification of the way the color string is stretched from the quark to one of the gluons, to the other gluon, and finally to the anti-quark. As discussed in the preceding section, however, the second-order cross section comes from two matrix elements, which may have different specifications for the ordering of the color string. A solution to this mismatch is to introduce a further parameter specifying the relative probability for the ordering of the two gluons in the color string. This type of problem is to be expected because the perturbative QCD is carried out quantum mechanically while the fragmentation is semi-classical.

Finally, we add the remark that in QCD the value of α_s depends on the renormalization scheme. In all these QCD models, the $\overline{MS}$ scheme (modified minimum subtraction scheme) is used. Therefore α_s always means α_s in this $\overline{MS}$ scheme.

Procedures to Determine α_s. It is seen from Table 4 that the first determination of α_s by the MARK J Collaboration gives a result significantly larger than those obtained by the other Collaborations using Hoyer et al. or independent fragmentation models. The reason remains unclear; it is doubtful that the method of energy flow is to be blamed.

In the comparison of experimental data with the QCD models for the purpose of determining α_s, the following point deserves emphasis. Both the incident electron and the incident positron may emit a photon, leading to e^+e^- annihilation at a lower energy. This initial state radiation must be taken into account, as discussed in ref. 76.

We give here briefly three examples of the procedures used in the determination of α_s.

(A) TASSO[41]. The TASSO Collaboration used the partial second order of Ali et al.[71]. The determination is based on the S-A (sphericity-aplanarity) distribution of their high sphericity data ($S > 0.25$). These data are analyzed by two different procedures.

In the first procedure, α_s is determined without any assumptions on the parameters of jet fragmentation. As shown in Table 5 the value of α_s is totally insensitive to the values of the fragmentation parameters, and the result is $\alpha_s = 0.16 \pm 0.04$ (statistical).

In the second procedure, this statistical error is reduced by using the fragmentation parameters of the first chapter. The resulting value of α_s is the one given in Table 4.

(B) JADE[67]. The JADE Collaboration uses the second order QCD result of Fabricius et al.[74]. The determination is based on the distributions as functions of x_1 and $x_\perp$, defined as follows. x_1 is the largest fractional energy of the jets

$$x_1 = \frac{2}{W} \max_\tau E^{(\tau)} , \tag{93}$$

where the $E^{(\tau)}$ are given by (54). Remember that, in the derivation of (54), the jet masses are neglected, and therefore $E^{(\tau)}$ is the same as the magnitude of the corresponding momentum $\vec{P}^{(\tau)}$. With this notation, $x_\perp$ is defined to be $2/W$ times the magnitude of the component of momenta of the other two jets perpendicular to the direction of x_1.

The comparison of the experimental and theoretical distribution in these two variables x_1 and $x_\perp$ is shown in Fig. 47. The final value of α_s is the one given in Table 4.

(C) MARK J[69]. The MARK J Collaboration uses the calculation of Ali et al.[71] and completes it by doing a Monte Carlo integration of the second-order virtual contributions computed by Ellis et al.[72]. The determination is based on the asymmetry in the energy-energy correlation function defined by[77]

$$A(\cos\chi) = \frac{1}{\sigma}\left\{\frac{d\Sigma}{d\cos\chi}(\pi-\chi) - \frac{d\Sigma}{d\cos\chi}(\chi)\right\} \tag{94}$$

with

$$\frac{1}{\sigma}\frac{d\Sigma}{d\cos\chi} = \frac{1}{N}\sum_{\text{event}}^{N}\sum_{i,j}\frac{E_i E_j}{E_{vis}^2}\,\delta(\cos\chi_{ij} - \cos\chi), \tag{95}$$

where the sum is over all hadronic events, E_i is the energy measured in a given solid angle element i, E_{vis} is the total event energy,

Table 5. Fitted values of α_s and σ_q obtained by TASSO[41] for different input values of a_F and P/(P+V). The fits are performed with the S-A distribution of the high sphericity data (S > 0.25).

P/(P+V) / a_F		0.1	0.3	0.5	0.7	0.9
0.1	α_s	=0.17±0.03	0.17±0.03	0.17±0.03	0.16±0.03	0.16±0.03
	σ_q	=0.44±0.11	0.46±0.10	0.47±0.10	0.48±0.09	0.48±0.08
0.3	α_s	=0.17±0.04	0.16±0.04	0.16±0.04	0.15±0.04	0.15±0.04
	σ_q	=0.42±0.12	0.44±0.11	0.46±0.10	0.47±0.09	0.48±0.08
0.5	α_s	=0.17±0.04	0.16±0.04	0.16±0.04	0.15±0.04	0.14±0.04
	σ_q	=0.35±0.12	0.38±0.12	0.41±0.12	0.43±0.10	0.44±0.09
0.7	α_s	=0.17±0.03	0.17±0.04	0.16±0.04	0.15±0.05	0.14±0.05
	σ_q	=0.28±0.09	0.30±0.10	0.33±0.10	0.36±0.10	0.39±0.09
0.9	α_s	=0.17±0.03	0.17±0.04	0.16±0.04	0.15±0.04	0.14±0.05
	σ_q	=0.21±0.08	0.23±0.08	0.26±0.08	0.30±0.09	0.33±0.08

and χ_{ij} is the angle separating the directions of the energy depositions. This definition used by MARK J differs from the usual one in the appearance of E_{vis} instead of W. The energy-energy correlation function has previously been used for the determination of α_s by PLUTO, MARK II, MAC and CELLO[78].

The comparison of the experimental and theoretical energy-energy correlation function is shown in Fig. 48. The final value of α_s is the one given in Table 4.

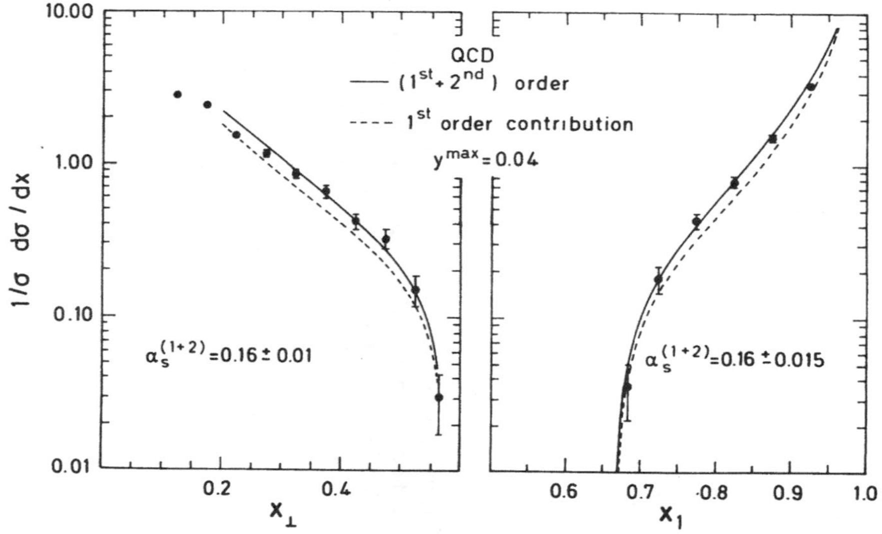

Fig. 47. The JADE[67] corrected x_1 and $x_\perp$ distributions together with the second order QCD best fit. The first order contribution of this fit is indicated by the broken curve. The fit does not include data above $x_1 = 0.85$ for x_1 - fit and below $x_\perp = 0.30$ for the $x_\perp$ - fit.

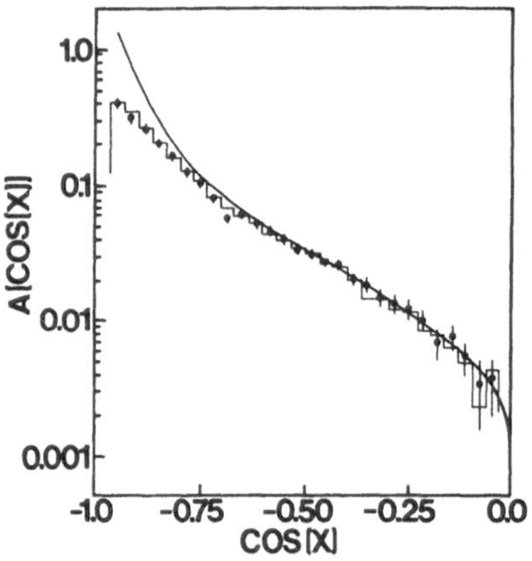

Fig. 48. MARK J[69] asymmetry data compared with predictions at parton level (curve) for α_s = 0.13 and predictions for the two fragmentation models (Lund, Ali et al.; histogram) for the best fit values of α_s. These two histograms are indistinguishable.

Comparison of Hoyer et al. and Lund Models. The question of the influence of the fragmentation model used, Hoyer et al. or independent fragmentation vs. Lund or string, on the value of α_s has been studied in first order by both the JADE[67] and CELLO[68] Collaborations. Their analyses reach different conclusions. JADE uses both fragmentation models to determine the corrected x_1 and $x_\perp$ distributions, and has found that the corrections are, within the errors, independent of the fragmentation scheme used[67]. On the other hand, CELLO has found that, depending on the distribution used, the Lund model gives α_s between 28% to 67% higher than the model of Hoyer et al. The two extreme cases are :

Smallest difference : three-jet fraction from events with oblateness larger than 0.3

Lund $\alpha_s = 0.255 \pm 0.050$

Hoyer et al. $\alpha_s = 0.200 \pm 0.035$ ratio = 1.28

Largest difference : energy-energy correlation function

Lund $\alpha_s = 0.25 \pm 0.04$

Hoyer et al. $\alpha_s = 0.15 \pm 0.02$ ratio = 1.67

In particular, this result from CELLO implies that, contrary to the earlier hope, the energy-energy correlation function is sensitive to fragmentation.

Properties of the Gluon

Spin of the Gluon. Since QCD is a gauge theory, its gluon must have spin 1. Experimentally, the gluon spin has been determined to be 1 by three Collaborations at PETRA : TASSO[45], PLUTO[52], and CELLO[62].

Before fragmentation, the $q\bar{q}g$ Dalitz plot can be described by the fractional energy variables

$$x_i = E_i/E_b \ , \tag{96}$$

where E_i is the energy of quark or gluon i and $E_b = 1/2\ W$ is the incident beam energy, so that

$$x_1 + x_2 + x_3 = 2. \tag{97}$$

We choose to order them such that

$$x_3 \leq x_2 \leq x_1, \tag{98}$$

which implies

$$2/3 \leq x_1 \leq 1 \ . \tag{99}$$

If the quarks and the gluon have negligible masses relative to E_B, the x_i are determined by the angles θ_i shown in Fig. 27

$$x_i = \frac{2 \sin \theta_i}{\sin \theta_1 + \sin \theta_2 + \sin \theta_3} \ . \tag{100}$$

[The x_1 here is the same as the x_1 of (93), and (100) is essentially (54) rewritten in different notations, where the θ_i are the $\phi^{(i)}$ rearranged such that (98) is satisfied.]

Fig. 27 (b) shows the angle $\tilde{\theta}$ suggested by Ellis and Karliner[44] to discriminate between vector and scalar gluons. In this figure the $q\bar{q}g$ system has been Lorentz boosted to the center-of-momentum frame of partons 2 and 3. Assuming negligible quark and gluon masses, $\cos \tilde{\theta}$ is given by

$$|\cos \tilde{\theta}| = \frac{x_2 - x_3}{x_1} = \frac{\sin \theta_2 - \sin \theta_3}{\sin \theta_1} \ . \tag{101}$$

The distribution functions for the x_i in QCD and in the scalar gluon model, after averaging over the production angles relative to the incident e^+e^- beams, are given by[39]

$$\text{vector} : \frac{1}{\sigma_0}\left(\frac{d\sigma}{dx_1 dx_2}\right)_V = \frac{2\alpha_s}{3\pi}\left[\frac{x_1^2 + x_2^2}{(1-x_1)(1-x_2)} + \begin{array}{l}\text{cyclic}\\ \text{permut.}\\ \text{of } 1,2,3\end{array}\right] \tag{102}$$

$$\text{scalar} : \frac{1}{\sigma_0}\left(\frac{d\sigma}{dx_1 dx_2}\right)_S = \frac{\tilde{\alpha}_s}{3\pi}\left[\frac{x_3^2}{(1-x_1)(1-x_2)} + \begin{array}{l}\text{cyclic}\\ \text{permut.}\\ \text{of } 1,2,3\end{array}\right] \ . \tag{103}$$

To avoid a number of difficulties associated with the singularities of these distributions at $x_1 = 1$, the TASSO Collaboration[45] uses the kinematic region defined by $x_1 < 0.9$. In the three-jet region so defined, the distributions are not strongly peaked either for vector or scalar gluons, making the dependence on fragmentation smearing small. As a further precaution, only distributions normalized to the number of events in this kinematic region are used. This means that the distinction between vector and scalar gluons is made only on the basis of the difference in shape of the two distributions in the three-jet region. In this way one eliminates, on the parton level, all dependence of the spin analysis on the values of the strong coupling constant α_s and $\tilde{\alpha}_s$ for vector and scalar gluons, respectively.

TASSO has applied the Wu-Zobernig[32] three-jet analysis to 16,000 hadronic events, and computed the x_i using (100). After a cut of $x_1 < 0.9$, which implies a minimum angle of 70° between the jets, 1600 events survive. In Fig. 49 these events, which include higher statistics than those given in ref. 45, are plotted as a function of the Ellis-Karliner angle, and compared with the Monte Carlo predictions under the two assumptions of spin 1 and spin 0. Spin 0 is clearly ruled out.

An alternative method of analysis[79] uses the variable

$$\xi = 2\left[\frac{x_3^2}{(1-x_1)(1-x_2)} + \text{permut.}\right] \Big/ \left[\frac{x_1^2 + x_2^2}{(1-x_1)(1-x_2)} + \text{permut.}\right]$$

$$= 2\left(\frac{x_1^2 + x_2^2 + x_3^2}{x_1^3 + x_2^3 + x_3^3} - 1\right) \qquad (104)$$

which is obtained from the right-hand sides of (102) and (103).

Using the cluster analysis method, the PLUTO Collaboration[52] has plotted the x_1 distributions of the three-jet events as shown in Fig. 50 (a) and (b). The solid curve in the Fig. 50 (a) is the prediction from first order QCD for vector gluons. A fit of the scalar gluon prediction to the data points yields a χ^2/D.F. = 9.1/4 as shown in Fig. 50 (b) (dashed curve). If one averages the x_1 distribution for $2/3 < x_1 < 0.95$, the predictions for vector and scalar gluons are 0.891 and 0.871 respectively. For the data obtained by PLUTO $\langle x_1 \rangle = 0.893 \pm 0.005$. Hence the hypothesis of scalar gluons is disfavored. Similar plot from the CELLO data[62] is shown in Fig. 51.

Angular Distribution of Particles in Three-Jet Events. The JADE Collaboration[43] has found an interesting and subtle effect in the distribution of hadrons in the three-jet events as a function of the angular variable in the event plane. Their study is motivated by a comparison between the independent fragmentation model of Hoyer et al.[70] and the string model of the Lund group[75]. In the model of Hoyer et al., the quark, the antiquark, and the gluon fragment independently of each other in essentially identical ways, producing final state mesons with limited momentum transverse to the directions of the original partons.

In the Lund model the fragmentation proceeds along the color flux lines as the primary partons move apart. For $q\bar{q}g$ events, these flux lines are not strung between quark and antiquark directly, but via the gluon as intermediary (see Figs. 46 and 52), because the quark is a color triplet while the gluon is a color octet. For example, if the quark is red (r) while the antiquark is anti-blue ($\bar{b}$), then in order to get a color singlet, the gluon must be antired-blue ($\bar{r}b$). Thus the red color line runs from the quark to the gluon, while in the same way the blue color line runs from the gluon to the

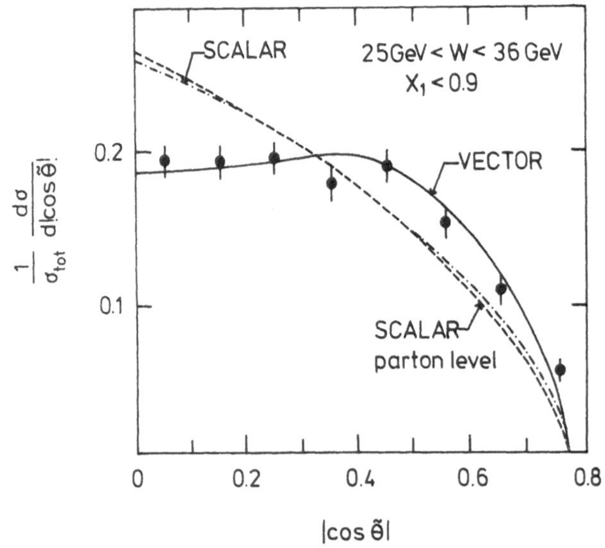

Fig. 49. Observed distribution of the TASSO data[45] in the region $x_1 < 0.9$ as a function of the Ellis-Karliner angle $\tilde{\theta}$. The solid line shows the QCD Monte Carlo prediction, dashed line the prediction for the scalar gluons (- - - for Monte Carlo scalar model; —•—•— for scalar model of parton level). All curves are normalized to the number of observed events.

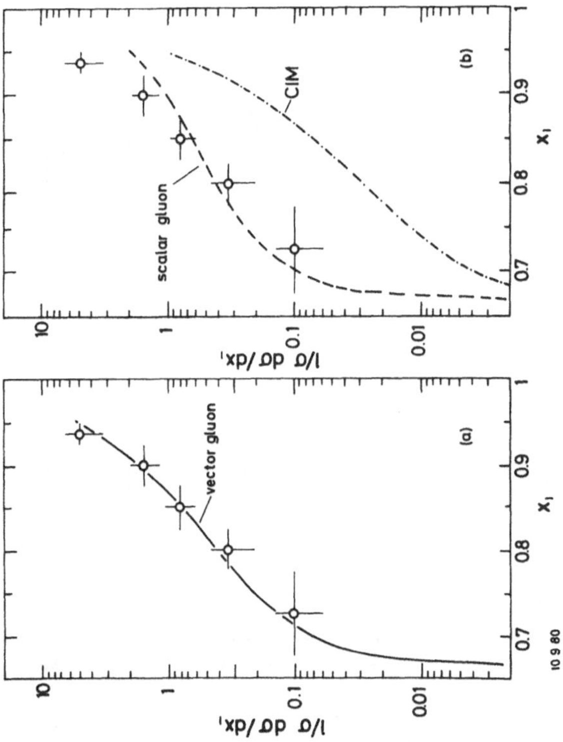

Fig. 50. PLUTO[52] distribution of the relative energy of the fastest parton (x_1). The data points are corrected for detector acceptance, radiation and hadronisation. The curves are (a) first order QCD, (b) dashed : scalar gluon hypothesis and dash-dotted : CIM (Constituent Interchange Model).

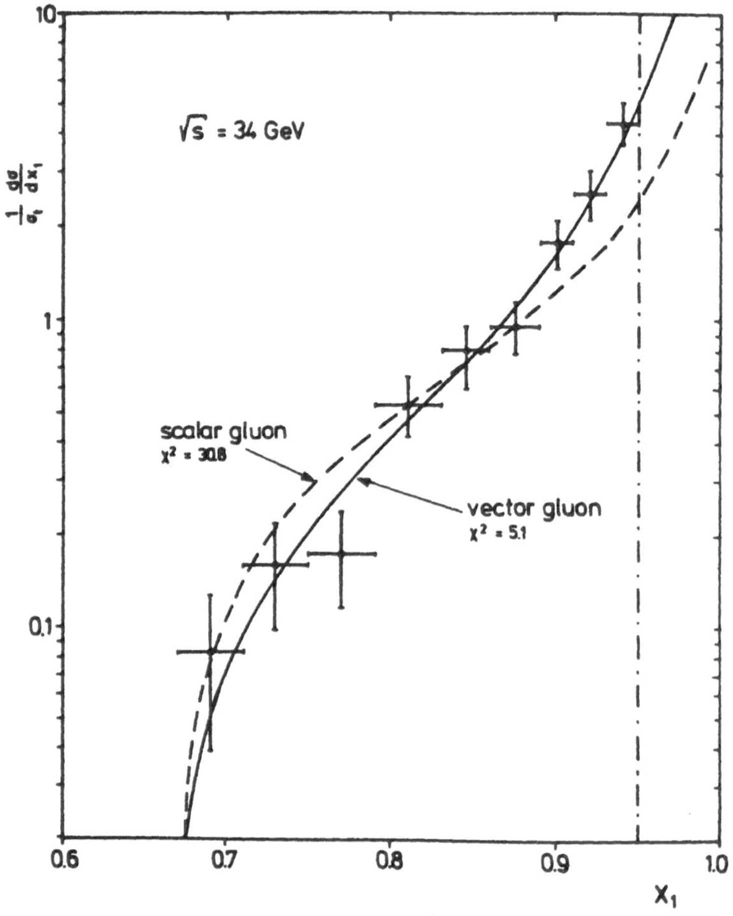

Fig. 51. CELLO[62] differential cross section of three-jet events with respect to the energy fraction x_1 carried by the most energetic parton. Data are compared to the QCD prediction of vector gluons (full curve) and a scalar gluon model (dashed curve).

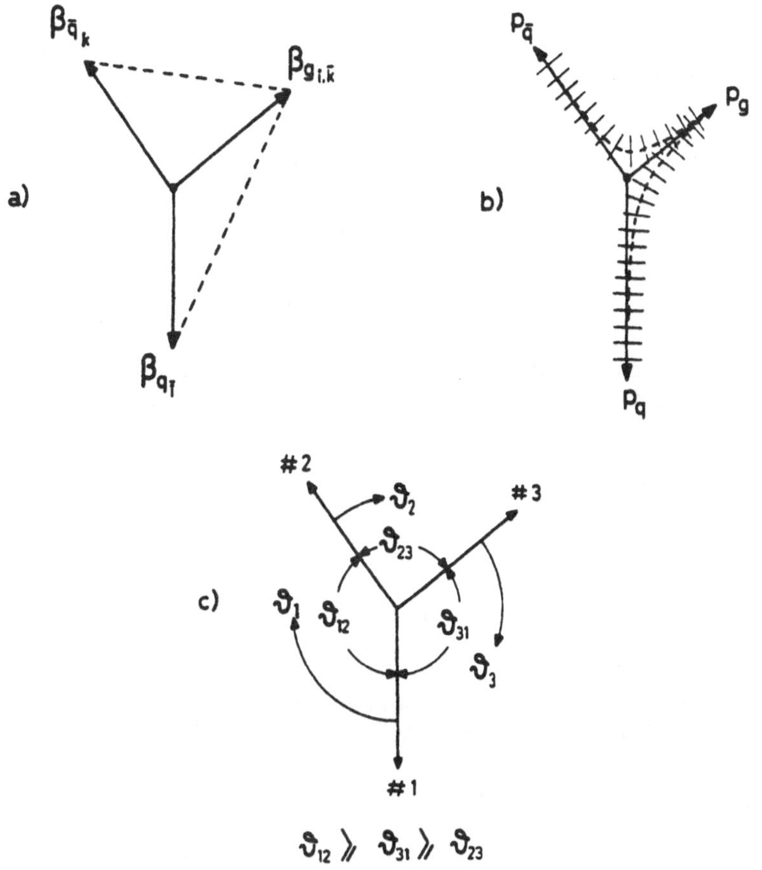

Fig. 52. Sketch of the quark and gluon velocities and of the colour-anticolour axes (a). Fragmentation along these axes, neglecting transverse momenta, yields particles of the same mass distributed in momentum space along two hyperbolas, as indicated in (b). The broadening due to different masses and transverse momenta is also indicated. The ordering scheme of the observed jets is sketched in (c). θ_{ik} is the angle between the jet axes #i and #k projected onto the event plane. This figure is from JADE[43].

antiquark. The Lund model is formulated in terms of strings and is kinematically equivalent to a treatment of the gluon as a colinear quark-antiquark pair (q', $\bar{q}'$) with the momentum shared between q' and $\bar{q}'$. Each of the two gluon components form a $q\bar{q}'$ or $q'\bar{q}$ two-jet system with the primary $\bar{q}$ or q, and these two-jet systems are treated in their respective center-of-mass systems, not in the laboratory system (which is also the $q\bar{q}g$ center-of-mass system). Aside from this Lorentz boost, the mesons within these jets are distributed according to the prescription of Field and Feynman[59], a special treatment being made only for the leading meson at the gluon corner. Neglecting transverse momenta with respect to the $q\bar{q}'$ and $q'\bar{q}$ jet axes final state particles of the same mass are distributed along hyperbolas in the overall c.m. momentum space as sketched in Fig. 52 (b). Therefore the Lund model predicts the production of more particles in the angular region between the gluon and the quark or the antiquark than between the quark and the antiquark.

Fig. 53 (a) shows the JADE result[43] on the angular distributions of the particles from their three-jet events. The average number of charged and neutral particles per event is plotted as a function of the normalized projected angle θ_i and θ_{ik} are defined in Fig. 52 (c). The two model predictions are also shown in Fig. 53 (a), and compared in Fig. 53 (b).

Apart from the region between jets No. 1 and No. 2, both models describe the data reasonably well. In this region, which is the region between the two quark jets for the majority of events, the model of Hoyer et al. predicts more particles than the Lund model, and the experimental data are in better agreement with the Lund model than with the model of Hoyer et al.

Difference between Quark and Gluon Jets. Ever since the first observation of gluon jets, an intriguing question is : how do we distinguish a gluon jet from a quark jet? Although some differences are expected from QCD, these two kinds of jets turn out to be remarkably similar. For example, JADE has measured the ratio of photon energy to charged particle energy separately for the three jets, and the results are[80]

Jet 1 : 0.47 ± 0.01 ,

Jet 2 : 0.46 ± 0.01 ,

Jet 3 : 0.44 ± 0.02 ,

where the errors are statistical but the systematic errors are expected to be small.

Recently, the JADE Collaboration has found experimental evidence

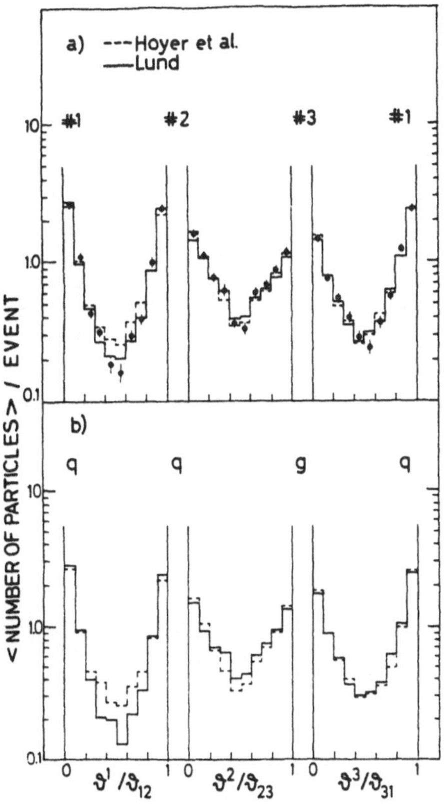

Fig. 53. The average number of particles per event between the indicated jet axes versus the normalized projected angle. The data together with the corresponding model predictions are shown in (a), the model predictions, ordered for quark and gluon jets, in (b). The data are from JADE[43].

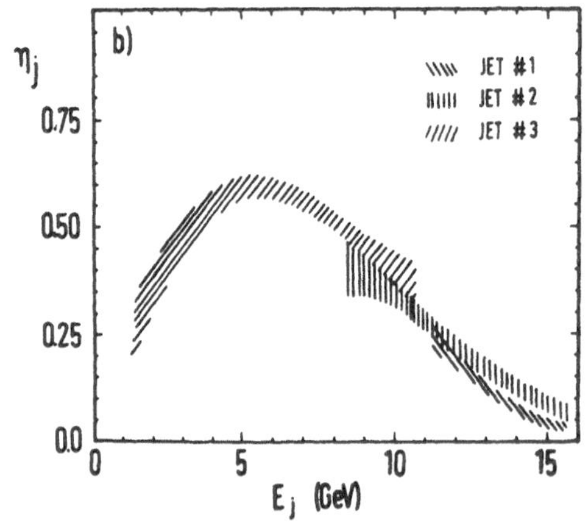

Fig. 54. The probability η_j given by JADE[80] for jet j being closest to the gluon direction as a function of the reconstructed jet energy E_j obtained from model calculations at a fixed c.m. energy of 33 GeV. The widths of the shaded areas indicate the statistical errors.

that the mean transverse momentum $\langle p_T \rangle$ of particles within a jet is larger for the gluon jet than for the quark jet[81]. Since the spin of the gluon is 1, jet 3, which is the least energetic one, is most likely to be the gluon jet. More precisely, on the basis of Monte Carlo calculation at center-of-mass energies of 33 GeV (22 GeV) the probability for jet #1, #2, and #3 being closest to the gluon direction is 12% (9%), 22% (20%), and 51% (34%), respectively. The probabilities do not add up to 100%; residual $q\bar{q}$ events faking three-jet structures account for the difference. Fig. 54 shows the probabilities η_j for jet j being closest to the gluon direction as a function of $E_j = x_j W/2$ at W = 33 GeV. Note that

$$W/3 \leq E_1 < W/2 \ ,$$

$$W/4 \leq E_2 < W/2 \ ,$$

and $$E_3 \leq W/3 \ . \qquad (105)$$

Therefore, for any given W, the overlap between E_1 and E_2 is from W/3 to W/2, that between E_2 and E_3 from W/4 to W/3, but E_1 and E_3 never overlap.

On the basis of 2048 planar three-jet events (out of 18424 hadronic events) at W between 29 and 36.4 GeV and 307 planar three-jet events (out of 1945 hadronic events) at W = 22 GeV, the JADE result for the average transverse momentum of a jet is shown in Fig. 55 (a). The g=q model calculation using Hoyer et al.[70] shown in Fig. 55 (b) does not give any larger transverse momentum for jet 3, indicating that there is no significant bias in the selection of events. The experimentally observed larger transverse momentum for jet 3 can be reproduced by increasing the σ_q for the gluon from 330 MeV/c to 500 MeV/c (independent of energy), or by using the Lund model[75]. In this Fig. 55, the overlaps of the energy ranges for the three jets are much more than those kinematically allwed for any given W, as discussed in the preceding paragraph and Fig. 54. The reason is of course that the values of W ranges from 22 GeV to 36.4 GeV. In particular, the five data points indicated by arrows in Fig. 55 (a) are from W = 22 GeV. Accordingly, as emphasized by the JADE Collaboration, the comparison is made between jet 3 at W = 29 to 36.4 GeV and jet 2 at W = 22 GeV. For this comparison, jet 3 has significantly larger average transverse momentum than jet 2 for the same jet energy.

Four-Jet Events. The three-jet events have been analyzed in various other ways[82], including the utilization of such variables as total transverse momentum and jet mass. Instead of going into these further analyses, we now turn our attention to four-jet events.

An interesting variable introduced by Nachtmann and Reiter[83] and applied by the JADE Collaboration[84] to distinguish the case of four jets against those of two and three jets is the tripodity D_3 :

$$D_3 = 2 \max \left(\sum_i |\vec{p}_i^{\,T}| \cos^3 \measuredangle (\hat{n}, \vec{p}_i^{\,T}) \ / \ \sum_i |\vec{p}_i|\right) \ , \qquad (106)$$

where the $\vec{p}_i^{\,T}$ are the particle or parton momenta projected onto the plane perpendicular to the event thrust axis, and $\hat{n}$ is a unit vector in this plane, oriented such that the quantity in brackets is maximized.

D_3 measures the symmetry of the momentum distribution in the plane normal to the thrust axis. If this distribution is symmetric

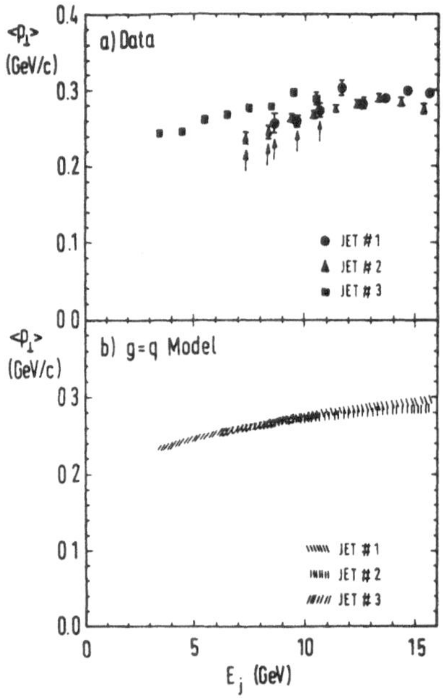

Fig. 55. Average p_T distribution from JADE [81] for the highest energy jets (E_1), medium energy jets (E_2), and lowest energy jets (E_3) as a function of the jet energy E_j calculated from the angles between the three jets. The center-of-mass energy is between 22 and 36.7 GeV. Fig. 55 (a) is from the experimental data, where the arrows indicate the data points at W = 22 GeV. Fig. 55 (b) is the prediction of the q = g Monte Carlo model.

with respect to the interaction point, $D_3 = 0$. The allowed range of D_3 is most easily understood for events without fragmentation of quarks and gluons. For two- and three-jet events, D_3 vanishes due to a symmetric distribution. Four-jet events fall into two separate classes : events in class I have two parton momenta on each side of the plane normal to the thrust axis, which leads to a symmetric distribution and vanishing D_3. Events in class II have one high momentum in one hemisphere and the three remaining ones in the opposite one, leading to $D_3 \geqq 0$ with a maximum value of 0.324[82].

In Fig. 56 the JADE tripodity and acoplanarity distributions are shown and compared with the theoretical expectations based on a combination of Ali et al.[74] and Lund[75]. Because of fragmentation, initial state radiation, and resolution as well as the fact that four axes are fitted to events with two or three jet structures, the expectation for two- and three-jet events alone (curves labelled L23) extends to large values of D_3. Inclusion of four-jet events (curves labelled L234) gives a significantly better fit to the data for W = 33 GeV. The conclusion is therefore reached that, for center-of-mass energy W above 30 GeV, the experimental distributions of the jet parameters acoplanarity and tripodity show significant deviations from the expectations for two- and three-jet events alone, and that the inclusion of four-jet events removes these deviations.

Examples of four-jet events observed by JADE and TASSO are shown in Fig. 10 and 57 respectively.

ELECTROWEAK INTERACTION

Introduction

We now come to one of the most exciting results from PETRA, the observation of weak and electromagnetic interference. Indeed, it was one of the main original purposes of building PETRA to observe such effects, especially the presence of forward-backward asymmetry in electron-positron annihilation[85].

The JADE[86] and TASSO[87] Collaborations are respectively the first and second groups to observe such an asymmetry. The process they studied is the creation of μ pairs :

$$e^- e^+ \rightarrow \mu^- \mu^+ .$$

Since the asymmetry increases rapidly with energy, such experimental results require both high energy and high statistics. In as much as the published results do not include the entire data set available now, in this chapter we shall update the experimental data whenever possible.

As of now, the statistically most significant data on weak and

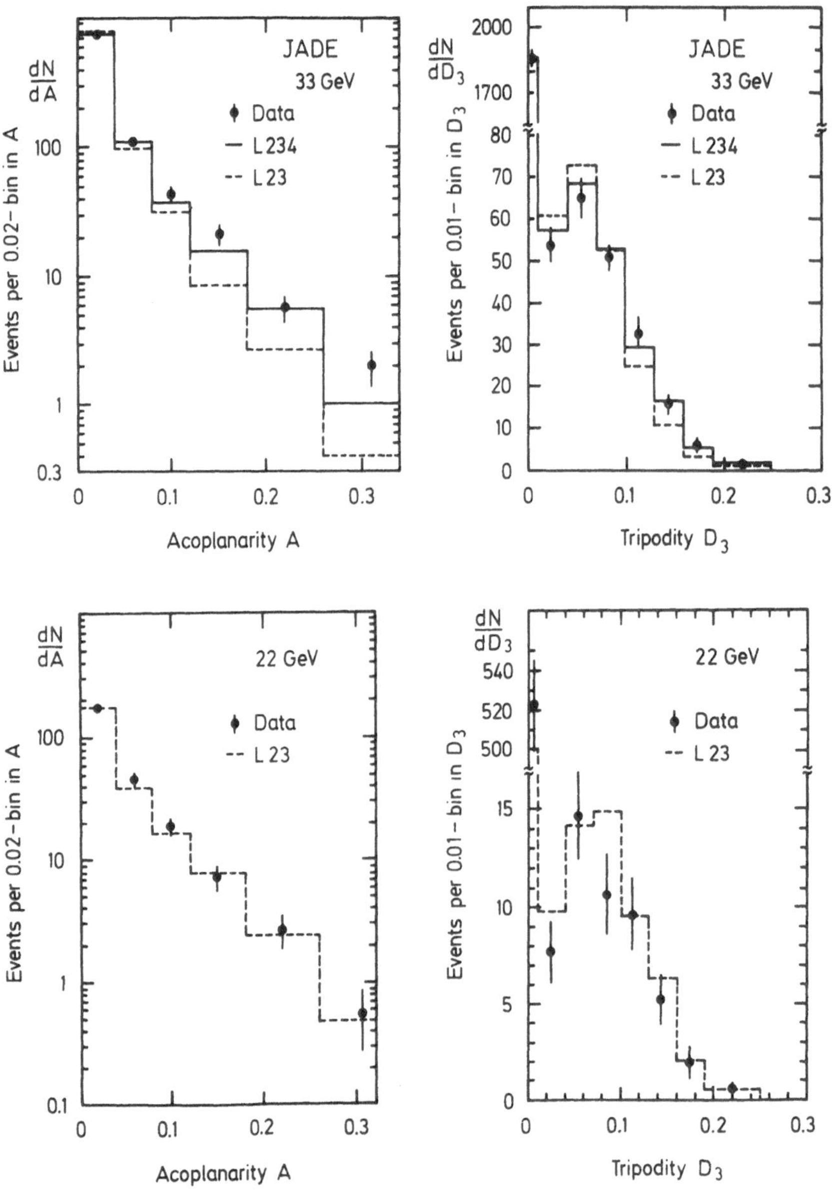

Fig. 56. JADE[83] experimental D_3-distribution for all events and A-distribution for the events with $D_3 < 0.01$. The variables D_3 and A were calculated from four reconstructed axes. (a) and (b) are for W = 33 GeV, (c) and (d) for W = 22 GeV. L23 is the model expectation for two- and three-jet events alone. L234 includes in addition four-jet events (5% at generation). Both L23 and L234 are normalized to the total number of events.

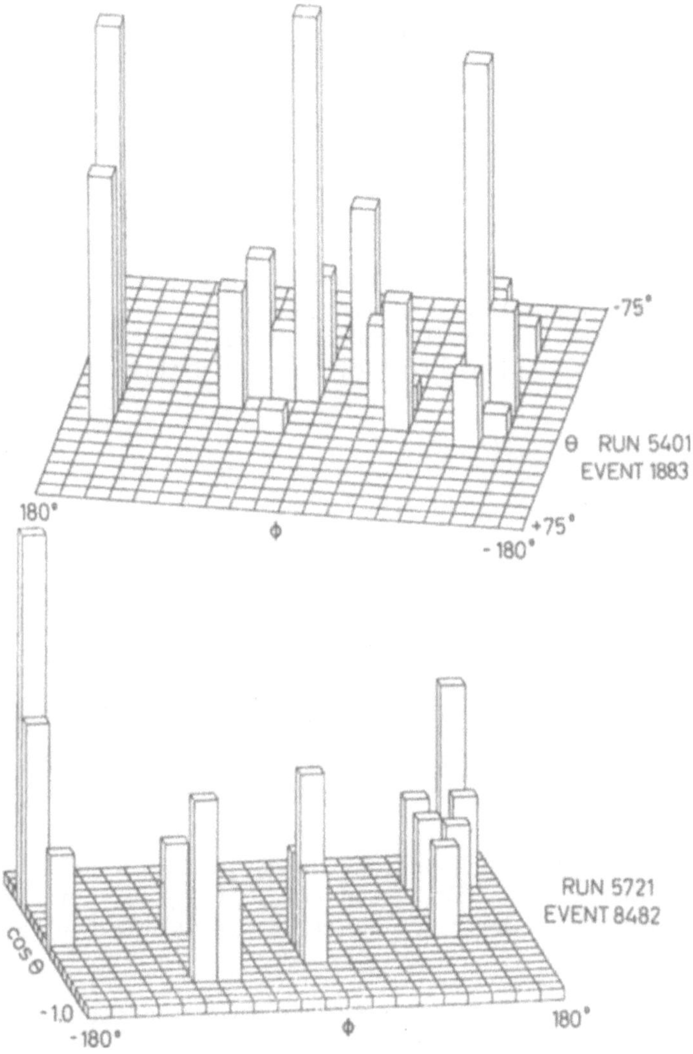

Fig. 57. Two examples of the many four-jet events observed at PETRA. These examples are from TASSO.

electromagnetic interference are still from this process of the creation of μ pairs. The processes $e^-e^+ \to \tau^-\tau^+$[88] and $e^-e^+ \to c\bar{c}$[89,90] have also been measured. The asymmetry measurements on these three processes will be described in this chapter later on.

Since the neutral intermediate vector boson Z^0[89,91] has not only axial vector but also vector coupling to fermions, this forward-backward asymmetry is not the only manifestation of electroweak interaction for electron-positron annihilation. In particular, there are also modifications on the total rate and on the angular distributions, which can be used to determine the vector coupling Z^0. Results from leptons are also summarized in this chapter.

Standard Model

Before describing the experimental procedure, we review briefly the standard model, which is the elegant scheme for the unification of the weak and electromagnetic interactions developed by Glashow, Weinberg and Salam[89]. In this model there are three families of quarks and leptons :

$$\begin{pmatrix} u \\ d \\ \nu_e \\ e \end{pmatrix} \begin{pmatrix} c \\ s \\ \nu_\mu \\ \mu \end{pmatrix} \begin{pmatrix} t \\ b \\ \nu_\tau \\ \tau \end{pmatrix}$$

and the electromagnetic and weak interactions are transmitted by the vector bosons γ, Z^0 and $W^\pm$. Of course, this model is also applicable if the number of families is more than three.

The process

$$e^-e^+ \to f\bar{f}$$

can proceed not only through one-photon annihilation (Fig. 58 a), but also through one-Z^0 annihilation (Fig. 58 b) where f can be any fundamental fermion in the three families, including the yet unobserved top quark t.

The couplings of the fermion f, including the electron e, to the photon and the Z^0 are explicitely shown in Fig. 58. While the coupling to γ involves as usual only the vector current, that to the Z^0 is a weak coupling and has both a vector and an axial vector part. The interference between the axial vector coupling to Z^0 and the vector coupling to both γ and Z^0 is responsible for the forward-backward asymmetry. The values of the vector coupling v_f and the axial vector coupling a_f as defined in Fig. 58 for various fermions are given in Table 6 in terms of the Weinberg angle θ_W[89]. Since the Weinberg angle is fairly accurately known from other experiments,

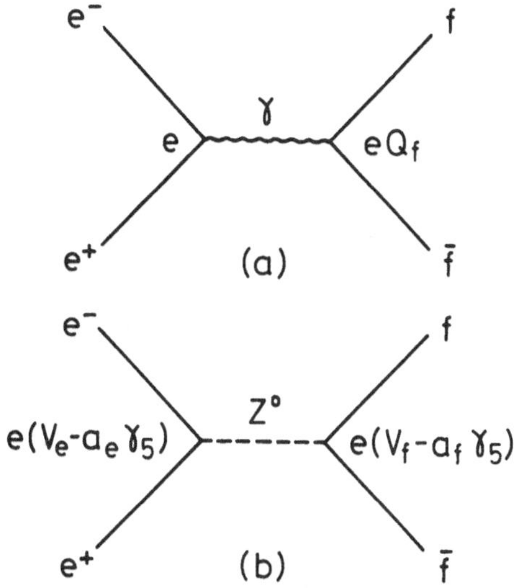

Fig. 58. The process $e^-e^+ \to f\bar{f}$ in the Standard Model.

the standard model gives unambiguous predictions for this forward-backward asymmetry. Thus a measurement of this asymmetry gives a direct test of the standard model.

Table 6. Values of v_f and a_f in the standard model

f	e^-, μ^-, τ^-	ν_e, ν_μ, ν_τ	d, s, b	u, c, t
Q_f	-1	0	$-\frac{1}{3}$	$\frac{2}{3}$
v_f	$-1+4\sin^2\theta_W$	1	$-1+\frac{4}{3}\sin^2\theta_W$	$1-\frac{8}{3}\sin^2\theta_W$
a_f	-1	1	- 1	1

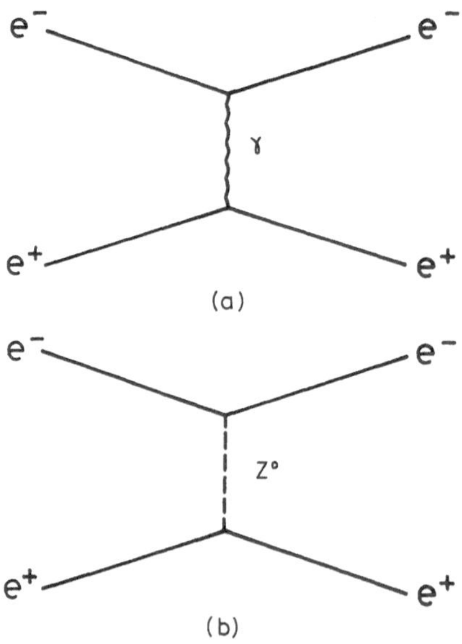

Fig. 59. Further second-order diagrams for Bhabha scattering $e^+e^- \to e^+e^-$ in the standard model. These are in addition to those of Fig. 58.

Which fermion is most suitable for the measurement of this asymmetry? The neutrinos are clearly out, since they have no electromagnetic coupling. The quarks are also less suitable because, in order to identify the flavor of the underlying quark from the observed jets, there is a large loss in statistics. Thus there are only three possible choices : e, μ, or τ. Of these three, e cannot be easily used because the Bhabha process[92] $e^+e^- \to e^+e^-$ also proceeds via Møller scattering[93], i.e. the exchange of a photon or a Z^0 in the cross channel as shown in Fig. 59, which gives a huge asymmetry. We are therefore left with just two candidates, μ and τ.

There is, however, an additional complication. The lowest-order radiative correction to the diagram of Fig. 58 (a) contains a part due to two-photon annihilation. The two diagrams of this process are shown in Fig. 60. The interference of the diagrams of Fig. 58 (a) and of Fig. 60 gives also a forward-backward asymmetry. Fortunately, this purely QED asymmetry is completely and reliably calculable and hence can be subtracted. Furthermore, this QED asymmetry is about 1.5%, significantly smaller than the expected asymmetry at PETRA energies from the standard model.

On the basis of the two diagrams of Fig. 58, the differential cross section for $e^-e^+ \to f\bar{f}$ is, in the standard model with the quark mass neglected[94]:

$$\frac{d\sigma(e^-e^+ \to f\bar{f})}{d\cos\theta}$$

$$= \frac{\pi\alpha^2}{2s}\left\{Q_f^2(1+\cos^2\theta) - \frac{2Q_f\, gs\,(s/M_z^2 - 1)\,[v_e v_f(1+\cos^2\theta)+2a_e a_f\cos\theta]}{(s/M_z^2 - 1)^2 + \Gamma_z^2/M_z^2}\right.$$

$$\left. + \frac{s^2g^2\,[(v_e^2 + a_e^2)(v_f^2 + a_f^2)(1+\cos^2\theta) + 8v_e a_e v_f a_f\cos\theta]}{(s/M_z^2 - 1)^2 + \Gamma_z^2/M_z^2}\right\}, \qquad (107)$$

where M_z and Γ_z are the mass and width of Z^0, and

$$g = \frac{\sqrt{2}\,G_F}{4e^2} = \frac{\sqrt{2}\,G_F}{16\pi\alpha} \qquad (108)$$

in terms of the Fermi weak coupling constant G_F. Integration over θ gives in particular :

$$R_f = Q_f^2 - \frac{2Q_f\, gs(s/M_z^2 - 1)\, v_e v_f - s^2g^2(v_e^2 + a_e^2)(v_f^2 + a_f^2)}{(s/M_z^2 - 1)^2 + \Gamma_z^2/M_z^2}. \qquad (109)$$

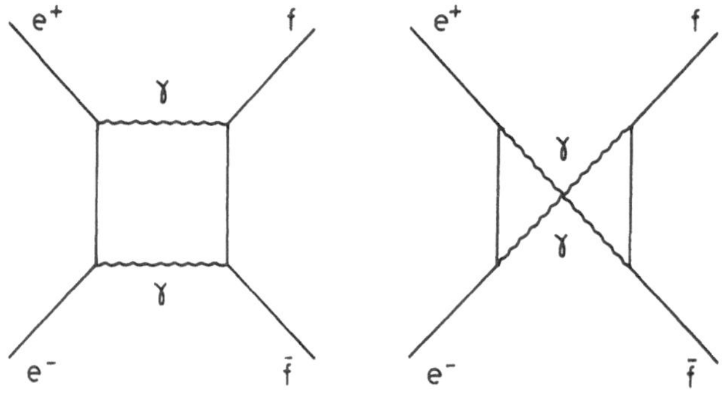

Fig. 60. Two-photon annihilation diagrams for $e^+e^- \to f\bar{f}$. The interference of the contributions from these diagrams and from those of Fig. 58 gives an asymmetry of known amount.

In writing down these formulas, the diagrams of Fig. 59 have not been included. Therefore, they are not valid for Bhabha scattering. As they stand, (107) and (109) are valid for μ and τ. They are also valid for the quarks provided that the right hand sides are multiplied by the color factor 3.

The asymmetry $A(\theta)$, as a function of θ, is defined by

$$A_f(\theta) = \frac{\left.\frac{d\sigma(e^-e^+ \to f\bar{f})}{d\cos\theta}\right|_{\theta} - \left.\frac{d\sigma(e^-e^+ \to f\bar{f})}{d\cos\theta}\right|_{\pi-\theta}}{\left.\frac{d\sigma(e^-e^+ \to f\bar{f})}{d\cos\theta}\right|_{\theta} + \left.\frac{d\sigma(e^-e^+ \to f\bar{f})}{d\cos\theta}\right|_{\pi-\theta}} . \qquad (110)$$

It therefore follows from (107) that the angular dependence of $A(\theta)$ is very simply

$$A_f(\theta) = A_f(0)\,\frac{2\cos\theta}{1+\cos^2\theta} , \qquad (111)$$

and $A_f(0)$, the asymmetry at $\theta = 0$, is explicity

$$A_f(0) = \frac{-2Q_f \, g \, s \, (s/M_z^2 - 1) \, a_e a_f + 4s^2 g^2 v_e a_e v_f a_f}{Q_f^2 [(s/M_z^2 - 1)^2 + \Gamma_z^2/M_z^2] - 2Q_f gs(s/M_z^2 - 1)v_e v_f + s^2 g^2 (v_e^2 + a_e^2)(v_f^2 + a_f^2)} \, .$$

$$\text{(112)}$$

If the acceptance is 4π, then by (111) the average asymmetry is

$$\langle A_f \rangle = \frac{\sigma(\theta < \pi/2) - \sigma(\theta > \pi/2)}{\sigma(\theta < \pi/2) + \sigma(\theta > \pi/2)} = \frac{3}{4} A_f(0) \, . \tag{113}$$

Since the acceptance is actually less than 4π, $\langle A_f \rangle$ is somewhat smaller in magnitude. The formulas (111), (112), and (113) are valid for all fundamental fermions, both leptons and quarks, with the exception of the electron.

Using the measured Weinberg angle of ref. 95,

$$\sin^2 \theta_w = 0.228 \tag{114}$$

this average asymmetry $\langle A_f \rangle$ of (113) is plotted in Fig. 61, in the PETRA energy range for the various fermions. Note that, for these fermions, the total cross section is more sensitive to v than to a, while the forward-backward asymmetry is more sensitive to a than to v.

To get a better physical feeling for this asymmetry, we get a simple approximate formula of high accuracy from (112) for PETRA energies. First, since the Fermi coupling constant is

$$G_F = 1.1663 \times 10^{-5} \text{ GeV}^{-2},$$

it follows from (108) that

$$g = 4.4967 \times 10^{-5} \text{ GeV}^{-2} \, .$$

Thus, for example, at the PETRA energy of $\sqrt{s} = 36.6$ GeV, gs ~ 0.06 is quite small. In the standard model, the Z^0 mass is given by

$$M_z = (2\sqrt{g} \, \sin 2\theta_w)^{-1} \, , \tag{115}$$

and hence the value (114) of $\sin \theta_w$ gives (radiative corrections increase M_z by a few percent)

$$M_z = 88.9 \text{ GeV.} \tag{116}$$

In the standard model, Z^0 has a width Γ_z of about 3 GeV. Thus the term Γ_z^2/M_z^2 in (112) can be omitted. Finally, because of (114), v_e from Table 6,

$$v_e = -0.088 \tag{117}$$

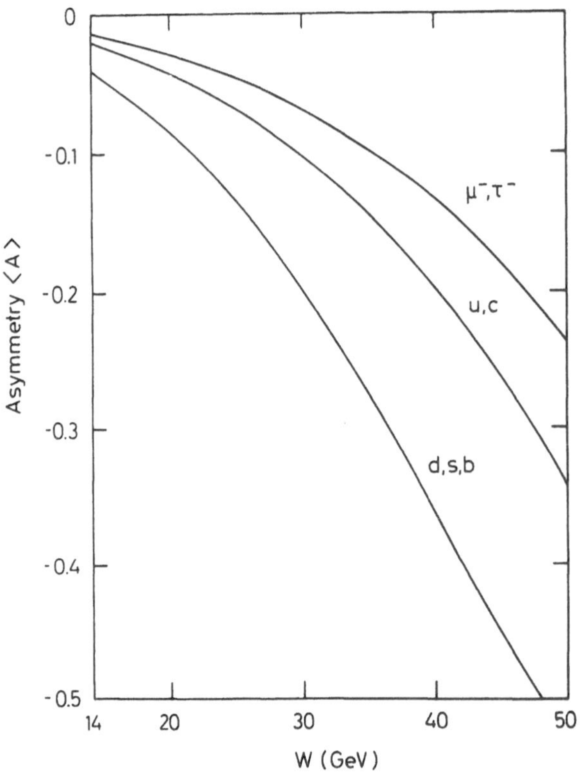

Fig. 61. Average asymmetry on the basis of the standard model.

is quite small. Neglecting $2gsv_e \sim -0.01$, (112) simplifies considerably to

$$A_f(0) \sim \frac{2\ g\ s\ a_e a_f}{Q_f(1 - s/M_z^2)} \ . \tag{118}$$

The appearance of Q_f in the denominator means that the asymmetry is larger for the quarks than for the charged leptons, while the factor s favors higher energies. In particular, for μ and τ, (118) is simply

$$A_\mu(0) = A_\tau(0) \sim - \frac{9 \times 10^{-5}\ s}{1 - s/7900} \ , \tag{119}$$

with s in GeV^2. From equation (113), the average asymmetry is

$$\frac{3}{4} A_\mu(0) = \frac{3}{4} A_\tau(0).$$

For completeness, we give here the differential cross section for Bhabha scattering taking into account all four second-order diagrams of Fig. 58 and Fig. 59[96]:

$$\frac{d\sigma(e^-e^+ \to e^-e^+)}{d\cos\theta} = \frac{\pi\alpha^2}{4s}[4B_1 + B_2\ (1-\cos\theta)^2 + B_3(1+\cos\theta)^2], \tag{120}$$

where

$$B_1 = (\tfrac{s}{t})^2\ |1 + (v_e^2 - a_e^2)\ R_t|^2 \ , \tag{121}$$

$$B_2 = |1 + (v_e^2 - a_e^2)\ R_s|^2 \ , \tag{122}$$

$$B_3 = \tfrac{1}{2}[|1 + \tfrac{s}{t} + (v_e+a_e)^2(R_s + \tfrac{s}{t}R_t)|^2 + |1 + \tfrac{s}{t} + (v_e-a_e)^2(R_s + \tfrac{s}{t}R_t)|^2], \tag{123}$$

$$R_s = \frac{g\ s\ M_z^2}{s - M_z^2 + i\Gamma_z M_z} \ , \tag{124}$$

and

$$R_t = \frac{g\ t\ M_z^2}{t - M_z^2 + i\Gamma_z M_z} \ , \tag{125}$$

with

$$t = -\frac{1}{2}\ s\ (1 - \cos\theta). \tag{126}$$

It is interesting to note that (107) with $Q_f = -1$ can be obtained from the above formula for Bhabha scattering by taking the limit $t \to \infty$. The reason is that, in this limit of $t \to \infty$, the contributions from the diagrams of Fig. 59, which involve the exchange of a photon or a Z^0 in the t channel, become zero. For both (107) and (120), the electron and the positron beams are assumed to be unpolarized.

Bhabha Scattering $e^-e^+ \to e^-e^+$

We begin the discussion of the various experimental results. Bhabha scattering, both at small angles and at large angles, is often used as a luminosity monitor. The selected events are required to have two oppositely charged tracks of sufficiently high momenta within acceptance and with an acollinearity angle of typically less than 10°. The tracks have to originate close to the intersection point, and cosmic ray backgrounds are further reduced by measurement of the time of flight. Contaminations from μ pairs and τ pairs are subtracted statistically.

Even at the highest energies the loss resulting from events having two tracks with the same charge assignment is small, less than 1% in the various experiments. This problem will be discussed in further detail later on. A much more serious problem is that the electrons and positrons from the Bhabha events tend to produce secondary particles due to showering in the beam pipe and other materials. For the TASSO detector for example, this material before the tracking chambers amounts to 0.13 radiation length. Therefore the restriction to events with exactly two charged tracks leads to losses which are around 13% and further are dependent on the polar angle θ. Taking into account radiative effects and shower formation, the resulting acceptance is found to be almost flat for $|\cos\theta| < 0.5$ with a value of 0.80 and falls to 0.72 at $|\cos \theta| = 0.7$. Monte Carlo simulation is used to correct for this and other effects, such as the detector efficiency.

All PETRA data on Bhabha scattering are in good agreement with QED[97-110]. As an example, the JADE[101] results are shown in Fig. 62. For the purpose of looking for weak effects, it is more useful to plot instead the ratio

$$\frac{d\sigma^{exp}/d\cos\theta}{d\sigma^{QED}/d\cos\theta} . \tag{127}$$

Such plots from CELLO[98] and TASSO[110] are shown in Figs. 63 and 64 respectively and compared with the standard model of Glashow, Weinberg and Salam. It is seen that these data on Bhabha scattering, taken by themselves, are not sufficiently accurate to differentiate between pure quantum electrodynamics and the standard model. They will, however, be used together with those from $e^-e^+ \to \mu^-\mu^+$ and $e^-e^+ \to \tau^-\tau^+$ to determine the vector and axial vector couplings.

For Bhabha scattering the ratio (127) is more sensitive to v_e

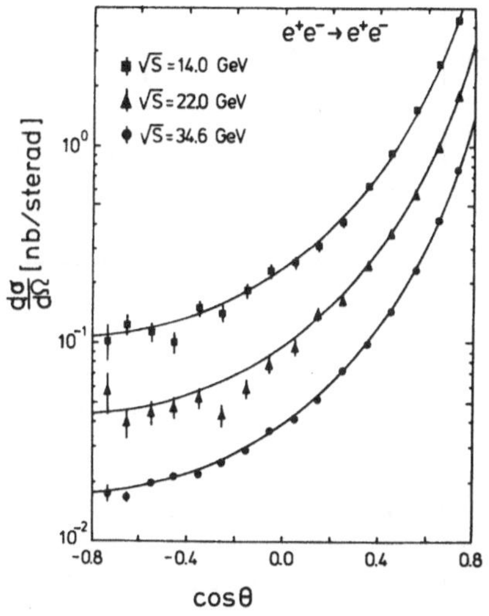

Fig. 62. JADE[101] differential cross sections for the reaction $e^+e^- \to e^+e^-$ for c.m. energies 14, 22 and 34.6 GeV. The lines show the QED expectations.

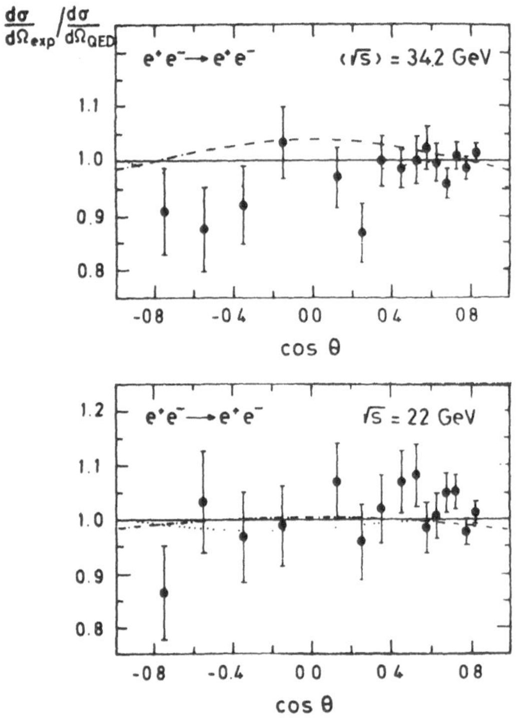

Fig. 63. The CELLO[98] differential cross section $d\sigma/d\Omega$ for Bhabha scattering at $\langle\sqrt{s}\rangle = 34.2$ GeV and $\sqrt{s} = 22$ GeV normalized to the QED cross section. The full line is the QED prediction. The dotted line represents the best fit for a^2 and v^2. The dotted-dashed line shows the prediction for the second solution for neutrino-electron scattering ($v^2 = 1.08$, $a^2 = 0.0$).

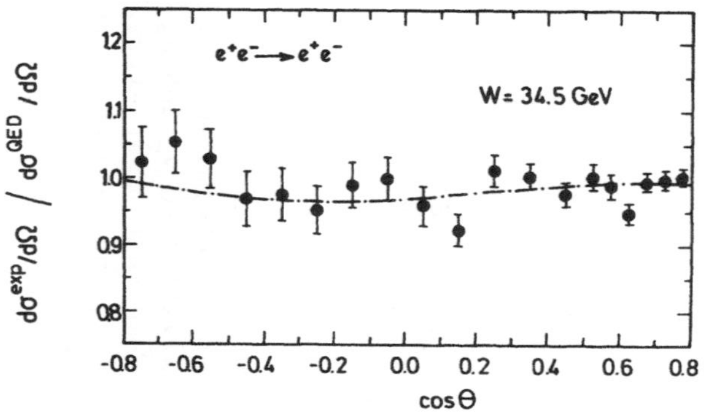

Fig. 64. The TASSO[110] differential cross section divided by the QED expectation for the reaction $e^+e^- \to e^+e^-$. The curve shows the fit to the data of the Glashow-Weinberg-Salam theory with $\sin^2\theta_W = 0.26$.

than a_e, while for pair production it is more sensitive to the axial-vector couplings than to the vector couplings. This is most clearly seen by considering the rough approximation where we keep gs only to the first order and neglect both s and Γ_z^2 as compared with M_z^2. In this approximation, it follws from eqs. (107) and (120) that the ratio can be expressed in the form

$$(128)$$

$$\frac{d\sigma(e^-e^+ \to f\bar{f})/d\cos\theta}{d\sigma^{QED}(e^-e^+ \to f\bar{f})/d\cos\theta} = 1 + 2Q_f^{-1} gs[v_e v_f + a_e a_f \frac{2\cos\theta}{1 + \cos^2\theta}]$$

and

$$\frac{d\sigma(e^-e^+ \to e^-e^+)/d\cos\theta}{d\sigma^{QED}(e^-e^+ \to e^-e^+)/d\cos\theta} = 1 + 2gs[v_e^2 S_v(\cos\theta) + a_e^2 S_a(\cos\theta)] \quad , (129)$$

where

$$S_v(\cos\theta) = \frac{3\sin^2\theta}{3 + \cos^2\theta} \quad , \quad (130)$$

and

$$S_a(\cos\theta) = \frac{-\sin^2\theta (5 - 8\cos\theta - \cos^2\theta)}{(3 + \cos^2\theta)^2} \quad . \quad (131)$$

Accordingly, in this approximation, for pair production the vector couplings of Z^0 changes only the normalization while the axial-vector couplings give the asymmetry in angular distribution. For Bhabha scattering, the effects are more complicated. The functions S_v and S_a, which give the sensitivities to v_e and a_e respectively, are plotted in Fig. 65. It is seen that S_v is everywhere larger than S_a in absolute values, but, except near the forward direction, the magnitudes are not very different.

Total Cross Section of μ Pair Creation

Figure 66 shows the results obtained for the measurements of R_μ, as a function of s, by the four Petra groups[111] : CELLO, JADE, MARK J and TASSO. Again, no deviation from QED is observed.

From the standard model, the total cross section for μ pair production is given by (109) with f = μ :

$$R_\mu = 1 + gs \frac{2(s/M_z^2 - 1) v_e v_\mu + gs (v_e^2 + a_e^2)(v_\mu^2 + a_\mu^2)}{(s/M_z^2 - 1)^2 + \Gamma_z^2/M_z^2} . \quad (132)$$

Since gs is small at PETRA energies, deviation from 1 gives a measure of the quantity $v_e v_\mu$. Therefore the data shown in Fig. 66 imply that $v_e v_\mu$ is small.

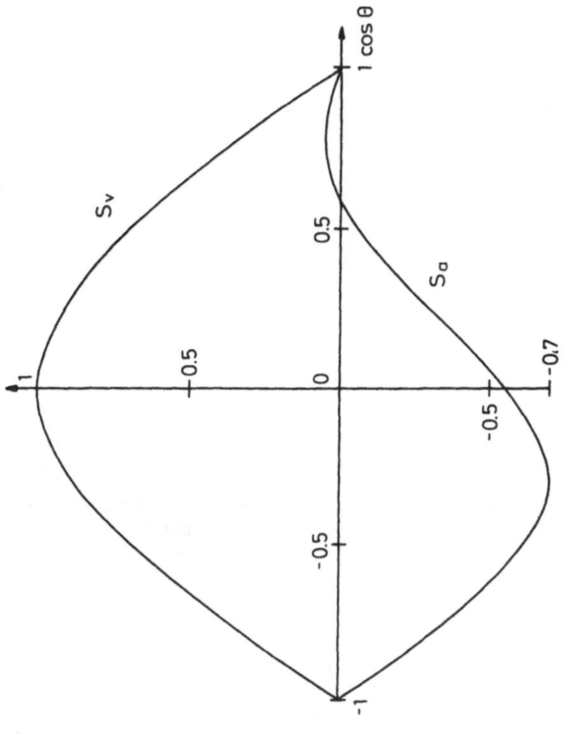

Fig. 65. Sensitivity of the Bhabha scattering cross section divided by the QED prediction to the vector and axial-vector couplings of Z^0.

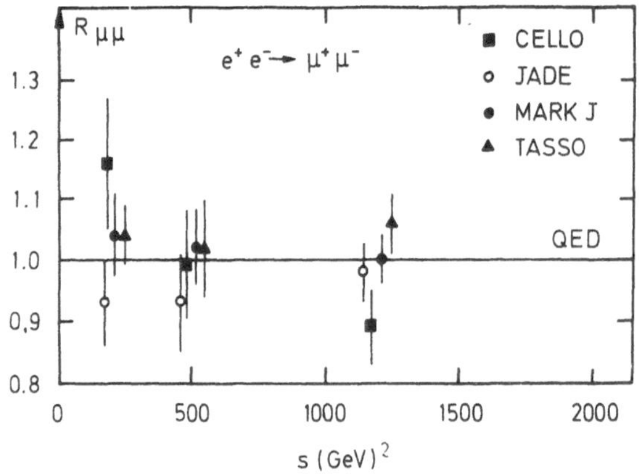

Fig. 66. Measured R_μ for $e^-e^+ \rightarrow \mu^-\mu^+$ from CELLO, JADE, MARK J, and TASSO[111].

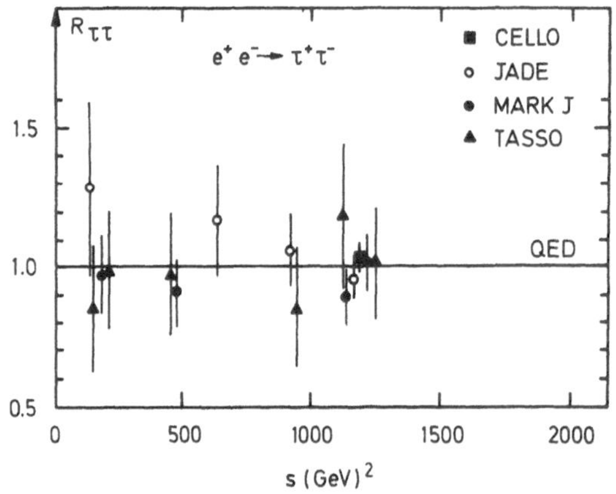

Fig. 67. Measured R_τ for $e^-e^+ \rightarrow \tau^-\tau^+$ from CELLO, JADE, MARK J, and TASSO[111].

Total Cross Section for $e^-e^+ \to \tau^-\tau^+$ and Branching Ratios

Figure 67 shows the results obtained at PETRA[111] for the measurements of R_τ as a function of s. Again, no deviation from QED is observed. The standard model again gives (132) for R_τ, provided that the couplings v_μ and a_μ are replaced by v_τ and a_τ respectively.

Among the detectors at PETRA, CELLO has the largest acceptance for $e^-e^+ \to \tau^-\tau^+$[88,112]. We therefore discuss in more detail the results from CELLO[113,114].

Event candidates are selected on the basis of the relatively low multiplicity and the distinct back-to-back topology of $\tau^-\tau^+$ events. More precisely, the cuts are as follows :

- For all events : at least two tracks satisfying
  (a) $p_T > 350$ MeV/c with respect to the beam axis.
  (b) $|\cos\theta| < 0.86$.

- For two-prong events
  (c) $|\vec{p}_1| + |\vec{p}_2| > 5$ GeV/c.
  (d) Total momentum transverse to the beam > 300 MeV/c.
  (e) Acollinearity in space > 20 mrad and < 0.7 rad.
  (f) Angle between the prongs in the projection containing the beam direction > 3.5 mrad.
  (g) Angle between the prongs in the projection transverse to the beam > 10 mrad.

- For ≧ four-prong events
  (h) $\Sigma|\vec{p}| > 4$ GeV/c.
  (i) At least one track in two back-to-back sectors 80° wide in the projection transverse to the beam direction.
  (j) Number of tracks ≦ 8.

The acolinearity cuts (e), (f), (g) suppress contributions from Bhabhas, μ pairs and cosmic muons. Background due to 2γ processes is removed by the cuts (c), (d) and (e). Multihadronic events are significantly rejected by the multiplicity cut (j), residual beam-gas events by the back-to-back requirement (i) and energy cut (h). The polar angle cut ensures a high efficiency for the charged trigger and also that the tracks enter the sensitive region of the liquid argon barrel calorimeter.

In addition, most Bhabha events are suppressed by the requirement of an energy deposition < 15 GeV in at least one of the two back-to-back calorimeter modules corresponding to the direction of the final particles.

After these candidates are double scanned in order to remove the remaining background events, a total of 526 $\tau^-\tau^+$ events survive, corresponding to the CELLO integrated luminosity of 11.2 pb^{-1} at a center-of-mass energy of 34 GeV. The following decay channels have been studied :

(1) $\tau \rightarrow \rho\nu$, characterized by one-prong decay with one or two photons in the same hemisphere,
(2) $\tau \rightarrow \mu\nu\bar{\nu}$, where a particle momentum larger than 2 GeV/c and an associated hit in the muon chambers are required,
(3) $\tau \rightarrow e\nu\bar{\nu}$, where the e, with momentum larger than 1 GeV/c, is identified by the characteristic shower pattern in the calorimeter; and
(4) $\tau \rightarrow \pi\nu$, where the shower pattern is required to be consistent with a minimum ionizing particle.

For these four decay channels, the efficiencies and background contributions are tabulated in Table 7.

The resulting branching ratios are given in Table 8 and compared with the world average[95]. From the momentum spectra of the observed decay products, CELLO has initiated the study of τ polarization asymmetry. No such asymmetry has yet been observed, the result being (1 ± 22)%.

Forward-Backward Asymmetry in $e^-e^+ \rightarrow \mu^-\mu^+$

We now come to the most exciting topic of the observation of weak and electromagnetic interference in the process $e^-e^+ \rightarrow \mu^-\mu^+$. This process has the great advantage that the systematic error in the asymmetry is certainly less than 1%, because the backgrounds (due to misidentified Bhabha events, cosmic rays, two-photon process, and τ pairs) can be kept small and do not produce artificial asymmetry.

As already mentioned earlier, the loss resulting from events having two tracks with the same charge assignment is small. In Fig. 68, the MARK J plot[115] is shown on the measured normalized reciprocal momentum p_{beam}/p_μ for the forward-going muon versus that of the backward-going muon. The sign of this reciprocal momentum is defined as the charge determined from the track curvature. The events are concentrated in the vicinities of the two points (1, -1) and (-1, 1), corresponding to the two cases of μ^+ forward, μ^- backwards and vice versa. It is seen from this Figure that the percentage of events misidentified as $\mu^+\mu^+$ and $\mu^-\mu^-$ is indeed small.

Higher order α^3 QED corrections[116], due to the interference of contributions from the diagrams of Fig. 60 with those from Fig. 58, are applied to the experimental data. Within the acceptance of most detectors, $|\cos\theta| < 0.80$ or slightly more, this correction corresponds to a positive asymmetry of $\sim$ 1.5%. Checks of this procedure have been performed by investigating distributions which are genera-

Table 7. Summary of the four decay samples and their respective efficiencies and background contributions, from CELLO[114].

	ρ	π	e	μ
No. of events	101	34	60	47
Efficiency	0.45±.04	0.48±.04	0.73±.04	0.70±.06
Contamination from :				
μ	0	1.4±1.4	0	-
e	0	0.5±0.5	-	0
π	0	-	0.7±0.7	2.0±1.0
ρ	-	5.4±1.2	0	0
$\pi\pi^0\pi^0$	15.4±8.4	0	0	0
K,K^*	2.3±1.1	3.6±1.4	0	0

Table 8. τ branching ratio in %.

	CELLO[114]	World Average[95]
$BR(\tau\to\rho\nu)$	22.8±2.5±2.1	21.6±3.6
$BR(\tau\to\mu\nu\bar{\nu})$	17.6±2.6±2.1	18.5±1.2
$BR(\tau\to e\nu\bar{\nu})$	18.3±2.4±1.9	16.2±1.0
$BR(\tau\to\pi\nu)$	9.9±1.7±1.3	10.7±1.6

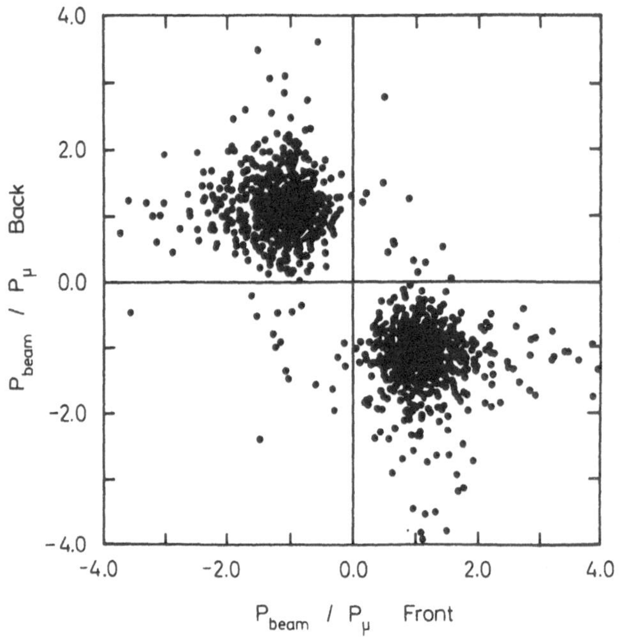

Fig. 68. The MARK J[115] normalized reciprocal momentum p_{beam}/p_μ for the forward and the backward muons in the reaction $e^-e^+ \to \mu^-\mu^+$.
The sign of this reciprocal momentum is defined as the charge determined from the track curvature.

Table 9. PETRA data on μ-pair asymmetry[120]

Experiment	$\sqrt{s}$ (GeV)	$\langle A_\mu \rangle$ (%)	$\langle A_\mu \rangle_{GWS}$ (%)
CELLO	34.2	-6.4±6.4	- 9.2
JADE	34.4	-11.0±1.8	- 9.3
MARK J	34.6	-11.7±1.7	- 9.5
PLUTO	34.7	-12.4±3.1	- 9.5
TASSO	34.5	-9.1±2.3	- 9.5
combined	34.5	-10.8±1.1	- 9.4

ted by radiative effects, for example, the acolinearity distribution between the two muons. No correction has been applied in the data for higher order weak-electromagnetic effects. The angular distributions for $e^-e^+ \to \mu^-\mu^+$ as presented by the various PETRA Collaborations are given in Fig. 69[117]. It is seen from these high-energy data that there is a distinct deviation from the symmetrical angular distribution predicted by QED.

Table 9 summarizes the results on the μ-pair asymmetry at PETRA with the data from CELLO[118], JADE[86], MARK J[119], PLUTO[118] and TASSO[110] but updated to the time of the 1983 Cornell Conference[120]. Each of the three high-statistics experiments JADE, TASSO and MARK J shows an effect of four or more standard deviations, in agreement with the standard model. If the five results are averaged according to their statistical significance, then the result at W = 34.5 GeV is[120]

$$\langle A_\mu \rangle_{av} = (-10.8 \pm 1.1)\%,$$

to be compared with the predictions -9.4% of the standard model. Note that the experimental value has not been corrected for radiative effects from diagrams containing Z^0.

Thus the weak-electromagnetic interference has been observed

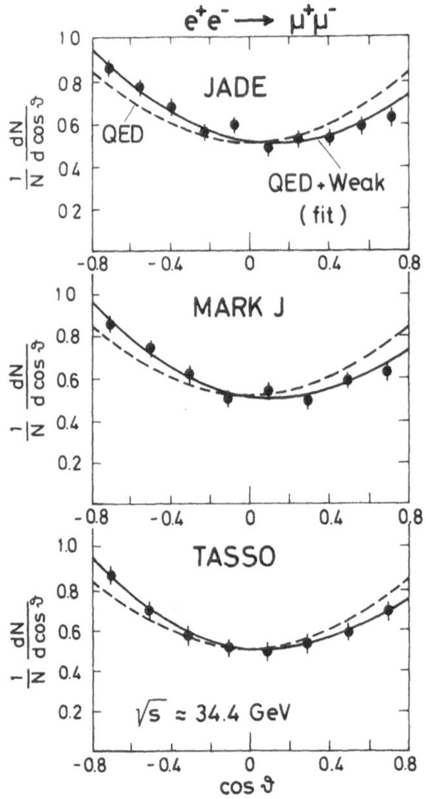

Fig. 69. Angular distribution for μ pair production $e^-e^+ \to \mu^-\mu^+$ measured by the PETRA experiments[117]. The data have been corrected for QED radiative effects. The dashed lines are the lowest-order QED prediction of $1 + \cos^2\theta$, while the solid lines are the fits to the data using eq. (107) of the standard model.

unambiguously at PETRA. For the sake of visual display, the combined fit to the angular distributions is shown in Fig. 70[120].

Forward-Backward Asymmetry in $e^-e^+ \to \tau^-\tau^+$

In the standard model, the Z^0 couplings to μ and τ[88] are identical, as shown in Table 6. Therefore, the asymmetries are theoretically identical for the processes $e^-e^+ \to \mu^-\mu^+$ and $e^-e^+ \to \tau^-\tau^+$.

The experimental problems of measuring these two asymmetries, however, are quite different. The selection and identification of events for $e^-e^+ \to \tau^-\tau^+$ have already been discussed. Compared with the μ pairs, the backgrounds for τ pairs are larger, arising from Bhabha events with bremsstrahlung in the beam pipe, two-photon processes, multi-hadron events, etc. The contamination due to Bhabha scattering requires special attention, since it produces a positive asymmetry, thereby counteracting the expected negative asymmetry due to weak electromagnetic interference. The effects of these backgrounds depend on the ability to identify correctly the final states. The resulting systematic uncertainty is about 2%. Although this is significantly larger than the <1% for μ pairs, it is not a serious problem at PETRA because it is still less than the statistical uncertainty.

The experimental data are treated in a way similar to that for $e^-e^+ \to \mu^-\mu^+$, described in the preceding section. There are only two journal publications from PETRA on this asymmetry; they are :

TASSO[87] $(-4 \pm 6)\%$

and

CELLO[109] $(-10.3 \pm 5.2)\%$.

The combined fit to the updated angular distributions from the PETRA groups is shown in Fig. 71 taken from the 1983 Cornell Conference. Taken individually, none of these results can be said to have established a significant asymmetry. However, every one is in the expected direction in the sense of having a negative value, and a 4 standard-deviation effect is obtained if the values presented at the Cornell Conference[120] are averaged statistically. The combined experimental asymmetry is

$$\langle A_\tau \rangle_{av} = (-7.6 \pm 1.9)\%$$

to be compared with the theoretical prediction of -9.4%.

Overall Fit of the Leptonic Vector and Axial-Vector Weak Couplings

As already discussed, the best evidence at PETRA for weak and

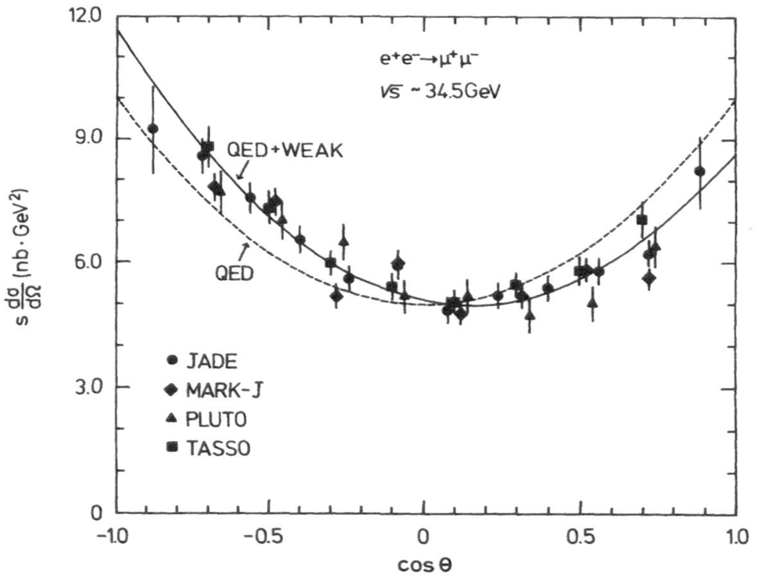

Fig. 70. Compilation[120] of PETRA high-energy data for the angular distribution of $e^+e^- \to \mu^+\mu^-$ at $\sqrt{s} \sim 34.5$ GeV. The data are corrected for effects α^3. The full curve shows a fit to the data allowing for an asymmetry; the dashed curve is the symmetric QED prediction.

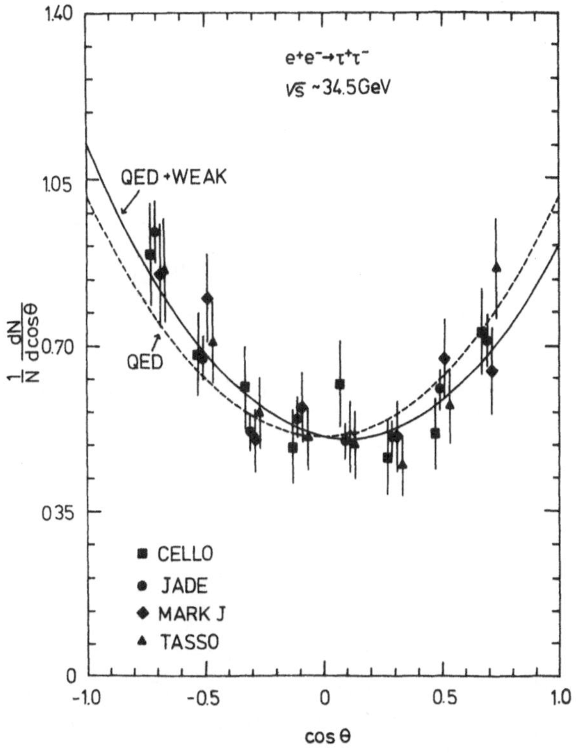

Fig. 71. Compilation[120] of PETRA high-energy data for the angular distribution of $e^+e^- \to \tau^+\tau^-$ at $\sqrt{s} \sim 34.5$ GeV. The data are corrected for effects α^3. The full curve shows a fit to the data allowing for an asymmetry; the dashed curve is the symmetric QED prediction.

Table 10. Overall fit of $e^+e^- \to$ lepton pairs.

Experiment	Data Used	a^2	v^2	$\sin^2\theta_W$
CELLO[98]	ee, μμ, ττ	1.22±.47	−.12±.33	$.21^{+.14}_{-.19}$
JADE[111]	μμ	1.17±.24	.20±.32	
MARK J[111]	ee, μμ, ττ	1.12±.24	−.08±.20	.26±.09
TASSO[110]	ee, μμ	0.88±.22	−.14±.21	.26±.07
Average		1.06±.13	−.06±.12	

electromagnetic interference comes from the forward-backward asymmetry in μ-pair production $e^-e^+ \to \mu^-\mu^+$. All observations are consistent with the prediction of the standard model that

$$v_e = v_\mu = v_\tau , \tag{133}$$

and

$$a_e = a_\mu = a_\tau . \tag{134}$$

On the assumption of this universality, overall fits have been carried out by the various PETRA Collaborations to the leptonic data on $e^-e^+ \to e^-e^+$, $e^-e^+ \to \mu^-\mu^+$ and $e^-e^+ \to \tau^-\tau^+$. Each Collaboration uses its own data; the choice of data and the results of the overall fits are tabulated in Table 10, where v and a mean the quantities (133) and (134) respectively. The CELLO data are from ref. 98, the JADE and MARK J data from refs. 86 and 119 respectively but updated to the time of the Paris Conference[111] and the TASSO data from ref. 110. Table 10 gives results from two separate overall fits, one for the values of v^2 and a^2, while the other is a direct fit for $\sin^2\theta_W$ on the basis of the standard model.

The errors on the vector coupling and the axial-vector coupling constants are correlated. As an example of this correlation, the TASSO data is shown in Fig. 72 (a) with 95% confidence-level contours using the definition of $g_V = v/2$ and $g_A = a/2$. In that figure, the shaded areas show the two solutions of νe experiments. It is an impressive feat of the PETRA experiments to obtain a determination

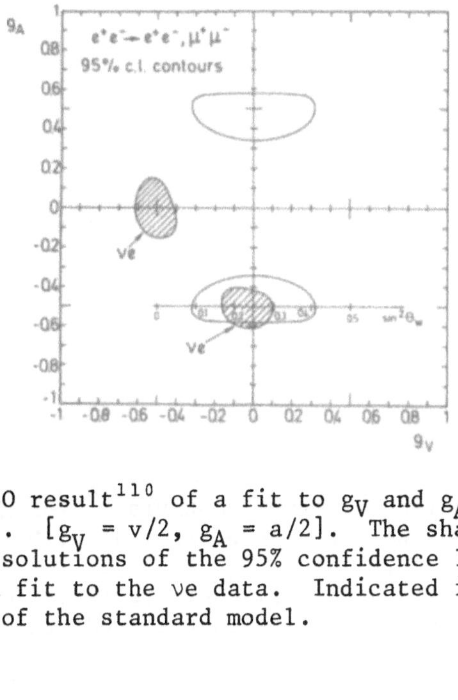

Fig. 72. (a) TASSO result[110] of a fit to g_V and g_A with 95% c.l. contours. [$g_V = v/2$, $g_A = a/2$]. The shaded areas show the two solutions of the 95% confidence level contours from a common fit to the νe data. Indicated is also the prediction of the standard model.

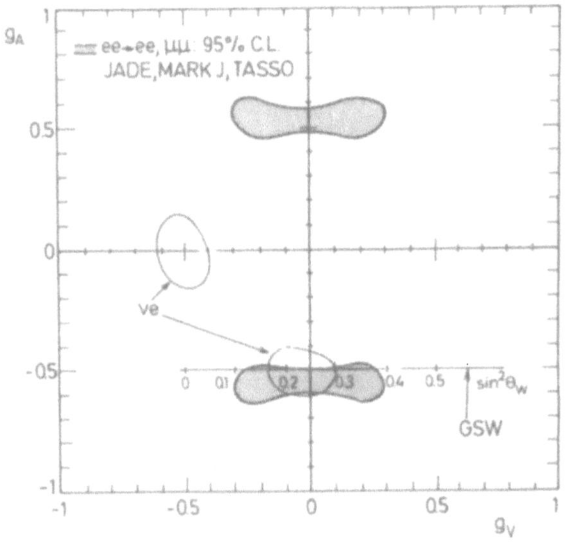

Fig. 72. (b) 95% confidence level contours[120] for g_V and g_A from $e^+e^- \to \mu^+\mu^-$ and $e^+e^- \to e^+e^-$ from JADE, MARK J, and TASSO (shaded area) [$g_V = v/2$, $g_A = a/2$]. The open areas are the 95% confidence level contours from a common fit to all νe data. Indicated is also the prediction of the standard model.

of the weak neutral-current coupling constants at these high energies with an accuracy comparable to that from the neutrino-electron scattering experiments at much lower energies. The combined data from the PETRA groups are shown in Fig. 72 (b)[120].

Determination of the Weinberg Angle θ_W from the Total Hadronic Cross Section

So far in this chapter, only the experimental data on the production of lepton pairs have been discussed. In this and the next section, attention is focused on weak effects in the hadronic events at PETRA.

In the standard model, when eq. (109) is summed over the various quarks, the result for the total hadronic cross section is

$$R = 3[\sum_q Q_q^2 - \frac{2gs\,(s/M_Z^2 - 1)v_e \Sigma_q Q_q v_q - s^2 g^2 (v_e^2 + a_e^2)\Sigma_q (v_q^2 + a_q^2)}{(s/M_Z^2 - 1)^2 + \Gamma_Z^2/M_Z^2}]. \quad (135)$$

For the PETRA energy range, the quarks that should be taken into account are : d, u, s, c, and b. In (135) radiative corrections due to gluons have not been included. For energies away from the thresholds of production for new quark flavors and much below M_Z, the inclusion of these radiative corrections leads approximately to an additonal factor of $1 + \alpha_s/\pi$.

The precise measurement of R at PETRA has already been presented in the first chapter, Fig. 6 for TASSO[121], and 7 for JADE[122]. One of the main motivations for these precise measurements is that, as seen from (135) with this factor of $1 + \alpha_s/\pi$, R as a function of the center-of-mass energy gives unambiguous determination[105,121-123], independent of fragmentation, of the quark-gluon coupling constant α_s and the Weinberg angle θ_W for electroweak interaction. MARK J[105] obtains

$$\sin^2\theta_W = 0.27 \begin{array}{l} + 0.34 \\ - 0.08 \end{array}$$

independent of the value of α_s. Both TASSO[121] and JADE[122] Collaborations have performed simultaneous fits to the data on the total cross section by varying the two fundamental constants α_s and θ_W. The TASSO results are

$$\alpha_s = 0.18 \pm 0.03 \pm 0.14 \,, \quad \text{for } s = 1000 \text{ GeV}^2,$$

and

$$\sin^2\theta_W = 0.40 \pm 0.16 \pm 0.02.$$

The resulting R with these parameters are shown as the solid line in Fig. 6. The more accurate JADE results are

$$\alpha_s = 0.20 \pm 0.08 \quad \text{for } s = 900 \text{ GeV}^2,$$

and

$$\sin^2\theta_W = 0.23 \pm 0.05,$$

where the errors include both statistical and systematic contributions, and give the one-standard deviation limits when the other parameters are left free. The resulting R with these parameters are shown as the solid line in Fig. 7.

Forward-Backward Asymmetry in $e^-e^+ \to c\,\bar{c}$

From the inclusive spectrum for $D^{*\pm}$, we identify events from $e^-e^+ \to c\bar{c}$, and can, in particular, give an experimental determination of the forward-backward aysmmetry for these events.

The decays used for D^{*+} by TASSO[124] are

$$D^{*+} \to D^0\pi^+,$$

with two modes for D^0 :

$$D^0 \to K^-\pi^+ ,$$

and

$$D^0 \to K^-\pi^+\pi^0 ,$$

through either ρ^+ or K^{*-} with the neutral pion not detected. These two modes for D^0 are referred to as D^0 and S^0. The corresponding charge-conjuguate modes are used for D^{*-}.

For the determination of the angular distribution of the $D^{*\pm}$, the D^0 and S^0 candidates are required to satisfy :

a) $1.50 \text{ GeV} < M(K^-\pi^+) < 1.984 \text{ GeV}$;

b) $M(K^-\pi^+\pi^+) - M(K^-\pi^+) < 0.15 \text{ GeV}$;

and

c) $E(K^-\pi^+\pi^+) / E_{beam} > 0.5$.

In order to enhance the sensitivity to the weak contribution, only events with W > 34 GeV are accepted, the average W being 35 GeV. The background under the D^0 and S^0 peaks is reduced by requiring the

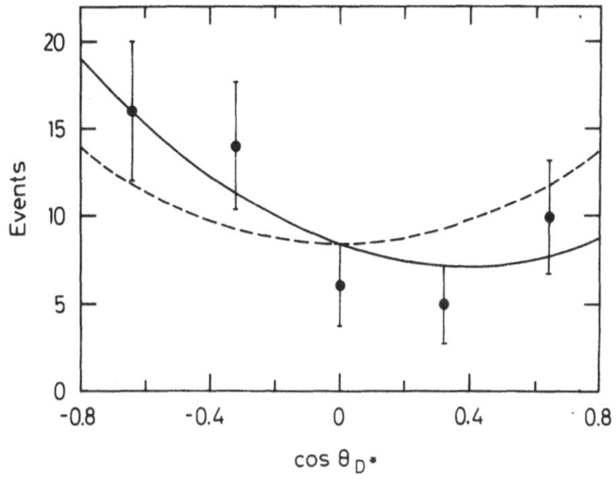

Fig. 73. The TASSO[124] $D^{*\pm}$ production angular distribution; θ is the angle between the e^- beam and the $K^-\pi^+\pi^+(K^+\pi^-\pi^-)$ system. The full curve indicates the fit $d\sigma/d\cos\theta \sim 1 + a\cos\theta + \cos^2\theta$. The dashed cuve is proportional to $1 + \cos^2\theta$.

momenta of each of the particles forming the $(K^-\pi^+)$ sytem to have $p > 1.4$ GeV/c. In total there are 51 D^* candidates with an estimated background of 5. The angle θ is taken to be the angle between the incoming e^- and the $K^-\pi^+\pi^+$ system. Figure 73 shows the D^{*+} angular distribution. The acceptance is uniform over the range $|\cos\theta| < 0.8$. The angular distribution is consistent with the form

$$d\sigma/d\cos\theta \sim 1 + a\cos\theta + \cos^2\theta.$$

Fitting this form to the data for $|\cos\theta| < 0.8$ and extrapolating to $\cos\theta = \pm 1$ yields for the asymmetry $A_c = 3/8\ a = -0.28 \pm 0.13$. No correction is applied for the background. Assuming the latter to be forward-backward symmetric would change A_c to -0.31 after correction. The fraction of $D^{*\pm}$ mesons coming from bottom hadrons is estimated to be ∿4% and gives a negligible contribution to the measured asymmetry. This measured asymmetry is to be compared with the standard-model value of - 0.14, which is 50% larger than A_μ because of the smaller quark charge [see Eq. (118)].

This result is the first indication for a weak-current contribution to charm production $e^-e^+ \to c\bar{c}$. From the value of A_c = -0.28 ± 0.13, it follows that $a_e a_c$ = -1.96 ± 0.92. This is consistent with the standard-model value of -1. If, furthermore, the experimental value of a_e^2 = 1.06 ± 0.13 from Table 10 is used, then $|a_c|$ = 1.90 ± 0.90, to be compared with the standard-model value of 1.

QED Tests and Search for a Second Neutral Intermediate Vector Boson

We briefly discuss in this section two unrelated topics : QED tests and comparison of the experimental data with models with two neutral intermediate vector bosons.

QED predictions are based on the validity of Maxwell's equations and on the assumption that leptons are point-like objects without excited states. The procedure used at PETRA to compare the data with the QED predictions is as follows :

(a) the data are corrected for weak effects on the basis of the standard-model.
(b) the data are also corrected for radiative effects and effects due to hadron vacuum polarization.
(c) the corrected cross sections are compared to the QED predictions, and deviations are parametrized in terms of form factors[125]

$$F_\pm = 1 \mp \frac{q^2}{q^2 - \Lambda_\pm^2} . \tag{136}$$

Table 11 gives the 95% confidence level lower limits for $\Lambda_\pm$ for various processes. On the basis of the largest value in that table, QED is found to be valid to the extremely short distance of 10^{-16} cm.

Finally, we discuss the possible modification of the standard model by including more than one neutral intermediate vector boson, say Z_1 and Z_2. Since copious production of Z_1 (where the convention is that $M_{Z_1} < M_{Z_2}$) via the process

$$e^-e^+ \to \gamma Z_1 \tag{137}$$

as shown in Fig. 2 (b) has not been seen at the highest PETRA energy,

Table 11. QED parameters : 95% confidence lower limits in GeV/c

Experiment	$e^-e^+ \to e^-e^+$ Λ_+	Λ_-	$e^-e^+ \to \mu^-\mu^+$ Λ_+	Λ_-	$e^-e^+ \to \tau^-\tau^+$ Λ_+	Λ_-	$e^-e^+ \to \gamma\gamma$ Λ_+	Λ_-
CELLO	83	155	186	101	142	121	59	48
Ref.	97	97	117	117	113	113	97	97
JADE	178	200	142	126	111	93	61	57
Ref.	101	101	131	131	131	131	101	101
MARK J	128	161	192	163	100	127	58	38
Ref.	115	115	126	119	126	126	127	131
PLUTO	80	234	107	101	79	63	46	36
Ref.	106	106	129	129	129	129	130	130
TASSO	155	251	222	187	104	189	34	42
Ref.	110	110	110	110	131	131	108	108

Table 12. 95% confidence upper limit for the parameter C

CELLO[98]	0.031
JADE[101]	0.031
MARK J[109]	0.021
PLUTO[107]	0.060
TASSO[110]	0.012
Combined	0.010

the value of M_{Z_1} must be at least 42 GeV. A more refined comparison is based on the fact that the presence of two or more Z's leads to a change of the vector coupling constant[132]:

$$v^2(Z_1,Z_2) = v^2 + 16C, \tag{138}$$

where v^2 is the value for the standard model. Table 12 gives the

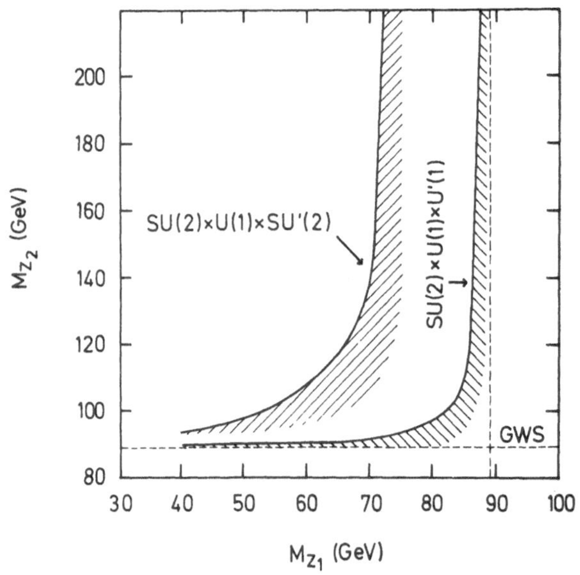

Fig. 74. TASSO[108] limits on the Z^0 masses for models with two Z^0 bosons. The allowed masses of Z_1 and Z_2 are confined to the region (partially shaded) between the solid curve and the dashed lines.

upper limits of C as obtained by the various PETRA Collaborations.

The relation between the value of C and the masses of Z_1 and Z_2 is model dependent. On the basis of the TASSO data[110], for example, the mass limits are shown in Fig. 74 for the two models SU(2)xU(1)x U'(1) and SU(2)xU(1)xSU'(2). In both of these models, Z_1 is lighter than, while Z_2 is heavier than, the mass of Z^0 in the standard model.

ACKNOWLEDGEMENTS

I wish to thank R. Gastmans for his hospitality at the Summer Institute at Cargèse. I am grateful to J. Ellis, E. Lohrmann, P. Söding, V. Sörgel, B. Wiik, G. Wolf, T.T. Wu and G. Zobernig for numerous discussions and constructive comments. This work is in part supported by the U.S. Department of Energy contract number DE-AC02-76ER00881.

REFERENCES

1. H.J. Bhabha, Proc. Roy. Soc. (London), A154 (1935) 195
2. S.L. Glashow, Nucl. Phys. 22 (1961) 579
   S. Weinberg, Phys. Rev. Lett. 19 (1967) 1264
   A. Salam, Proceedings of the Eighth Nobel Symposium, May 1968, ed. : N. Svartholm (Wiley, 1968) 367
   S.L. Glashow, J. Iliopoulos, and L. Maiani, Phys. Rev. D2 (1970) 1285
3. M.L. Perl et al., Phys. Rev. Lett. 35 (1975) 1489 and Phys. Lett. 63B (1976) 466
   G.J. Feldman et al., Phys. Rev. Lett. 38 (1977) 117
4. UA1 Collaboration, G. Arnison et al., Phys. Lett. 126B (1983) 398
   UA2 Collaboration, G. Bagnaia et al., Phys. Lett. 129B (1983) 130
5. P. Söding and G. Wolf, Ann. Rev. Nucl. Part. Sci. 31 (1981) 231
   TASSO Collaboration, M. Althoff et al., DESY 83-130 (1983) and Z. Phys. C-Particles and Fields (to be published.)
   TASSO Collaboration, M. Althoff et al., DESY 84-001 (1984) and Physics Lett. (to be published.)
6. A. Quenzer, Thesis, Orsay Report LAL (1977) 1294
   A. Cordier et al., Phys. Lett. 81B (1979) 389
   J. Perez-Y-Jorba, Proc. XIX Int. Conf. on High Energy Physics, Tokyo (1978) 277
   R.F. Schwitters, XVIII Int. Conf. on High Energy Physics, Tbilisi (1976) B34
   J. Burmester et al., Phys. Lett. 66B (1977) 395
   R. Brandelik et al., Phys. Lett. 76B (1978) 361
   C. Bacci et al., Phys. Lett. 86B (1979) 234
   P. Bock et al., Z. Phys. C - Particles and Fields 6 (1980) 125
7. J.J. Aubert et al., Phys. Rev. Lett. 33 (1974) 1404
   J.-E. Augustin et al., Phys. Rev. Lett. 33 (1974) 1406
   C. Bacci et al., Phys. Rev. Lett. 33 (1974) 1408
   G.S. Abrams et al., Phys. Rev. Lett. 33 (1974) 1453 and 34 (1975) 1181
   P. Rapidis et al., Phys. Rev. Lett. 39 (1977) 526
8. S.W. Herb et al., Phys. Rev. Lett. 39 (1977) 252
   W.R. Innes et al., Phys. Rev. Lett. 39 (1977) 1240

Ch. Berger et al., Phys. Lett. 76B (1978) 243
C.W. Darden et al., Phys. Lett. 76B (1978) 246
D. Andrews et al., Phys. Rev. Lett. 44 (1980) 1108 and 45 (1980) 219
T. Böhringer et al., Phys. Rev. Lett. 44 (1980) 1111
G. Finocchiaro et al., Phys. Rev. Lett. 45 (1980) 222
9. CELLO Collaboration, H.J. Behrend et al., DESY 81-029 (1981)
10. JADE Collaboration, W. Bartel et al., Phys. Lett. 88B (1979) 171
JADE Collaboration, W. Bartel et al., Phys. Lett. 89B (1979) 136
JADE Collaboration, W. Bartel et al., Phys. Lett. 91B (1980) 152
JADE Collaboration, W. Bartel et al., Phys. Lett. 100B (1981) 364
11. JADE Collaboration, W. Bartel et al., Phys. Lett. 129B (1983) 145
12. MARK J Collaboration, D.P. Barber et al., Phys. Rev. Lett. 42 (1979) 1113
MARK J Collaboration, D.P. Barber et al., Phys. Rev. Lett. 43 (1979) 901
MARK J Collaboration, D.P. Barber et al., Phys. Lett. 85B (1979) 463
MARK J Collaboration, D.P. Barber et al., Phys. Rev. Lett. 44 (1980) 1722
MARK J Collaboration, Phys. Rev. Lett. 50 (1983) 799
13. PLUTO Collaboration, Ch. Berger et al., Phys. Lett. 81B (1979) 410
PLUTO Collaboration, Ch. Berger et al., Phys. Lett. 86B (1979) 413
14. TASSO Collaboration, R. Brandelik et al., Phys. Lett. 83B (1979) 261
TASSO Collaboration, R. Brandelik et al., Z. Phys. C, Particles and Fields 4 (1980) 87
TASSO Collaboration, R. Brandelik et al., Phys. Lett. 88B (1979) 199
15. TASSO Collaboration, R. Brandelik et al., DESY 82-010 (1982) and Phys. Lett. 113B (1982) 499
16. G. Hanson et al., Phys. Rev. Lett. 35 (1975) 1609
17. M. Gell-Mann, Phys. Lett. 8 (1964) 214;
G. Zweig, CERN-Report TH 401 and TH 412 (1964)
18. C.N. Yang and R.L. Mills, Phys. Rev. 96 (1954) 191
19. D.J. Gross and F. Wilczek, Phys. Rev. Lett. 30 (1973) 1343
H.D. Politzer, Phys. Rev. Lett. 30 (1973) 1346
T. Appelquist and H. Georgi, Phys. Rev. D8 (1973) 4000
A. Zee, Phys. Rev. D8 (1973) 4038
20. J.D. Bjorken and S.J. Brodsky, Phys. Rev. D1 (1970) 1416
21. PLUTO Collaboration, Ch. Berger et al., Phys. Lett. 82B (1979) 449
K. Hagiware, Nucl. Phys. B137 (1978) 164
22. Sau Lan Wu and Georg Zobernig, Z. Phys. C - Particles and Fields 2 (1979) 107
23. E. Farhi, Phys. Rev. Lett. 39 (1977) 1587
24. S. Brandt et al., Phys. Lett. 12 (1964) 57
25. G. Sterman and S. Weinberg, Phys. Rev. Lett. 39 (1977) 1436
26. K.H. Mess and B.H. Wiik, DESY Report 82-011 (1982) and Les

Houches, Session XXXVIII, 1981, Gauge theories in high energy physics, edited by M.K. Gaillard and R. Stora and published by North-Holland Publishing Company, 1983
27. R.D. Field and R.P. Feynman, Nucl. Phys. B136 (1978) 1
28. T. Meyer, Program Write-Up (1979) for TASSO Collaboration, unpublished.
29. TASSO Collaboration, R. Brandelik et al., Phys. Lett. 94B (1980) 437
30. TASSO Collaboration, R. Brandelik et al., Phys. Lett. 100B (1981) 357
31. Sau Lan Wu, DESY Report 81-003 (1981) and Physica Scripta 25 (1981) 212
32. Sau Lan Wu and Georg Zobernig, Z. Phys. C, Particles and Fields 2 (1979) 107
Sau Lan Wu and Georg Zobernig, TASSO Note No. 84, June 26, 1979
33. B.H. Wiik, Proceedings of the International Conference on neutrinos, weak interactions and cosmology, Bergen, Norway, 18-22 June 1979, p.113
34. P. Söding, Proceedings of the European Physical Society International Conference on High Energy Physics, Geneva, Switzerland, 27 June - 4 July, 1979, p. 271
35. TASSO Collaboration, R. Brandelik et al., Phys. Lett. 86B (1979) 243 [received on August 29, 1979]
36. MARK J Collaboration, D.P. Barber et al., Phys. Rev. Lett. 43 (1979) 830 [received on August 31, 1979]
37. PLUTO Collaboration, Ch. Berger et al., Phys. Lett. 86B (1979) 418 [received on September 13, 1979]
38. JADE Collaboration, W. Bartel et al., Phys. Lett. 91B (1980) 142 [received on December 7, 1979]
39. J. Ellis, M.K. Gaillard and G.G. Ross, Nucl. Phys. B111 (1976) 253 [Errata : B130 (1977) 516]
T.A. DeGrand, Y.J. Ng and S-H. Tye, Phys. Rev. D16 (1977) 3251
A. DeRújula, J. Ellis, E.G. Floratos and M.K. Gaillard, Nucl. Phys. B138 (1978) 387
J. Ellis, Comments Nucl. Part. Phys. 9 (1981) 153
40. G. Hanson et al., Phys. Rev. Lett. 35 (1975) 1609
41. TASSO Collaboration, R. Brandelik et al., Phys. Lett. 94B (1980) 437
42. S. Brandt and H.D. Dahmen, Z. Phys. C, Particles and Fields 1 (1979) 61
43. JADE Collaboration W. Bartel et al., Phys. Lett. 101B (1981) 129
44. J. Ellis and I. Karliner, Nucl. Phys. B148 (1979) 141
45. TASSO Collaboration, R. Brandelik et al., Phys. Lett. 97B (1980) 453
For higher statistics in gluon spin analysis from TASSO, see Sau Lan Wu, DESY Report 83-007 and Proceedings of Summer Institute on Particle Physics, August 16-27, 1982, SLAC, Stanford, California.
46. M.R. Anderberg, Cluster Analysis for Applications, Academic Press, New York, 1973

47. K. Lanius, DESY 80-36 (1980)
48. J. Dorfan, Z. Phys. C, Particles and Fields 7 (1981) 349
49. H.J. Daum, H. Meyer, and J. Bürger, Z. Phys. C, Particles and Fields 8 (1981) 167
50. K. Lanius, H.E. Roloff, and H. Schiller, Z. Phys. C, Particles and Fields 8 (1981) 251
51. C.T. Zahn, IEEE Trans. Computers C-20 (1971) 68
52. PLUTO Collaboration, Ch. Berger et al., Phys. Lett. 97B (1980) 459
53. JADE Collaboration, W. Bartel et al., Phys. Lett. 119B (1982) 239
54. A. De Rújula et al., Nucl. Phys. B138 (1978) 387
55. MARK J Collaboration, D.P. Barber et al., Phys. Lett. 108B (1982) 63
56. Sau Lan Wu, Z. Phys. C, Particles and Fields 9 (1981) 329
57. TASSO Collaboration, M. Althoff et al., Phys. Lett. 126B (1983) 493
58. JADE Collaboration (presented by S. Orito)
MARK J Collaboration (presented by H. Newman)
PLUTO Collaboration (presented by Ch. Berger)
TASSO Collaboration (presented by G. Wolf)
Proceedings of the 1979 International Symposium on Lepton and Photon Interactions at High Energies, Fermilab, Batavia, Illinois, August 23-29, 1979.
59. R.D. Field and R.P. Feynman, Nucl. Phys. B136 (1978) 1
A. Ali, J.G. Körner, G. Kramer, and J. Willrodt, Z. Phys. C, Particles and Fields 1 (1979) 203 and 269;
T. Meyer, Program Write-Up for TASSO Collaboration (1979)
60. MARK J Collaboration, Physics Report 63 (1980) 337. The claim that the "first statistically relevant results, establishing the 3-jet pattern from $q\bar{q}g$ of a sample of hadronic events, were presented by the MARK J Collaboration" is not justified [see last paragraph of the subsection on the MARK J analysis]
61. P. Söding and G. Wolf, Ann. Rev. Nucl. Part. Sci. 31 (1981) 231
62. CELLO Collaboration, H.J. Behrend et al., Phys. Lett. 110B (1982) 329
63. MARK J Collaboration, D.P. Barber et al., Phys. Lett. 89B (1979) 139
64. PLUTO Collaboration, Ch. Berger et al., Phys. Lett. 99B (1981) 292
65. CELLO Collaboration, H.J. Behrend et al., Phys. Lett. 113B (1982) 427
66. CELLO Collaboration, H.J. Behrend et al., Z. Phys. C, Particles and Fields 14 (1982) 95
67. JADE Collaboration, W. Bartel et al., Phys. Lett. 119B (1982) 239
68. CELLO Collaboration, H.J. Behrend et al., Nucl. Phys. B218 (1983) 269
69. MARK J Collaboration, B. Adeva et al., Phys. Rev. Lett. 50 (1983) 2051

70. P. Hoyer, P. Osland, H.G. Sander, T.F. Walsh and P.M. Zerwas, Nucl. Phys. B161 (1979) 349
71. A. Ali, E. Pietarinen, G. Kramer and J. Willrodt, Phys. Lett. 93B (1980) 155
72. R.K. Ellis, D.A. Ross, and A.E. Terrano, Phys. Rev. Lett. 45 (1980) 1226 and Nucl. Phys. B178 (1981) 421
73. K.J.F. Gaemers and J.A.M. Vermaseren, Z. Phys. C, Particles and Fields 7 (1980) 81
J.A.M. Vermaseren, K.J.F. Gaemers and S.J. Oldham, Nucl. Phys. B187 (1981) 301
74. K. Fabricius, G. Kramer, G. Schierholz, I. Schmitt, Phys. Lett. 97B (1981) 431 and Z. Phys. C, Particles and Fields 11 (1982) 315
A. Ali, J.G. Körner, G. Kramer, Z. Kunszt, E. Pietarinen, G. Schierholz, and J. Willrodt, Phys. Lett. 82B (1979) 285, and Nucl. Phys. B167 (1980) 454
75. B. Andersson, G. Gustafson, and T. Sjöstrand, Z. Phys. C, Particles and Fields 6 (1980) 235
T. Sjöstrand, Lund Preprint, LU TP 80-3 (1980), LU TP 82-3 (1982)
B. Andersson, G. Gustafson, and T. Sjöstrand, Nucl. Phys. B197 (1982) 45
76. L.W. Mo and Y.S. Tsai, Rev. Mod. Phys. 41 (1969) 205
G. Bonneau and F. Martin, Nucl. Phys. B27 (1971) 381
F.A. Berends, K.J.F. Gaemers and R. Gastmans, Nucl. Phys. B57 (1973) 381
F.A. Berends, K.J.F. Gaemers and R. Gastmans, Nucl. Phys. B63 (1973) 381
Y.S. Tsai, Rev. Mod. Phys. 46 (1974) 815
F.A. Berends and G.J. Komen, Phys. Lett. 63B (1976) 432
F.A. Berends and R. Kleiss, Nucl. Phys. B178 (1981) 141
77. C.L. Basham, L.S. Brown, S.D. Ellis, and S.T. Love, Phys. Rev. Lett. 41 (1978) 1585; Phys. Rev. D17 (1978) 2298; Phys. Rev. D19 (1979) 2018; and Phys. Lett. 85B (1979) 297
L.S. Brown and S.D. Ellis, Phys. Rev. D24 (1981) 2383
78. PLUTO Collaboration, Ch. Berger et al., Phys. Lett. 99B (1981) 292
D. Schlatter et al., Phys. Rev. Lett. 49 (1982) 521
CELLO Collaboration, H.J. Behrend et al., Z. Phys. C, Particles and Fields 14 (1982) 95
79. Sau Lan Wu, DESY Report 81-003 (1981) and Physica Scripta, 25 (1981) 212
80. JADE Collaboration, W. Bartel et al., Z. Phys. C, Particles and Fields 9 (1981) 315
81. JADE Collaboration, W. Bartel et al., Phys. Lett. 123B (1983) 460
82. PLUTO Collaboration, Ch. Berger et al., Phys. Lett. 100B (1981) 351
PLUTO Collaboration, Ch. Berger et al., Z. Phys. C, Particles and Fields, 12 (1982) 297
83. O. Nachtmann and A. Reiter, Z. Phys. C, Particles and Fields

16 (1982) 45
A. Reiter, QCD-Untersuchungen zur Elektron-Positron Annihilation in 4 jets, University of Heidelberg Report, HD-THEP-81-10 (1981)
84. JADE Collaboration, W. Bartel et al., Phys. Lett. 115B (1982) 338
85. CELLO Collaboration, Proposal for a 4π Magnetic Detector for PETRA
JADE Collaboration, Proposal for a Compact Magnetic Detector at PETRA (1976)
MARK J Collaboration, A Simple Detector to measure e^-e^+ Reactions at High Energies (1976)
PLUTO Collaboration, Proposal for Experiments at PETRA with PLUTO (1976)
TASSO Collaboration, Proposal for a Large 4π Magnetic Detector for PETRA (1976)
86. JADE Collaboration, W. Bartel et al., Phys. Lett. 108B (1982) 140
87. TASSO Collaboration, R. Brandelik et al., Phys. Lett. 110B (1982) 173
88. M.L. Perl et al., Phys. Rev. Lett. 35 (1975) 1489; Phys. Lett. 63B (1976) 466
G.J. Feldman et al., Phys. Rev. Lett. 38 (1977) 117
89. S.L. Glashow, Nucl. Phys. 22 (1961) 579
S. Weinberg, Phys. Rev. Lett. 19 (1967) 1264
A. Salam, Proceedings of the Eighth Nobel Symposium, May 1968, ed. : N. Svartholm (Wiley, 1968) 367
S.L. Glashow, J. Iliopoulos, and L. Maiani, Phys. Rev. D2 (1970) 1285
90. M.K. Gaillard, B.W. Lee and J.L. Rosner, Rev. Mod. Phys. 47 (1975) 277
91. UA1 Collaboration, G. Arnison et al., Phys. Lett. 126B (1983) 398
UA2 Collaboration, P. Bagnaia et al., Phys. Lett. 129B (1983) 130
92. H.J. Bhabha, Proc. Roy. Soc. (London), A154 (1935) 195
93. C. Møller, Ann. Phys. 14 (1932) 531
94. N. Cabibbo and R. Gatto, Phys. Rev. 124 (1961) 1577
T. Kinoshita, J. Pestiau, P. Roy, and H. Terazawa, Phys. Rev. D2 (1970) 910;
J. Godine and A. Hankey, Phys. Rev. D6 (1972) 3301;
R. Budny, Phys. Lett. 45 (1973) 340
95. Particle Data Group, M. Roos et al., Review of Particle Properties, Phys. Lett. 111B (1982)
96. R. Budny, Phys. Lett. 55B (1975) 227
97. CELLO Collaboration, H.J. Behrend et al., Phys. Lett. 103B (1981) 148
CELLO Collaboration, H.J. Behrend et al., Phys. Lett. 123B (1983) 127
98. CELLO Collaboration, H.J. Behrend et al., Z. Phys. C, Particles and Fields 16 (1983) 301
99. JADE Collaboration, W. Bartel et al., Phys. Lett. 92B (1980) 206

100. JADE Collaboration, W. Bartel et al., Phys. Lett. 99B (1981) 281
101. JADE Collaboration, W. Bartel et al., Z. Phys. C, Particles and Fields 19 (1983) 197
102. MARK J Collaboration, D.P. Barber et al., Phys. Rev. Lett. 42 (1979) 1110
103. MARK J Collaboration, D.P. Barber et al., Phys. Rev. Lett. 43 (1979) 1915
104. MARK J Collaboration, D.P. Barber et al., Phys. Lett. 95B (1980) 149
105. MARK J Collaboration, D.P. Barber et al., Phys. Rev. Lett. 46 (1981) 1663
106. PLUTO Collaboration, Ch. Berger et al., Z. Phys. C, Particles and Fields 4 (1980) 269
107. PLUTO Collaboration, Ch. Berger et al., Z. Phys. C, Particles and Fields 7 (1981) 289
108. TASSO Collaboration, R. Brandelik et al., Phys. Lett. 94B (1980) 259
TASSO Collaboration, R. Brandelik et al., Phys. Lett. 117B (1982) 365
109. A. Böhm, DESY Report 82-084 and Proceedings of the SLAC Summer Institute on Particle Physics, Stanford, California, August 1982
110. TASSO Collaboration, M. Althoff et al., DESY 83-089 (1983) and Z. Phys. C-Particles and Fields 22 (1984) 13
111. M. Davier, Proceedings of the XXI International Conference on High Energy Physics, Paris 26-31, July 1982, p. C3-471
112. Y.S. Tsai, Phys. Rev. D4 (1971) 2821;
H.B. Thacker and J.J. Sakurai, Phys. Lett. 36B (1971) 103
113. CELLO Collaboration, H.J. Behrend et al., Phys. Lett. 114B (1982) 282
114. CELLO Collaboration, H.J. Behrend et al., Phys. Lett. 127B (1983) 270
115. J.G. Branson, MIT-LNS Technical Report Number 133 and Lectures at the International School of Subnuclear Physics, Erice (1982)
116. F.A. Berends, K.J.F. Gaemers and R. Gastmans, Nucl. Phys. B63 (1973) 381 and B68 (1974) 541
F.A. Berends and R. Gastmans, Nucl. Phys. B61 (1973) 414
F.A. Berends and R. Kleiss, Nucl. Phys. B178 (1981) 141
F.A. Berends, R. Kleiss and S. Jadach, Nucl. Phys. B202 (1982) 63
M. Böhm and W. Hollik, DESY 83-060 (1983)
117. A. Böhm, Proceedings of the International Europhysics Conference on High Energy Physics, Brighton, UK, 20-27 July 1983
118. CELLO Collaboration, H.J. Behrend et al., Z. Phys. C, Particles and Fields 14 (1982) 283
PLUTO Collaboration, Ch. Berger et al., Z. Phys. C, Particles and Fields 21 (1983) 53
119. MARK J Collaboration, B. Adeva et al., Phys. Rev. Lett. 48 (1982) 1701

120. B. Naroska, Proceedings of the 1983 International Symposium on Lepton and Photon Interactions at High Energies, Cornell University, Ithaca, NY, August 4-9, 1983
121. TASSO Collaboration, R. Brandelik et al., Phys. Lett. 113B (1982) 499
122. JADE Collaboration, W. Bartel et al., Phys. Lett. 129B (1983) 145
123. JADE Collaboration, W. Bartel et al., Phys. Lett. 101B (1981) 361
124. TASSO Collaboration, M. Althoff et al., Phys. Lett. 126B (1983) 493
125. R.P. Feynman, Phys. Rev. 76 (1949) 769
V.B. Berestetskii, O.N. Krokhin and A.K. Khlebnikov, ZETF 30 (1956) 788
[English translation : Soviet Physics - JETP 3 (1956) 761]
S.D. Drell, Ann. Phys. 4 (1958) 75
B. De Tollis, Nuovo Cimento 16 (1960) 203
J.A. McClure and S.D. Drell, Nuovo Cim. 37 (1965) 1638.
126. MARK J Collaboration, D.P. Barber et al., Phys. Rev. Lett. 45 (1980) 1904
127. MARK J Collaboration, B. Adeva et al., Phys. Rev. Lett. 48 (1982) 967
128. MARK J Collaboration, Physics Report 63 (1980) 337
129. PLUTO Collaboration, Ch. Berger et al., Phys. Lett. 99B (1981) 489
130. PLUTO Collaboration, Ch. Berger et al., Phys. Lett. 94B (1980) 87
131. K.H. Mess and B.H. Wiik, DESY Report 82-011 (1982) and Les Houches, Session XXXVII, 1981, Gauge theories in high energy physics, edited by M.K. Gaillard and R. Stora and published by North-Holland Publishing Company, 1983
132. P.Q. Hung and J.J. Sakurai, Nucl. Phys. B143 (1978) 81
J.D. Bjorken, Phys. Rev. D19 (1979) 335
E.H. de Groot, G.J. Gounaris and D. Schildknecht, Phys. Lett. 85B (1979) 399; 90B (1980) 427; Z. Phys. C, Particles and Fields 5 (1980) 127
D. Schildknecht, Proc. Intern. School of Subnuclear Physics (Erice, 1980); Bielefeld preprint BI-TP 81/12
V. Barger, W.Y. Keung and E. Ma, Phys. Rev. Lett. 44 (1980) 1169; Phys. Rev. D22 (1980) 727; Phys. Lett. 94B (1980) 377

e^+e^- PHYSICS AT CESR

Paolo Franzini

Columbia University

INTRODUCTION

The study of high energy electron positron interactions has, in the past decade, provided us with a wealth of experimental information about all aspects of particle physics. The charmed quark, or c-quark, was simultaneously discovered in hadron interactions and in e^+e^- annihilations. It was however mostly at e^+e^- colliders that its properties were (and still are) investigated. The measurements of the annihilation cross section into hadrons has given tangible evidence about the existence of color and the continued study of these processes at high energies has confirmed many qualitative predictions of Quantum Chromo Dynamics (QCD), the best candidate at present for a complete theory of the strong interactions. The discovery, in hadron collisions, of the upsilons in 1978, required the existence of yet another quark, the b-quark, where b stands sometimes for beauty or bottom, a new flavor of quarks. Because of its large mass, $\sim$ 5 GeV, the b quark has greatly helped in clarifying many aspects of strong interactions, especially in the spectroscopy of heavy particles. The study of the b quark, carried out only at electron positron colliders, has given new proofs that quarks carry color and fractional charge (1/3e for the b quark) and that gluons exist, have spin 1 and carry color, as required by QCD. In the field of weak interactions further proof of the validity of the "standard model" has been obtained and strong constraints on the mixing angles have been established. In the last three years most of the study of the b quark has been carried out at the Cornell Electron Storage Ring, CESR, because of a fortunate set of circumstances. In these lectures I will present what we have learned from this study at CESR [1], where the focus has been and continues to be on the Υ resonances [2,3], from which we learn about heavy quark spectroscopy, and on the weak interactions of the b-quark, through the study of B-meson properties.

CESR, CLEO, CUSB AND THE UPSILONS

The Cornell Electron Storage Ring

CESR was first proposed just before the discovery of the T's, as a high luminosity storage ring operating in the beam energy range of 5 to 8 GeV or total center of mass energy W of 10 to 16 GeV. Construction was greatly accelerated, following the discovery of the T, and first collisions were observed in the fall of '79. The machine has undergone many modifications and improvements since that time and can reach peak luminosities of 1.5×10^{31} cm^{-2} s^{-1} at present, with single bunch operation and mini-β inserts. Extensive modifications during the summer of 1983, including multi-bunch operation, improved electron and positron filling etc., should result in further increase of the machine luminosity, to values higher than those achieved by the rebuilt DORIS at DESY.

Because of the great interest in the physics of the T resonances CESR was never operated at energies higher than W $\sim$ 11.5 GeV. Figure 1 shows a plan view of the tunnel housing the 10 GeV Cornell electron synchrotron, which is used as an injector for CESR, and the storage ring. The two experimental areas housing the detectors CLEO and CUSB are also indicated.

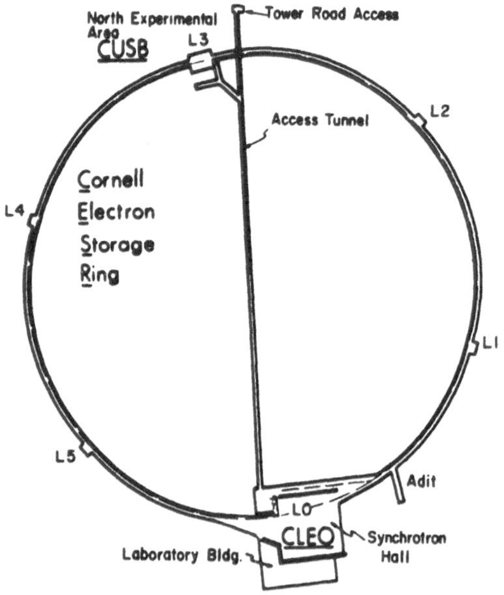

Fig. 1. Plan of CESR, showing the location of the two detectors CLEO and CUSB.

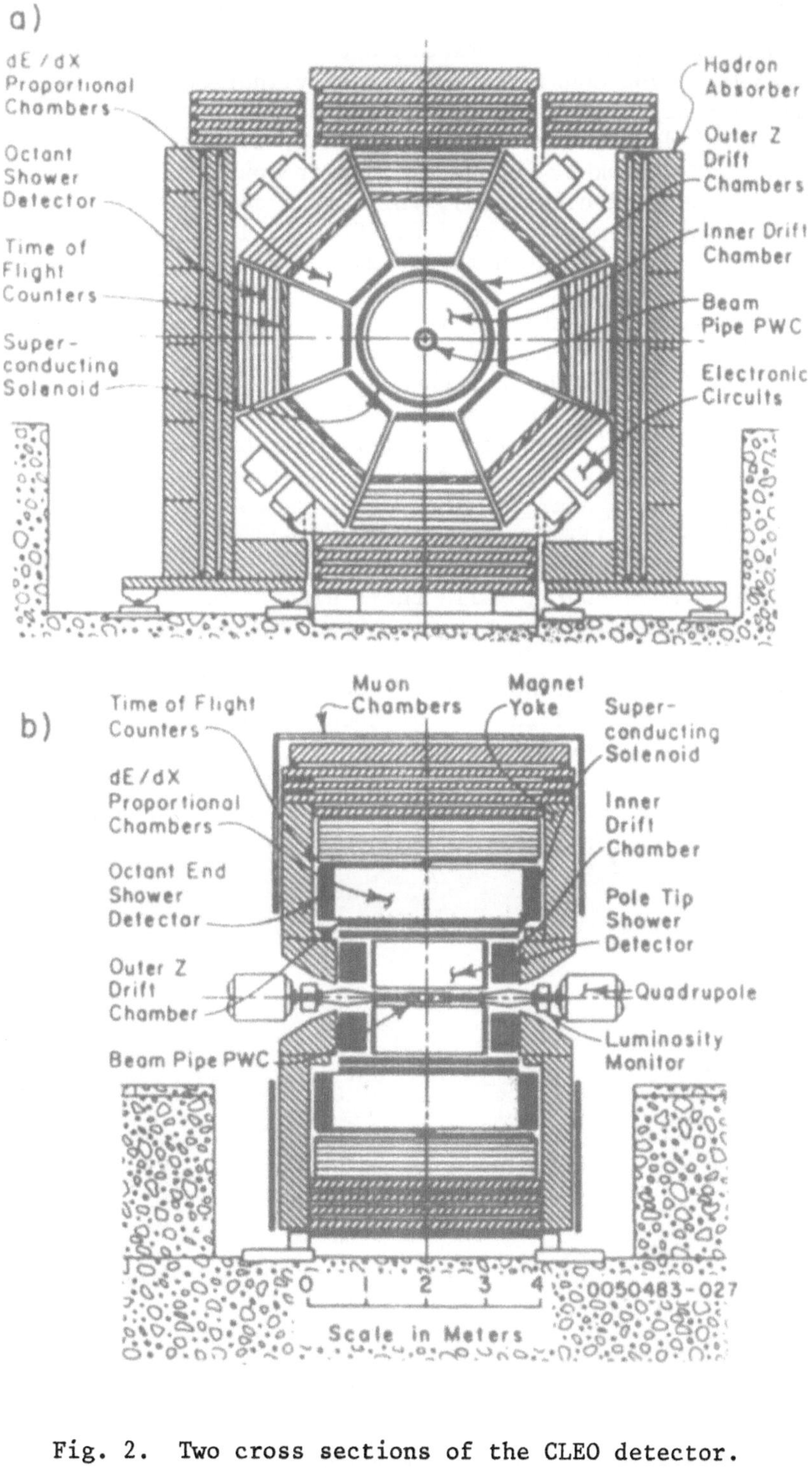

Fig. 2. Two cross sections of the CLEO detector.

The CESR detectors

Two basic types of detectors are used at e^+e^- storage ring. The most popular type consists of a large cylindrical tracking chamber around the beam pipe, in a magnetic field parallel to the beam direction. Particle identification is provided by time of flight, dE/dx measurements and Cerenkhov counters. Shower counters measure the energy of photons (and electrons), while iron absorbers typically provide filtering for muon detection. CLEO [3a] is an example of such detectors and is shown in figure 2.

The other type of detectors is based on "calorimetry", i.e. measurements of the total energy carried by the products of e^+e^- annihilations. CUSB, shown in figure 3, is an example of this kind. It consists of four layers of tracking chambers around the beam pipe, surrounded by $\sim$ 15 radiation lengths (X_0) of active converter material, where photons are completely absorbed and hadrons deposit a significant fraction of their energy. The active radiator consists of four thin layers of NaI ($\sim$ 1.1 X_0), followed by four more X_0 of NaI and 7 X_0 of lead glass blocks. The radial layering of the NaI permits recognition of e.m. showers, while their directions are obtained from four cathode readout chambers inserted between layers.

The very first result obtained at CESR in the fall of 1979 [4,5] was the resolving of the T", at about 10350 MeV, evidence for which had been claimed by the original Fermilab experiment [6] which discovered the T's, figure 4. Figure 5 shows the first energy scan at CESR, where the first two T's, first resolved at DORIS [7], and the T" are all fully resolved.

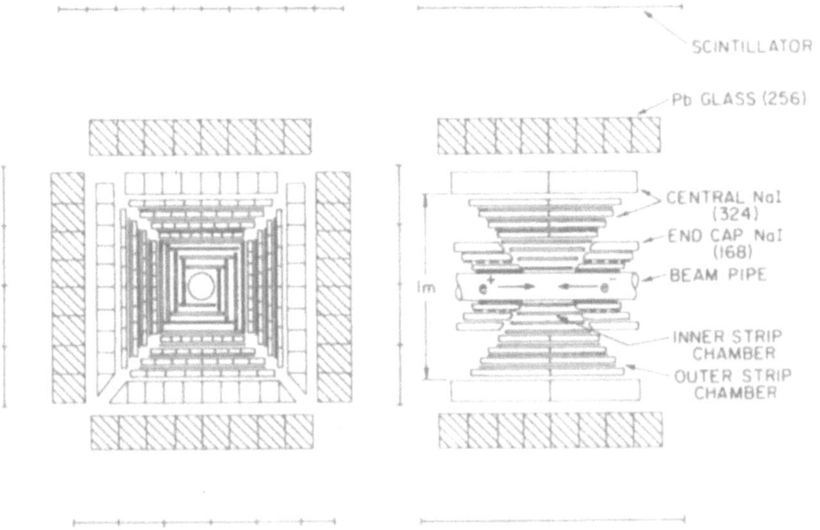

Fig. 3. The CUSB detector.

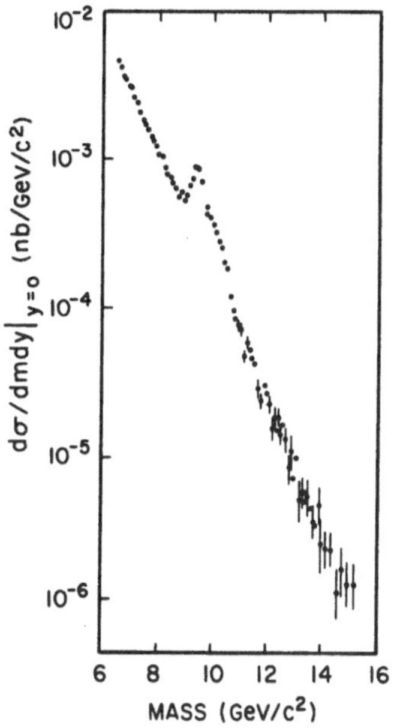

Fig. 4. Discovery of the Υ at Fermilab. A peak is observed at ∿ 10 GeV above the smoothly falling dimuon mass spectrum. Muons are produced in p-Be collisions at 400 GeV.

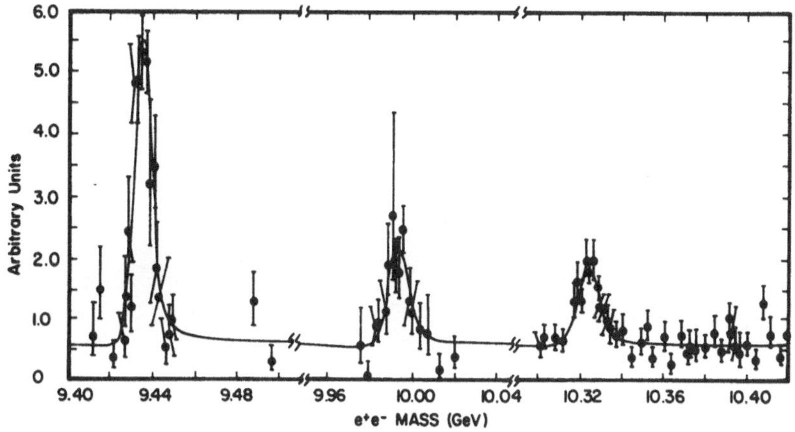

Fig. 5. First scan at CESR in 1979, showing the first three upsilons fully resolved. (CUSB)

Precision measurement of the Υ mass

At e^+e^- colliders the mass of a resonance like the Υ is given by twice the beam energy at which the peak of the cross section is observed. The beam energy is usually determined by estimates of ∫Bdl of the machine guide field, obatined from measurements of field and excitaion curves of the magnets. This method results in accuracies of about 0.1 %. Thus the early results for the Υ mass were 9460 and 9434 MeV at DORIS and CESR respectively, with estimated errors of 10 to 30 MeV.

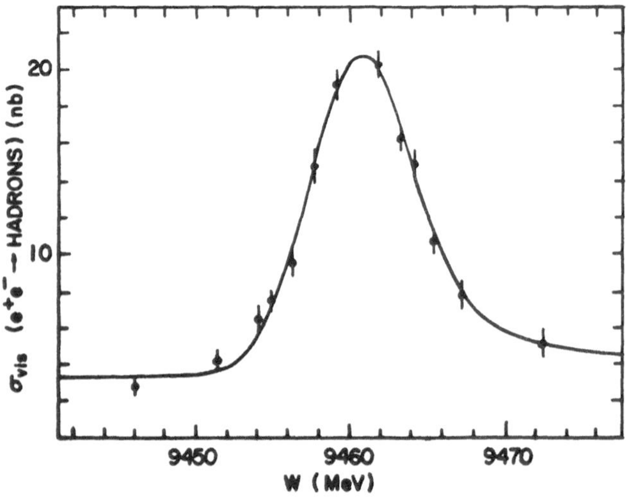

Fig. 6. Scan of the Υ for the precision measurement of its mass. (CUSB)

An elegant and very precise method for obtaining the energy of a stored electron beam, used first at Novosibirsk [8], consists of measuring the difference between spin precession frequency ω_p and cyclotron frequency ω_c, due to the anomalous part of the electron magnetic moment. Precession and circulating frequencies are related by

$$\omega_p = \omega_c(1 + \gamma(g-2)/2),$$

where $\gamma = E/m_e$ is the electron Lorentz factor and $(g - 2)/2 = 0.001159652209(13)$. ω_p is determined by applying a small periodic perturbation which produces resonant depolarization of the stored beam. Measurements at VEEP-4 resulted in a value of the Υ mass of 9459.7 ± 0.6 MeV [8] and this summer at CESR we obtained M(Υ) = 9459.95 ± 0.1 MeV [9]. Figure 6 shows the scan of the Υ peak at CESR with the energy scale determined by the resonance method described. It is truly remarkable that the mass of the ϒ is known to ∿ 10 ppm in terms of the mass of the electron. (The mass of the electron is known to ∿ 2.7 ppm).

UPSILON SPECTROSCOPY

e^+e^- annihilations into hadrons in the ϒ region

Figure 7 shows measurements of $\sigma(e^+e^- \to \text{hadrons})$ at CESR around the known ϒ's. The first three upsilons are prominently visible and a fourth small enhancement in the cross section is observed around 10.6 GeV. This enhancement, discovered at CESR in early 1980 [10], is called the ϒ'''. While the first three ϒ's are observed with a width of ∿ 4 MeV r.m.s., which is consistent with the machine energy spread, the ϒ''' has a total width of ∿ 20 MeV, indicating that it lies above the flavor threshold, providing us with a copious source of b-flavored meson, about which we will say much more, later.

In the quark-parton model, in the absence of resonances, the total cross section for e^+e^- annihilation into hadrons is given by

$$\sigma(e^+e^- \to \text{had}) = 3(\Sigma_i\ q_i^2)\sigma(ee \to \mu\mu) ,$$

where q_i is the electric charge of the ith quark in units of the electron charge and $\sigma(ee \to \mu\mu)$ is the cross section for production

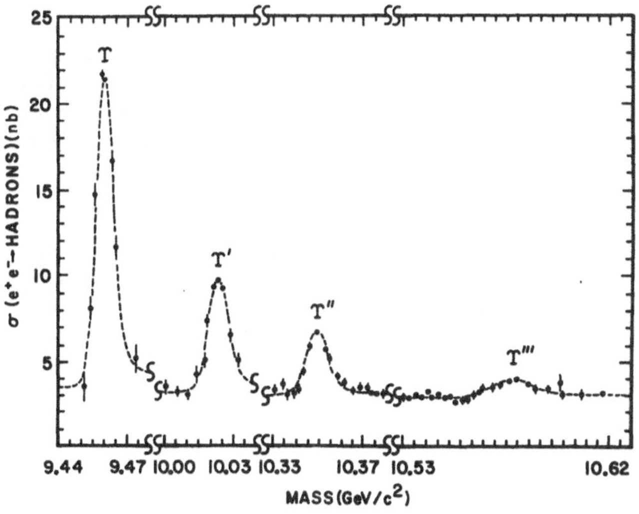

Fig. 7. Cross section for e^+e^- annihilations into hadrons in the ϒ region. More than 700 000 events appear in this plot. Three narrow resonances are observed plus a fourth one, whose width is significantly wider than the machine energy spread. (CUSB)

of $\mu^+\mu^-$ pairs. To first order in QED this cross section is given by :

$$\sigma(ee \to \mu\mu) = 86.8/s \text{ nb, s in GeV}^2,$$

where $s = W^2$ is the square of the total c.m. energy. The factor 3 in the hadronic cross section accounts for color, which is carried by all quarks, independently of their flavor. It is customary to define the dimensionless ratio $R = \sigma_{had}/\sigma_{\mu\mu}$ which, below the Υ where we know of the u, d, c and s quarks, is predicted to be (to lowest order) :

$$R_o = 3(4/9 + 1/9 + 4/9 + 1/9) = 3.33 \ .$$

Higher order QCD corrections give [11] :

$$R = R_o(1 + \alpha_s/\pi + 1.5(\alpha_s/\pi)..) \ ,$$

from which, for $\alpha_s \sim 0.2$, $R = 3.56$, below the upsilon region.

If the total energy W is very close to the mass M of a $J^{PC}=1^{--}$ bound $q\bar{q}$ state, such as a vector meson V, then the total cross section acquires additional resonant contributions given by [12] :

$$\sigma_i = \frac{3\pi}{M^2} \frac{\Gamma_{ee}\Gamma_i}{(W-M)^2 + \Gamma_{tot}^2} \ ,$$

where σ_i is the cross section to channel i, Γ_{ee} and Γ_i are respectively the partial widths for V decay to e^+e^- and to channel i, and Γ_{tot} is the total decay width of the meson V. Integrating this additional contribution and summing over all channels one obtains a relation between Γ_{ee} and the total resonant cross section :

$$\int\sigma_{tot,res}\, dW = (6\pi^2/M^2)\Gamma_{ee} \ .$$

We wish however to express the leptonic width in terms of the resonant hadronic cross section only. Defining $B_{\mu\mu} = \Gamma_{\mu\mu}/\Gamma_{tot}$ and assuming lepton universality, i.e. $B_{ee} = B_{\mu\mu} = B_{\tau\tau}$ we can write

$$\Gamma_{tot} = \Gamma_{had} + 3B_{\mu\mu}\Gamma_{tot} = \Gamma_{had}/(1 - 3B_{\mu\mu}) \ ,$$

obtaining

$$\Gamma_{ee} = (M^2/6\pi^2) \int\sigma_{res,had}\, dW/(1 - 3B_{\mu\mu}) \ .$$

From measurements of $\sigma_{had}(W)$ and $B_{\mu\mu}$ we can therefore obtain Γ_{ee} and $\Gamma_{tot} = \Gamma_{ee}/B_{\mu\mu}$.

Measurements of R, leptonic widths and masses

CESR has delivered so far ∿ 130,000 nb^{-1} to each of the two interaction regions corresponding to some of 3/4 million hadronic events collected in each detector. From this large amount of data the properties of the upsilon have been extracted, as well as measurements of R below and above the b-flavor threshold. Table 1 gives the results from CLEO and CUSB for R below threshold and the change across the threshold. The values obtained at CESR for R are in good agreement with those obtained at DORIS at 7.4 - 9.4 GeV. The quoted systematic errors are quite large at present. This is mostly due to the fact that both CLEO and CUSB are rather young detectors and absolute measurement of the cross section requires good knowledge of efficiency, luminosity, effects of radiative corrections which are detector dependent etc. Most of these uncertainties cancel out in the measurement of ΔR, as indicated in the table below, for measurements done with the same detectors, at the same machine. The value for ΔR obtained at CESR is a very direct proof that the b quark charge is 1/3e, for which, from the formula above one expects ΔR = 0.36. Also the measured values of R above the threshold, R = 3.99 from CUSB and R = 4.11 from CLEO are in very good agreement with values from PETRA and PEP at much higher energies (20 to 35 GeV), proving that no new odd particles (technipions ?..) are produced in the 10 to 35 GeV energy range.

Table 2 summarizes the properties of the upsilons as measured at CESR and reviewed in ref. 2, corrected for the new measurement of the Υ mass. Both masses and mass differences are given, as well as leptonic widths and their ratios. Mass differences are given because they are directly relevant to potential model calculations.

Table 1. R and ΔR in the Υ region

W (GeV)	R^a	Experiment[b]
10.4-10.5	3.77±0.06 (±0.28)	CLEO
10.4-10.5	3.63±0.06 (±0.37)	CUSB
ΔR across flavor threshold		
	0.34±0.09 (±0.11)	CLEO
	0.36±0.09 (±0.03)	CUSB

[a]The errors in parenthesis are estimates of systematic uncertainties.
[b]As summarized in reference 2.

Table 2. Properties of the upsilon resonances

Parameter	Υ	Υ'	Υ''	Υ'''
M(MeV)	9459.95±0.1	10020.5±0.7	10350.0±0.7	10576±3
$M-M_{\Upsilon}$(MeV)		560.8±0.4	890.3±0.4	1116±3
Γ_{ee} [a]	1.22±0.03	0.52±0.02	0.38±0.02	0.29±0.03
$\Gamma_{ee}/\Gamma_{ee}(\Upsilon)$		0.42±0.02	0.31±0.02	0.24±0.03
$B_{\mu\mu}$(%)	2.83±0.28	1.9±0.3	3.3±1.3	$\sim 10^{-5}$

[a]Systematic uncertainty is ∿ 10 %

This also applies to the leptonic width ratios. In addition, experimental systematic uncertainties in their values, of order of 10%, cancel out in the ratios.

Υ'and Potential Models

The Υ's are $^3S_1$ bound states of a $b\bar{b}$ pair. Because of the massiveness of the b quark (∿ 5 GeV) the level structure of these states can be calculated using an effective static potential in the Schroedinger equation. This approach, which was first used for the lighter charmonium states, has been extended to the Υ family by several authors. QCD suggests that the potential should be of the form $-(4/3)(\alpha_s/r)$ for small distances, where the exchange of hard gluons results in a coulomb-like potential. At distances of the order of 1 fermi, confinement sets in, resulting in a constant force between quarks or potential linear in r, with a "string tension" related to the slope of Regge trajectories. The simplest form of the potential is therefore the combination of a linear term and a coulomb term, as proposed by the Cornell group [13]. More sophisticated calculations take into account asymptotic freedom [14] and higher order QCD corrections [15]. Alternative approaches consist of assuming a simple power law for the potential [16,17], such as $a+br^{\alpha}$, with the parameters determined from experimental data. The quark masses, as well as the absolute scale for the potential are also free parameters. Excellent agreement is obtained with measurements of the level spacing for both $c\bar{c}$ and $b\bar{b}$ states, thus confirming the basic requirement of QCD that the interquark forces be flavor independent. Because of the freedom in the quark masses the ground states cannot be predicted in potential models. Potential models, in general, cannot compute accurately the value of the wave function at zero quark separation and therefore the leptonic width which is directly related to its magnitude square. Ratios of leptonic widths are however reliably calculable, since many uncertainties cancel out. Table 3 shows a comparison of the measured properties of the upsilons with the results of four typical potential model calculations.

Table 3. Comparison of prediction and measurements for the Υ's

Quantity	Eichten [13]	Krasemann [14]	B.G.T. [15]	Martin [16]	Experiment
M(2S)-M(1S) (MeV)				565	561
M(3S)-M(1S) (MeV)	898	862	890	900	890
M(4S)-M(1S) (MeV)	1170	1108	1180	1142	1116
$\Gamma_{ee}(1S)$ (keV)		1.05	1.07		1.22±0.03
$\Gamma_{ee}(2S)/\Gamma_{ee}(1S)$	0.39	0.43	0.44	0.41	0.42±0.02
$\Gamma_{ee}(3S)/\Gamma_{ee}(1S)$	0.27	0.31	0.32	0.35	0.31±0.02
$\Gamma_{ee}(4S)/\Gamma_{ee}(1S)$	0.22	0.25	0.26	0.27	0.24±0.03

The agreement between data and calculations is extremely good in all cases, confirming the identification of Υ', Υ'' and Υ''' as the 2^3S_1, 3^3S_1 and 4^3S_1 $b\bar{b}$ states. The predictions for the excitation energy of the fourth Υ is somewhat poorer than for the lower levels. This is however expected, since the Υ''' has strong decay channels available, as discussed in A. Martin lectures. That there should be just three bound Υ's is also a prediction of potential models.

Decay of the Υ

A remarkable property of the heavy vector mesons is their very small total decay width. While the ρ-meson has a mass of ∿ 750 MeV and a total width of ∿ 150 MeV, the Υ has a mass of ∿ 10 000 MeV and a total width of only ∿ 0.04 MeV. The explanation of this fact, first given by Appelquist and Politzer [18], is one of the outstanding successes of QCD. Color confinement and charge conjugation invariance require, to lowest order, the Υ to decay into hadrons by annihilation of the $b\bar{b}$ pair into three gluons according to the amplitude of figure 8 (in analogy to the decay of ^3S positronium into

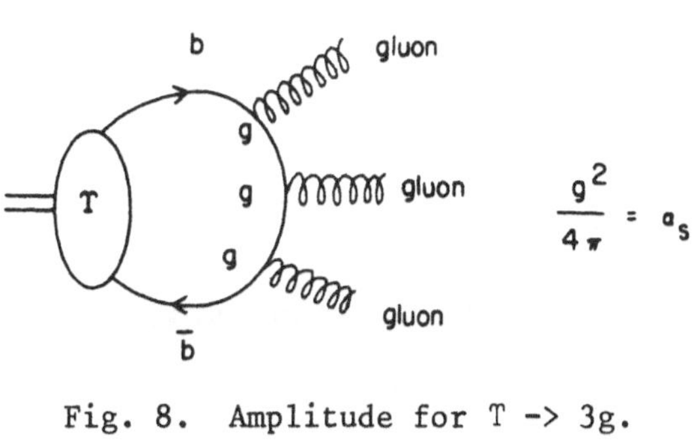

Fig. 8. Amplitude for Υ -> 3g.

three photons) followed by the evolution of the gluons into hadrons with unit probability.

Properly accounting for the fact that the three gluons must be in a color singlet state, one obtains

$$\Gamma_{ggg} = (160/81)\alpha_s^3(\pi^2-9)(|\psi(0)|^2/M^2) ,$$

which, using current values of α_s gives $\Gamma_{ggg}\sim 30$ keV. Because of the large suppression of the strong decay, electromagnetic processes can favorably compete. The annihilation of the $b\bar{b}$ pair into a virtual photon leads to e^+e^- pairs, the inverse of the Υ production reaction in e^+e^- annihilations, $\mu\mu$ and $\tau\tau$ pairs, as well as quark anti-quark pairs, according to the amplitude of figure 9. By the definition of R and assuming lepton universality, we have

$$\Gamma_{\ell\bar{\ell},q\bar{q}} = (3+R)\Gamma_{ee} \sim 8 \text{ keV} .$$

The leptonic decay width for the amplitude of figure 9 has been computed by several authors and is given by

$$\Gamma_{ee} = 16\pi\alpha^2 q_b^2(|\psi(0)|^2/M^2) ,$$

where q_b is the b quark charge. Note that both Γ_{ggg} and Γ_{ee} are proportional to the modulus square of the wave function at zero quark separation, as expected for processes invoving annihilation of the $q\bar{q}$ pair. We also note that the decay of the Υ by annihilation into a single virtual gluon is not forbidden by energy and momentum conservation, as in the case of positronium via one photon (because there are no charged particles lighter than the electron). This process would lead to decay widths for the Υ of order of 100 MeV. That this is not the case is a very direct proof that gluons carry color and of "color confinement".

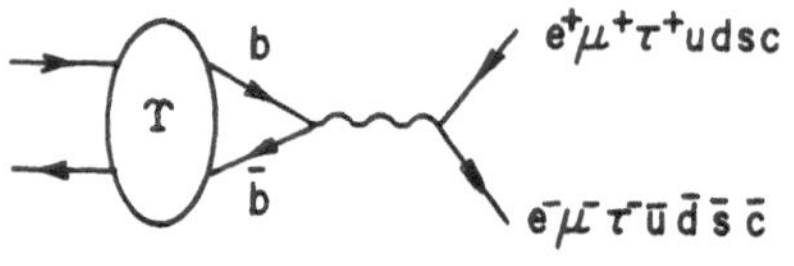

Fig. 9. Amplitude for Υ annihilation via virtual photon.

Decay Dynamics

A direct proof that vector mesons decay into three gluons has been obtained for the first time by observing the three jet structure in Υ decay. While many global parameters may be employed to characterize the structure of a many body final states, the one used in various forms by all groups and most amenable to calculations in QCD, is the quantity called thrust, defined as

$$T = \mathrm{Max}(\sum_i |\vec{p}_i \cdot \hat{n}| / |\vec{p}_i|) ,$$

where $\vec{p}_i$ are the particle momenta (energy clusters in a calorimeter) and the unit vector $\hat{n}$ is rotated until a maximum is found. For a collinear two jet final state $\hat{n}$ is the jet axis, for a three jet event $\hat{n}$ is close to the axis of the most energetic jet. The thrust distribution dN/dT can be calculated for two quarks and three gluon [19] final states. The thrust axis angular distribution, for three gluons from vector mesons also depends on the gluon spin. The angular distribution for the polar angle of the unit vector $\hat{n}$ is given by

$$1 + \alpha_T \cos^2\theta ,$$

where $\alpha_T = 1.0$ for two quark jets (hadronic events from continuum e^+e^-annihilations), $\alpha_T = 0.39$ for Υ decays into three spin 1 gluons and $\alpha_T = -1.0$ for decays into three spin 0 gluons [20]. The thrust distributions obtained by CLEO, as reported in ref. 3, and CUSB[21], figure 10, change significantly between continuum ($q\bar{q}$) and Υ (mostly three gluons) and the data agree well with predictions for three gluon decays. The parameter α_T has been measured by CLEO [3]. They obtain a value of 0.32 ± 0.11. CLEO has also measured the angular distribution of the normal to the event plane as shown in figure 11. All these results together are strong support for the original explanation of the decay mechanism of the heavy vector mesons as a third order process in QCD. In addition they provide evidence for the vector nature of the gluon.

Υ' and Υ'' Decays

The decay widths of Υ' and Υ'' via annihilation of the $b\bar{b}$ pair are expected to be smaller than that of the ground state Υ, because the excited states have larger radii and therefore the value of $|\psi(0)|^2$ is smaller. This fact is reflected in the decreasing values of Γ_{ee}. Because of the smallness of the annihilation width, decays of the excited Υ's to other $b\bar{b}$ states can contribute significantly to the total decay widths. For the ψ's, for instance, we know that $\Gamma(\psi) \sim 60$ keV and $\Gamma(\psi') \sim 220$ keV. 50% of the latter is due to $\psi' \to \psi\pi\pi$ and 26 % to $\psi' \to \gamma\, \chi_c$, where χ_c are the three triplet p-wave (^3P_J) $c\bar{c}$ states. The study of similar transitions for "beautium" is more complex because there are three ^3S and two ^3P bound states.

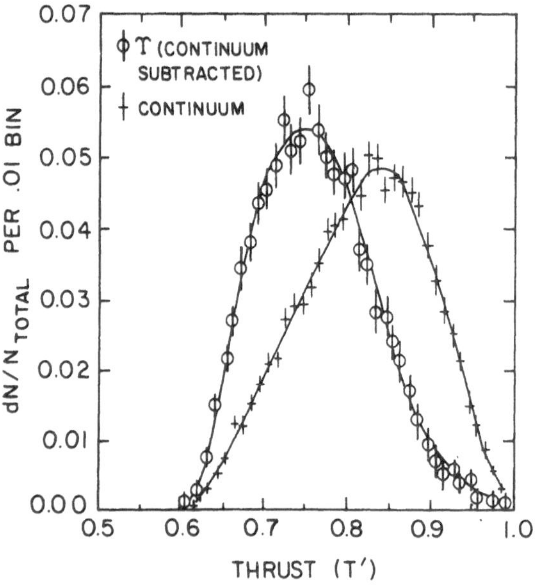

Fig. 10. Thrust distribution for continuum subtracted Υ events (circles) and pure continuum events (crosses). Upsilon decays are almost pure three gluon final states and continuum events are two-quark states. The observed Υ thrust distribution is in excellent agreement with QCD calculations. (CUSB)

(Singlet S and P states, as well as D states are much harder to reach).

The ππ Transitions

The first inter-upsilons transition observed at CESR in 1980 was $\Upsilon' \rightarrow \Upsilon\pi^+\pi^-$ [22]. This reaction is uniquely identified in both CUSB and CLEO by detecting the inclusive events from the decay chain: $\Upsilon' \rightarrow \Upsilon\pi^+\pi^-$, $\Upsilon \rightarrow e^+e^-$ or $\mu^+\mu^-$. CLEO in addition can compute the mass recoiling against any $\pi^+\pi^-$ pair and observes a clear peak at the mass of the Υ, as shown in figure 12. From measurement at CESR of the branching ratio (BR) for $\Upsilon' \rightarrow \Upsilon\pi^+\pi^-$ and using isospin invariance, one obtains for the Υ' a partial width $\Gamma_{\pi\pi} \sim 9$ keV, to be compared with $\Gamma_{\pi\pi} \sim 110$ keV for the ψ' case, although the Q-values for the two cases are practically identical. The transition $\Upsilon' \rightarrow \Upsilon\pi\pi$ is supposed to be due to the emission of two gluons by the $b\bar{b}$ bound system, the two gluons appearing as the two pions, as indicated in figure 13. Perturbative QCD cannot calculate this amplitude. Gottfried [23] suggested that soft gluon emission amplitudes could be

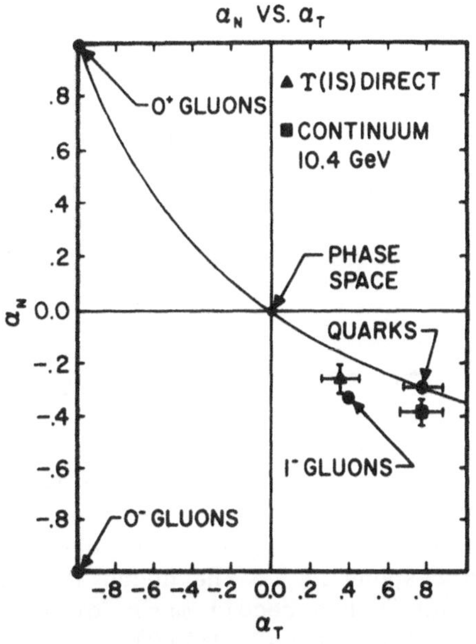

Fig. 11. Predicted (circles) and measured (squares, triangles) coefficients for the angular distributions $1+\alpha_T\cos^2\theta$ and $1+\alpha_N\cos^2\theta$ of the thrust axis and the normal to the event plane. The solid line gives the correlation between thrust axis and event plane normal when the latter is isotropic around the first. (CLEO)

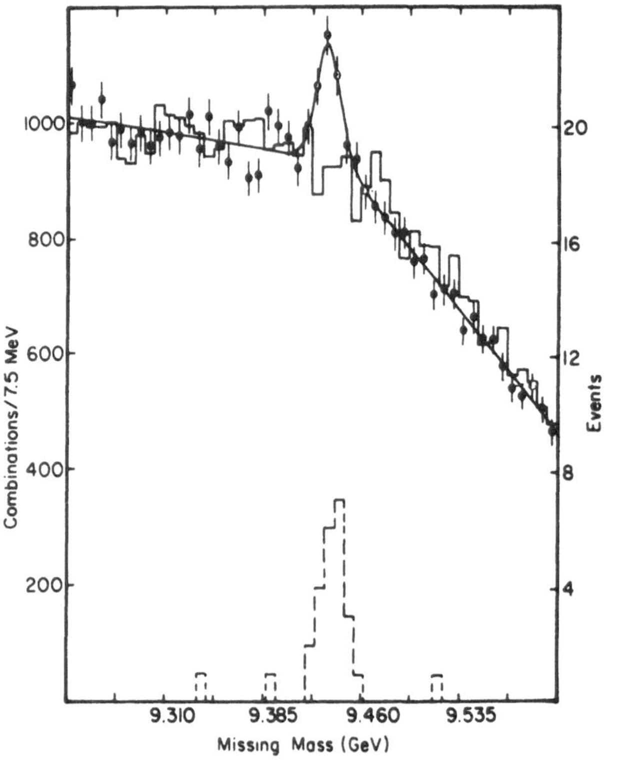

Fig. 12. Distribution of the mass recoiling against opposite sign dipions (data points) and same sign pion pairs (solid line histogram) in Τ' decays. A clear peak at the Τ mass is observed corresponding to the decay $\Upsilon' \to \Upsilon\pi\pi$. The dashed line histogram is the recoil mass for events of the type $\Upsilon' \to \Upsilon\pi\pi$, $\Upsilon \to ee$ or $\mu\mu$. (CLEO)

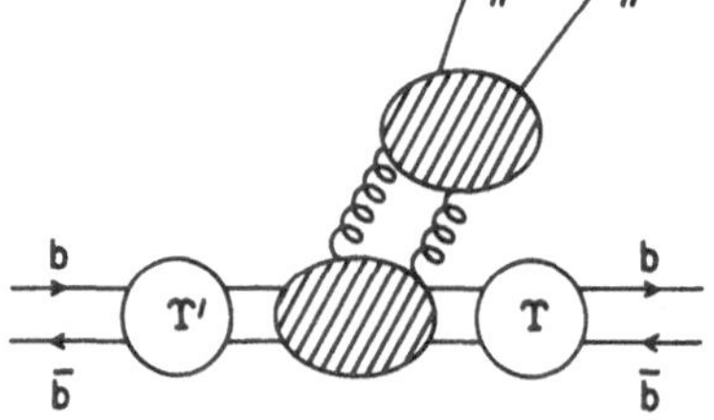

Fig. 13. Amplitude for $\Upsilon' \to \Upsilon\pi\pi$.

understood in terms of a multipole expansion of the "color field". In this picture the emission of two gluons $\Upsilon' \to \Upsilon$ transition corresponds, to lowest order, to two "color-electric" dipole transitions. Since the electric dipole transition element squared is of the order of the r.m.s. size of the system, $\langle r^2 \rangle$, one expects $\Gamma_{\pi\pi} \sim \langle r^2 \rangle^2$. Using values for $\langle r^2 \rangle$ from potential models, one obtains

$$\frac{\Gamma(\psi' \to \psi_{\pi\pi})}{\Gamma(\Upsilon' \to \Upsilon\pi\pi)} \sim \frac{\langle r^2 \rangle^2}{\langle r^2_\Upsilon \rangle^2} \sim \frac{1}{10} ,$$

in good agreement with observation ($\sim$9/110). Spin zero gluons would be emitted with an amplitude proportional to the color charge rather than the color-electric dipole moment. In this case there would be no suppression of $\Upsilon' \to \Upsilon\pi\pi$. The observed suppression is a particularly beautiful proof that gluons carry spin 1.

The decays $\Upsilon'' \to \Upsilon'\pi\pi$ and $\Upsilon'' \to \Upsilon\pi\pi$ have also been observed at CESR [24]. The first reaction was observed only by CUSB, because of the very small Q-value ($\sim$ 50 MeV), which results in very slow pions, not detectable in CLEO. CUSB has also detected all the pion transitions with π°'s. The observed rates for the π° transitions are $\sim$ 1/2 of those for $\pi^\pm$ transitions, from which experimental confirmation is obtained for the assumption that the two pion systems are in an I=0 state. The combined CESR results for the branching ratios for pion (charged and neutral) transitions of the Υ's are given below, compared to the calculations of Kuang and Yan [25]:

$\Upsilon' \to \Upsilon\pi\pi$	$\Upsilon'' \to \Upsilon'\pi\pi$	$\Upsilon'' \to \Upsilon\pi\pi$	
29% ± 6%	4.6% ± 3.0%	6.9% ± 1.1%	CESR
25.1% to 27%	2.3% to 3.5%	2.1% to 5.1%	K + Y

Photon Transitions and P-Wave States

Triplet P-wave states have $J^{PC} = 2^{++}, 1^{++}, 0^{++}$ and cannot be produced directly in e^+e^- annihilations. The transition $^3S_1 \to {}^3P_{2,1,0}$ is an electric dipole (E1) transition, with an amplitude given by $q\langle f|r|i\rangle$, which can compete with decays by annihilation of the $b\bar{b}$ pair. For the Ψ' case the E1 partial width is $\sim$ 50 keV. Simple scaling using the ratio of the quark charges squared and the mean squared radii leads to $\Gamma_{E1}(\Upsilon') \sim 4$ keV, or a branching ratio of the order of 10%. The situation is slightly more favorable for the transition $3^3S_1 \to \gamma + 2^3P_J$ because of the larger size of the Υ'', the smaller annihilation rate and the smaller pion transition rates leading to an expected branching ratio of the order of 30%. (The first ^3P_J $b\bar{b}$ states are also called χ_b and the second ones χ_b'). The first indirect evidence for a rather large BR for Υ'' decays to χ_b' states was obtained by CUSB [21], which observed an anomalous high thrust for Υ'' decays, compared to Υ decays. Because of their positive C value the $^3P_{2,0}$ states decay by annihilations into two gluons, resulting in a two-jet-like final state, with thrust distributions

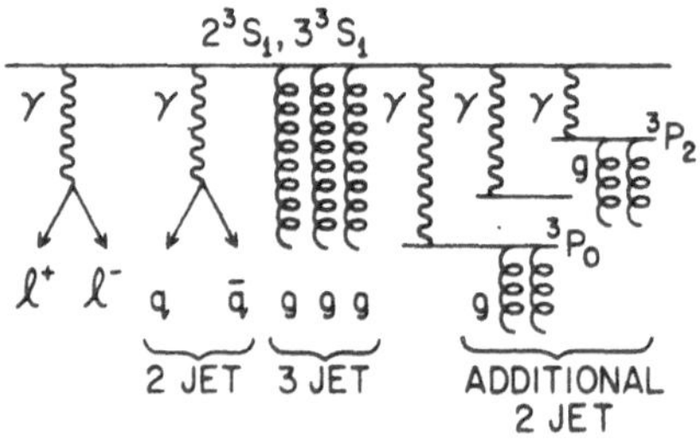

Fig. 14. Schematic representation of ^3S state decays showing the excess 2 jet contribution via the ^3P states.

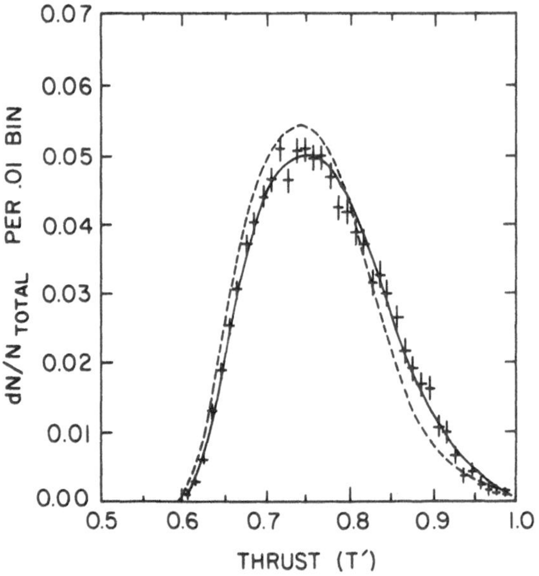

Fig. 15. Thrust distribution for Υ'' decays. The dashed line is the expected distribution for the Υ''. The continuous line is a linear superposition of the Υ and contiuum distributions, corresponding to a 20% excess two-jet-like events in Υ'' decays. (CUSB)

peaked at high values, as for continuum events. This is shwon schematically in figure 14. Figure 15 shows the thrust distribution from Υ'' decays, compared with that of the Υ and with a fit to a superposition of continuum and Υ distributions corresponding to a 20% excess two-jet-like fraction in Υ'' decays attributed to $\Upsilon'' \to \gamma + \chi_b \to 2$ gluons $\to$ two jets. This result, obtained first with a few thousand Υ'' decays is a good example of how powerfull a tool thrust can be in the study of complicated multiparticle final states.

Direct evidence for E1 transitions of both the Υ' and Υ'' to the corresponding χ_b and χ_b' was obtained by CUSB much later when sufficient luminosity became available at CESR. In 1981-82 in a three month runs, CESR delivered 14 pb^{-1} integrated luminosity at the Υ'' peak. CUSB collected during this run 65 000 hadronic events, 37 000 of which are resonance decays. In 1982-83 a similar length run gave 30 pb^{-1}, yielding 230 000 hadronic events at the Υ' peak of which 150 000 are resonance decays.

Evidence for E1 transitions for both Υ' and Υ'' was obtained in two ways. A quasi monochromatic signal is observed in the inclusive photon spectrum from all hadronic events at the two resonance peaks. An alternate method consists in searching for exclusive decay channels containing only two photons and an ee or $\mu\mu$ pair in the final state, corresponding to the decay chain

$$\Upsilon^n \to \gamma_1 + \chi_b^{n-1} \to \gamma_1 + \gamma_2 + \Upsilon^{n-1}, \quad \Upsilon^{n-1} \to e^+e^- \text{ or } \mu^+\mu^- .$$

The decay $\chi_b \to \gamma + \Upsilon$ is also an E1 transition, which is expected to have large branching ratio for the J=1 state, small for J=2 and very small for J=0, according to the decay rate into gluons computed [25] from potential model wave functions and the formulae of Barbieri and collaborators [26].

Figure 16 and 17 show the inclusive photon spectrum, after subtraction of the very large background of photons from π^0's, for the Υ' and Υ'' [27,28]. In figure 16 a broad signal is observed in the region 90 to 160 MeV. A narrow signal is also seen at 427 MeV. Referring to the level scheme of figure 18, the low energy signal, which shows some evidence for structure and is considerably wider than the experimental resolutions, is interpreted as being due to the three lines labelled 1, 2, 3. The signal at 427 MeV results from merging together of lines 4 and 5, with relative weights of 2 and 5. Adding the weighted average of the energies of lines 1 and 2 with the observed energy of the higher energy signal and the recoil energy (10.5 MeV) gives 560.0 MeV in excellent agreement with the Υ'-Υ mass difference of 560.8 MeV. The relative strength of the low energy and high energy signals ($\sim$ 15 to 4) proves that the low energy photons are from Υ' decays, while the more energetic photons are due to transitions to the ground states. These in turn remove possible ambiguities in the mass of the χ_b's. The inclusive photon

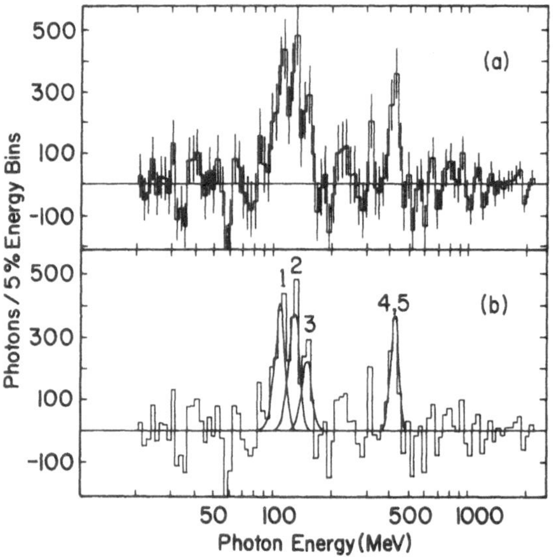

Fig. 16. Inclusive photon signal from Υ' E1 transitions (background subtracted.). (CUSB)

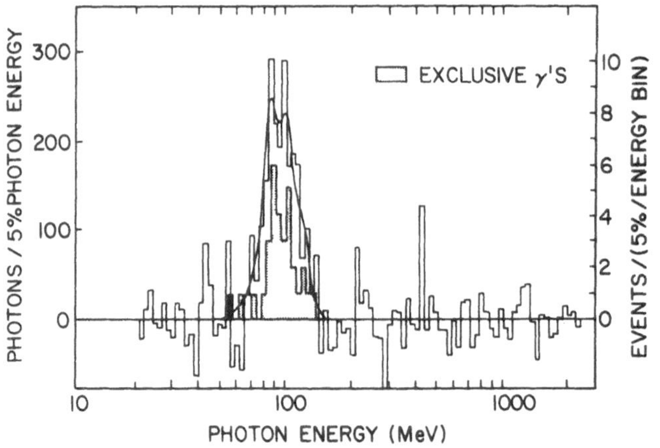

Fig. 17. Inclusive photon signal from Υ'' E1 transitions (background subtracted). (CUSB)

signal from Υ'' decays in figure 17 is more prominent and is approximately 20 MeV lower than for Υ' case, as expected since the level separation decreases for higher excitation. There is in this case no hint of fine structure, partly because it is smaller, partly because the energy resolution of CUSB was slightly worse in our earlier run. Still, the observed signal is considerably broader than the

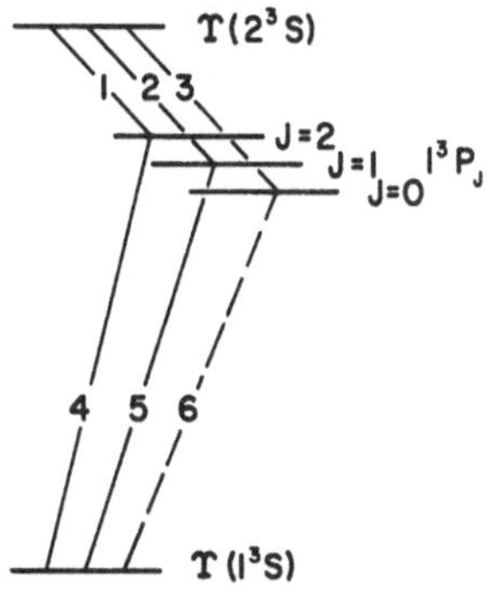

Fig. 18. Level scheme for 1^3S, 2^3S and $1^3P_{1,2,3}$.

experimental resolution and the data can be fitted to three lines of arbitrary position and intensities smeared by the resolution, to obtain the energy of the lines. No significant signals are observed for decays of the χ_b' to the Υ' and Υ. This is partly due to the much smaller sample of events available for the Υ'' and partly to the smaller branching ratios for the chain decays. The results of the study of the inclusive spectra are presented in tables 4 and 5.

Confirmation of these results is obtained from the search in the exclusive channels. The results of these searches [29,30] are shown in figures 19 and 20. Events are displayed as points in a scatter plot of the more energetic photon ($E_{\gamma,hi}$) versus the energy of the softer one ($E_{\gamma,\ell o}$). For the Υ' there is a clear clustering of events around ($E_{\ell o}$; E_{hi}) $\sim$ (120, 430 MeV), corresponding, including 10 MeV of recoil energy, to the Υ'-Υ mass difference of 560.8 MeV. Figure 21 shows the spectrum of the lower energy photons from Υ' decays, with the two lines corresponding to $\Upsilon' \to 1^3P_2 + \gamma$ and

T' → 1^3P_1 + γ well resolved. This allows the determination of the individual product BR's for T' → Tγγ via the J=2 and J=1 states. The results from the analysis of the exclusive decay channels of the T' are also given in table 4. For the plane T" case, the events cluster

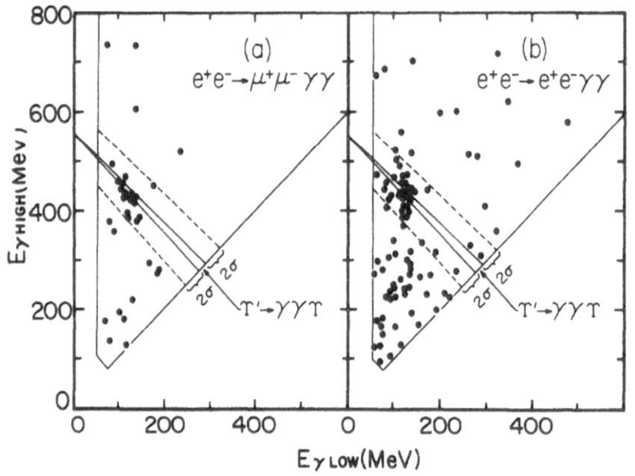

Fig. 19. Scatter plots of the higher energy photon versus the lower energy photon for T' decays. Two photon transitions to the T are expected to lie in the indicated diagonal band. (CUSB)

in two regions of the energy correlation plane, corresponding to the decay chains :

T" → χ'γ → T'γγ and T" → χ'γ → Tγγ .

The results from the analysis of the exclusive decays via two photons of the T" are given in table 5. As can be seen in tables 4

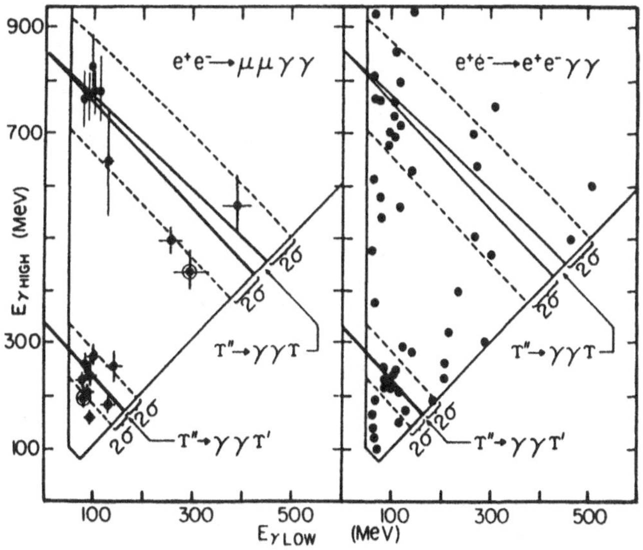

Fig. 20. Scatter plots of the higher energy photon versus the lower energy photon for Υ'' decays. The upper diagonal band contains two photon transitions to the Υ and the lower to the Υ'. (CUSB)

Table 4. Experimental Results for $2^3S \to 1^3P \to 1^3S$

	E_γ, inclusive	E_γ, exclusive	B.R., inclusive
$2^3S_1 \to 1^3P_2$	108.2±.3 MeV	107±2	6.1±1.4 %
$\to 1^3P_1$	128.1±.4	128±2	5.9±1.4
$\to 1^3P_0$	149.4±.7	-	3.5±1.4
$\to 1^3P_{2,1,0}$			15.5±2.5 (+5,-2)
$1^3P_2 \to 1^3S_1$	427.0±1		20 ± 5
$1^3P_1 \to 1^3S_1$			47 ± 18

$2^3S \to 1^3P \to 1^3S$	$\Sigma(\Pi(\text{B.R.}))$	= 4.0 ± 1 % (inclusive)
		= 3.6 ± .9 (exclusive)
$2^3S \to 1^3P_2 \to 1^3S$	$\Pi(\text{B.R.})$	= 1.2 ± 0.7
$\to 1^3P_1 \to$		= 2.4 ± 0.8

Center of Gravity (1^3P_J) = 9900 ± 3 MeV

$$\frac{M(J=2) - M(J=1)}{M(J=1) - M(J=0)} = 0.93 \pm 0.1 \ (\pm 0.2)$$

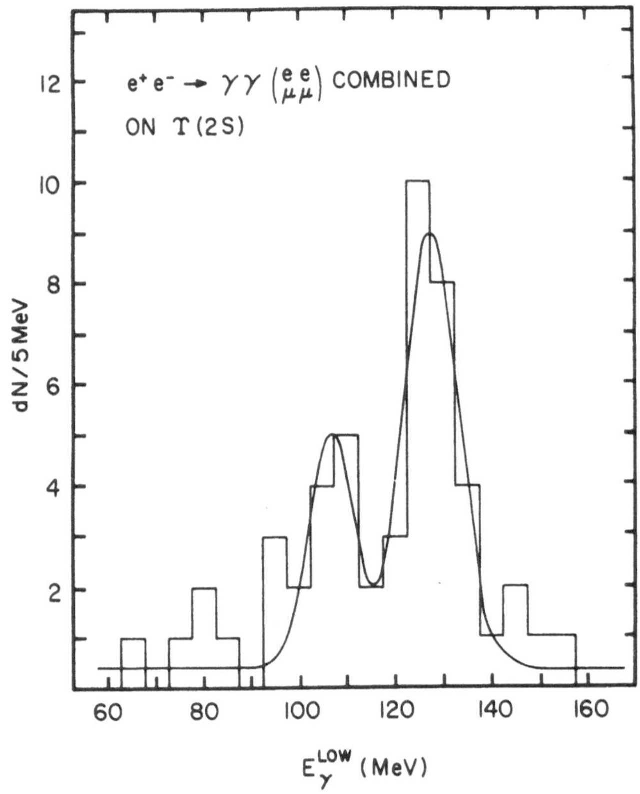

Fig. 21. Photon spectrum for the lower energy photon in the accepted band of figure 16, from Υ' decays. Line 1 and 2 of figure 18 are clearly resolved. (CUSB)

Table 5. Experimental Results for $3^3S \to 2^3P \to (2^3S$ or $1^3S)$

	E_γ, inclusive	E_γ, exclusive	B.R., inclusive
$3^3S1 \to 2^3P_2$	84.5±2 MeV	84±3	15.9±1.6 %
$\to 2^3P_1$	99.5±3.2	99±2	12.5±1.3
$\to 2^3P_0$	117.2±5	-	6.4±1.1
$\to 2^3P_{2,1,0}$			34.0±3.0 (3.0)
$3^3S \to 2^3P \to 2^3S$		$\Sigma(\Pi(B.R.)) = 5.9 \pm 2.1$ %	
$3^3S \to 2^3P \to 1^3S$		$\Sigma(\Pi(B.R.)) = 3.6 \pm 1.2$	

Center of Gravity (2^3P_J) = 10256 ± 5 MeV

$$\frac{M(J=2) - M(J=1)}{M(J=1) - M(J=0)} = 0.85 \pm 0.1 \quad (\pm 0.3)$$

and 5 the two methods employed in the study of T' and T" radiative decays are in excellent agreement both for line energies and, where possible, for the product branching ratios for cascade decays. Also, for the case of the χ_b it is possible, combining inclusive and exclusive results, to obtain the E1 BR's for the 1^3P_2 and 1^3P_1 states.

While thus far we have assumed that the observed photons correspond to E1 transitions to ^3P_J $b\bar{b}$ states, the data support this assumption because the observed rates to the J=2,1,0 states are, within errors, proportional to $k^3(2J+1)$ as expected for E1 transitions, k being the photon energy. Also, the measured center of gravity (cog) of the observe states agree well with potential model predictions, thus confirming their identification with the first and second ^3P_J $b\bar{b}$ systems.

Partial Rates for T's Decays

As described previously a knowledge of $B_{\mu\mu}$ is necessary to obtain Γ_{total}. $B_{\mu\mu}$ is known at present with reasonable accuracy only for the T. For the other two narrow T's it is possible however to obtain Γ_{tot} indirectly, and therefore $B_{\mu\mu}$, by using the fact that all annihilation channel widths are proportional to Γ_{ee}, together with measurements of the branching ratios for decays without annihilation of the $b\bar{b}$ pair. In addition to the annihilation of the $b\bar{b}$ pair into three gluons one must also include annihilation into a virtual photon, leading to lepton pairs and quark pairs. For three lepton flavors and by definition of R, the partial width for decays via a virtual photon is given by $(3+R)\Gamma_{ee}$. We can therefore write $\Gamma_{total} = \Gamma_{ggg} + (3+R)\Gamma_{ee} + \Gamma_{other}$, where Γ_{other} is the decay width for all

Table 7. Branching ratios and partial rates for the bound 's

Parameter	Υ	Υ'	Υ''
Measured			
Γ_{ee} (keV)	1.22±0.03	0.52±0.02	0.38±0.02
$B_{\mu\mu}$ (%)	2.83±0.28		
$B_{\pi\pi}$ (%)		0.29±0.04	0.115±0.033
B_{E1} (%)		0.15±0.05	0.34±0.03
Derived			
Γ_{tot} (keV)	43.1±4.3	32.8±4.6	24.6 ± 3.4
Γ_{ggg} (keV)	35.0±4.3	14.9±1.8	10.9 ± 1.3
$\Gamma_{\pi\pi}$ (keV)		9.5±1.4	2.8 ± 0.9
Γ_{E1} (keV)		4.9±1.8	8.4 ± 1.4
$B_{\mu\mu}$ (%)		1.58±0.24	1.54 ± 0.22

channels not requiring annihilation (Γ_{ggg} as defined here includes a ∿ 4 % [31] contribution from $\Upsilon \to \gamma gg$). We thus obtain

$$\Gamma^n_{tot} = \Gamma^n_{ee}/(B_{\mu\mu}(\Upsilon)(1-B^n_{other})),$$

where n refers to the nth Υ and B_{other} is the branching ratio for all decays without annihilation. Using the values given previously we derive $B_{other}(\Upsilon') = 0.49 \pm 0.06$ and $B_{other}(\Upsilon'') = 0.46 \pm 0.04$ (Searches for other decay modes of the Υ' and Υ'' have given limits of < 10 % [21] for their branching ratios). The results in Table 7 are derived using the outlined procedure. The directly measured quantities are given first, followed by the derived ones.

P-wave States and E1 Transitions Versus Theory

The discovery of the P wave $b\bar{b}$ states brings a wealth of new information to be compared with predictions of potential models as well as with results of QCD sum rules. We can compare three quantities : the center of gravity (cog) of the level masses, the E1 transition rates and the fine structure splitting, using for instance the ratio :

$$r = M(J=2)-M(J=1) \ / \ M(J=1)-M(J=0).$$

The cog's are perhaps casier to predict, because of the good knowledge of binding potential. The transition rates and splittings are sensitive to relativistic effects and there is no satisfactory theory of the effective spin dependence of the forces in quarkonium. While all these problems are expected to be less severe for the $b\bar{b}$ case, the predictions for the ratio r for the χ'_b range from 0.48 to 1 and the splitting themselves range from 14 to 24 MeV for the two

Table 8. Comparison of predictions and measurements for ^3P levels

Author	1^3P cog (MeV)	2^3P cog (MeV)	$\Gamma(2^3S\to1^3P)^a$ (keV)	$\Gamma(3^3S\to2^3P)^a$ (keV)
Büchmuller and Tye [15]	9890	10250	4.2	6.1
Büchmuller [32]			4.1	6.8
Eichten et al. [13,25]	9924	10271	4.4	6.1
Martin [16]	9861	10242		
Quigg and Rosner [33]	9888	10244	4.3	7.2
Krasemann [19,34]	9936	10271	3.1	4.8
Gupta et al. [35]	9898	10256		
Byers and McClary [36]	9923	10267	4.4	7.6
Voloshin [37]	9835±30	10267		
Bertlman [38]	9803±10			
Experiment	9900±5	10256±5	4.9±1.6	8.4±1.4

^{a}Adjusted using experimental cog's, see text.

upper levels and from 17 to 40 for the two lower ones. The spread of the predictions is even larger for the χ_b case. Since r is poorly determined at present we only compare measurements and calculations for masses and transition rates. Similarly, the products of branching ratios given earlier are consistent with expectations [25] although the experimental accuracy is very limited. Results of many authors are shown in table 8 for the ^3P level masses and transition rates. Since the rate for E1 transitions is proportional to k^3, where k is the photon energy, we have rescaled all E1 widths using the experimental value $k_{cog}(1P)$ = 119 MeV and $k_{cog}(2P)$ = 93 MeV.

The agreement between measurements and calculations is in fact impressive for potential models and very poor for the QCD sum rules predictions of the lowest ^3P state mass.

Conclusions on the Spectroscopy of the Υ's

The present experimental status of upsilon spectroscopy is best summarized by the level diagram of figure 22, where the observed levels and transitions are illustrated (solid lines). We have quantitatively verified that the Υ decays into 3 gluons by measuring the partial width and by observing the three jet structure in its decay. The spin of the gluon was shown to be one, both from studying the Υ thrust distribution and from the study of hadronic cascades. R and ΔR are fine to lowest order in QCD the charge of the b quark is $|1/3|$ and there are no vibrational states visible.

Potential models cannot unambiguously predict the fine structure of the ^3P states, nor the hyperfine splitting between the $^3S_1(\Upsilon$'s) and $^1S_0(\eta_b$'s) states. Experimental information is necessary to es-

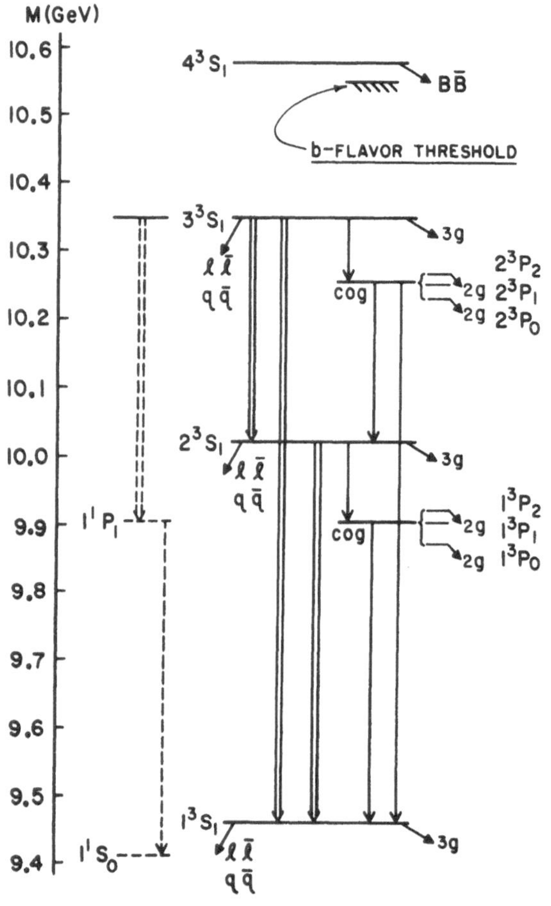

Fig. 22. A summary of our present knowledge of the $b\bar{b}$ system. Solid lines indicate observed levels and transitions. Unlabelled single lines stand for photon transitions and double lines for $\pi\pi$ transitions. The dashed portion of the diagram refers to states and transitions which could be observed in the not to far future.

tablish the form of the effective spin-spin and spin-orbit interactions responsible for these effects. The next generation of experiments both at CESR and DORIS will concentrate in this area, which require very fine resolution, very high machine intensities, and luck. In the Υ system there appears to be an alternate path to reach the ^1P and the η_b which is experimentally promising, as shown on the left side of figure 22 in dashed lines. This is the decay chain $\Upsilon'' \to \pi\pi 1^1P$, $1^1P \to \eta_b + \gamma$. We have designed an upgraded CUSB which will capitalize on this fortunate circumstance.

ABOVE THE FLAVOR THRESHOLD

The Fourth Upsilon and the B Meson

As we have mentioned before, the fourth Υ is observed at CESR with a total width of $\sim$ 20 MeV. This fact is interpreted as proof for the Υ''' having strong decay channels, not requiring annihilation of the $b\bar{b}$ pair. The $b\bar{b}$ pair has enough energy to escape the binding potential, its energy going into creating a pair of light quarks $u\bar{u}$ or $d\bar{d}$. Light and heavy quarks bind together to form two color neutral mesons. The pseudoscalar $b\bar{u}$ etc. bound states are called B mesons, while the vector states are called B*. Since the light quark and the b quark carry different flavors, which are conserved by the e.m. and strong interactions, the B meson is stable except for the weak interaction, which, at the parton level, induces $b \to c\ell^-\bar{\nu}$ (or $b \to c\bar{u}d$ etc.) and $b \to u\ell^-\bar{\nu}$ transitions. Conclusive proof of the existence of a new meson, of mass around 5 GeV, the B meson, which is stable except for weak decay, was obtained by CUSB and CLEO[39,40] in 1980, by observing a large increase in the yield of electrons of 1 to 2.5 GeV at the Υ''' peak. In the so called spectator model the B meson decay is due to the amplitude of figure 23 (the light quark just sits there). Simple counting (with color not forgotten and taking into account kinematic factors) results in estimates for the branching ratios for $B \to e^-\bar{\nu}X$ or $B \to \mu^-\bar{\nu}X$ of 10 to 15 %. Since $M_B \sim 1/2 M_\Upsilon \sim 5$ GeV, the electrons from B decays have momenta up to $\sim$ 2.5 GeV while in e^+e^- hadronic annihilations at $W \sim 10$ GeV, very

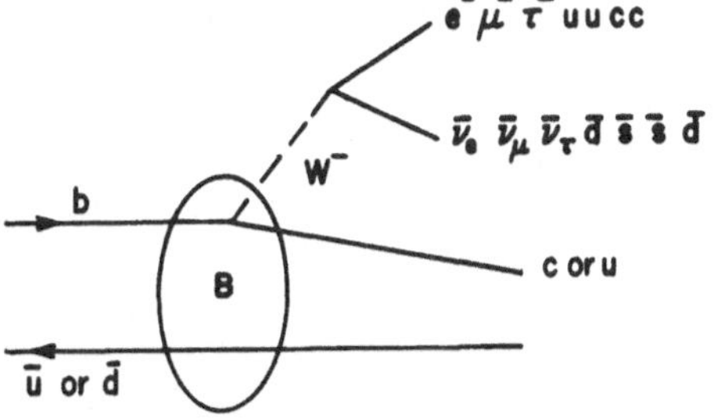

Fig. 23. Spectator model amplitude for B meson decay.

few electrons with momenta greater than 1 GeV are produced.

The Mass of the B and B* Mesons

As for ρ-π, K*-K and D^*-D, the vector states containing b-flavor are expected to be heavier than the pseudoscalar ones. From scaling arguments in potential models [13,16] the B^*-B mass difference is expected to be $\sim$ 50 MeV. The actual mass of B and B^* mesons are not however predictable to tens or even hundreds of MeV, because of the intrinsic freedom in choosing quark masses and the depth of the potential. It is not therefore possible to predict whether the Υ''' decays into B or B^* mesons. An obvious bound for the B (and B^*) mass is $M_{\Upsilon''} < 2M_B < M_{\Upsilon'''}$. For very low momenta of B and B^* the decay rate for Υ''' are proportional to β^3 with statistical spin factors of 1 for BB, 4 for BB^* and 7 for B^*B^*. In addition one must take into account the overlap between initial and final states, essentially given by the Υ''' wave function in momentum space at the B or B^* momenta [41]. In particular the nodes of the radial wave functions of the $\Upsilon(4^3S)$ result in zeros in momentum space. Thus for increasing $\Upsilon'''-2M_{B,B^*}$ mass difference, decays to the lighter states might be suppressed in favor of BB^* or B^*B^* channels. This is in fact the case for the ψ(4030,4160,4415), which decay mostly into D^* mesons.

For the expected B-B^* mass difference of $\sim$ 50 MeV, the B^* decays with a very short lifetime via $B^* \to B + \gamma$, resulting in monochromatic photons in the B^* system, which become slightly doppler broadened in the laboratory. Thus if $M_{\Upsilon'''} - 2M_B > 50$ MeV one expects the presence of a strong photon signal. This photon signal (in the 40 to 70 MeV range) has been searched for by CUSB with negative results [42]. Figure 24 shows the inclusive photon spectrum from Υ''' decays, after subtraction of the (large) contribution from the continuum. Also shown is the signal expected for BR($\Upsilon''' \to$ monochromatic photon) = 1.0. From the negative results, for $M_{B^*}-M_B \sim 50$ MeV, one can put more stringent limits on the B meson mass of :

$$5263 < M_B < 5278 \text{ MeV}.$$

Exclusive Decays of B Mesons

According to the amplitude of figure 23, final states consisting only of quarks can be reached in the decay of b quarks resulting in decays of the B meson into lighter mesons. Assuming, as will be shown later that $b \to c$ is dominant, the allowed weak couplings lead to the decay chain $b \to cW^-$, $c \to sW^+$ which in terms of known mesons, corresponds to the possible decays :

$B^- \to D^\circ\pi^-$
$\bar{B}^\circ \to D^\circ\pi^+\pi^-$
$\bar{B}^\circ \to D^{*+}\pi^-$
$B^- \to D^{*+}\pi^-\pi^-$ and their charge conjugates

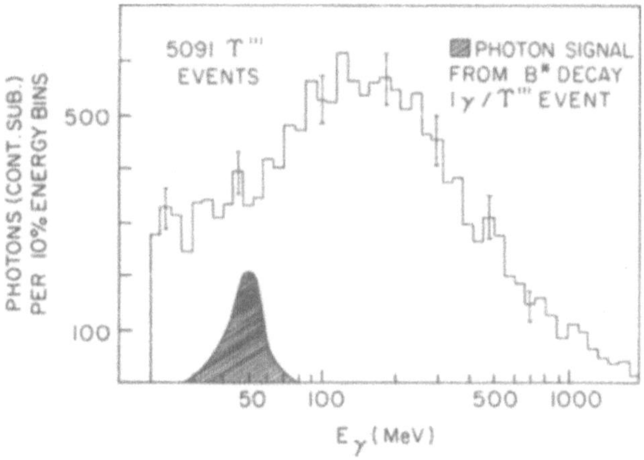

Fig. 24. Inclusive photon spectrum from ϒ''' decays. Also shown is the photon signal expected for $BR(\Upsilon'''\to BB^*) = 100\%$ (CUSB).

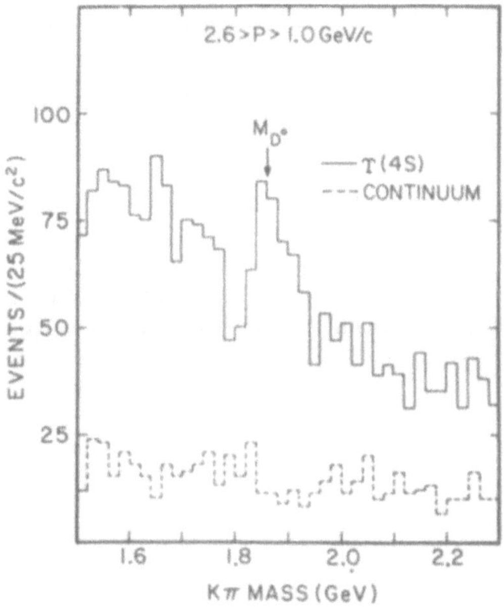

Fig. 25. D meson signal in the $K^{\pm}\pi^{\mp}$ mass spectrum. (CLEO)

with branching ratios estimated to be of the order of a few per cent. In all the above decays D or D^* are present and they are the beginning point for the search of B decays. D° mesons can be identified via their decays into $K^{\pm}\pi^{\mp}$, as shown in figure 25, and $D^{*\pm}$ are found by searching for $(K^{\pm}\pi^{\mp})\pi^{\pm}$ triplets for which

$$M[(K\pi)\pi] - M(K\pi) = M_{D^*} - M_D \quad ,$$

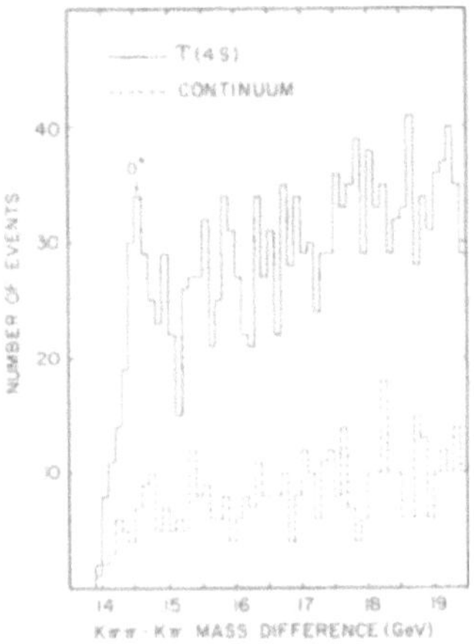

Fig. 26. D^* meson signal in the spectrum of $M[(K\pi)\pi] - M(K\pi)$, see text. (CLEO)

figure 26. Finally the D,D^* candidates are combined with one or two additional pions and energy-momentum constraints are applied, corresponding to the decay chain $T''' \to B\bar{B}$, $B \to D$ or D^*, $D \to K\pi$ or $D \to K\pi\pi$.

The results of this program, as carried out recently by CLEO [43], is shown in figure 27, where a clear peak in the reconstructed B mass is observed around 5272 MeV. This peak contains 18 events, with an estimated background of 4 to 7 events. The mass of the neutral B and the charged B mesons are given as 5274.2±1.9±2.0 and 5270.8±2.3±2.0. The mass difference between neutral and charged B's is not quite determined at the moment but is consistent with estimates of 2 to 4 MeV [41], with the correct sign. This experiment

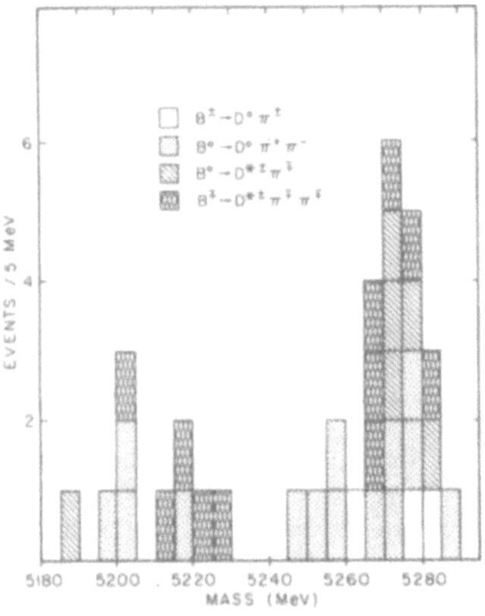

Fig. 27. Mass distribution of reconstructed B mesons. (CLEO)

determines the Q-value for $\Upsilon''' \rightarrow B\bar{B}$ to be 32.4±3.0±4.0 MeV in good agreement with the CUSB bounds of 20 to 50 MeV [42]. The CLEO results are however insensitive to having missed a 50 MeV photon because of the imposed kinematical constraints, this means that the reconstructed states of 5272 MeV mass could be either B's or B^*'s. The second possibility is excluded by the null results of the CUSB search for B^* decay photons. See the lecture by J. Lee-Franzini, for more on B and B^* meson masses.

An Aside on the F Meson

There certainly is ample evidence that the $D°$ $(c\bar{u})$ and the D^+ $(c\bar{d})$ mesons exists and are stable (together with their charge conjugate states). Evidence for the third stable c-flavored meson F $(c\bar{s})$, is less concrete. In carrying out its program of searching for exclusive decays of charmed and beautiful mesons CLEO has searched for the F meson through one of its clearest signatures, the decay $F^{\pm} \rightarrow \phi\, \pi^{\pm}$, $\phi \rightarrow K^+K^-$, among the hadrons in e^+e^- annihilations, where F production is expected to be an appreciable fraction of the total cross section. First all identified K^+K^- are paired and if their mass is within ± 5 MeV of the ϕ mass, the kaon pair is combined with all $\pi^{\pm}$ in the event. This results in the mass spectrum of figure 28 where a very sharp peak is observed at $M(\phi\pi)$ = 1970±5±5 MeV [44]. 109 events are observed in this peak above a background of 176 events. Production cross section, decay angular distribution, observed width (consistent with being due entirely to the experimental resolution) and BR for the detected mode, all make a very convincing argument that the strange-charmed stable pseudoscalar meson has indeed been observed by CLEO.

WEAK INTERACTION OF THE b QUARK

The Standard Model

With all these preliminaries disposed of, let us turn now to

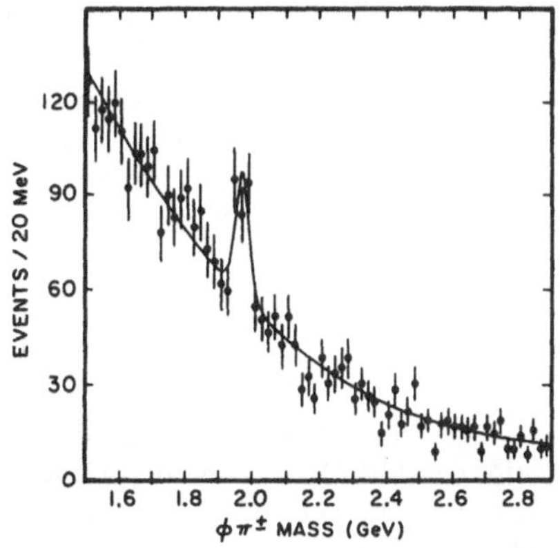

Fig. 28. F meson signal in the $\pi^{\pm}$ mass spectrum. (CLEO)

the description of weak interactions of quarks and leptons. In the standard $SU(2)_L$ model, with three quark generations, the weak interaction is due to a gauge coupling between the gauge bosons and left handed fermion weak isospin doublets :

$$\begin{pmatrix} e \\ \nu_e \end{pmatrix}_L \begin{pmatrix} \mu \\ \nu_\mu \end{pmatrix}_L \begin{pmatrix} \tau \\ \nu_\tau \end{pmatrix}_L \begin{pmatrix} u \\ d' \end{pmatrix}_L \begin{pmatrix} c \\ s' \end{pmatrix}_L \begin{pmatrix} t \\ b' \end{pmatrix}_L ,$$

allowing for the not yet discovered t-quark. The weak quark eigenstates d', s' and b' are related to the mass eigenstates d, s and b by a unitary matrix, first introduced by Kobayashi and Maskawa [45]. This mixing matrix generalizes the idea of Cabibbo universality [46] and maintains the validity of the GIM [47] cancellation of flavor changing neutral currents. It also allows for the introduction of a CP violating phase, although by no means requires it. In symbolic notation we write :

$$\begin{pmatrix} d' \\ s' \\ b' \end{pmatrix} = \begin{pmatrix} V_{ud} & V_{us} & V_{ub} \\ V_{cd} & V_{cs} & V_{cb} \\ V_{td} & V_{ts} & V_{tb} \end{pmatrix} \begin{pmatrix} d \\ s \\ b \end{pmatrix} .$$

Thus the decay of the b quark is induced by the couplings $(ud')_\alpha W_\alpha = ..V_{ub}(ub)_\alpha W_\alpha$ and $(us')_\alpha W_\alpha = ..V_{cb}(cb)_\alpha W_\alpha$. $(ub)_\alpha$ stands here for the standard V-A current : $\bar{u}\gamma_\alpha(1-\gamma_5)b$ and W_α is the weak interaction gauge boson (whose existence has finally been recently proven at CERN) [See C. Rubbia lecture]. As an example, the β-decay amplitude for the b quark is :

$$[V_{ub}(ub)_\alpha W_\alpha + V_{cb}(cb)_\alpha W_\alpha][W_\beta(e\nu)_\beta] .$$

From the amplitude of figure 23, taking into account phase space factors for the masses of quarks and leptons one obtains the total decay rate as [48] :

$$\Gamma(B) \sim \Gamma(b) \sim [(M_b{}^5 G_F{}^2)/(192\pi^3)](2.8|V_{cb}|^2 + 7.7|V_{ub}|^2).$$

Prior to the recent results on the study of the B decays, the relevant K-M matrix elements where only constrained by $|V_{ub}| < 0.09$ and $|V_{cb}| < 0.78$, both from the unitarity requirements [49]. This in turn reflects in the bounds $\tau_B > 6.6\times10^{-15}$ s and $0 < BR(b\to u)/BR(b\to c) < \infty$.

What if There Is No Top

Since the top quark has not been found yet, we must ask what would the weak interaction of the b quark be if the t quark does not exist. Within the context of the standard model a b quark without a charge 2/3 partner would have to be assigned to a weak isospin singlet. In this case the GIM cancellation is not possible and the b-quark decay is mediated by both the W's and the Z°, resulting in the possibility of decays like $b \to ce^+e^-$ etc. Kane and Peskin [50] have shown, under very general assumptions, that it is possible in this case to establish a lower bound for $\Gamma(b \to \ell^+\ell^-X)/\Gamma(b\to \ell\ \bar{\nu}\ X)$ of 0.125. CLEO [51] has found for the same ratio an upper limit of 0.08 (90% c.l.), thus proving that, in the standard model, the b-quark must have have a partner with which to form a doublet. Several models [52] have also been proposed for the decay of the b quark which are not mediated by the weak gauge bosons. All proposed models results in a large fraction of the initial energy disappearing in T''' decays, either because it is carried away by neutrinos or because of copious baryon production resulting in an apparent missing energy of the order of the proton mass. All these models have been excluded by both CUSB and CLEO by measurements of the average visible energy in direct T''' decays as compared to continuum and resonance decays or to Monte Carlo calculations. Figure 29 shows the CUSB results [53] for the average of E_{vis}/W at the bound upsilons, in the continuum and for T''' decays, after continuum subtraction. The lower value observed at the T''' is in very good agreement with being due to B semileptonic decays as predicted by the standard model and provides strong evidence against the proposed new couplings. Quantitatively one can establish that the various 'exotic' couplings proposed contribute to less than 3 to 6% of all B decays (90 % c. l.). Correlations between visible energy and semileptonic branching ratios have been computed by Monte Carlo methods by CLEO for all proposed models. They are displayed in figure 30 together with

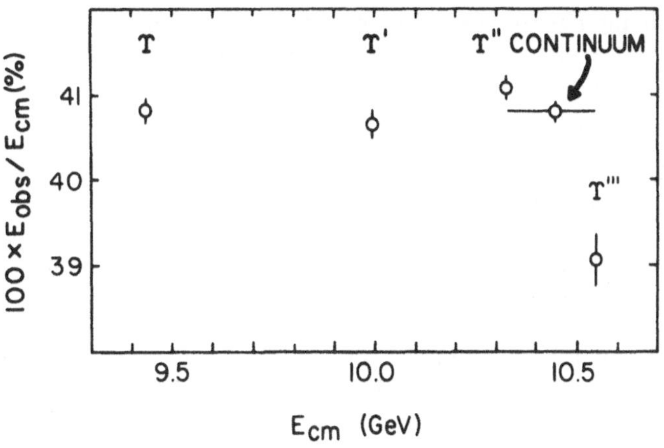

Fig. 29. Fractional visible energy observed in CUSB. The lower value observed at the T''' is due to energy carried away by neutrinos from B decays, in good agreement with expectations in the standard model.

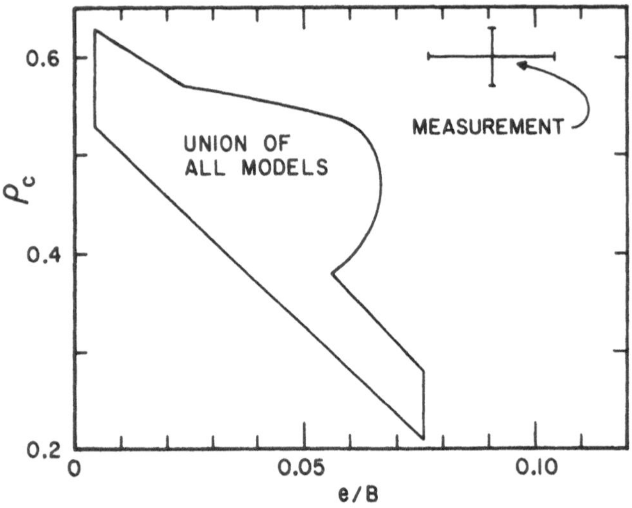

Fig. 30. Charged energy fraction ρ_C versus electron yield e/B as measured in CLEO at the Υ''' and as calculated for all proposed exotic models, ref. 52.

their measurements, again giving strong evidence against exotic couplings [54].

Semileptonic Decays of the B Meson

The branching ratio (BR) for $B \to e\nu X$ has been measured both by CUSB and CLEO and the corresponding BR for muons has also been obtained by CLEO. The three results are :

$BR(B \to eX) = 13.6 \pm 1.5$	CUSB	[55]
$BR(B \to eX) = 12.7 \pm 1.7$	CLEO	[55]
$BR(B \to \mu X) = 12.4 \pm 1.7$	CLEO	[55] ,

giving an average value of 13.0±0.9. A comparison of these results with theory is complicated by the uncertainties in the calculations of the corrections for the decay rate into non leptonic channels to which W exchange and quark annihilation amplitudes can contribute. We recall that in the case of the D meson these corrections are very large and result in the total decay rates differing by a factor of 2 to 5 five for neutral and charged meson. Moreover these estimates have had a strong tendency to follow the difference in the measured lifetimes of neutral and charged D meson, whose ratio appears finally to have settled around 2 [57]. An additional problem is due to the fact that the CESR measurements mentioned above are from an unknown mixture of charged and neutral B's, which naively is expected to be 40-60, using a mass difference of 4 MeV and phase space for $\Upsilon''' \to B\bar{B}$. However it is widely accepted that such corrections should be smaller for the B case, because of the much higher mass of the b quark with respect to the c quark. Also, the charged and neutral B total decay rates are expected to differ by only tens of per cent. Various calculations [58] give for the semileptonic BR of the B mesons values of 10 to 15 %, in reasonable agreement with the measured value above. It will certainly be valuable to measure independently the BR's of neutral and charged B's to directly confront calculations. The values of the semileptonic branching ratios are in principle independent of the magnitude of the quark mixing parameters V_{ub} and V_{cb} and are very insensitive to their ratio. In the following we shall show how information on the latter can be obtained from the lepton spectrum.

The Lepton Spectrum and Bounds on $|V_{ub}/V_{cb}|$

The V-A coupling for $b \to e\nu q$ (q stands for any quark) has the same fermion ordering as in the case of μ-decay. The electron spectrum is therefore given, ignoring the masses, by $d\Gamma/dx = 12x^2(3-2x)$ where $x = p_e/p_{max}$, $0<x<1$, $p_{max} = M_b/2$. The sharp drop of this spectrum at its end point is softened by (QCD) radiative correction and finite masses, while the end point is lowered by masses. Moreover in the decay of B mesons, rather than bare quarks, one has to estimate the final masses via the rather poorly understood mechanism of 'hadronization'. As a heuristic example we shall calculate the difference in end points for $b \to c$ and $b \to u$. The end point energy is given by $E_{max} = (M_b^2 - M_q^2)/(2M_b)$. Using $M_b = 5.2$ GeV, $M_c = 1.8$ GeV and $M_u = 0$ (because of the quadratic dependence of the end point on the quark mass, we can use $M_u = 300$ MeV without changing the result) we obtain $E_{max} = 2288$ MeV for $b \to c$ and $E_{max} = 2600$ MeV for $b \to u$, a very significant difference with respect to the experimental resolution of ~ 60 MeV (for CUSB). We can in fact argue that the above values for the end points are also valid for the B meson decays. The main contributions to X in $B \to e\nu X$ are D and D^* meson for $b \to c$ and π and ρ mesons for $b \to u$. Since the B, D, π and ρ masses are very similar to the ones used in the example above one obtains essentially the same answers. A self consistent approach to this problem has been used by Altarelli and collaborators [59] in the spirit of the spectator model. They consider the B meson as a bound state of a b and a light quark, moving with a gaussian distributed fermi motion. This motion is convoluted with the decay spectrum of the slightly off-the-mass-shell b quark including QCD radiative corrections. The motion of the B meson in the laboratory (T=16 MeV or β=0.078) is finally folded in. The results of this model are used by CUSB [55] to compare with experimental results. Figure 31 gives the electron spectrum obtained at the peak of the Υ'''. This spectrum is due to : i) B meson decays, B mesons being approximately 30% of all the events, ii) the tail of fast moving D's from B decays, iii) the tail of directly produced

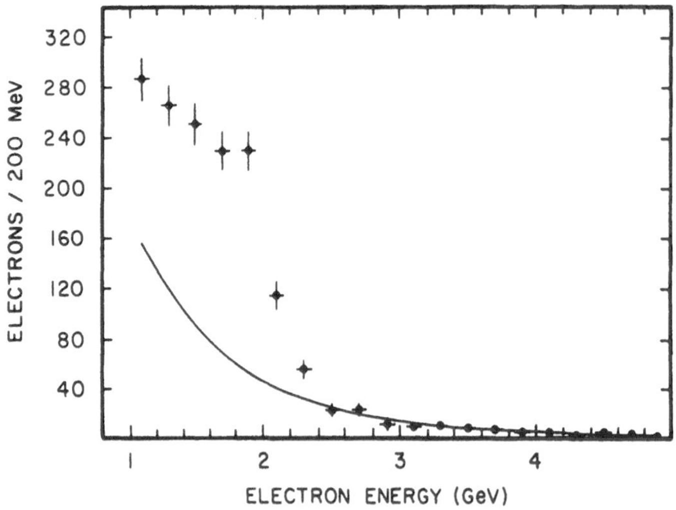

Fig. 31. Observed electron spectrum at the Υ''', data points. The continuous curve is the contribution from contnuum events. (CUSB)

D mesons, which are faster in the lab than the previous ones and are about 30% of all the events, iv) a ∿ 2% background due to the overlap of a π° and a charged pion, which is incorrectly called an electron. The contributions from iii) and iv) can be directly obtained from data in the continuum and at the Υ. In this way the continuous curve for the background is obtained. The electron spectrum after subtraction is shown in figure 32, together with the expected spectra for $b \to c$ and $b \to u$. The experimental spectrum agrees quite well with the curve for $b \to c$ and very little if any contribution from $b \to u$ is present. From these data CUSB concludes that $\Gamma(b \to u)/\Gamma(b \to c) < 0.055$ (90% c.l.) [55]. Using the appropriate phase space factors this limit corresponds to $|V_{ub}/V_{cb}|^2 < 0.026$ or $|V_{ub}/V_{cb}| < 0.16$. Last year CLEO [51] reported limits for $\Gamma(b \to u)/\Gamma(b \to c)$ of 10% for both decays with electrons and muons. Preliminary analysis of new data has brought down the limit to ∿ 5% [60]. Since both CUSB and CLEO do not observe any sign of $b \to u$ transitions, one could combine the above results to obtain $|V_{ub}/V_{cb}| < 0.09$. To be conservative we will use later $|V_{ub}/V_{cb}| < 0.15$.

Further Evidence for $b \to c$

Looking back at figure 23, if we assume that $c \to s$ dominantly and that for every strange quark at the parton level we should observe a K meson in the final state, we expect more K's in B decays if the (bc)W coupling dominates. Both CLEO [51] and CUSB [61] have

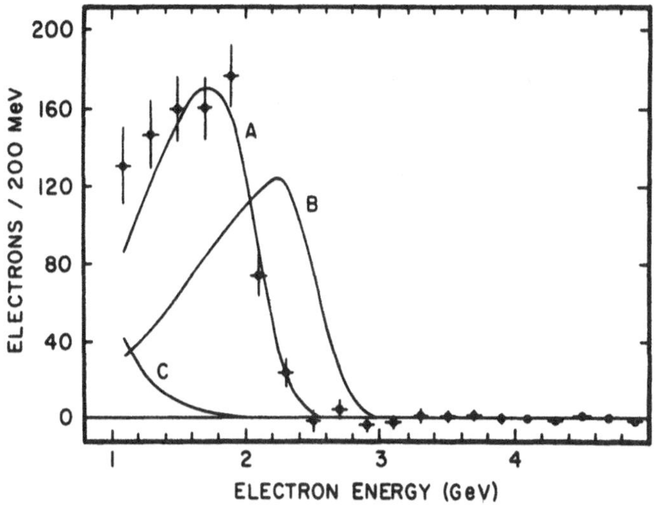

Fig. 32. The electron spectrum from B decay. The continuous curves are : A) the predicted spectrum for $b \to c$, B) same for $b \to u$ and C) the contribution from electrons from D's from B decays. (CUSB)

observed a large yield of K's in B decays, inconsistent with the coupling (ub)W being dominant. Unfortunately, it is not possible to convert the observed kaon yield into a stringent limit for $|V_{ub}/V_{cb}|$ because of uncertainties in strong interaction corrections and the poor understanding of the process of quark dressing.

Recently CLEO [62] has directly established that D mesons are abundantly produced in B meson decays, with a momentum spectrum appropriate to that expected from the amplitude of figure 23. They report that the D meson fraction from B decays is 0.8±0.2±0.2. While this results is of limited statistical significance, it lends further support, together with the shape of the momentum spectrum, to the conclusion that B mesons decay mostly according to $B \to (D \text{ or } D^*) + X$, as expected if $b \to c$ is dominant.

The Quark Mixing Matrix

While the study at CESR of the B mesons has led to a very complete picture in good agreement with expectations of the standard model and has allowed us to put strong bounds on the ratio of two of the elements of the mixing matrix, one additional measurement is necessary to obtain values for V_{ub} and V_{cb} as mentioned earlier. Only through the measurement of the lifetime this becomes possi - ble. Recently two experiments at PEP have reported results for the lifetime of the B meson. MAC [63] gives :

$$\tau_B = (1.8 \pm 0.6 \pm 0.4)\ 10^{-12}s \quad ,$$

and MARKII [64] obtains :

$$\tau_B = (1.2\ {}^{+\,0.45}_{-\,0.36} \pm 0.3)\ 10^{-12}s \ ,$$

while some time ago JADE had reported an upper limit of $1.4 \cdot 10^{-12}$s [65]. If we take as an example 10^{-12} seconds for the B meson lifetime, then, from the formula for the total rate given earlier and using $|V_{ub}/V_{cb}|^2 < 0.02$ from CESR, we obtain $V_{cb} \sim 0.05$ and $V_{ub} < 0.008$.

These results to a large extent define most of the mixing matrix which, apart from signs, looks now like :

$$V \sim \begin{pmatrix} 0.974 & 0.225 & <0.008 \\ -0.225 & 0.972 & 0.05e^{i\delta} \\ 0.01e^{i\delta}-(<0.008) & -0.05e^{i\delta}-(<0.008) & >0.999 \end{pmatrix} ,$$

a vast improvement with respect to expectations of only one year ago [49]. Many physical quantities reflect upon the mixing matrix elements. Ginsparg, Glashow and Wise [66] have recently derived a relation between the CP-non conservation parameter ε for K° mesons, the mixing angles and the top quark mass. For the above values this corresponds to a bound for the top mass of $M_{top} > 40$ GeV. Likewise $B\bar{B}$ mixing and CP violations depend on the values of the mixing angles and the top quark mass. Present estimates tend to give small mixing and very small CP violation. Theoretical uncertainties in the calculations of box diagrams and matrix elements of local quark operators make however these predictions not totally unambiguous. More measurements are therefore necessary before we can improve our knowledge of the mixing matrix.

Conclusions and Outlook for B Meson Physics

The study of B mesons at CESR has confirmed the validity of the standard model in the b quark sector and together with the lifetime measurements has almost completed our knowledge of the K-M matrix, except for the value of the CP violating phase δ. The mixing matrix appears to be almost diagonal and it should be noted that if any mixing angles were to vanish, the phase δ can be rotated away, thus removing the only 'natural' way of introducing CP violation in the standard model. It is therefore important to continue to improve the measurements of V_{ub}/V_{cb} to verify whether it is zero or just very small. In addition to the value of V_{ub} there are three other outstanding questions to be explored in B meson physics:

I) - B° - $\bar{B}$° mixing.
II) - If I allows, CP violation in B° decays.

III) - Is $\Gamma(B^{charged}) = \Gamma(B^{neutral})$?

for which the theoretical expectations are:

I) - Small.
II) - Very small.
III) - Decay rates should be nearly equal.

Experiments will have the last word but they are not going to be easy.

REFERENCES

1. P. Franzini and J. Lee-Franzini, Phys. Rep. 81 (1982) 239
2. P. Franzini and J. Lee-Franzini, Ann. Rev. Nucl. Part. Sci. 33 (1983) 1
3. K. Berkelman, to be published in Phys. Rep. (1983)
3a. D. Andrews et al., Nucl. Inst. Meth. 211 (1983) 47
4. T. Böhringer et al., Phys. Rev. Lett. 44 (1981) 1111
5. D. Andrews et al., Phys. Rev. Lett. 44 (1981) 1108
6. S. W. Herb et al., Phys. Rev. Lett. 39 (1977) 252
K. Ueno et al., Phys. Rev. Lett. 42 (1979) 486
7. Ch. Berger et al., Phys. Lett. 76B (1978) 243 and 78B (1978) 176
C. W. Darden et al., Phys. Lett. 76B (1978) 246 and 78B (1978) 364
J. K. Bienlein et al., Phys. Lett. 78B (1978) 360
8. A. S. Artamonov et al. Phys. Lett. 118B (1982) 225
9. W. W. MacKay et al., to be published, (1983)
10. D. Andrews et al., Phys. Rev. Lett. 45 (1980) 1108
G. Finocchiaro et al., ibid. p. 1111
11. R. M. Barnett et al., Phys. Rev. D22 (1980) 594
12. J. M. Blatt and V. F. Weiskopf, Theoretical Nuclear Physics (J. Wiley, New York, 1952) p. 423
13. E. Eichten et al., Phys. Rev. D17 (1978) 3090 and D21 (1980) 203
14. K. H. Krasemann and S. Ono, Nucl. Phys. B154 (1979) 283
15. W. Büchmuller et al., Phys. Rev. Lett. 45 (1980) 103, 587(E) also
W. Büchmuller and S.-H. H. Tye, Phys. Rev. D24 (1981) 132
16. A. Martin, Phys. Lett. 93B (1980) 338 and 100B (1981) 511
17. C. Quigg and J. L. Rosner, Phys. Rep. 56 (1979) 167
18. T. Appelquist and H. D. Politzer, Phys. Rev. Lett. 34 (1975) 43
19. K. Koller and T. F. Walsh, Nucl. Phys. B140 (1978) 449
20. K. Koller and K. H. Krasemann, Phys. Lett. 88B (1979) 119
21. D. Peterson et al., Phys. Lett. 114B (1982) 277
22. G. Mageras et al., Phys. Rev. Lett. 46 (1981) 1115
J. Mueller et al., Phys. Rev. Lett. 46 (1981) 1181
23. K. Gottfried, in Proc. Int. Symp. on Lepton and Photon Interactions at High Energy, Hamburg 1977, Ed. F. Gutbrod (DESY, Hamburg)
24. G. Mageras et al., Phys. Lett. 118B (1982) 453
J. Green et al., Phys. Rev. Lett. 49 (1982) 617
25. Y.-P. Kuang and T.-M. Yan, Phys. Rev. D24 (1981) 2874, also
T.-M. Yan, Phys. Rev. D22 (1980) 1652

26. R. Barbieri, R. Gatto and R. Kögler, Phys. Lett. 60B (1976) 183
R. Barbieri, R. Gatto and E. Remiddi, ibid. 61B (1976) 465
27. C. Klopfenstein et al., Phys. Rev. Lett. 51 (1983) 160
28. K. Han et al., Phys. Rev. Lett. 49 (1982) 1612
29. F. Pauss et al., Phys. Lett., (1983) in print.
30. G. Eigen et al., Phys. Rev. Lett. 49 (1982) 1616
31. M. Chanowitz, Phys. Rev. D12 (1975) 918
32. W. Büchmuller, in Proc. Moriond Workshop on New Flavors, ed. J. Tran Thanh Van, L. Montanet, Gif-sur-Ivette, Editions Frontière (1982) p. 91
33. C. Quigg and J. L. Rosner, Phys. Rev. D23 (1981) 2625; also C. Quigg, H. B. Thacker and J. L. Rosner, ibid. D21 (1980) 234
34. K. H. Kraseman, CERN rep. Th. 3036 (1981)
35. S. N. Gupta et al., Phys. Rev. D26 (1982) 3305
36. R. McClary and N. Byers, UCLA Rep. UCLA/82/TEP/12 (1982)
37. M. Voloshin et al., ITEP Rep. ITEP-21 (1980)
38. R. A. Bartelman, CERN Rep. TH-3192 (1981)
39. K. Chadwick et al., Phys. Rev. Lett. 46 (1981) 88
40. L. J. Spencer et al. Phys. Rev. Lett. 47 (1981) 771
41. E. Eichten, Phys. Rev. D22 (1980) 1819
42. R. D. Schamberger et al., Phys. Rev. D26 (1982) 720
43. S. Behrends et al., Phys. Rev. Lett. 50 (1983) 881
44. A. Chen et al., Phys. Rev. Lett. 51 (1983) 634
45. M. Kobayashi and T. Maskawa, Prog. Theor. Phys. 49 (1973) 652
46. N. Cabibbo, Phys. Rev. Lett. 10 (1963) 531
47. S. L. Glashow, J. Iliopoulos and L. Maiani, Phys. Rev. D2 (1970) 1285
48. M. K. Gaillard and L. Maiani, in Proc. of the 1979 Cargèse Summer Institute on Quarks and Leptons, Ed. M. Levy et al., Plenum Press, New York, 1979, p. 433
49. L. Maiani, J. Phys. (Paris) Colloq. 43 (1982) C3-631
50. G. L. Kane and M. E. Peskin, Nucl. Phys. B195 (1982) 29
51. B. Gittelman, J. Phys. (Paris) Colloq. 43 (1982) C3-110
52. H. Georgi and M. Machacek, Phys. Rev. Lett. 43 (1979) 1639
E. Derman, Phys. Rev. D19 (1979) 319
R. N. Mohapatra, Phys. Lett. 82B (1979) 101
H. Georgi and S. L. Glashow, Nucl. Phys. B167 (1980) 173
53. P. Franzini, J. Phys. (Paris) Colloq. 43 (1982) C3-114
54. A. Chen et al., Phys. Lett. 122B (1983) 317
55. C. Klopfenstein, Phys. Lett., in print. (1983)
56. K. Chadwick et al., Phys. Rev. D27 (1983) 317
57. G. Kalmus, J. Phys. (Paris) Colloq. 43 (1982) C3-431
58. J. Leveille, ref. 32, p. 191
59. G. Altarelli et al., Nucl. Phys. B208 (1982) 365
60. B. Gittelman and S. Stone, private communication
61. L. J. Spencer et al., Nucl. Phys. B206 (1982) 1
62. J. Green et al., Phys. Rev. Lett. 51 (1983) 347
63. E. Fernandez et al., Preprint SLAC-PUB-3154
64. N. S. Lockyer et al., Preprint SLAC-PUB 3165 (1983)
65. W. Bartel et al., Phys. Lett. 114B (1982) 71

66. P. H. Ginsparg, S. L. Glashow and M. B. Wise, Phys. Rev. Lett. 50 (1983) 1415

SELECTED TOPICS IN UPSILON PHYSICS

Juliet Lee-Franzini

State University of New York at Stony Brook

INTRODUCTION

In this past week you have heard much about e^+e^- physics, in particular from Paolo Franzini on the triplet S $b\bar{b}$ bound states with their hadronic and photonic transitions, as well as on the properties of the b-flavored mesons; from Sau Lan Wu on the arduous task of extracting the strong coupling constant α_s at PETRA; and from André Martin on the theory of quarkonia. What I am going to talk about at this last lecture of the section devoted to Janus' backward glance are some selected topics in upsilon physics, or bringing up the rear (bottom) while forecasting what information is likely to be available the next time the Cargèse summer institute on particles and fields convenes.

I intend to cover the following topics: a) search for other states produced in e^+e^- annihilations, b) further tests of the QCD multipole expansions, c) the B^*-B mass difference, d) further implications of the P state measurements, e) the determination of α_s from $\Gamma_{\gamma gg}/\Gamma_{ggg}$ and from $\Gamma_{\mu\mu}/\Gamma_{ggg}$, f) searches for axions and Higgs.

SEARCH FOR OTHER STATES PRODUCED IN e^+e^- ANNIHILATIONS

Paolo Franzini (PF) has presented in his figure 7 the CUSB measurements of $\sigma(e^+e^- \to \text{hadrons})$ at CESR around the known Υ's, measurements of the cross sections in the intervening regions are shown in the bottom half of figures 1 through 3. No obvious structure is seen. In order to quantify that negative result we use a maximum likelihood method to obtain the 90% confidence level upper limit (90% cl) for the leptonic width of a resonance whose total width is given by the machine energy spread. This result as a function of

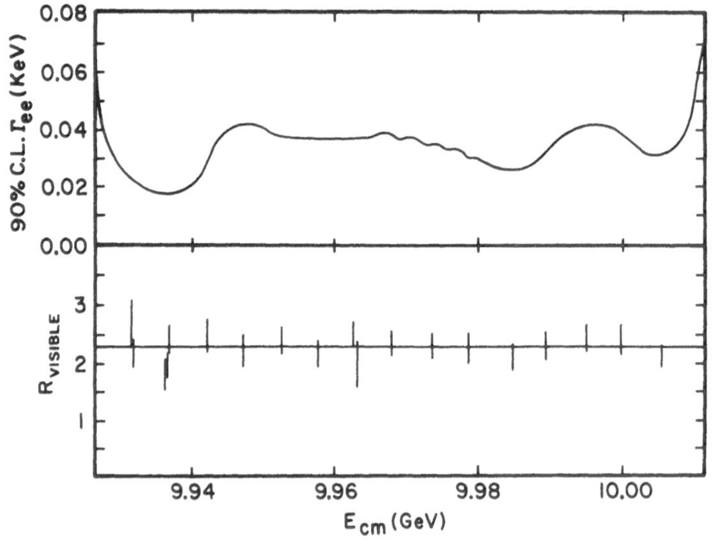

Fig. 1. 90% cl upper limit on Γ_{ee} vs c. of m. energy from 9.927 to 10.008 GeV.

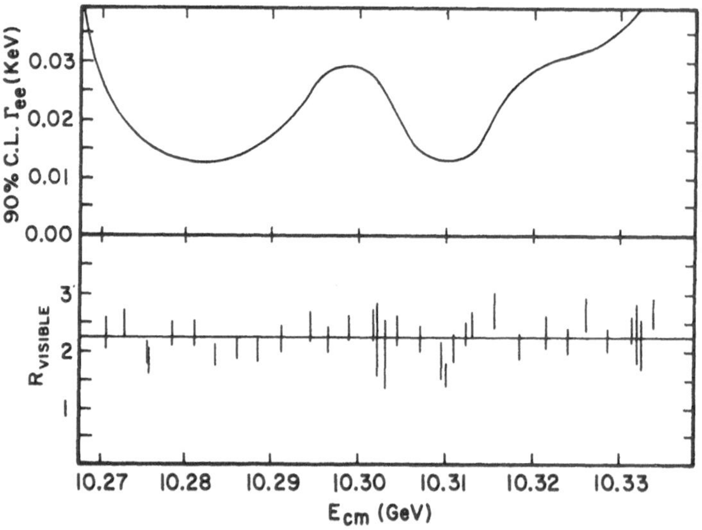

Fig. 2. 90% cl upper limit on Γ_{ee} vs c. of m. energy from 10.27 to 10.33 GeV.

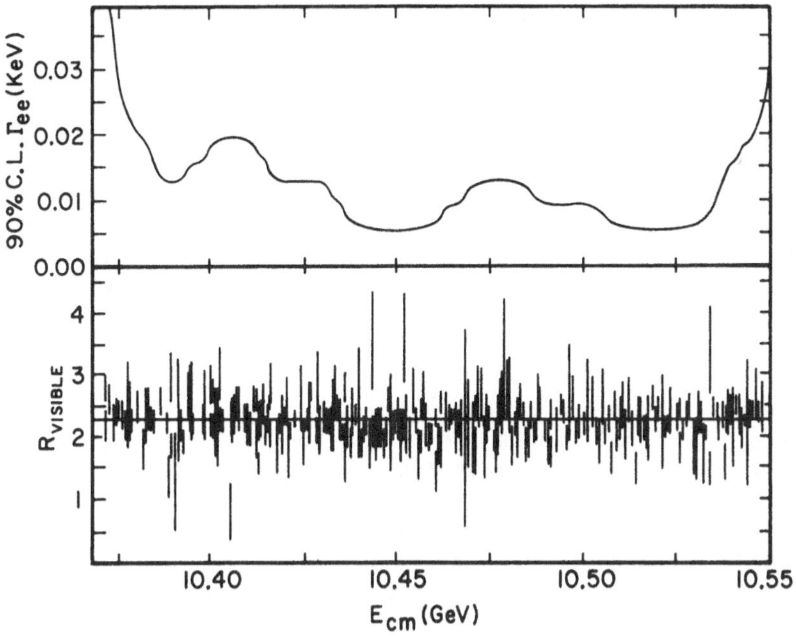

Fig. 3. 90% cl upper limit on Γ_{ee} vs c. of m. energy from 10.38 to 10.55 GeV.

mass (or W, total energy in the center of mass) is shown at the top of figures 1-3. They range around ∿ 32 eV in the mass interval of 9.927 to 10.008 GeV (figure 1), ∿ 24 eV in the mass interval of 10.27 to 10.33 GeV (figure 2), and ∿ 14 eV in the mass interval of 10.38 to 10.55 GeV (figure 3). These limits are an order of magnitude smaller than those expected for a bound triplet S state, therefore confirming that only three bound triplet S exist below flavor threshold.

We had expected to see the string vibrational states, predicted by Buchmuller and Tye to lie in the mass region between 10.37 to 10.46 GeV, with leptonic widths of the order of 70 to 270 eV[1]. However, to make these predictions they had to take a phenomenological approach of identifying either the $\psi(3.96)$ or $\psi(4.03)$ as a $(c\bar{c})$ vibrational state. Therefore our negative finding can mean either vibrational states do not exist or that the extrapolation scheme is inadequate. The latter possibility is more likely since Kuang and Yan[2] use string intermediate states for their rate computations of hadronic transitions which are in good agreement with experimental measurements.

Two triplet D-wave states are expected to lie below the b-flavor threshold, the centers of gravity (cog) of which are predicted almost uniformly by various potential models at ∿ 10.15 GeV for the

T(1D) and $\sim$ 10.43 GeV for the T(2D). Their expected leptonic widths are highly model dependent because the D-wave states can be seen in e^+e^- annihilations only through S-D mixing. In particular, the D-wave leptonic widths are given by the formula

$$\Gamma(^3D_1 \to \ell^+\ell^-) = (200/64m_b{}^6)\alpha^2 e_b{}^2 |R_D''(0) + (2\sqrt{2}/5)m_b{}^2 R_S(0)|^2,$$

where R are the radial wave functions and m_b, e_b are the b quark mass and electric charge respectively[3]. Moxhay and Rosner[4] find them to be 1.5 eV for the T(1D) and 2.7 eV for the T(2D), both values too low for present experimental sensitivity. The present conclusion is that no unusually large S-D mixing has been observed.

Above the free b-flavor threshold the radial excitations are expected to have total widths wider than the machine energy spread, as is true of the T(4S). The leptonic widths are expected to be somewhat less than that of the T(4S) since Γ_{ee} roughly scales as 1/n where n is the principal quantum number. At the bottom of figure 4 we show the resonance search for the T(5S, 6S, 7S) which most potential models predict to lie in the energy region between 10.6 to 11.3 GeV, spaced $\sim$ 200 MeV apart[5]. At the top of the figure we display the 90% cl upper limit for the leptonic width assuming that the total width of the resonance is 40 MeV ($\sim$ twice that of the T(4S))[6]. We see from these limits of $\sim$ 60 eV's ($\sim 1/4\Gamma_{ee}$ of the T(4S)), that the higher T(nS) are not narrow like the T(4S) but are probably broad shallow structures overlaying the continuum.

FURTHER TESTS OF THE QCD MULTIPOLE EXPANSIONS

The transition $T' \to T\pi\pi$ is observed in the CUSB detector by seeing two charged tracks or four γ's (from two π°'s) accompanying a high energy dilepton pair resulting from the decay of the T. During our recent run we have increased our sample of such events by over an order of magnitude relative to the 1980 run. We now have 221 $\pi^+\pi^- ee$ events, 147 $\pi^+\pi^-\mu\mu$ events and 44 $\pi^\circ\pi^\circ\ell\ell$ events. Our present branching ratio (BR) BR($T' \to \pi^+\pi^- T$) = (19.4±2.9)% is fully in agreement with our previous determination of (20±7)%, the error in the present case is dominated by the measurement uncertainty of BR($T \to \mu\mu$), $B_{\mu\mu}$. From the observed number of $\pi^+\pi^-\ell\ell$ events we expect 37±10 $\pi^\circ\pi^\circ\ell\ell$ events if the $\pi\pi$ system were in an I=0 state, 0 such events for the $\pi\pi$ in an I=1 state and 148 such events for the $\pi\pi$ in an I=2 state. The observed number is 44 events, greatly favoring I=0 as is required by isospin invariance in this transition. Of course, the new partial rate for this transition in the ($b\bar{b}$) system is still 9 keV and its ratio to the corresponding one in the ($c\bar{c}$) system (110 keV) is still the best proof that these transitions are two "color-electric" dipole transitions and that the gluon has spin one.

Another scaling law obtainable from QCD multipole expansion predicts the relative rates of the $T' \to \eta T$ and $\psi' \to \eta\psi$ transitions to be

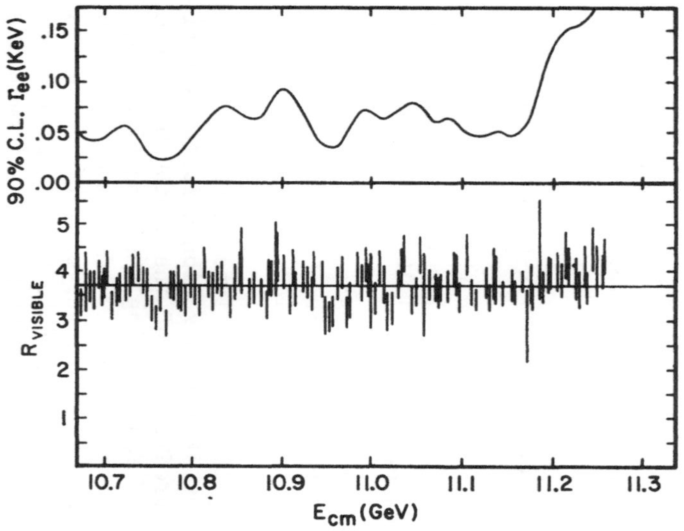

Fig. 4. 90% cl upper limit on Γ_{ee} vs c. of m. energy from 10.63 to 11.25 GeV.

1 to 400^2. Such transitions can be thought of as two M1 (or E1-M2) transitions whose matrix element scales inversely as the mass of the quark squared times the customary phase space factor. In other words

$$\Gamma(T' \to \eta T)/\Gamma(\psi' \to \eta\psi) \sim (m_c/m_b)^4 (P_\eta(T)/P_\eta(\psi))^3 \sim 1/400 \ ,$$

where the P_η's are the η momenta, their ratio cubed is 1/6, while the ratio of the quark masses to the fourth power is over ten times smaller, $\sim$ 1/67.

CUSB has searched for $T' \to \eta T$ transitions where again the T decays into a lepton pair and the η decays either into two photons or into $\pi^+\pi^-\pi^\circ$. The η is almost at rest, the minimum angle between the two γ's is 160° and their energies are 280±10 MeV. For the three pion decay mode we looked for two low energy charged pions and two photons accompanying the T decay dileptons. We found no candidates and obtain a 90% cl upper limit of

$$BR(T' \to \eta T) < 0.2\% \text{ and } \Gamma(T' \to \eta T) < 0.06 \text{ keV}.$$

These limits when combined with those from the ψ system $BR(\psi' \to \eta\psi) = (2.8\pm0.6)\%$ and $\Gamma_{tot}(\psi') = (215\pm40)$ keV[7] yield $\Gamma(T' \to \eta T)/\Gamma(\psi' \to \eta\psi)|_{experimental} < 1/100$.

We note that this limit is a factor 15 smaller than that expected from phase space suppression alone and goes a long way towards confirming the QCD multipole expansion picture.

While the scaling laws are derived in the QCD multipole expansion method independent of hadronization details, additional assumptions are needed for predictions of the dipion mass spectrum and angular distributions. Yan[8], noting that the pions involved in these hadronic transitions have relatively small momenta, adapted soft pion theorems where PCAC (partially conserved axial current)and current algebra constrain the variation of soft pion emission on pion momenta. The matrix element for $\pi\pi$ transitions from higher triplet quarkonia states to lower ones is proportional to $(A q_1^\mu q_{2\mu} + B q_1^\circ q_2^\circ)$ where q1 and q2 are the two pions' four momenta. Brown and Cahn[9] had noted in the study of the charmonium state transitions that the A term is dominant, and Yan had in his original formulation assumed the B term to be negligible. This assumption uniquely determines the dipion mass spectrum (to be peaked at high dipion mass) and requires the dipion to be isotropic around the colliding beam axis. With our recent data we have measured the dipion mass distribution, shown in figure 5. Curve a (b) is the spectrum resulting from the above transition amplitude with B=0 (A=0), the curves are normalized to the number of events. Our data is in excellent agreement with curve a, χ^2=9 for 13 degrees of freedom. A fit to the data for the best value of B/A gives B/A = - 0.02±0.09. The curve in figure 6 is for an isotropic distribution corrected for acceptance and normalized

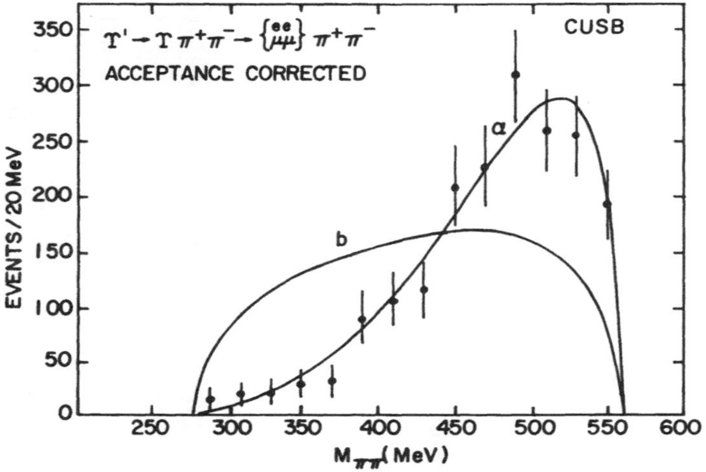

Fig. 5. Acceptance corrected $M_{\pi\pi}$ spectrum from $\Upsilon' \to \Upsilon\pi\pi$. See text for explanation of curves.

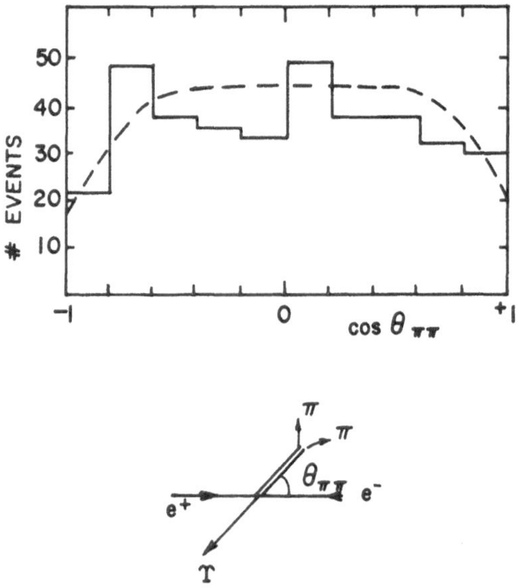

Fig. 6. Distribution of cos $\Theta_{\pi\pi}$, the angle of the dipion system with respect to the beam axis. The curve is an isotropic distribution corrected for acceptance.

to the number of events. The data are clearly consistent with s-wave production of the dipion system.

This outstanding confirmation of the QCD multipole expansion + PCAC predictions for the dipion mass and angular distributions also exists for the corresponding charmonium excited triplet S to ground triplet S state $\pi\pi$ transitions. However I would like to recall that in the decay $\Upsilon'' \rightarrow \pi\pi\Upsilon$ the dipion mass spectrum is not as expected : the CUSB spectrum is shown in figure 7 where the data is perfectly consistent with phase space[10] and CLEO had obtained $(B/A)^{-1} = -0.15 \pm 0.12$, that is, a spectrum reasonably consistent with phase space[11]. No explanation of this deviation from an otherwise highly successful theory has materialized yet.

THE B^*-B MASS DIFFERENCE

At the Υ(4S) CLEO has recently observed exclusive decay modes of b-flavored mesons[12] and obtained a mass of 5274.2±3.9 MeV for the neutral state and 5270.8±4.3 MeV for the charged state. Their particular method of identifying these b-flavored states is insensitive to the presence of additional low energy photons, therefore they can not tell whether they have reconstructed B's or B^*'s whose main decay mode would be $B^* \rightarrow \gamma+B$ as long as $M(B^*)-M(B) < m_\pi$. They have established however, that the threshold for producing a pair of such b-flavored mesons is 32.4±5.0 MeV below the Υ(4S). The question is then, is the Υ(4S) a $B\bar{B}$, $B\bar{B}^*$ or $B^*\bar{B}^*$ factory?

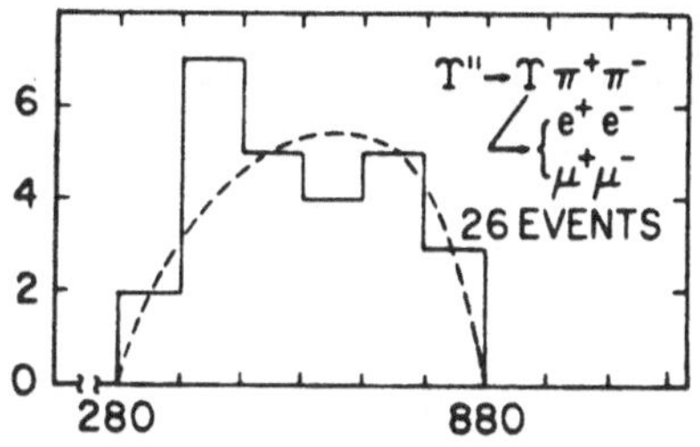

Fig. 7. $M_{\pi\pi}$ spectrum from $\Upsilon'' \rightarrow \Upsilon\pi\pi$.

The hyperfine splitting between the vector and scalar mesons is traditionally thought of as being due to a short range force resulting from a single hard gluon exchange, and thus is expected to scale inversely as the mass of the system[13]. Hence a simple estimate of the B^*-B mass difference is to scale it from D^*-D mass difference inversely as the masses of the heavy quarks (b,c) :

$M(B^*)-M(B) \sim \{m_c/m_b\}\{M(D^*)-M(D)\} \sim 50$ MeV.

CUSB had searched for such 50 MeV γ's at the Υ(4S) with null result[14]. However, aside from the range of numbers one might choose for the masses of the heavy quarks, several more refined models for calculating hyperfine splitting in heavy quarkonia have been extended to light-heavy and light-light quarkonia so that the B^*-B mass difference has been predicted to be even as low as 23 MeV[15] and as high as around 60 MeV[16].

We (CUSB) therefore made a renewed effort to search for low energy photons at the Υ(4S) with a special algorithm optimized for finding very low energy photons. For this algorithm the acceptance and efficiency combined is $\sim \{6.2 + 2.2\mathrm{x}((E_\gamma-25)/20)\,\%$ where E_γ is measured in MeV and the resolution is given by $\sigma(E_\gamma) = 0.9\mathrm{x}(E_\gamma + 5)^{1/2}$ MeV. The photon spectrum from 50,000 Υ(4S) decays (obtained from 170,000 hadronic events taken at the Υ''' peak and subtracting 120,000 continuum events) is shown in figure 8. The curves a, b and c are the expected contribution if there were one 25 MeV, 50 MeV and 70 MeV γ respectively per each Υ(4S) decay, we clearly see no such signal. Using the maximum likehood method on the unsubtracted photon spectrum we obtained 90% cl upper limit for the branching ratio BR(Υ(4S)→monochromatic γ) as a function of $\Delta M = M(B^*)-M(B) = E_\gamma$. It is tabulated below:

E_γ(MeV)	20	25	30	35	40	45	50	55	60	65	70
BR (%)	0	0	0	0	0	0	0	0	0	0	0
90% cl upper limit	10%	10%	10%	9%	8%	8%	8%	9%	11%	11%	9%

To realize the implications of the upper limits we need to know the expected partial rates for $B\bar{B}$, B^*B and B^*B^* on the Υ(4S) as a function of ΔM. These partial rates also depend on whether the CLEO b-flavored particle pair production threshold is the $B\bar{B}$, $B\bar{B}^*$, or $B^*\bar{B}^*$ threshold. Since the Υ(4S) is so close to these thresholds, no extra particles are produced and we can write

$$\#\text{ of }\gamma\text{'s}/\Upsilon(4S) = \{R(B\bar{B}^*)+2R(B^*\bar{B}^*)\} \;/\; \{R(B\bar{B})+R(B\bar{B}^*)+R(B^*\bar{B}^*)\}.$$

The R's are given by $\sigma(e^+e^-\to B_aB_b)/\sigma(e^+e^-\to\mu^+\mu^-) = p_{ab}{}^{2\ell+1}|M(p_{ab})|^2S_{ab}$ where (a,b) refers to B or B^*, p_{ab} is their relative momentum, $M(p_{ab})$ is a matrix element that describes the decay and S_{ab} is a statistical factor which depends on the spin of the final state. It is 1:4:7 for $B\bar{B}:B\bar{B}^*+\bar{B}B^*:B^*\bar{B}^*$[17]. Since both the B and B^* have odd intinsic parity, so does the Υ(4S), ℓ has to be odd and most likely equals 1 as we are near threshold. Therefore R_{ab} reduces to $p_{ab}{}^3|M(p_{ab})|^2S_{ab}$. The momentum p_{ab} can be expressed simply in terms of the number of B^*'s contained in CLEO's reconstructed b-flavored pairs (i=0,1,2), the number of B^*'s produced per Υ(4S) (j=0,1,2) and $\Delta M=E_\gamma$ in MeV : $p_{ij} \sim \{5272(32.4-(i-j)E_\gamma)\}^{1/2}$. We assume that the

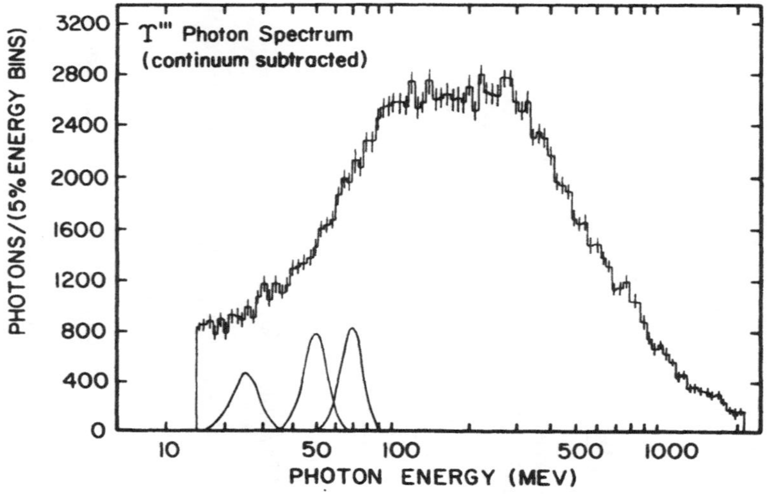

Fig. 8. Inclusive photon spectrum for 48000 Υ(4S) events after continuum subtraction. The solid lines show the expected photon signal for one photon per Υ(4S) decay for energies of (a) 25 MeV, (b) 50 MeV, (c) 70 MeV.

neutral and charged B mass difference can be neglected (in fact including the estimated mass difference of 2 to 4 MeV did not alter our results). The $R_{aa,ab,bb}$ can be estimated in two ways (i) by assuming $|M(P_{ab})|^2=1$, (ii) by using the T(4S) decay amplitude from reference 5. Method (ii), while obviously superior to method (i), is not strictly correct since we are using the T(4S) decay amplitude alone without considering mixing with other triplet S states.

The expected number of γ's per T(4S) as a function of E_γ and CLEO threshold is shown in figure 9 as computed using method (i) and in figure 10 as computed from using method (ii), superimposed on both is the experimental 90% cl upper limit. In either case we conclude that CLEO's reconstructed B meson pairs are $B\bar{B}$'s. We also conclude that the B^*-B mass difference is > 29 MeV at 90% cl from method (i) and > 31 MeV at 90% cl from method (ii).

FURTHER IMPLICATIONS OF THE P STATE MEASUREMENTS

Hadronic widths of the χ_b's

The two triplet P-wave (bb) bound states have been discovered by CUSB in 1982 (χ_b') and 1983 (χ_b), and PF has discussed how well the experimental E1 rates agree with those predicted by nonrelativistic potential models. Of considerable interest to QCD are the hadronic widths of these χ states which are predicted to within 30% accuracy to be ∿190 keV, 60 keV and 750 keV for the J=2,1,0 states respectively[2,18]. Their explicit dependence, to first order, on α_s is as follows :

$$\Gamma(\chi_{b,J=2} \to gg) \sim (128/5)\alpha_s{}^2(M_P)^{-4}|R_{P'}(0)|^2 ,$$

$$\Gamma(\chi_{b,J=1} \to q\bar{q}g_{soft}) \sim (128/3\pi)\alpha_s{}^2(M_P)^{-4}|R_{P'}(0)|^2 \ln(M_P r_P) ,$$

$$\Gamma(\chi_{b,J=0} \to gg) \sim 96\alpha_s{}^2(M_P)^{-4}|R_{P'}(0)|^2 ,$$

where M_P is the mass of the P state, R_P denotes its radial wave function and r_P is its radius[19,18].
We note that these widths can potentially give an independent means to measure α_s, as well as a check on these QCD computations. Experimentally, however, these widths are not directly accessible because they are so narrow. If we assume the E1 rates predicted by potential models to be correct, we can get an estimate of the hadronic widths using the CUSB measured BR($\chi_b \to \gamma$+T)'s. From

$$\Gamma(\chi_b \to \text{hadrons}) = \Gamma(E1)(1/BR-1) ,$$

we have :

$$\Gamma(\chi_{b,J=2} \to \text{hadrons}) = 152 + 63 - 50 \text{ keV},$$

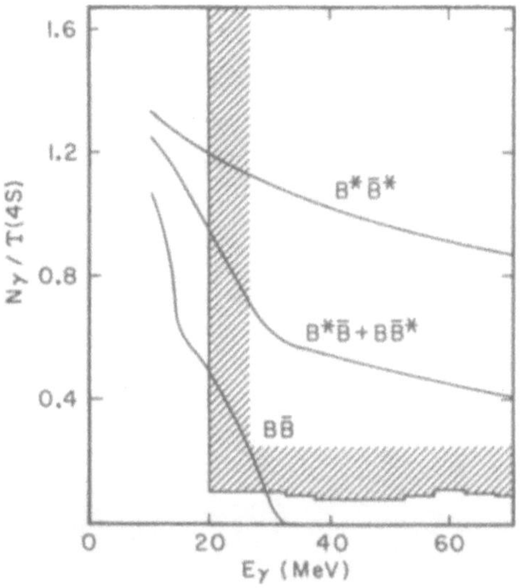

Fig. 9. Number of γ's expected per Υ(4S) as a function of E_γ, using method (i), see text. Region to the right of 20 MeV and above the CUSB 90% cl upper limit is excluded.

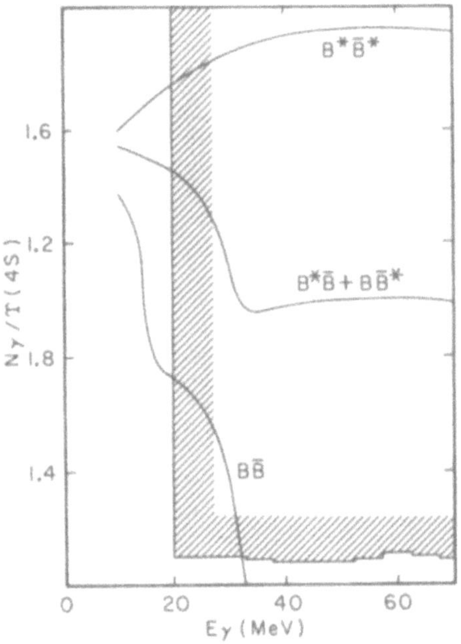

Fig. 10. Number of γ's expected per Υ(4S) as a function of E_γ, using method (ii), see text. Region to the right of 20 MeV and above the CUSB 90% cl upper limit is excluded.

and $\Gamma(\chi_{b,J=1} \to \text{hadrons}) = 41 + 40 - 20$ keV,

where the error is derived from measurement uncertainties only. Within the present errors these numbers are in agreement with the above quoted expectations. Clearly much more accuracy is needed both in the theoretical and experimental quantities.

P-states' centers of gravity and flavor independence

It has been shown by PF that at least four differently derived potential models describe the triplet S bound ($b\bar{b}$) level spacing and ratios of leptonic widths equally well. This is because all successful central potentials have the same radial dependence in the region probed by the above states, almost logarithmic, or if one were to parametrize the potential V(r) as Ar^v, v is ~ 0. Quigg and Rosner[20] have pointed out that ratios of level spacings can be directly related to the power v. Using the CUSB measured centers of gravity of the triplet bound P states of 9900 MeV for the 1^3P_1 and 10256 MeV for the 2^3P_1 and the curves of Quigg[21], we obtain $v \sim -0.15$ from figure 11 and $v \sim -0.22$ from figure 12. In figure 11 is also shown the corresponding point from the charmonium system[20] which yields $v \sim 0.15$. Note that all three values are close to zero, confirming that the potential is flavor independent. The small negative value of v of the ($b\bar{b}$) system indicates that the potential has become more Coulomb like at the shorter distances.

MEASUREMENT OF $\Gamma(\gamma gg)/\Gamma(ggg)$ IN Υ AND Υ' DECAYS

While the dominant decay mode for bound Υ's from $b\bar{b}$ annihilation is into three gluons, a finite fraction $\sim \alpha_{em}/\alpha_s$ could decay into a photon and two gluons (γgg). The ratio of these two partial rates, including first order QCD corrections is[22,23]:

$$\Gamma(\Upsilon \to ggg)/\Gamma(\Upsilon \to \gamma gg) = (5/36)(1/e_b^2)(1/\alpha_{em})\alpha_s(1-2.2\alpha_s/\pi).$$

Because of cancellations, the next term in the α_s/π expansion is expected to be small, hence this ratio could be the best source for determining α_s and the QCD scale parameter $\Lambda_{\overline{MS}}$, where the two quantities are related by[24]:

$$\alpha_s(Q^2) = \frac{4\pi}{\beta_0 \ln\{Q^2/\Lambda^2_{\overline{MS}}\}} - \frac{4\pi\beta_1 \ln\{\ln(Q^2/\Lambda^2_{\overline{MS}})\}}{\beta_0^3 \ln^2\{Q^2/\Lambda^2_{\overline{MS}}\}} ,$$

with $\beta_0 = 11 - (2/3)n_f$, $\beta_1 = 102 - (38/3)n_f$, $n_f = 4$, and the appropriate Q^2 to be used is $(0.157M_\Upsilon)^2$ [25].

The photon spectrum from the γgg decay is expected to peak towards the end of the kinematic spectrum. If we define $z = E_\gamma/E_{beam}$,

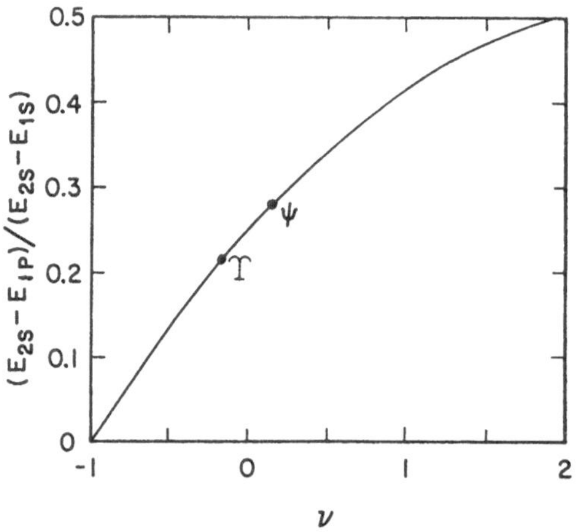

Fig. 11. Predicted dependence of level splitting ratio $\Delta E(2S-1P)/\Delta E(2S-1S)$ vs power of the potential[20]. The lower data point is the CUSB measured value.

the spectrum is given to 0th order by[26]:

$$dN/dz \sim \{2/(\pi^2-9)\}\{z(1-z)/(2-z)^2-2(1-z)^2\ln(1-z)/(2-z)^3+(2-z)/z+2(1-z)\ln(1-z)/z^2\},$$

which is essentially a straight line through the origin. Radiative corrections are expected to modify the shape of the end of the spectrum but not the integral from z_{min} to 1 of $\int(dN/dz)dz$ for reasonable z_{min}'s[27]; for example this integral is expected to be 0.75 for $z_{min}=0.5$. In contrast, the spectrum of π° decay photons resulting from the hadronization of the gluons of the three gluon decay mode is peaked towards very low energies ($z \sim 0.004$).

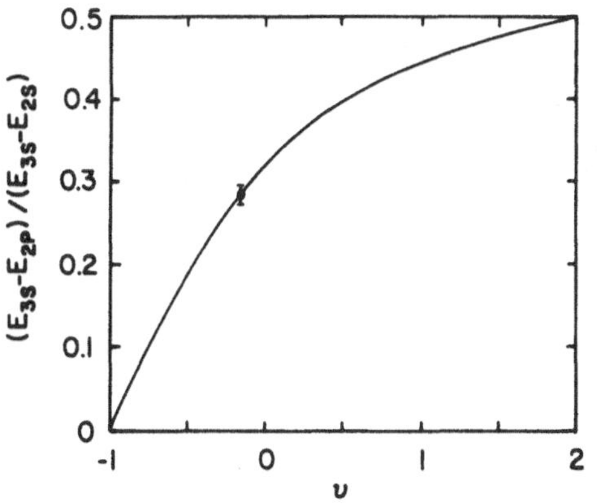

Fig. 12. Predicted dependence of level splitting ratio $\Delta E(3S-2P)/\Delta E(3S-2S)$ vs power of the potential[20]. The data point is the CUSB measured value.

Experimentally, the high energy neutral particle spectrum includes both π°'s and γ's since the opening angle between the two decay photons from the π° becomes at some energy smaller than the detector's angular resolution. For the CUSB detector, the π°'s are removed statistically using the properties of the radially layered active converters in the following way. We calculate the relative conversion probability of π°'s and direct γ's in each layer of sodium iodide (for a very thin layer that ratio is two). We count for each observed energy the fraction which had converted in the first layer, second layer,etc. From both pieces of information we obtain the fraction of direct photons contained in the spectrum as a function of photon energy and transform the neutral particle spectra taken at resonance peaks and on the continuum into the corresponding direct

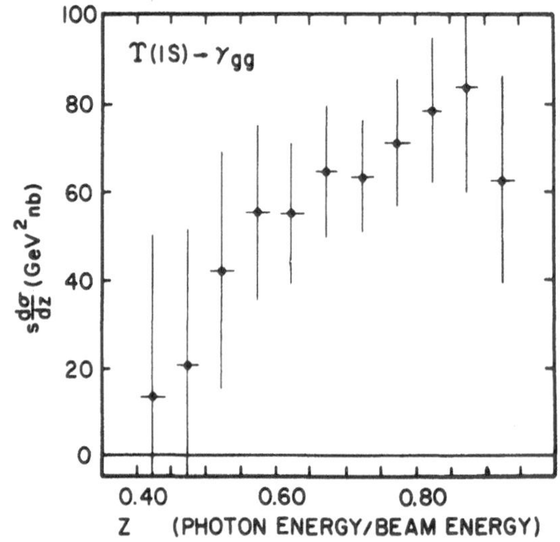

Fig. 13. Acceptance corrected spectrum of direct photons from Υ(1S) decays.

photon spectra. Since there are continuum (and continuum like, e.g. virtual γ's) contributions in each direct photon spectrum obtained at resonance peak energies, they are removed using the measured continuum direct photon spectrum weighted by the appropriate equivalent integrated luminosities. Finally, we have to fold in our hadronic event selection efficiency which is E_γ dependent and our high energy electromagnetic shower search code acceptance (very mildly E_γ dependent) to obtain the true direct photon spectra from the Υ(1S), figure 13 and from the Υ(2S), figure 14. We note that their shapes are

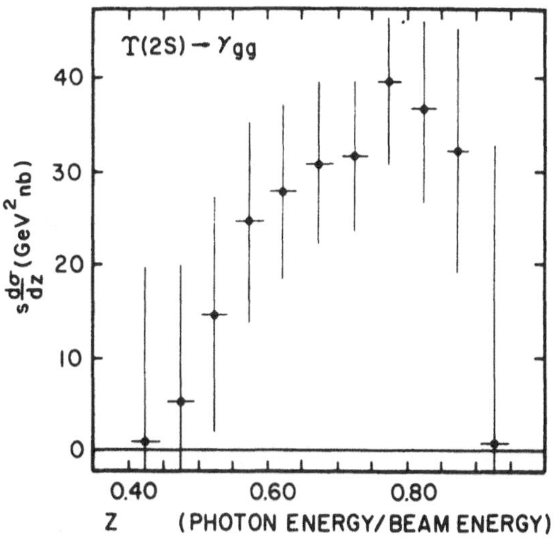

Fig. 14. Acceptance corrected spectrum of direct photons from Υ(2S) decays.

generally consistent with a linear dependence on z, modified by some rounding off at high z.

We note that these spectra are devoid of fine structure, which is consistent with the results of our exotic particle search (see later) and thus assume that they are due to γgg decays. The following tables summarize the luminosity, hadronic events observed, number of neutral particles, fraction of γ's in the spectrum averaged over the spectrum and number of direct photons obtained at various beam ener-

Table 1. Observed number of direct γ's

	Luminosity (pb^{-1})	Hadronic events observed	$N(\gamma+\pi^o)$	f_γ	n_γ
Υ region 9.452 GeV<E_{CM}<9.470 GeV	6.7	100,032	1079 ±33	0.31	334 ±27
Υ' region 10.012 GeV<E_{CM}<10.038 GeV	30.0	238,417	2507 ±50	0.30	750 ±50
continuum	24.2	59,563	866 ±29	0.16	139 ±27

Table 2. Number of γgg and ggg events

	N_γ	Hadronic events	$<\varepsilon_h(\gamma gg)>$	$<\varepsilon_\gamma>$	$N(\gamma gg)$	$N(ggg)$
Υ(1S) decays	275 ±28	74815	0.63	0.24	1958 ±198	89240
Υ(2S) decays	482 ±67	123792	0.60	0.22	3363 ±658	147662

gies (Table 1) and the number of direct γ's, hadronic events, average hadron and γ efficiencies, number of γgg and ggg decays from the Υ(1S) and Υ(2S) resonance decays (Table 2).
From the above numbers we obtain : $\Gamma(\gamma gg)/\Gamma(ggg) = 0.0292 \pm 0.0030 \pm 0.0029$ for Υ(1S) decays and $\Gamma(\gamma gg)/\Gamma(ggg) = 0.0328 \pm 0.0060 \pm 0.0066$ for Υ(2S) decays, where the first error is statistical, the second one systematic.
Using the formulae quoted at the beginning of this section we obtain :
$\alpha_s = 0.23 \pm 0.03$ and $\Lambda_{\overline{MS}}$ = 119 MeV (+52-34) MeV from Υ(1S) and
$\alpha_s = 0.20 \pm 0.05$ and $\Lambda_{\overline{MS}}$ = 81 MeV (+88-41) MeV from Υ(2S).

These determinations are in good agreement with those obtained from deep inelastic scattering data and are not subject to the uncertainties in hadronization details.

MEASUREMENT OF $\Gamma(\mu\mu)/\Gamma(ggg)$ FROM $\Upsilon(1S)$ AND $\Upsilon(2S)$ DECAYS

Using the CUSB detector we have measured $B_{\mu\mu}$ ($\Gamma_{\mu\mu}/\Gamma_{tot}$) for the $\Upsilon(1S)$ and $\Upsilon(2S)$, by measuring the rate of $e^+e^- \rightarrow \mu^+\mu^-$ both on and off resonance. This rate's increase on a resonance is due to the bound $(b\bar{b})$'s annihilation into a muon pair and is proportional to that resonance's $B_{\mu\mu}$.
Specifically what we measure is $\bar{B}_{\mu\mu} = \Gamma_{\mu\mu}/\Gamma_{hadrons}$ which is related through lepton universality to $B_{\mu\mu}$ $(= 3 + \bar{B}_{\mu\mu}^{-1})^{-1}$.

If we define $N_{\mu\mu}(tot)$ as the total number of $\mu\mu$ events occurring at the resonance energy, $N_{\mu\mu}(res)$ as the number of $\mu\mu$ events occurring at the resonance energy through resonance production, $N_{\mu\mu}(cont)$ as the number of $\mu\mu$ events occurring at the resonance due to continuum production, $\varepsilon_{\mu\mu}(res)$ as the efficiency for detecting $\mu\mu$ events due to resonance production, and $N_h(tot)$, $N_h(res)$, $N_h(cont)$ and $\varepsilon_h(res)$ the corresponding quantities for hadronic events then

$$B_{\mu\mu} = [N_{\mu\mu}(res)/\varepsilon_{\mu\mu}(res)] / [N_h(res)/\varepsilon_h(res)]$$

$$= \{[N_{\mu\mu}(tot)-N_{\mu\mu}(cont)]/\varepsilon_{\mu\mu}(res)\}/\{[N_h(tot)-N_h(cont)]/\varepsilon_h(res)\}$$

We determine $N_{\mu\mu}(cont)$ both at the $\Upsilon(4S)$ where $B_{\mu\mu}$ is zero and on the continuum between the $\Upsilon(4S)$ and $\Upsilon(3S)$, scaling these results by large angle Bhabhas to $\Upsilon(1S)$ and $\Upsilon(2S)$. $N_h(cont)$ is also determined from the measured rate of hadron production on the continuum between the $\Upsilon(4S)$ and $\Upsilon(3S)$ scaled by large angle Bhabhas to $\Upsilon(1S)$ and $\Upsilon(2S)$. When doing these scalings, we correct the Bhabha rate measured at the resonance energies for resonance production of e^+e^-. This correction is 3.3% for the $\Upsilon(1S)$ and 0.8% for the $\Upsilon(2S)$. We measured $\varepsilon_{\mu\mu}(res)$ by comparing the number of dimuon events seen on and below the $\Upsilon(4S)$ to the number calculated using the integrated luminosities determined by large angle Bhabhas and the continuum cross section. This continuum efficiency is then corrected, using the Monte Carlo of Berends and Kleiss[28], for initial state radiation effects, which occur in continuum production but not in resonance production. We measure

$$B_{\mu\mu}(1S) = (2.7 \pm 0.3 \pm 0.3)\% \text{ and } B_{\mu\mu}(2S) = (1.9 \pm 0.3 \pm 0.5)\% ,$$

where the first error is statistical and the second error is systematic whose dominant contribution is a 3% uncertainty in the relative Bhabha normalization used to scale $N_{\mu\mu}(cont)$. Using these values of $B_{\mu\mu}$ we can calculate $\Gamma_{ggg}/\Gamma_{\mu\mu}$[29]:

$$\Gamma_{ggg}/\Gamma_{\mu\mu} = (1/B_{\mu\mu})\{1-0.086/\pi-B_{\mu\mu}(3+R)-B_{\gamma gg}-B_{\pi\pi}-B_{E1}\} ,$$

where R = 3.63[6], $B_{\gamma gg}$'s were given in the previous section, $B_{\pi\pi}(1S) = B_{E1}(1S) = 0$ and $B_{\pi\pi}(2S) = 0.29 \pm 0.04$, $B_{E1}(2S) = 0.155 \pm 0.03$ [PF].

$$\Gamma_{ggg}(1S)/\Gamma_{\mu\mu}(1S) = 24.8 \ (+4.4-3.5),$$

$$\Gamma_{ggg}(2S)/\Gamma_{\mu\mu}(2S) = 19.6\ (+4.9-3.6),$$

where the errors are statistical only.

According to Brodsky, Lepage and Mackenzie[25], the appropriate scale Q^* to use in the Υ system for defining α_s is $0.157M_\Upsilon$, as we have done in the previous section when obtaining $\Lambda_{\overline{MS}}$ from $\Gamma_{ggg}/\Gamma_{\gamma gg}$. However, then

$$\Gamma_{ggg}/\Gamma_{\mu\mu} = \{10(\pi^2-9)/(81\ \pi e_b{}^2)/\alpha_{QED}{}^2\}\alpha_{\overline{MS}}{}^3(Q^*)$$
$$\times\{1-14.0\alpha_{\overline{MS}}(Q^*)/\pi + \ldots\}.$$

We note that the series is far from convergent, the correction term in the second bracket is almost equal to 1 and the next order term is surely large if this formula is to fit the above data. The authors suggest one should use the physical process described in the previous section to define α_D, then the correction term in the above bracket becomes $\{1- 7.4\alpha_D/\pi + \ldots\}$ and the new Q^* is $\sim 1.3\times0.157M_\Upsilon$. One notes that the second term is still of the order of 0.5, so the theoretical uncertainties are still large. We obtain from the above ratios

$$\alpha_D(1S) = 0.22 \pm 0.015,\ \Lambda_{\overline{MS}} = 130\ \text{MeV}\ (+54-24)\ \text{MeV from}\ \Upsilon(1S)$$

and

$$\alpha_D(2S) = 0.18 \pm 0.02,\ \Lambda_{\overline{MS}} = 75\ \text{MeV}\ (+56-34)\ \text{MeV from}\ \Upsilon(2S).$$

Despite the large experimental and theoretical uncertainties, it is very rewarding to see that the QCD scale parameter extracted using two completely different processes (from this section and the last section) agree so extremely well. It certainly adds credence to $\Lambda_{\overline{MS}}$ being around 100 MeV.

SEARCH FOR AXIONS AND HIGGS PARTICLES

All current models of the electroweak interactions require the existence of at least one pseudoscalar Higgs particle[30]. They could be almost massless, as the axions postulated by Weinberg[31] and Wilczek[32]. Or, if they acquire mass through radiative corrections, be quite light ($<M_\Upsilon$), like that of Nanopolous' $M_{H^\circ} \sim 4.4$ GeV[33]. The Higgs particle couples to quarks proportionally to the square of the quark mass which result in the prediction[34]

$$BR(\Upsilon\to\gamma+H)/BR(\Upsilon\to\mu\mu) = G_F M_\Upsilon{}^2/(4\sqrt{2}\alpha\pi)\times[1-M_H{}^2/M_\Upsilon{}^2]\ ,$$

where G_F is the Fermi coupling constant. Models with two Higgs fields modify this result by a factor $(v_1/v_2)^2$ for charge 1/3 quarks (Υ's) while the corresponding BR ratio for the Ψ is multiplied by

$(v_2/v_1)^2$, where v_1 and v_2 are the vacuum expectation values of the two Higgs fields[35].

The second model applies to the Weinberg-Wilczek axion, a, which is a very light (M<1MeV), long-lived, and semiweakly interacting pseudoscalar particle, distinguished by its 'invisibility'. Since the ratio of the two vev's ($X=v_1/v_2$) is not known, one cannot predict the BR's for either Υ or $\Psi\to\gamma+a$. However their product is given by :

$$BR(\Upsilon\to\gamma+a)/BR(\Psi\to\gamma+a) = B_{\mu\mu}(\Upsilon)B_{\mu\mu}(\Psi)G_F{}^2M_\Upsilon{}^2M_\Psi{}^2/\{32\alpha^2\pi^2\}$$
$$= (1.6 \pm 0.3) \times 10^{-8}.$$

We searched for these decays in CUSB, where the event signature is a single neutral electromagnetic shower with $E_\gamma \sim E_{beam}$, accompanied by nothing else in the detector since the axion does not interact and if its mass is less than 10 MeV, would decay outside the detector. We found no example of such events in an equivalent sample of $\sim$ 60000 Υ decays, giving an upper limit of 1.2×10^{-4} for the BR[36]. Combining this result with the limits of Edwards et al.[37] for $\Psi\to\gamma+a$ of 1.4×10^{-5}, we quote a limit of 0.6×10^{-9} for the above product of the BR's at 90% cl, an unambiguous result arguing against the existence of the axion.

We have also searched for Higgs from the decay $\Upsilon\to\gamma+H$ where the Higgs decays into hadrons. The inclusive photon spectrum from $\sim$80000 Υ's is shown in figure 15. It can be fit by a smooth polynomial function (the curve in the figure). A maximum likehood search was made for monochromatic photon signals, the 90% cl upper limit for the BR($\Upsilon\to\gamma H$) calculated as a function of the recoil mass is shown in figure 16. We are at present not sensitive to the expected rate for the single Higgs model (shown as the curve in figure 16). For models with two Higgs, the above BR is enhanced by a factor of X^2, and we can exclude a large region in the (X, M_H) plane, as is shown in figure 17.

FUTURE PROSPECTS IN Υ PHYSICS

It is irresistible for me to venture a forecast of our status in 1985, despite the risk of doing such in our fast moving field. On the experimental side I believe the fine splitting will have been well measured, providing strong contraints on model builders of the spin dependent interquark forces (spin-orbit and tensor interactions). We will probably not have found the η_b yet. We will have definitely established whether $M_{H^\circ}<M_\Upsilon$ even in the single Higgs models. On the theoretical side, I hope very much that G. Parisi[38] and company will have calculated the central potential responsible for quark binding. Perhaps it is too much to ask that he will have done the same for the spin-dependent ones as well.

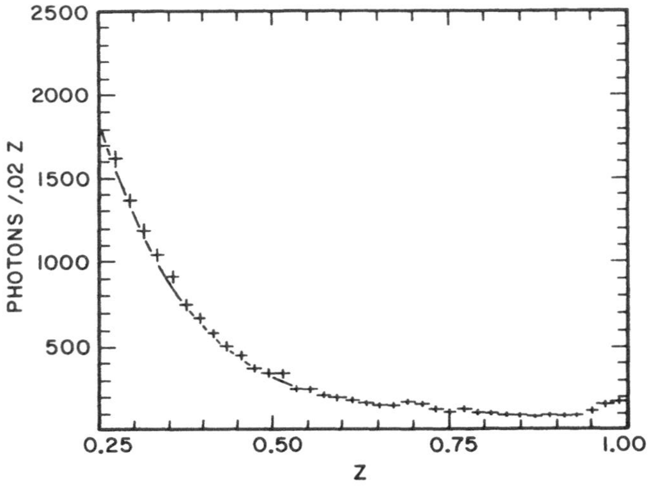

Fig. 15. Inclusive photon spectrum from 80000 T's, $z = E_{\gamma}/E_{beam}$. Curve is a polynomial fit to the data.

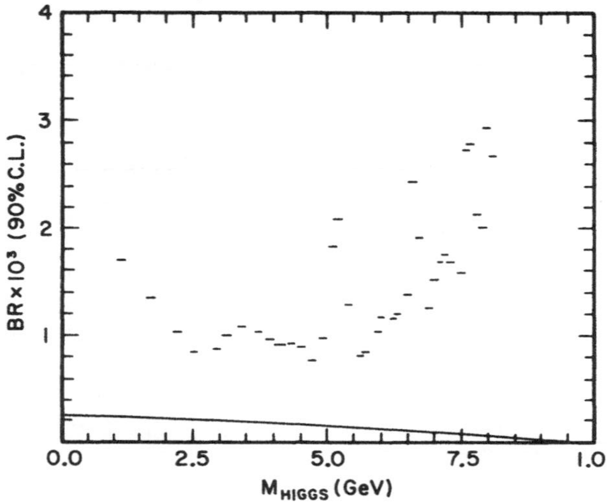

Fig. 16. CUSB 90% cl upper limit of BR($\Upsilon \to \gamma + H$) vs M_{Higgs}. Curve is theoretically predicted rate for one Higgs models.

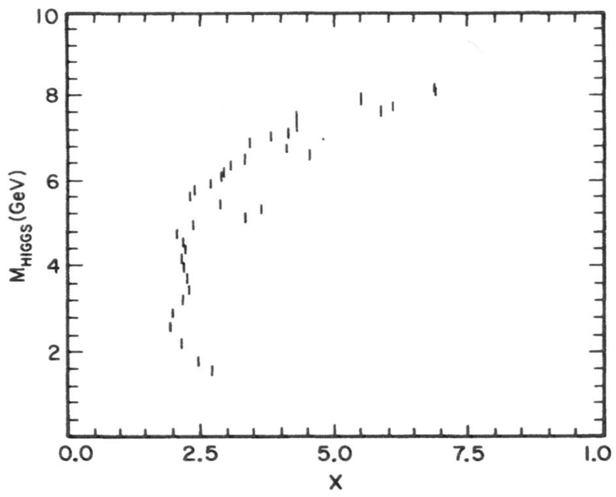

Fig. 17. Contour of excluded region for (M_H, X), where $X = v_1/v_2$.

REFERENCES

1. W. Buchmuller and S.-H. H. Tye, Phys. Rev. Lett. 44, 850 (1980).
2. Y. P. Kuang and T. M. Yan, Phys. Rev. D24, 2874 (1981).
3. V. A. Navikov et al., Phys. Rep. 41C, 1 (1978).
4. P. Moxhay and J. L. Rosner, Enrico Fermi Inst. preprint EFI 83-15.
5. E. Eichten, Phys. Rev. D22, 1819 (1980).
6. E. Rice et al., Phys. Rev. Lett. 48, 906 (1982) and new data.
7. Particle Data Group, Phys. Lett. 111B, 11 (1982).
8. T. M. Yan, Phys. Rev. D22, 1652 (1980).
9. L. S. Brown and R. N. Cahn, Phys. Rev. Lett. 35, 1 (1975).
10. G. Mageras et al., Phys. Lett. 118B, 453 (1982).
11. J. Green et al., Phys. Rev. Lett. 49, 617 (1982).
12. S. Behrends et al., Phys. Rev. Lett. 50, 881 (1983).
13. E. Eichten et al., Phys. Rev. D21., 203 (1980); Ibid., D22, 1819 (1980).
14. R. D. Schamberger et al., Phys. Rev. D26, 720 (1982).
15. S. N. Gupta et al., Wayne State Univ. preprint 83-02.
16. A. Martin, Phys. Lett. 103B., 55 (1980).
17. E. Eichten et al., Phys. Rev. D17, 3090 (1978).
18. J. L. Rosner, private communication.
19. R. Barbieri et al., Phys. Lett. 60B, 183 (1976); Ibid., 61B, 465 (1976).
20. C. Quigg and J. L. Rosner, Phys. Rep. 56C, 167 (1979).
21. C. Quigg, private communication.
22. M. Chanowitz, Phys. Rev. D12, 918 (1975).
23. P. B. Mackenzie and G. P. Lepage, in Perturbative Quantum Chromodynamics, ed. D. W. Duke and J. F. Owens (AIP, New York, 1981).
24. A. J. Buras et al., Rev. Mod. Phys. 52, 199 (1980).
25. S. Brodsky et al., Phys. Rev. D28, 228 (1983).
26. S. Brodsky et al., Phys. Lett. 73B, 203 (1978).
27. G. P. Lepage, private communication.
28. A. Berends and R. Kleiss, Nucl. Phys. B178, 141 (1981).
29. P. B. Mackenzie and G. P. Lepage, Phys. Rev. Lett. 47, 1244 (1981); F. J. Yndurain, Phys. Rev. Lett. 48, 897 (1982).
30. J. Ellis et al., Ann. Rev. Nucl. Part. Sci. 32, 443 (1982).
31. S. Weinberg, Phys. Rev. Lett. 40, 223 (1978).
32. F. Wilczek, Phys. Rev. Lett. 40,220 (1978).
33. D. Nanopolus et al., CERN preprint 83-TH.3651.
34. F. Wilczek, Phys. Rev. Lett. 39, 1304 (1977).
35. R. D. Pecceiand H. R. Quinn, Phys. Rev. Lett. 39, 1440 (1977).
36. M. Sivertz et al., Phys. Rev. D26, 717 (1982).
37. C. Edwards et al., Phys. Rev. Lett. 48, 903 (1982).
38. G. Parisi, private communication.

HARD PROCESSES INVOLVING REAL PHOTONS

D. Treille

CERN

Geneva, Switzerland

Photons, in the same way as leptons, give us a direct look at short-distance phenomena and, through them, access to the manifestation of the basic constituents of matter.

Photons can act as elementary partons directly coupled to the quark electric charge. For such behaviour the number of competing basic processes is smaller and the structure of the event is simpler than in ordinary hard hadronic interactions. A list of such reactions is given in Table 1. In processes (c) and (e) the photon transfers its full energy to large-p_T partons, and the absence of fragmentation products in the forward region gives a very striking topology.

As we know, real high-energy photons generally do not behave that way. They interact by fluctuating into a $q\bar{q}$ pair, and the relevant quantity is the fluctuation time, $\tau = E_\gamma / p_T^2$ for a real γ. (For a virtual one, $\tau = 1/\sqrt{-q^2}$, and such fluctuations are suppressed.) If p_T, the transverse momentum of a created parton, is of the order of the scale parameter Λ of QCD, the fluctuation time is long, the partons of the fluctuating system interact with each other, and globally the photon appears as a superposition of vector mesons with an ordinary (i.e. $\sim$ 1-x) quark content. If p_T is larger ($\Lambda << p_T << E_\gamma^*$), the photon behaves rather as a pair of quasi-independent quark and antiquark. This is quite similar to lepton pair creation. The quark content of the photon is harder than the previous situation and about flat in x. This corresponds to the well-known and badly named "anomalous" behaviour of the photon. At still larger p_T the point-like picture of the photon takes over.

In Section 1, I will briefly mention, following the excellent review by Zerwas[1], what could be the prompt-gamma physics in e^+e^- collisions. Except for $\gamma\gamma$ hard scattering (see subsections 1c and

Table 1. Large-p_T processes with photons in the initial or in the final state (Ref. 1)

a)	$e^+e^- \to \gamma X$	$e^+e^- \to q\bar{q}$ $\to \gamma$ brems.	
b)	$(Q\bar{Q}) \to \gamma X$	$(Q\bar{Q}) \to \gamma GG$ $\to \gamma GGG$	
c)	$\gamma N \to$ large-p_T jet + X	$\gamma q \to Gq$ $\gamma G \to q\bar{q}$	QCD Compton fusion
			↓
d)	Direct γ in hadron coll. $AB \to \gamma X$	$Gq \to \gamma q$ $q\bar{q} \to \gamma G$	inverse
e)	Deep inelastic Compton $\gamma N \to \gamma X$	$\gamma q \to \gamma q$ $\gamma G \to \gamma G$	QED Compton via box
f)	Large-p_T hadrons in $\gamma\gamma$ $\gamma\gamma \to$ large-p_T jets	$\gamma\gamma \to q\bar{q}$ $\gamma\gamma \to GG$	 via box
g)	Two high-p_T γ in hadron coll. $AB \to \gamma\gamma X$	$q\bar{q} \to \gamma\gamma$ $GG \to \gamma\gamma$	

3b) this domain is still lacking experimental results.

In section 2, I will treat, in parallel, the hard photoproduction and processes involving a prompt γ in the final state, and review most of the currently available results in that field.

Section 3 will deal with two-photon processes (QED Compton, $pp \to \gamma\gamma + X$, $\gamma\gamma \to$ jets) and discuss the problem of a direct measurement of quark charges. Finally, the application of another original feature of the photon - its ability to produce heavy quarks abundantly - will be described.

e^+e^- PHYSICS

For details, the reader should refer to the review by Zerwas[1] and the references therein.

Quark bremsstrahlung

Quark large-angle bremsstrahlung. This offers the possibility to measure the electroweak charges of the quarks, with a weighting between u and d types which is basically different from the one obtained in the annihilation cross-section measurement. Without going

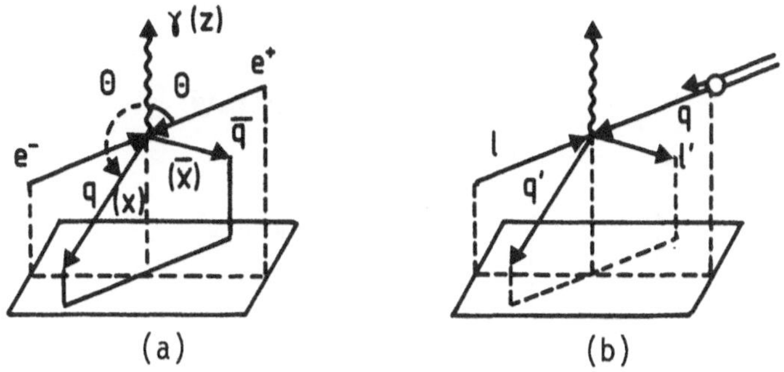

Fig. 1. Gamma wide-angle bremsstrahlung : (a) in e^+e^- collision ; (b) in lepton-hadron deep inelastic scattering.

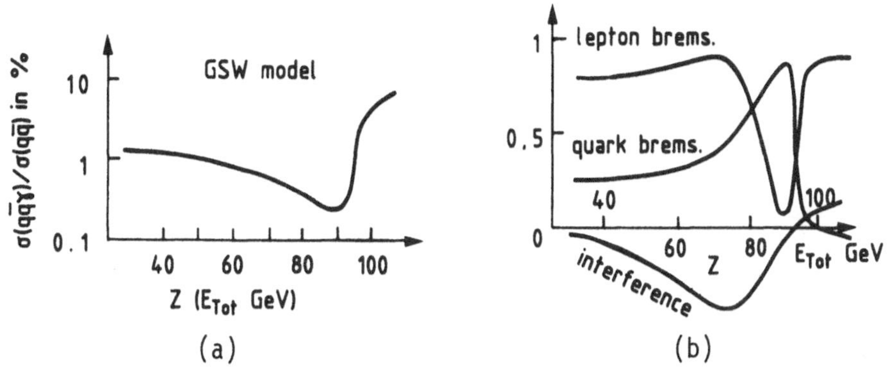

Fig. 2. (a) Cross-section for gamma production in e^+e^- collisions (from Ref. 1) : thrust ≤ 0.975; $z \leq x_q$ or $x_{\bar{q}}$; $45° \leq \theta \leq 135°$. See Fig. 1 for notation (drawn from Ref. 1). (b) Decomposition of the cross-section with respect to the γ sources (from Ref. 1).

into details (see Fig. 1a), the bremsstrahlung cross-section is proportional to

$$\sum_q e_q^2 R_q = 3 \sum_q e_q^2 \left| e_q - \frac{G_F\, q^2 (v_e - h_e a_e)(v_q - h_q a_q)}{8\sqrt{2}\,\pi\,\alpha (q^2 - m_Z^2 + i m_Z \Gamma_Z)/m_Z^2} \right|^2 .$$

The familiar coupling of a Z° to a quark can be recognized. Far below the Z° this factor reduces to $\sim \Sigma\, e_q^4$. With several kinematical cuts giving significance to the Born approximation and eliminating uninteresting configurations, the bremsstrahlung process represents about 1% of the total rate[1] (Fig. 2a). It can be split according to the origin of the photon (Fig. 2b) : at the Z°, bremsstrahlung from the quarks dominates, and there is, for instance, the interes-

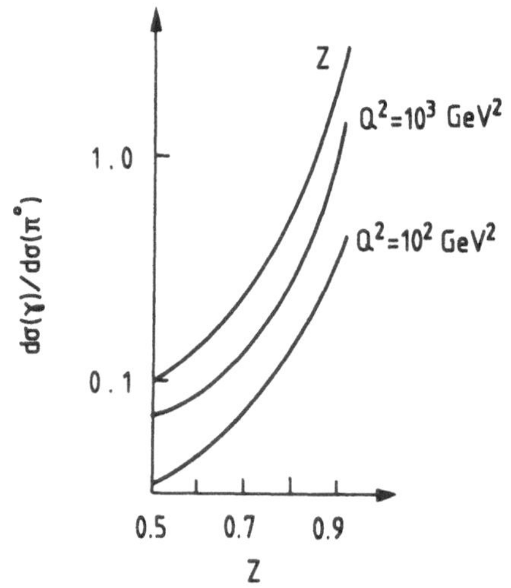

$$Z \equiv \frac{E_{\gamma,\pi^\circ}}{E_B}$$

Fig. 3. γ/π° ratio in e^+e^- annihilation (from Ref. 3).

ting relationship

$$\frac{\sigma(q\bar{q}\gamma)}{\sigma(q\bar{q})} = \frac{\Sigma\ e_q^2\ R_q}{\Sigma\ R_q}\ \frac{\sigma(\mu\mu\gamma)}{\sigma(\mu\mu)}\ ,$$

giving access to the original weighting of electroweak charges quoted above.

The background to this process is discussed in Ref. 1.

Smaller-angle bremsstrahlung. The smaller-angle bremsstrahlung (i.e. lower γ-q mass) brings us into the anomalous regime of the photon coupling. The hard-photon structure function[2] is here reflected into a hard fragmentation function of a quark into prompt gammas, for the same physical reason. We can therefore expect a spectacular rise of the relative yield of prompt γ at large fractional energies[3], as shown in Fig. 3. This should be measurable and complement the study of the γ structure function.

Charge asymmetry[1,4]. Figure 4 shows the origin of the asymmetry.

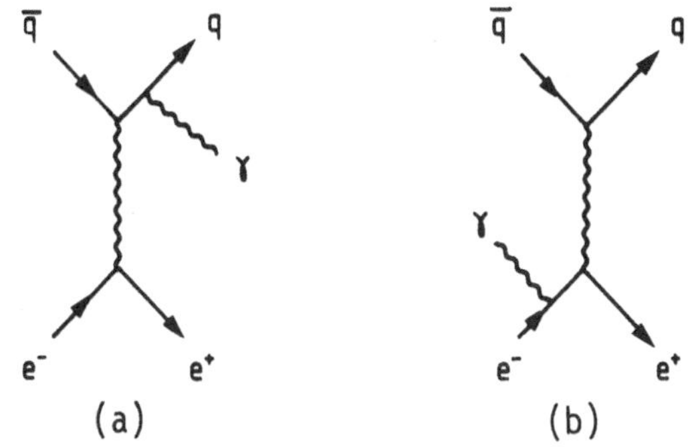

Fig. 4. The sources of prompt gammas.

It is due to the interference between diagrams 4a and 4b in which the hadronic final states have C = + 1 and C = - 1, respectively. At the quark level the asymmetry is quite large but there is no direct access to it. It is transmitted into an inclusive charge asymmetry, which can be expressed as

$$\rho = \frac{d\sigma(\gamma\pi^+) - d\sigma(\gamma\pi^-)}{d\sigma(\gamma\mu^+) - d\sigma(\gamma\pi^-)} ,$$

smaller but probably measurable. The asymmetry is proportional to $\Sigma\, e_q^3$ and to the difference between quark and antiquark fragmentation functions. Let us define also

$$\rho' = \frac{d\sigma(\pi^+) + d\sigma(\pi^-)}{d\sigma(\mu^+) + d\sigma(\mu^-)} ,$$

which gives the sum of these fragmentation functions. When the reduced momentum (p_π/p_q) goes to 1, it can easily be shown that the ratio ρ/ρ' should approach 3/5 in the case of standard fractional-charge quarks. Integer Han-Nambu quarks would give an integral ratio. However, the reader should refer to Section 3 for a "modern" version of the integer-charge assumption.

Bremsstrahlung in lepton-hadron deep-inelastic scattering. Figure 1b shows that for bremsstrahlung processes, e^+e^- collisions and deep-inelastic muon scattering processes are in a one-to-one correspondence. To get the interference term, we have now to change the sign of the incident lepton. This was done in the past at SLAC[5], but the result was not accurate enough to tell about quark charges.

The EMC Collaboration[6] has the prospect of extracting the interference term. The experimental problems are severe (for instance, the background of halo muons) and the result is not yet available.

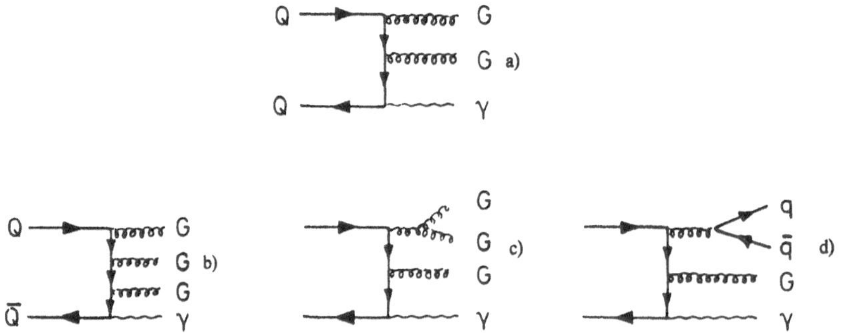

Fig. 5. Gamma decays of orthoquarkonium (Ref. 7).

Decay of orthoquarkonium

Again this is quite speculative since - a heavy mass being needed - it mostly applies to the still undiscovered toponium state.

The rate of the gamma decay of toponium $t\bar{t}$ is presumably large : O(10 - 20%).

Second-order processes (Fig. 5) should make it possible to see the gluon self-coupling of Fig. 5c. Figure 6[7] shows indeed that the properties of the hadronic final state (the mass of which can be tuned by selecting the gamma energy) are dominated by the gluon splitting.

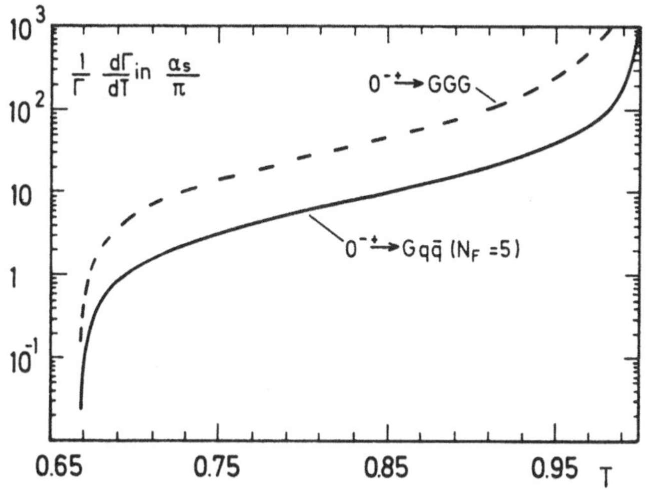

Fig. 6. Thrust distribution of the final hadronic state in gamma decay of orthoquarkonium (Ref. 7).

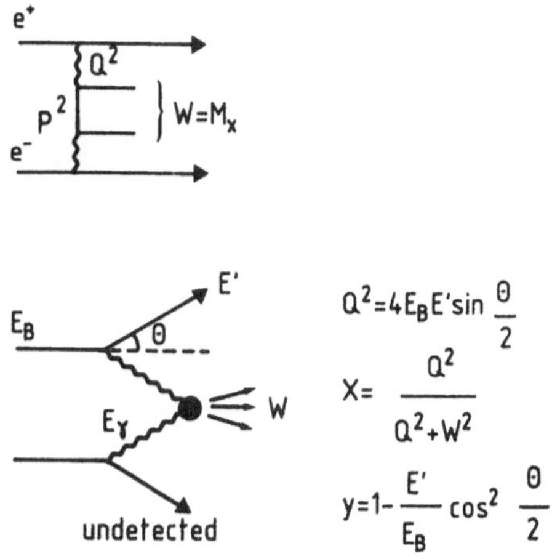

Fig. 7. Kinematics of $\gamma\gamma$ scattering.

As is well known, another imperative reason for looking at the toponium decays involving a gamma is the search for Higgs bosons. However, one has first to find the toponium and get reasonable sta-

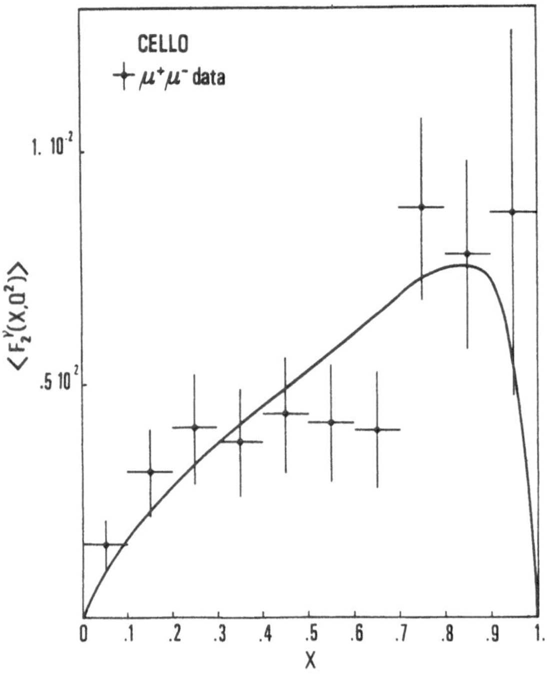

Fig. 8. $\gamma\gamma \to \mu^+\mu^-$.

tistics on it - and this is not an easy task.

<u>The photon structure function</u>

Here experimental information and excellent reviews are available[8].

The basic question is : When one probes a real gamma with a deeply virtual one (Fig. 7), does one find a hadron-like structure or, as expected from the parton model, something similar to the purely leptonic case[9] (Fig. 8)? The answer is clear[8,10]: the gamma structure function looks like the parton model expectation :

$$\frac{F_2(x,Q^2)}{x} = \frac{3\alpha\ \Sigma\ e_i^4}{\pi}\ [x^2 + (1 - x)^2]\ \ln \frac{Q^2}{m_q^2}\ .$$

Here QCD would rather suggest replacing m_q by Λ, and higher-order corrections are supposed to modify the low- and high-x regions. Figure 9a[11] seems to indicate that a clear distinction between QCD and the naïve parton model is in principle possible : however, Fig.9b shows that unfortunately, once the level of observed events is reached, the difference between the two models has shrunk to an at present unobservable value. We are still far from proving QCD with the gamma structure function measurement.

HARD PHOTOPRODUCTION AND PROMPT-GAMMA PHYSICS

This is a quite active sector of high-energy physics[12].

Figure 10 illustrates the similarity of the physical content in both situations. The basic processes (Fig. 10a) are the same. The background - an interesting one - has the same origin (Fig. 10b). Both have two-photon processes (Fig. 10c) and higher-twist processes (Fig. 10d) with quite noticeable topologies. Table 2 (from Ref. 12) evaluates the relative merits of these and other processes under various theoretical criteria.

However, the experimental conditions and methods are different. Figure 11 shows that several energy domains should be explored to get all types of parton scattering. Figure 12 reminds us that at SPS energies an E_T trigger is not efficient for uncovering jet structures[13], and hard scattering should rather be selected there by triggering on individual particles. This is also true for incident

Table 2. Theoretical merits of different processes (from R. Petronzio, Ref. 12). The stars have the usual Michelin meaning, $(*)^{-1}$ is a new concept.

	Subprocesses	K factors	Intrinsic p_T
DY	* * *	* * *	/
DY_{p_T}	* *	* * *	* * *
Direct γ	* *	* * (?)	* *
Photoproduction	* *	/	* * *
Deep Compton	* * *	* *	* * *
$hh \to jet/\pi^\circ$	*	/	$(*)^{-1}$

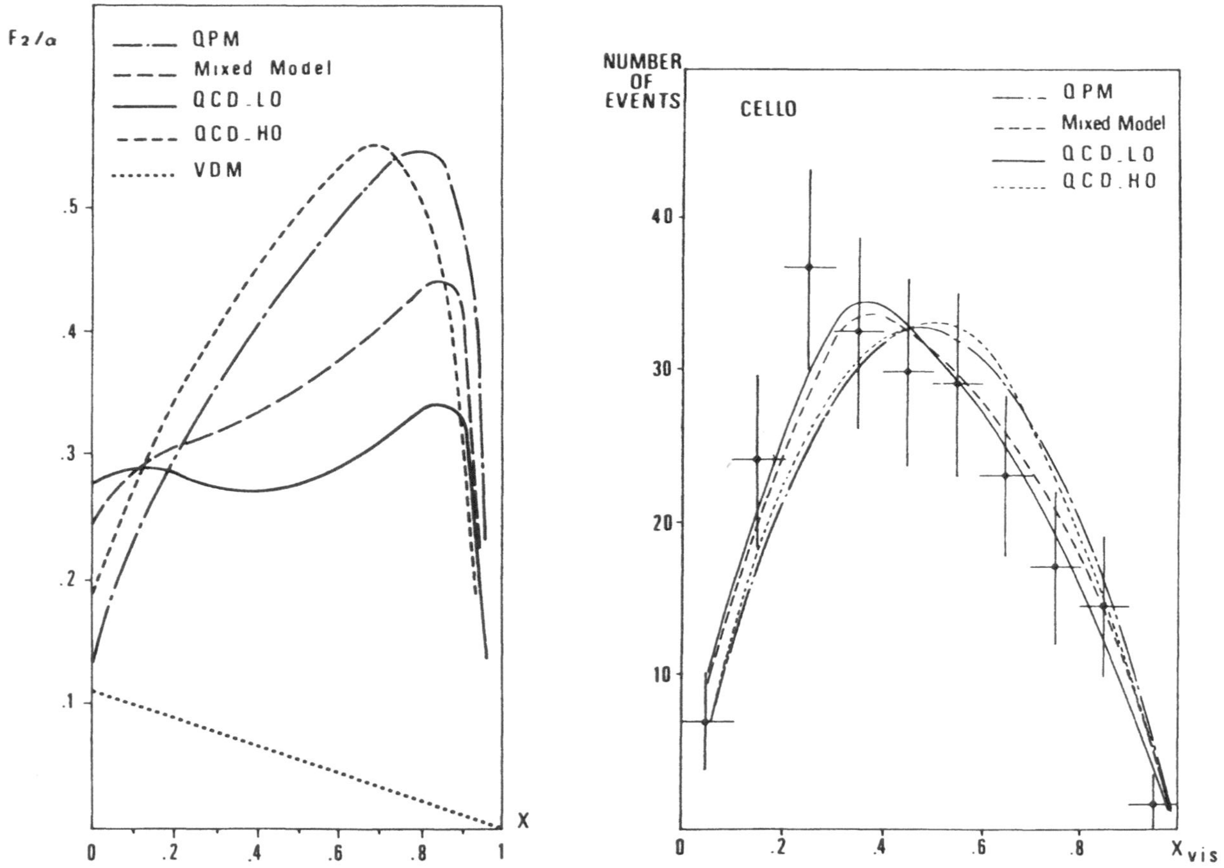

Fig. 9. The photon structure function : prediction of various models : (a) the structure function itself; (b) the number of observable events.

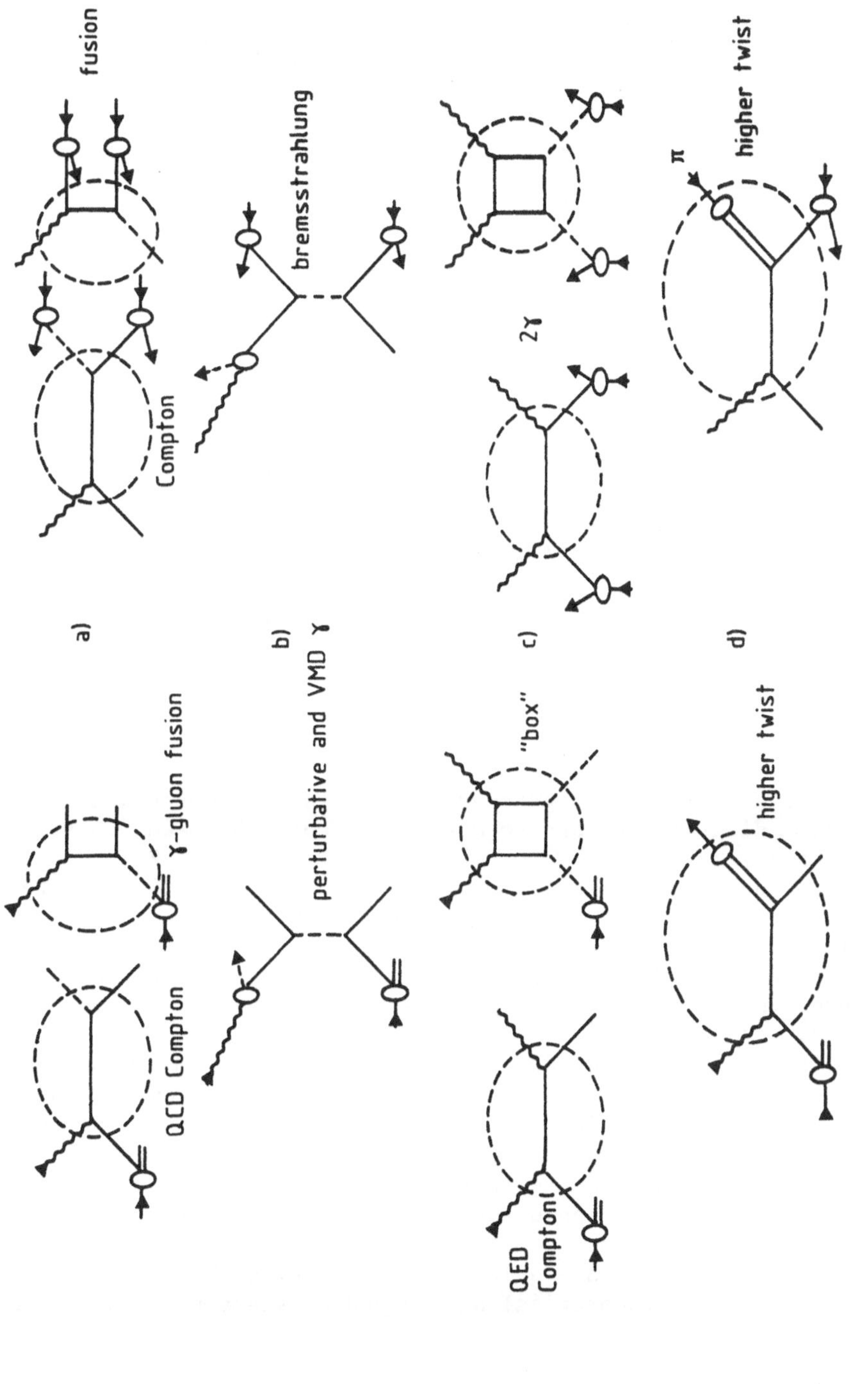

Fig. 10. The physics of hard photoproduction and prompt-gamma hadroproduction.

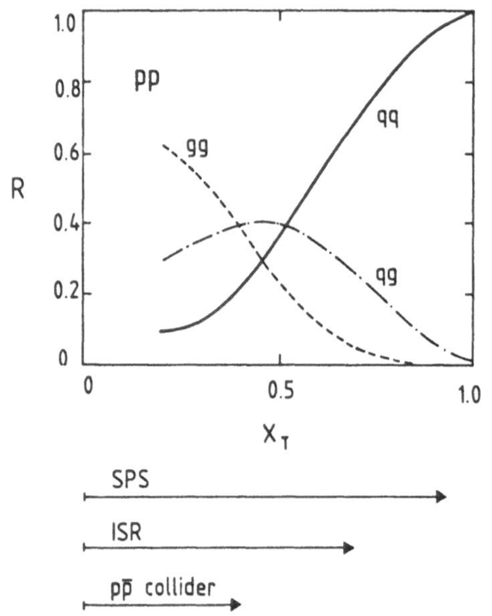

Fig. 11. The fraction of various parton-parton subprocesses predicted from a QCD model as a function of $x_T = 2p_T/\sqrt{s}$. The kinematic ranges covered by experiments studying jet physics at the SPS, ISR, and p$\bar{p}$ Collider are shown.

gammas, although they are a potentially richer source of jets[14], as shown in Fig. 13.

To isolate a given hard process, for instance Compton scattering or the fusion diagram, one has to make differences between cross-sections, namely :

- for photoproduction, the difference between inclusive final-state distributions for particles of both signs should be considered[15] (see Fig. 14);
- for prompt-gamma physics with hadrons, the differences between cross-sections obtained with projectiles of both signs should be used (Fig. 15).

Up to now the field has been dominated by ISR prompt-gamma measurements. These are described by Figs. 16[16]. The prompt gamma signal around 90° is clear (Fig. 16a). The prompt gamma prefers to be unaccompanied (Fig. 16b). The charge of the away system is rather positive, which seems to indicate that indeed inverse Compton scattering, predominantly involving u quarks, plays a role (Fig. 16c). The bremsstrahlung contribution seems to be weak, if not zero (Fig. 16d). Results at small angles[17] are quite compatible with the large-angle ones (Fig. 16e). The theoretical understanding seems to be satisfactory[18].

Results from fixed-target experiments[19] are still preliminary

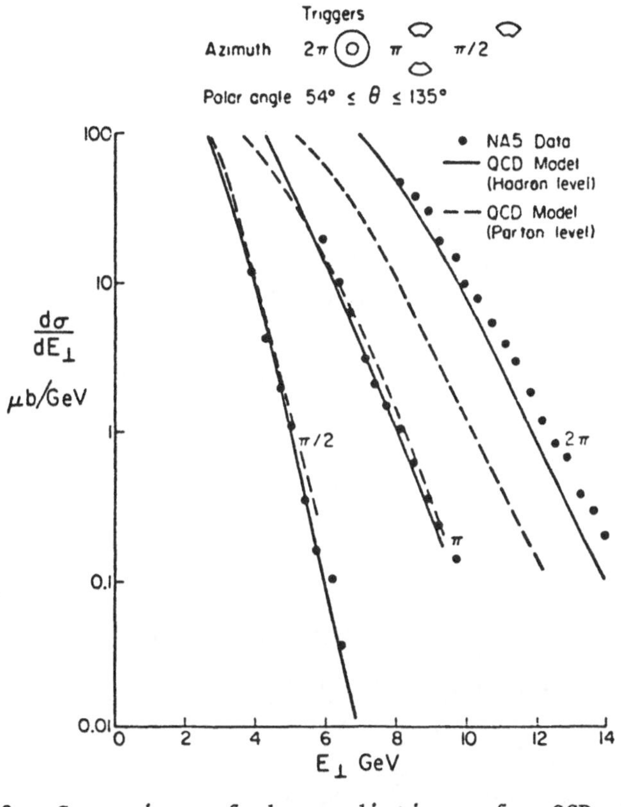

Fig. 12. Comparison of the predictions of a QCD parton-shower Monte Carlo model with data from NA5 for both small- and large-aperture calorimeter triggers. The solid and dashed curves are the predictions at the hadron and parton level, respectively.

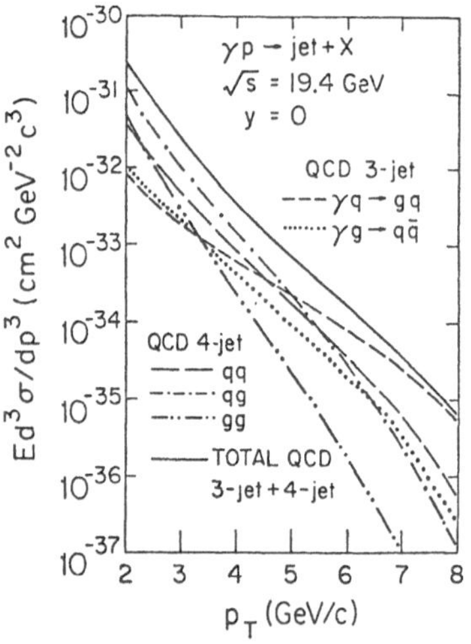

Fig. 13a. Cross-sections predicted by a QCD-model calculation for the reaction $\gamma p \rightarrow$ jet + X at $\sqrt{s}$ = 19.4 GeV.

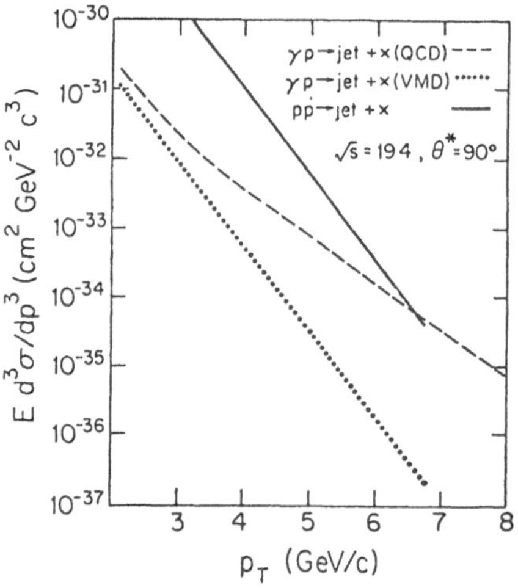

Fig. 13b. Cross-sections predicted by a QCD-model calculation for $\gamma p \rightarrow$ jet + X compared with pp $\rightarrow$ jet + X.

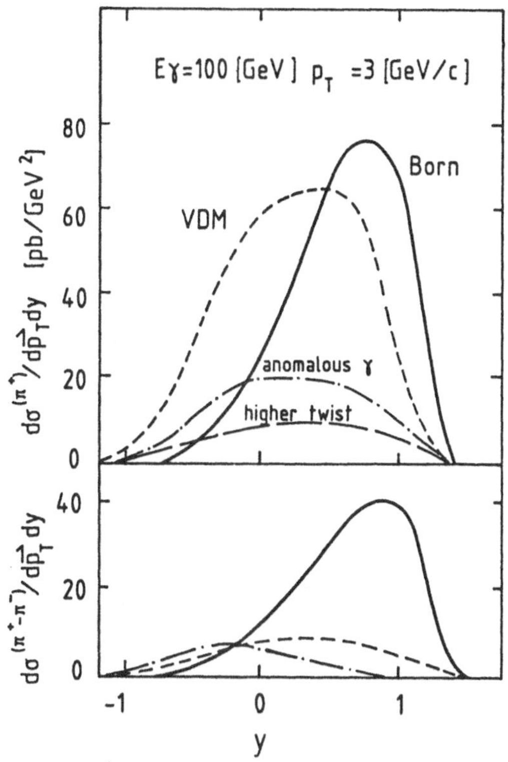

Fig. 14. QCD-model calculations of the inclusive π^+ cross-section and the difference of the inclusive cross-sections $\pi^+-\pi^-$ for the reaction $\gamma p \to \pi^+ X$ and $\gamma p \to \pi^- X$ are shown. The contribution of the point-like behaviour of the photon (Born term) is shown as a solid line, the anomalous in dashed-dotted, the hadronic or VDM in small-dashed, and higher twist in big-dashed lines.

(Fig. 17) but many groups are currently working along these lines (Table 3). These experiments are quite difficult because the identification of a prompt gamma (among π°/η and various halo backgrounds) is a delicate task.

In the case of photoproduction, preliminary results have come from the NA14 group at CERN[20]. This experiment at the Super Proton Synchrotron uses the most intense e^- beam available, the E12 (Fig. 18), which is a very pure, tagged beam ($\sigma_E \sim 3$ GeV). The NA14 (Fig. 19) is an open-geometry spectrometer with wide coverage (up to 135° c.m.) for charged particles and photons. The structure of the electromagnetic calorimeters is adequate enough to allow for an excellent γ-π° separation, and they are well protected against the background from halo muons. The π°'s and η's are measured in the calorimeters (Fig. 20) and their reconstruction efficiency is well understood.

Table 3. List of approved prompt-photon experiments at the CERN SPS and at Fermilab (from F. Costantini and K.P. Pretzl).

Expt.	Start of data-taking	Beam	Energy (GeV)	Physics goals stated in Proposal	Photon calorimeter	Photon calorimeter resolution $\sigma(E)/E$	Two-shower separation Δx (mm)	Hadron measurement	Sensitivity (events per nb per day)
SPS									
NA3	82	$\pi^{\pm}$, p	200	$hh \to \gamma X$, $hh \to \gamma\gamma X$, q and g fragmentation, g structure function.	Scintillator/lead + shower chamber	$< 0.21/\sqrt{E}$	< 50	Magnet spectrometer, Cherenkov.	600
NA24	83	$\pi^{\pm}$, p	200 300	"	Proportional tube/ lead + scintillator/ lead	$< 0.24/\sqrt{E}$	15	Segmented hadron calorimeter.	800
WA70	83	$\pi^{\pm}$, p	200 300	"	Liquid scintillator/ lead	$0.16/\sqrt{E}$	15	Ω' spectrometer, RICH counter.	800
UA6	84	$\bar{p}$, p	270	$hh \to \gamma X$, $\pi^0 X$, Drell–Yan e^+e^- pairs, and Λ and $\bar{\Lambda}$ production.	Proportional tube/ lead	?	$\gtrsim 15$	Magnet spectrometer, dE/dx, transition radiation.	40 ($\bar{p}p$) (if they are main user) and 300 (pp)
FNAL									
E705	84	p, $\bar{p}$, $\pi^{\pm}$	300	Charmonium $hh \to \chi X$, $hh \to \gamma X$, $hh \to \gamma\gamma X$.	Scintillating lead-glass	?	50	Magnet spectrometer Cherenkov.	~ 120
E706	85	$\pi^{\pm}$, p	400 530 800	$hh \to \gamma X$, $hh \to \gamma\gamma X$, Drell–Yan e^+e^- pairs, $hh \to \pi^0\gamma, \eta^0\gamma$, $\pi^0\eta^0 + X$.	Liquid argon	$0.14/\sqrt{E}$	?	Magnet spectrometer, calorimeter, RICH counter, silicon micro-strips.	$\sim 10^3$

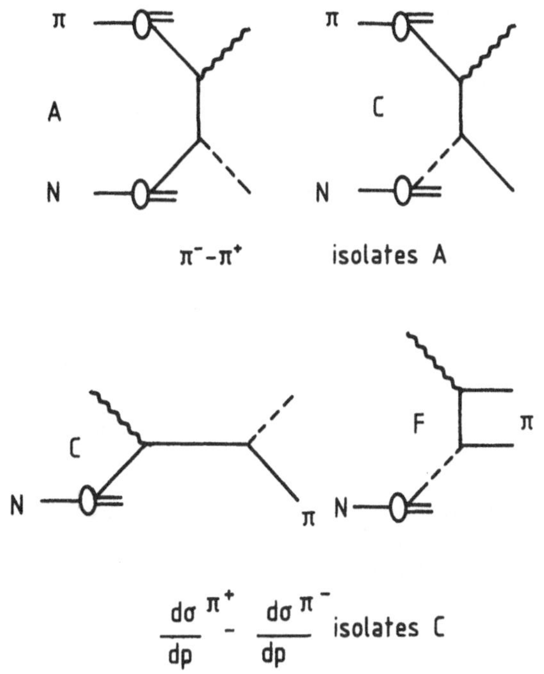

Fig. 15. How to isolate a given process by subtraction.

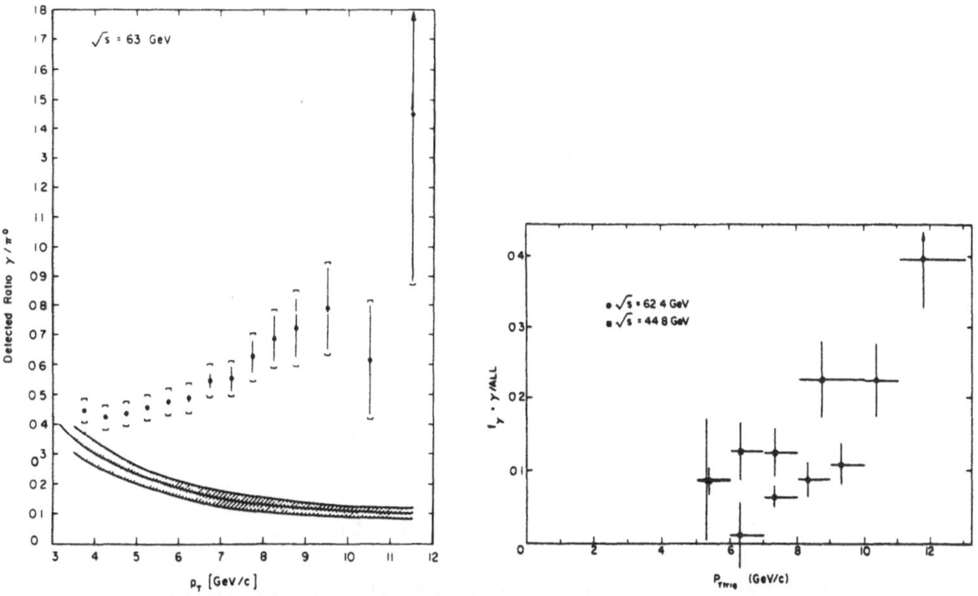

Fig. 16a. Left : Experiment R806. Observed ratio γ/π° as a function of p_T. Right : Experiment R108. Ratio $\gamma/$ all as a function of p_T.

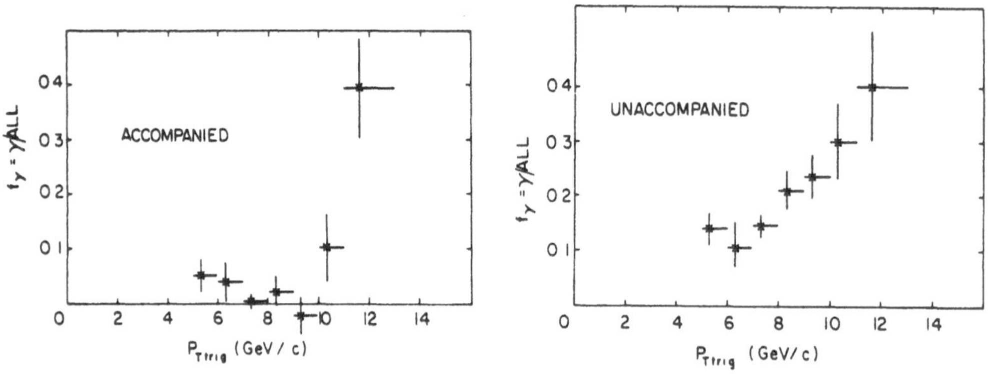

Fig. 16b. Experiment R108. Ratio γ / all for events with particles accompanying (left) and not accompanying (right) the photon in its hemisphere.

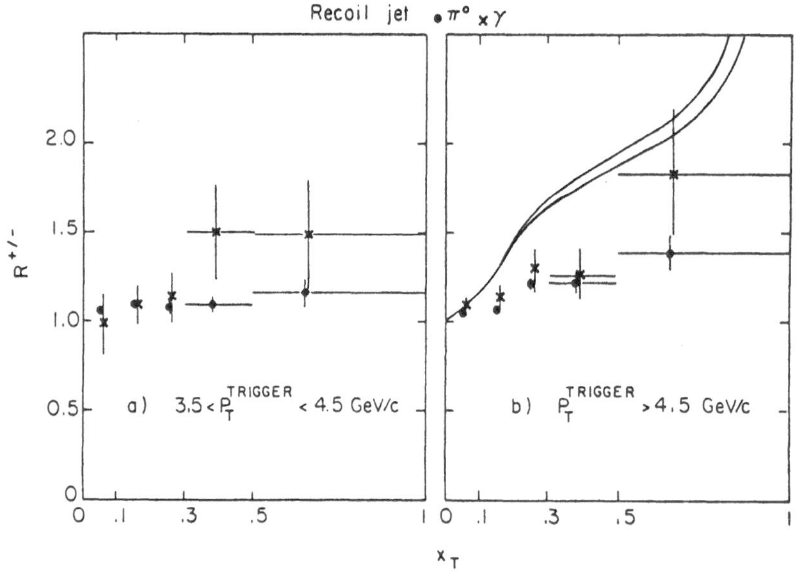

Fig. 16c. Experiment R807. Ratio R of positive to negative particles in the hemisphere opposite to the trigger and for two p_T intervals as a function of x_T. Solid lines are QCD predictions of R.

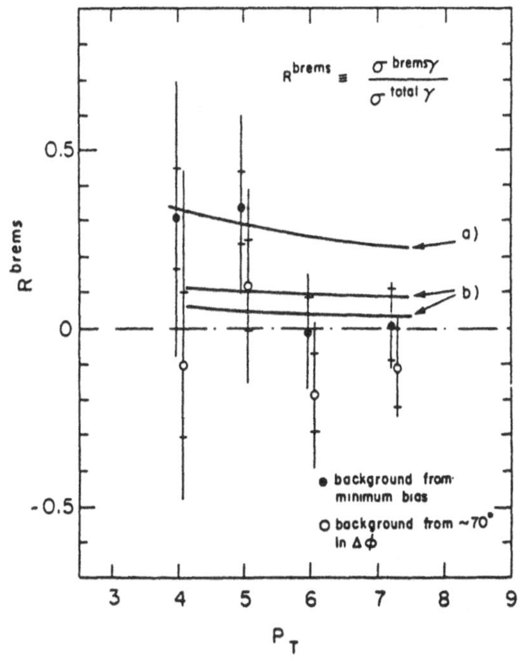

Fig. 16d. Experiment R807. Fraction of bremsstrahlung events as a function of the p_T of the direct γ. The calculation is made by assuming two kinds of level for the unassociated background : the minimum-bias events or the level at 70° in $\Delta\phi$ from the trigger. The solid lines are QCD predictions for this ratio.

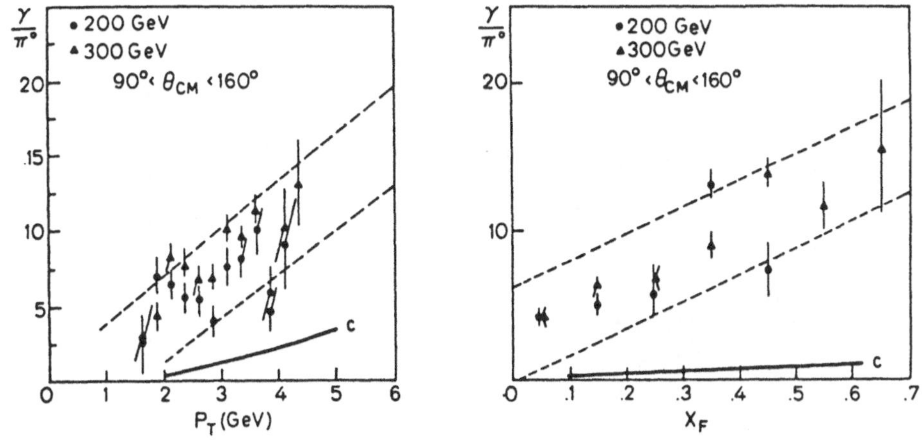

Fig. 17a. FNAL data on pp → γ + X (from Ref. 19a).

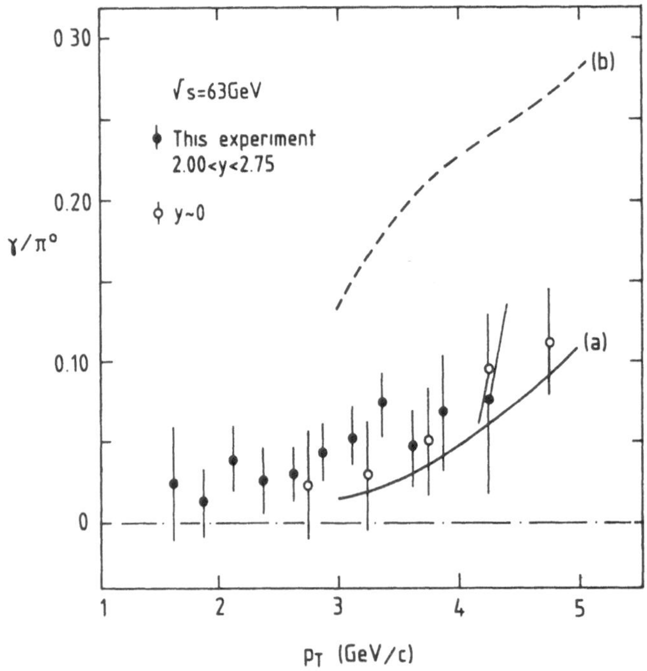

Fig. 16e. The γ/π° ratio at the ISR in the forward region.

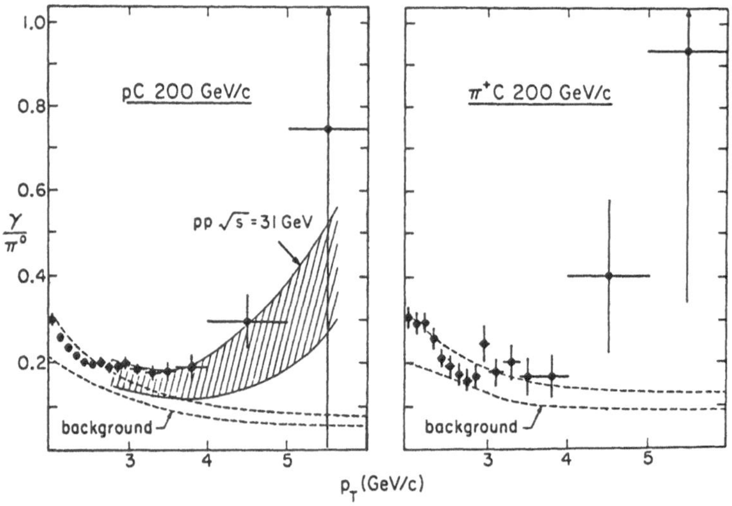

Fig. 17b. FNAL data on $pC \to \gamma + X$ and $\pi^{+}C \to \gamma + X$ at 200 GeV/c (from Ref. 19b).

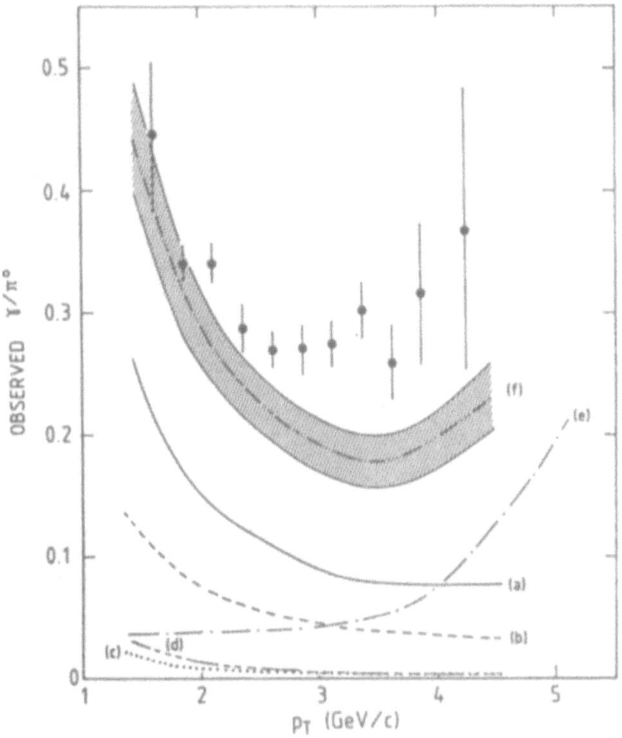

Fig. 16f. Detail of the background contributions to the observed γ/π° ratio. The curves represent : a) background from π°; b) from η; c) from η'; d) from ω; e) from merging π°; f) sum of (a) to (e), where the bound gives the statistical and systematic errors on the calculation.

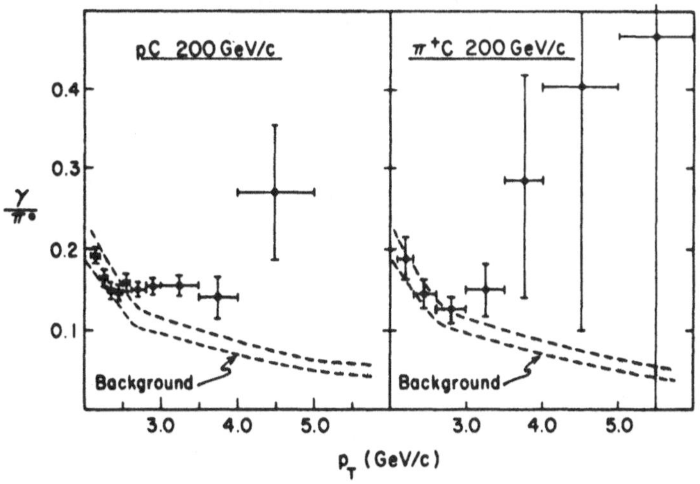

Fig. 17c. FNAL data on the ratio γ/π° (from Ref. 19c).

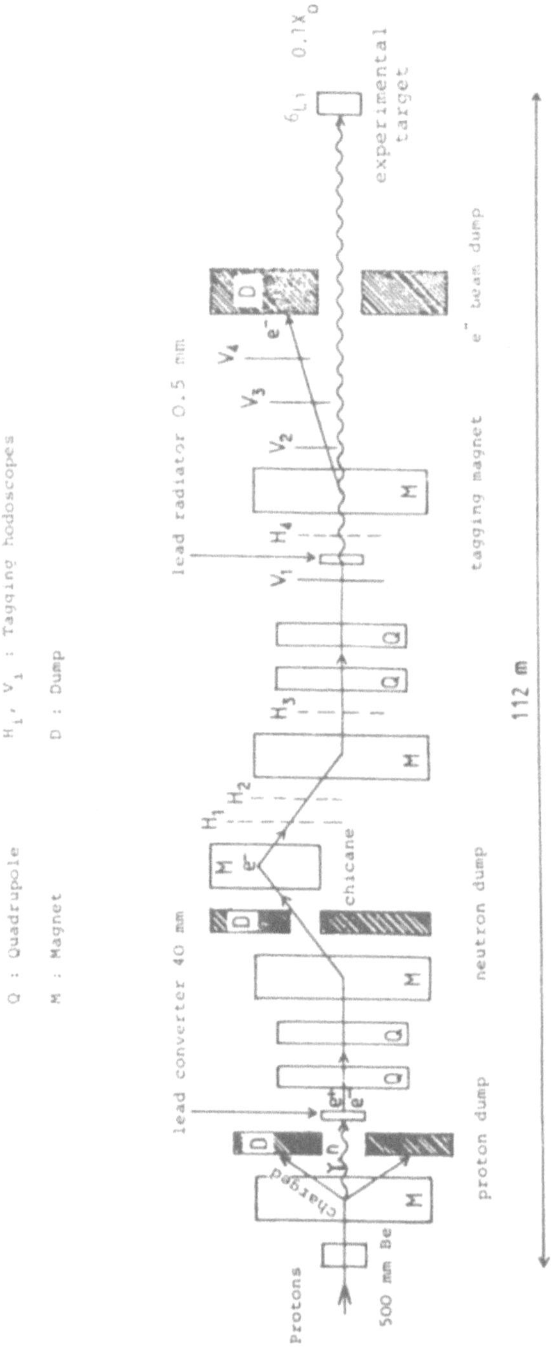

Fig. 18. The E12 beam layout.

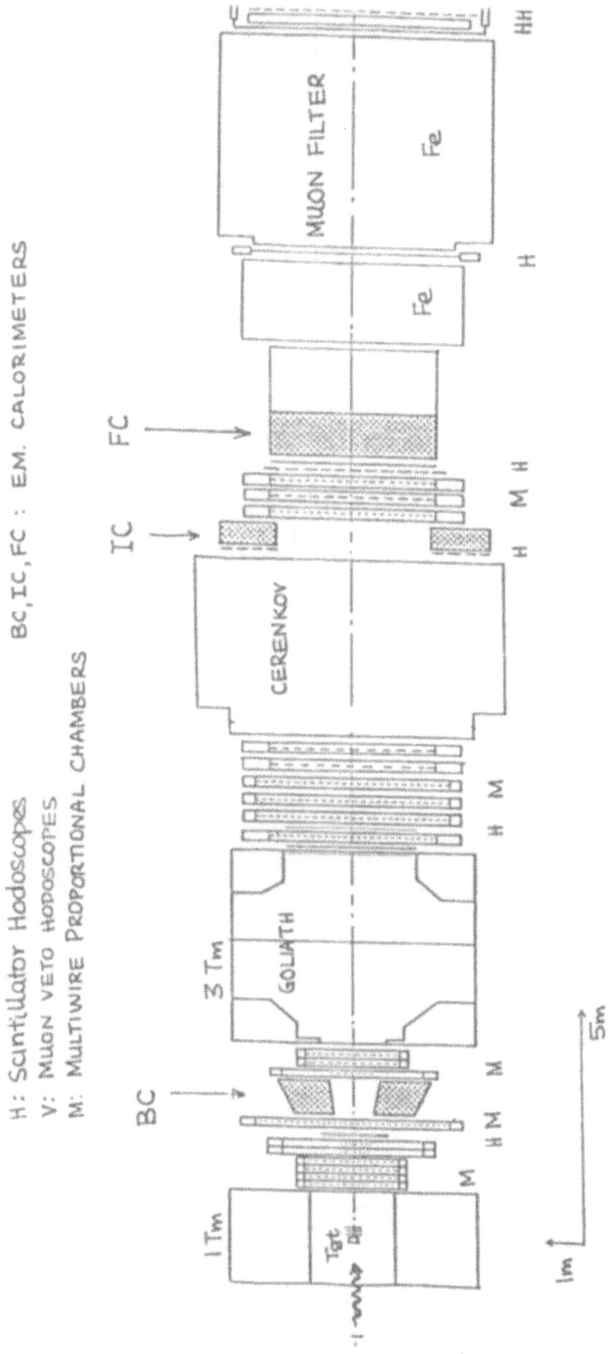

Fig. 19. The NA14 spectrometer.

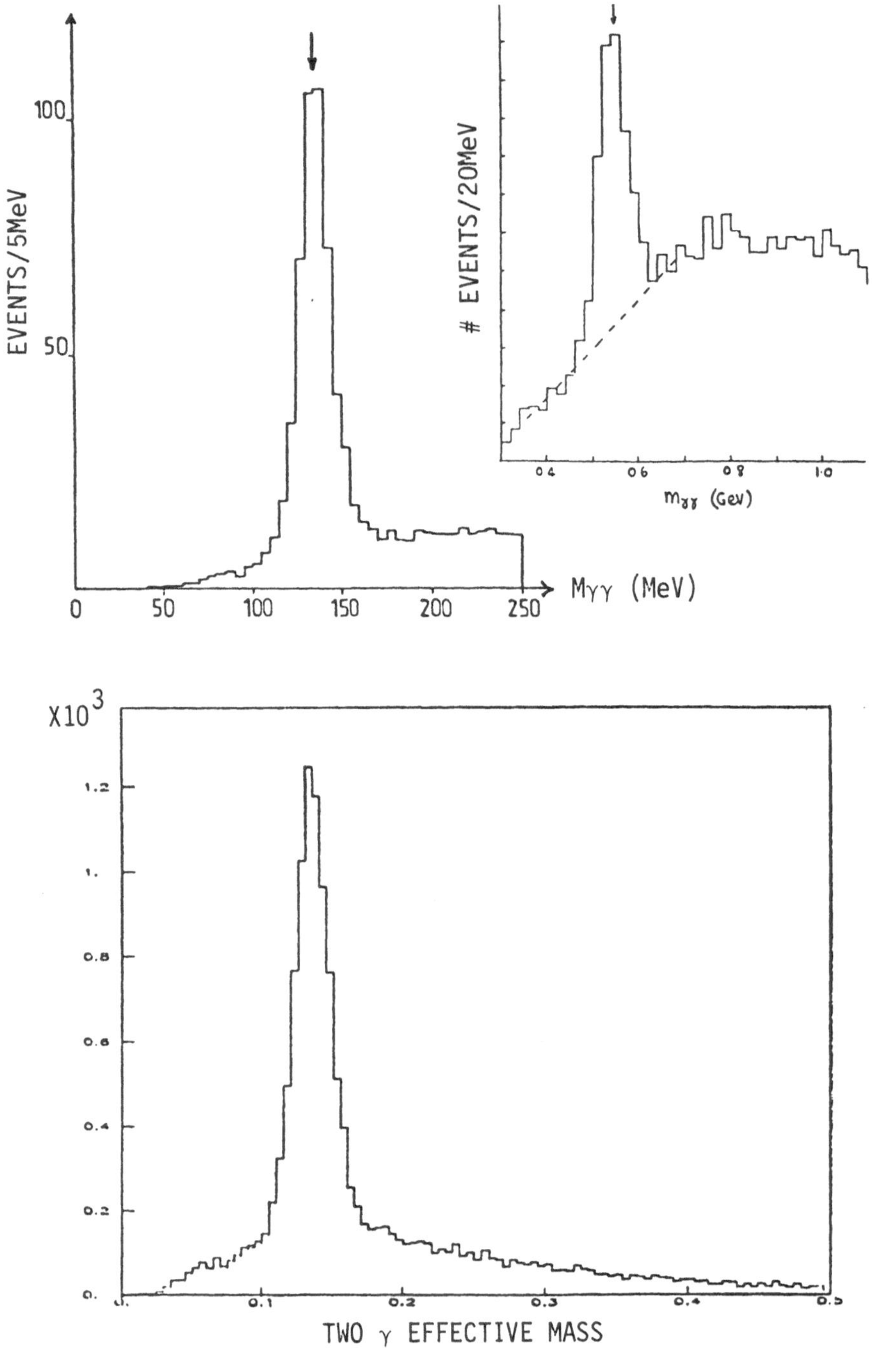

Fig. 20. γγ invariant masses in the NA14 calorimetry.

Inclusive π° distributions are shown in Fig. 21. The normalization obtained by several methods, which agree well (beam counting, Bethe-Heitler μ pairs, ψ, etc.), has an uncertainty of $\sim$ 20%. The sensitivity in Fig. 21 is above one event per picobarn. In addition to the normal hadronic contribution inferred from pion scattering[21], Fig. 21a and, to a lesser extent, Fig. 21b show that photon hard-scattering terms are needed. The total sum reproduces data quite well. A recent and preliminary computation[22] indicates that higher-order terms will modify the QCD estimate : one will go to the uppermost curve in Fig. 21. The fact that in the data of Fig. 21b, around 90°c.m., the QCD terms have a smaller relative contribution than in the forward region (Fig. 21a) is in good agreement with QCD prediction.

As already quoted, the right way to isolate QCD Compton scattering is by taking the difference between π^+ and π^- inclusive distributions. This work is under way. In 1983 the NA14 experiment will complete its data-taking and increase its statistics by a large factor. We will come back to this in subsection 3.3.

TWO-GAMMA PROCESSES

Two-gamma processes involve the fourth power of quark charges, and in principle are quite sensitive to the numerical value of these.

It may be appropriate here to recall that models with integer charges are still frequently advocated in the literature[23]. They are mostly based on an $SU(3)_c \times U(1)$ symmetry, with a spontaneous breaking of colour symmetry in the large distance (i.e. strong coupling) regime but without giving up confinement. The gluons get a non-zero Lagrangian mass. If this mass is small enough ($\leq$ 200 MeV or so), the authors[23] claim that no experiment, except two-gamma processes with real gammas, can tell the difference between this and the standard model, at least with the accuracies at present available. Indeed, the quark charge behaves as :

$$Q = Q_1^{\text{fract.charge}} + Q_8 \frac{m_g^2}{q^2 + m_g^2} \, .$$

At $q^2 = 0$, $Q_1 + Q_8$ reproduces the old Han-Nambu integer charge assignment. But at even moderate q^2 ($q^2 \gg m_q^2$), the octet part fades away and the fractional charge results are recovered. Furthermore, any single-photon process - since, apart from the "tinted" gamma, only colour singlets are involved - will always project the photon onto its singlet component. Quasi-reality ($q^2 \lesssim m_g^2$) and the presence of more than one gamma are therefore needed in order to feel the octet part of the charge. Discussions on controversial points can be found elsewhere[24,25]. Let us now turn to experiments.

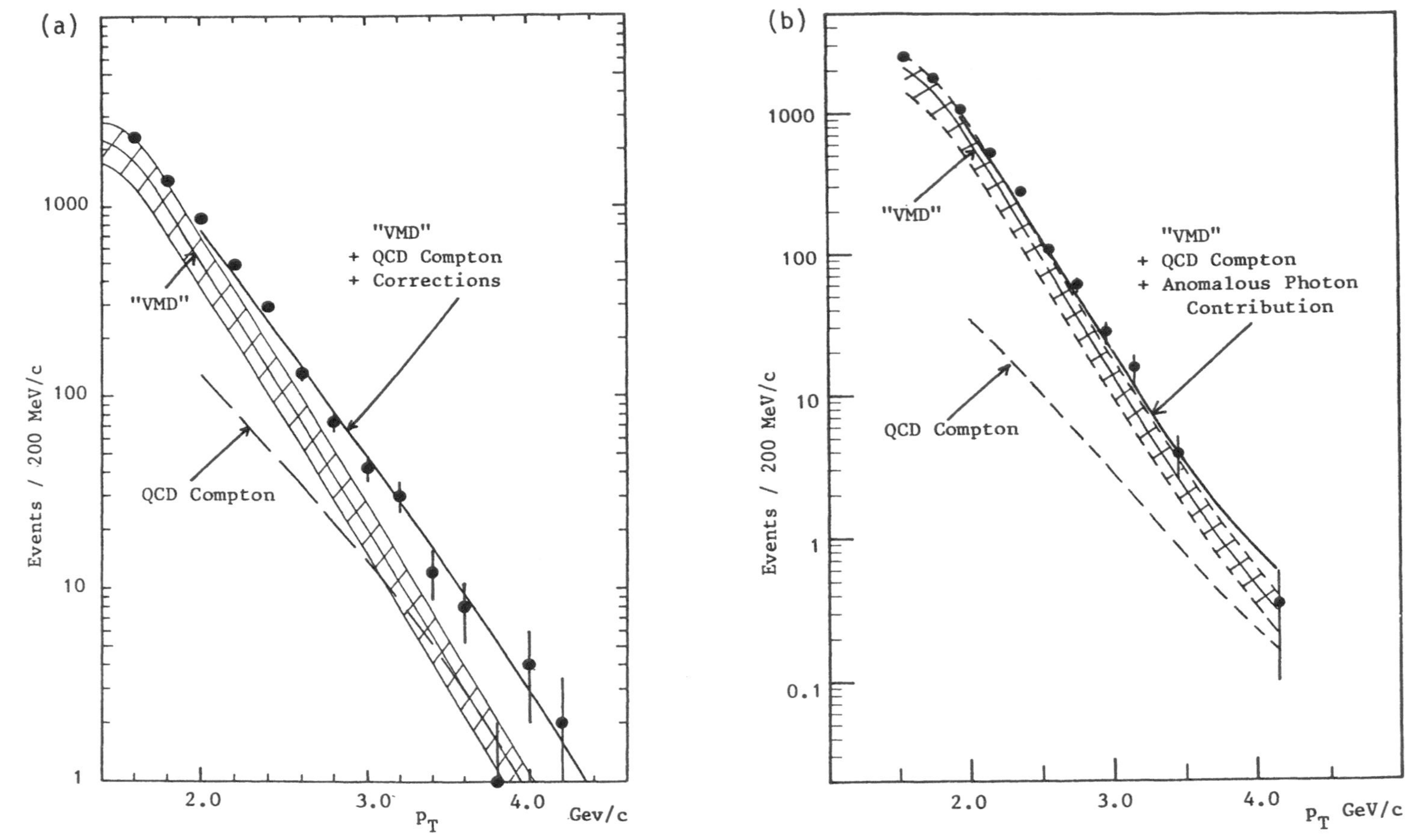

Fig. 21. π^0 inclusive spectra in NA14 : (a) forward calorimeter (Olga); (b) 90° c.m. calorimeter (IC).

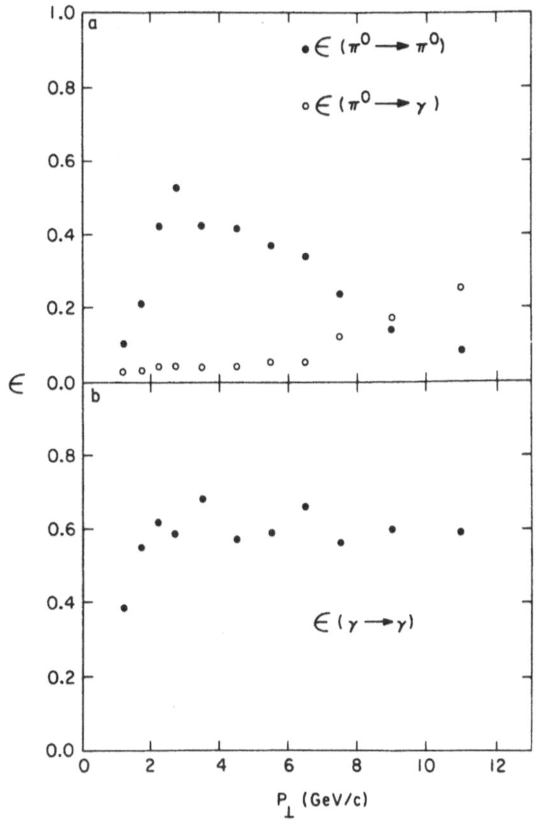

Fig. 22. Neutral identification in the ISR calorimeter. Monte Carlo calculated probability (including detection and reconstruction efficiency) for a) a π°, with both decay photons incident on a calorimeter, to be accepted as a single shower (open circles) or a π° (solid dots); b) a single photon to be identified as a single shower.

$pp \to \gamma\gamma X$

This process has been measured at the ISR in experiment R806[26]. Its very small cross-section and the value of the ratio $\gamma\gamma/\pi^\circ\pi^\circ \approx 10^{-3}$ indicate how difficult this is. The $\gamma\gamma$ cross-section is expected to be about equal to e^+e^- pair production by the Drell-Yan process, and the ratio of both cross-sections is a meaningful quantity to consider. Figure 22 shows the ability for neutral identification of the calorimeter : at high p_T (i.e. high p), merging neutral pions can simulate gammas. Figure 23 shows that if one compares the measured points with the Monte Carlo expectation in the absence of a $\gamma\gamma$ source, such a contribution is present. The error on the final result

$$\gamma\gamma/e^+e^- = 1.7 \pm 1.0$$

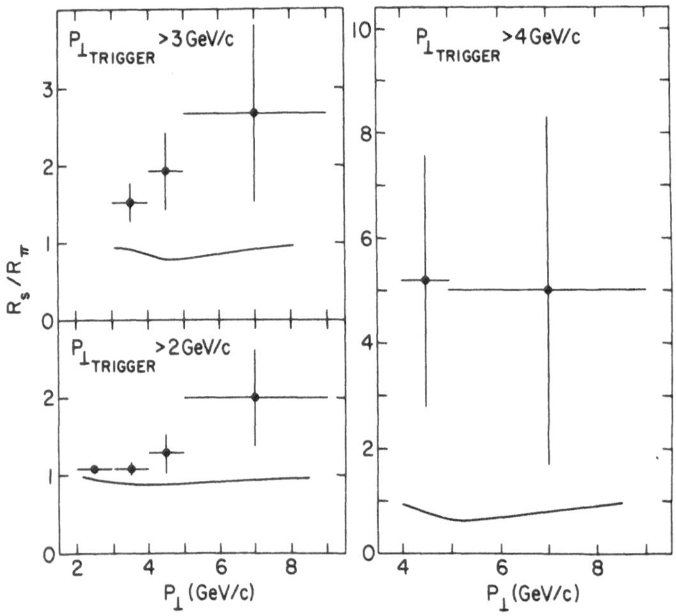

Fig. 23. Evidence for a $\gamma\gamma$ source : the observed and the Monte Carlo calculated R_S/R_π ratio for three different p_T thresholds.

reflects the statistical and systematic limitations of the measurement.

$\gamma\gamma \rightarrow$ jet-jet

Figure 24 summarizes the available data in this field of two-gamma physics.[27]

In particular, it can be seen that "no tag" results are now available[28].

When comparing these with various expectations, it should be remembered that

- isolating such a final state is not an easy experimental task, and therefore systematic errors could be difficult to avoid;
- in the colour-breaking model the q^2 domain of the exploration matters a lot. Jarayaman et al.[29] have analysed this problem and give the value of the increase of the two-jet cross-section to be expected in such a theory (Table 4).

Clearly, the experimental results tend to be above the simple standard QCD predictions. However, at this stage, both the experimental and the theoretical uncertainties, such as the contribution of non-leading terms and of different mechanisms in the standard model, preclude the drawing of conclusions and make premature the

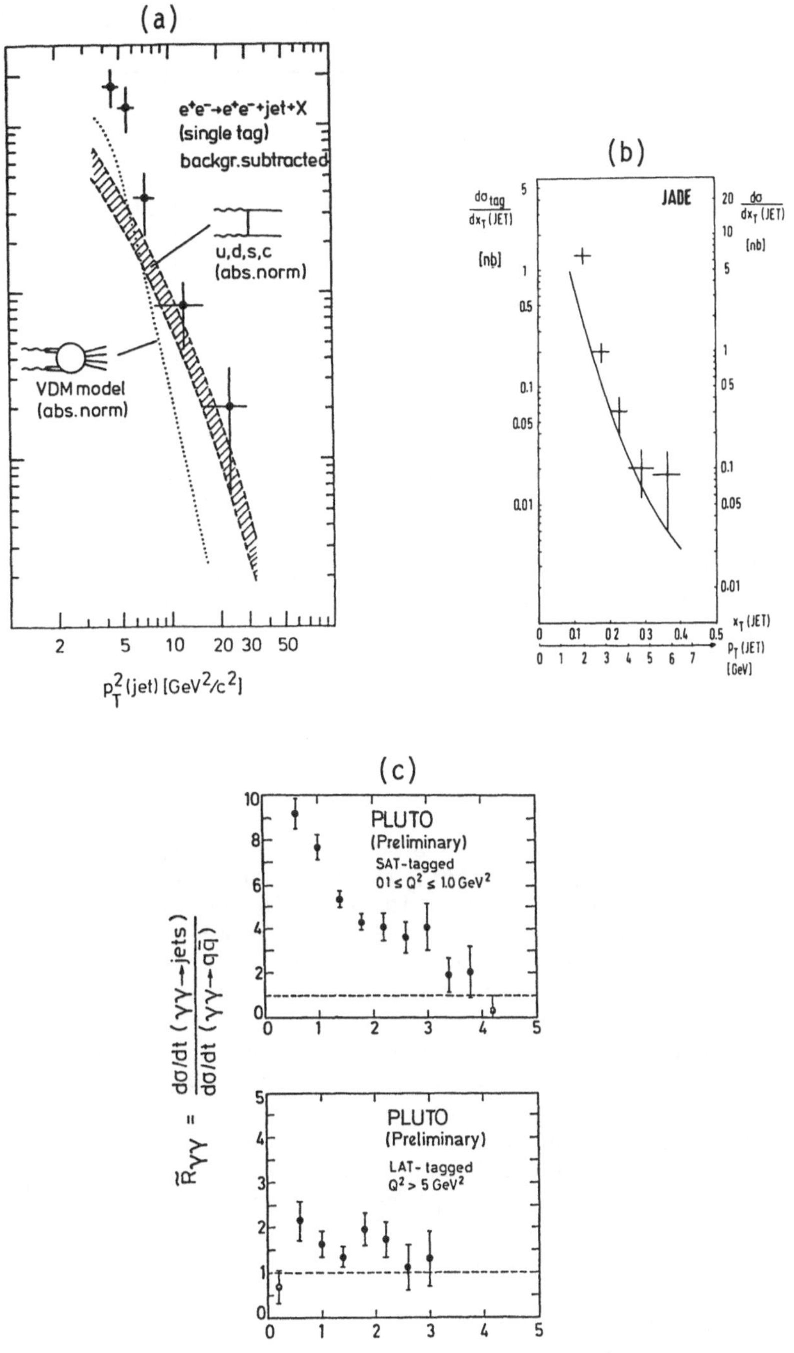

Fig. 24. $\gamma\gamma \to$ jet + X in various set-ups : a)TASSO,b)JADE,c)PLUTO.

Table 4. Predictions for $R_{\gamma\gamma}$ in the various cases considered in the text (Ref.29)

Model	$R_{\gamma\gamma}$
FCQ	1.26
ICQ (m_g = 0.2 GeV)[a]	1.82
ICQ (m_g = 0.1 GeV)[a]	1.45
ICQ (m_g = 0.05 GeV)[a]	1.34
ICQ (m_g = 0)[a]	1.31
ICQ with Higgs (m_g = m_H = 0)[a]	1.51
ICQ (m_g = 0) above threshold	2.19

[a]Below colour threshold.

reference to non-standard quark charges. It is nevertheless a crucial point to follow up.

QED Compton scattering

This process, which has long been advocated as a basic test, has been measured by the NA14 group. However, data are still preliminary[30].

The prompt-gamma spectrum in gamma hadronic final states, identified as such by the spectrometer, is shown in Fig. 25. The band represents the estimated contamination of π^o's and other neutral mesons. Since π^o's have been measured in the same exposure (Fig. 21), this contamination estimate is normalization-independent. The band width includes systematic uncertainties in the acceptance and identification, as well as statistical effects. Both the slope and the absolute level indicate that a prompt-gamma signal, increasing in relative value with p_T, is present. Adding the QED standard term, with higher-order corrections[31] and the box diagram (Fig. 10c) contribution, roughly reproduces the data. They have some tendency to lie above the expected curve, both at small and intermediate p_T. Subsequent measurements and a complete study of the exclusive features of the events will strengthen the reliability of the data and tell us whether this effect is confirmed.

HEAVY-FLAVOUR PHOTOPRODUCTION

At least one thing is known : more charm seems to be produced, in relative value, by photons than by hadrons. Heavy-flavour photoproduction is intersting in two respects :
i) for understanding the production mechanism;

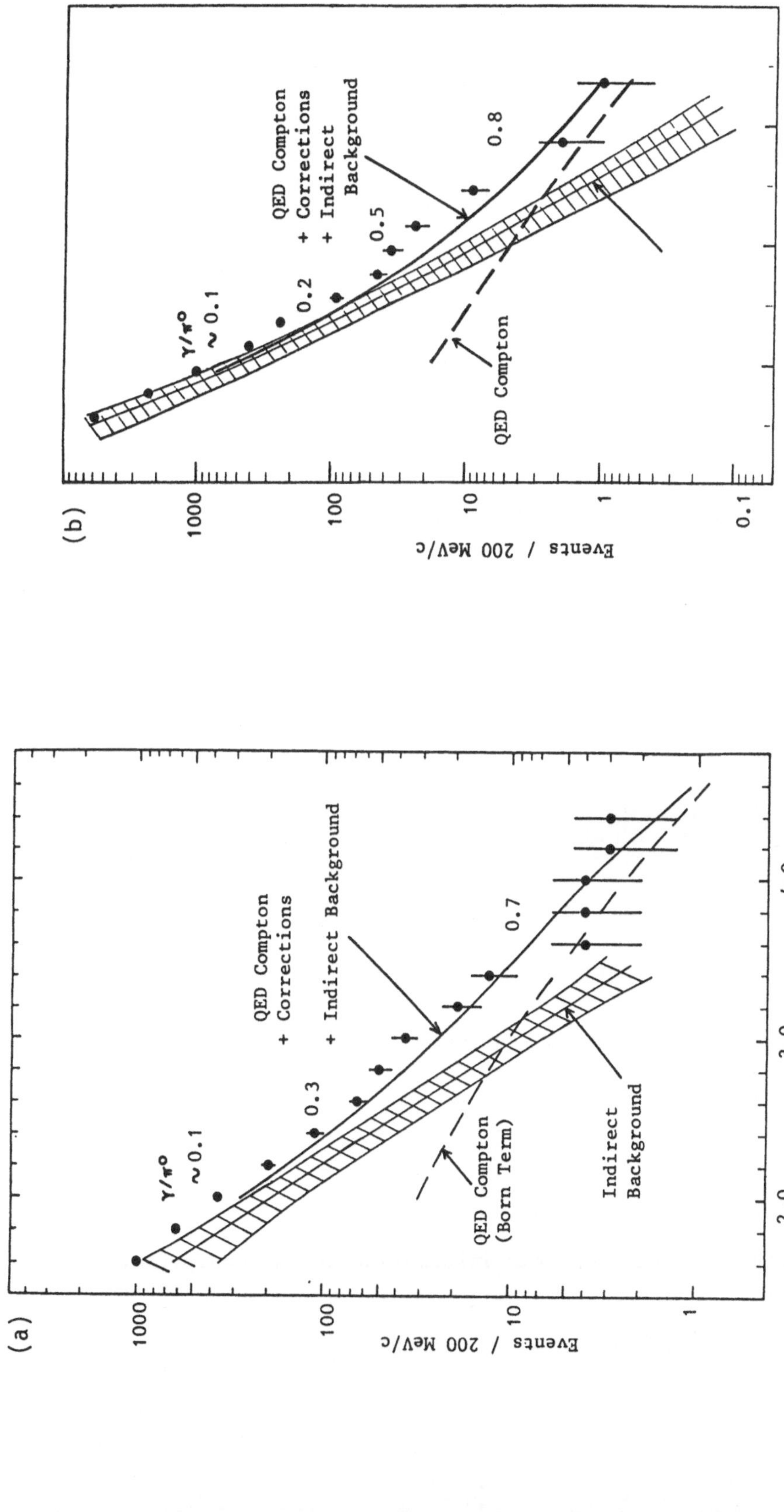

Fig. 25. Single-gamma inclusive spectra in NA14 : a) forward calorimeter (Olga): b) 90°c.m. calorimeter(IC).

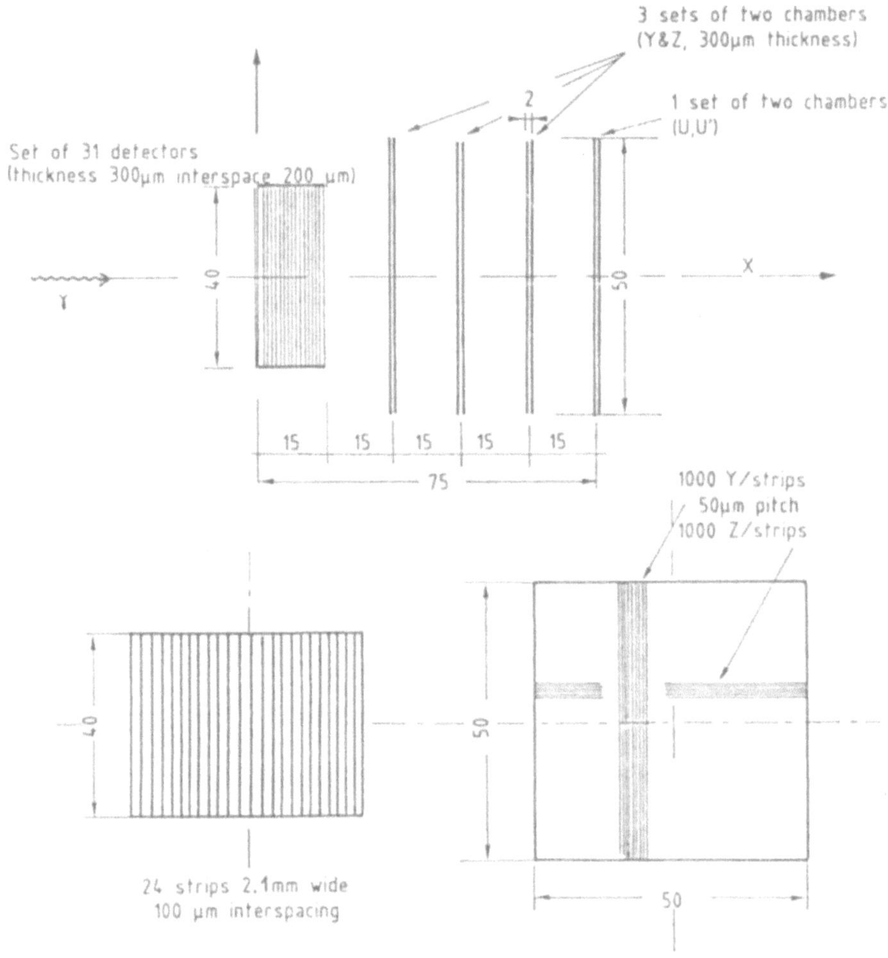

Fig. 26. The target arrangement in the NA14 proposal.

ii) for providing spectroscopic information on the heavy-flavoured states.

I will be very brief, since these points have been discussed in great detail at the recent Fixed-Target Workshop at CERN[12].

Production mechanism

Photon-gluon fusion is claimed to provide a relevant description of hidden and open charm photoproduction. Many reviews on the subject are available[32]. An interesting remark[33] is that if indeed the ψ excitation curve, for instance, measures the gluon distribution, it would be interesting, in the same experiment, to photoproduce ψ's on various materials to see whether differences due to the EMC effect are observed.

Information on heavy-flavoured states

This is relevant for the measurement of lifetimes and for pure spectroscopic information.

Clearly the e^+e^- rings, on the "charm factory", will be much better for D's than will any other set-up. But for F's and charmed baryons the combination of an active target and a powerful spectrometer can bring us much information. Table 5, taken from a CERN proposal[34], gives an idea of what can be achieved in a photoproduction experiment using the E12 beam. Compared to a hadronic exposure, the gain is obviously related to the better signal/background, which itself reflects the higher relative yield of charmed events.

With such an intense beam, which is necessary to reach the sensitivity quoted, the only viable solution for an active target is the use of silicon devices. The proposal of Ref. 34, following the pioneering work of other groups*), suggests the arrangement shown in Fig. 26, which is under test.

The recent claim that beauty has a long lifetime ($\sim 10^{-12}$s) makes us very hopeful of detecting it in such experiments. The present upper limits on its production cross-section are given in Fig. 27[32]. If the photoproduction cross-section is not too far below the quoted number - an occurrence which can be reasonably expected- an exposure of 1 event per picobarn could provide enough events for beauty identification and study.

ACKNOWLEDGEMENTS

I warmly thank the organizers of the Cargèse Summer Institute for their hopitality. I also thank C. Fabjan, J. Weyers and my colleagues of NA14 for a critical reading of the manuscript, and M. Millar, S. Vascotto and K. Wakley for their patience and efficiency in typing and correcting the report.

*Especially the NA1 and NA11 groups.

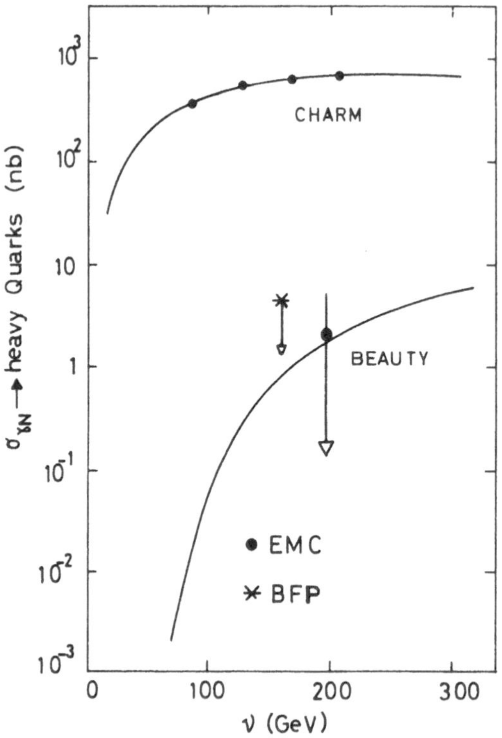

Fig. 27. Cross-section of $B\bar{B}$ pair photoproduction versus ν. Also shown are the data of the EMC for $C\bar{C}$ pairs. The curves represent the prediction of the γ-g fusion model (leading order, m_c = 1.5 and m_b = 5 GeV).

Table 5. Number of useful decays for lifetime measurement (S/B means signal / background). 20 M events registered with minimum bias trigger sensitivity $\sim$ 1 evt/pb.

	0.26 $F/\bar{F}$ per charmed event $F \rightarrow K^+K^-\pi$	0.2 $\Lambda_c/\bar{\Lambda}_c$ per charmed event $\overset{(-)}{\Lambda}_c \rightarrow pK^{\pm}\pi^{\mp}$
BR ?	5%	5%
for spectroscopy	260	130
S/B	0.25	0.3
for τ measurement	124	62
S/B	1.2	1.6
+ visible jump in active target	50	25

REFERENCES

1. P.M. Zerwas, Aachen report PITHA 81-12 (1981).
2. E Witten, Nucl. Phys. B120 (1977) 189.
3. K. Koller et al., Z. Phys. C2 (1979) 197.
4. S. Brodsky et al., Phys. Rev. D14 (1976) 2264.
5. D. L. Fancher et al., Phys. Rev. Lett. 38 (1977) 1980.
6. H. E. Montgomery, private communication.
7. K. Koller et al., DESY 80-132 (1980).
8. D. Cords, DESY 82-083 (1982).
9. H.J. Behrend et al. (CELLO Collaboration), DESY 83-017 (1983).
10. Ch. Berger et al. (PLUTO Collaboration), Phys. Lett. 107 (1981) 168.
11. H.J. Behrend et al. (CELLO Collaboration), DESY 83-018 (1983).
12. D. Treille, Proc. Workshop on SPS Fixed-Target Physics in the Years 1984-1989, Geneva, 1982 [CERN 83-02 (1983)], Vol. I, p.124.
K.P. Pretzl, ibid., Vol. II, p. 280.
R. Petronzio, ibid., Vol. II, p. 433.
13. C. de Marzo et al., Phys. Lett. 112B (1982) 173; Nucl. Phys. B211 (1983) 375.
14. J.F. Owens, Phys. Rev. D21 (1980) 54.
15. M. Fontannaz et al., Z. Phys. C6 (1980) 357; Phys. Lett. 89B (1980) 263.
16. A.L.S. Angelis et al., Phys. Lett. 94B (1980) 106.

M. Diakonou et al., Phys. Lett. 87B (1979) 292.
V. Cavasinni, preprint CERN-EP/82-175 (1982), to appear in Proc. Europhysics Study Conf. on Jet Structures from Quark and Lepton Interactions, Erice, 1982.
17. T. Åkesson et al., preprint CERN-EP/83-18 (1983).
18. W.J. Willis, preprint CERN-EP/81-45 (1981).
19. a) R.M. Baltrusaitis et al., Phys. Lett. 88B (1979) 372.
b) J. Biel et al., paper submitted to the 21st Int. Conf. on High-Energy Physics, Paris, 1982.
c) M. McLauglin et al., Phys. Rev. Lett. 51 (1983) 971.
20. P. Astbury et al. (CERN NA14 Collaboration), Inclusive π° photoproduction, paper submitted to the Int. Symp. on Lepton and Photon Interactions at High Energies, Cornell, 1983.
T. Virdee, Real photons in hard scattering, Proc. Int. Europhysics Conf. on High-Energy Physics, Brighton, 1983 (to be published).
E. Augé, Thèse de 3e cycle, Orsay LAL 83/09 (1983).
21. G. Donaldson et al., Phys. Lett. 73B (1978) 375.
22. M. Fontannaz, A. Mantrach, B. Pire and D. Schiff, Z. Phys. C6 (1980) 241.
P. Aurenche et al., Bielefeld preprint BI-TP 83/18 (1983).
23. A.V. Efremov, Dubna preprint JINR-E2 82-433 (1982).
A.Yu. Ignatev et al., Theor. Math. Phys. 47 (1981) 373.
G.M. Vereshkov et al., Yad. Fiz. 32 (1980) 227.
24. L.B. Okun'et al., Moscow preprint ITEP-79 (1979).
25. H. Lipkin, Nucl. Phys. B155 (1979) 104.
26. C. Kourkoumelis et al., Z. Phys. C16 (1982) 101.
27. W. Bartel et al. (JADE Collaboration), DESY 81-048 (1981); Phys. Lett. 107B (1981) 163.
R. Brandelik et al. (TASSO Collaboration), Phys. Lett. 107B (1981) 290.
28. TASSO Collaboration, Observation of hard processes in untagged $\gamma\gamma$ collisions, paper presented at the Int. Europhysics Conf. on High-Energy Physics, Brighton, 1983.
29. T. Jarayaman et al., Phys. Lett. 119B (1982) 215.
30. T. Virdee, Prompt γ photoproduction measurement, Proc. Int. Europhysics Conf. on High-Energy Physics, Brighton, 1983 (to be published).
31. D. Duke and J. Owens, Phys. Rev. D26 (1982) 1600; Erratum : Phys. Rev. D28 (1983) 1227.
M. Fontannaz and D. Schiff, Z. Phys. C14 (1982) 151.
32. W. Geist and S. Reucroft, Proc. Workshop on SPS Fixed-Target Physics in the Years 1984-1989, Geneva, 1982 [CERN 83-02 (1983)], Vol. II, p. 190.
J. Sacton, ibid., Vol. II, p.222.
33. H. Wahlen, Wuppertal Univ. report WUB 83-4 (1983).
34. P. Astbury et al. (NA14 Collaboration), A program of heavy-flavour photoproduction, CERN proposal CERN/SPSC/82-73; SPSC/P 109 Add. 2 (Oct. 1982).

MUON EXPERIMENTS AT HIGH ENERGY

G. Smadja

CEN-Saclay
DPhPE
91191 Gif-sur-Yvette Cedex, France

We present a selection of results from muon experiments at CERN and FNAL; from the EMC[1] and BCDMS[2] collaborations at CERN, the BFP[3] and CHIO[4] collaborations at FNAL.

After a description of the recent data on structure functions, we discuss two aspects of the fragmentation process measured in muon experiments : charm production and transverse momentum.

EXPERIMENTAL ENVIRONMENT

The beam line

The main elements of the beam line are seen on Fig. 1 : there are four important stages beyond the primary target :

- momentum selection of pions and kaons
- a decay channel of about 600 m
- absorption of the hadrons in a Beryllium block, leaving a contamination smaller than 10^{-4}
- momentum analysis of each muon.

The beam line is polarised and the mean value of the muon helicity is directly controlled by the choice of pion and muon momenta : $\langle\lambda\rangle = \cos\xi$, where ξ is the angle on Fig. 2 between the pion direction and the laboratory line of flight in the muon rest frame.

For a choice of pion and muon momenta of 220 and 203 GeV/c $\cos\xi = .81$. Beam hodoscopes located along the beam allow to measure the trajectory of each muon and give its momentum with an accuracy $\Delta p/p \sim 3\ 10^{-3}$. The timing information is supplied by the hodoscopes in a gate of 100 ns around each event.

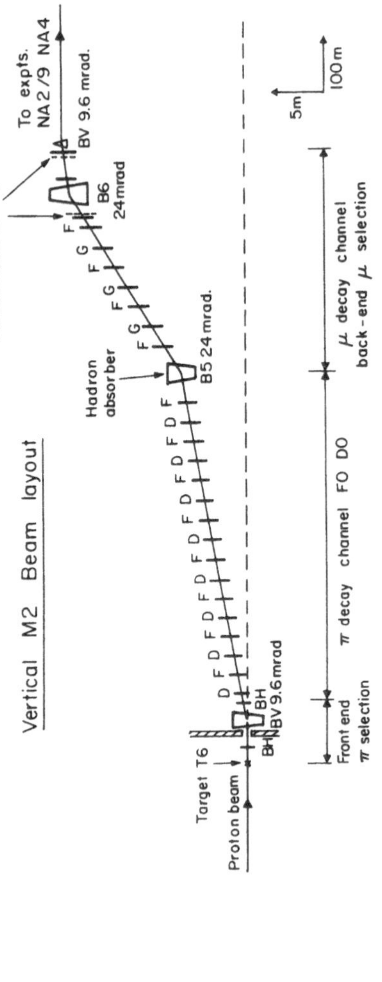

Fig. 1. The muon beam lay out.

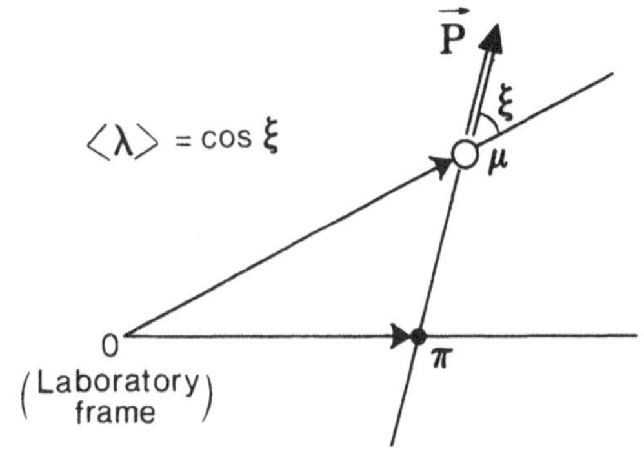

Fig. 2. Polarisation of the muon in π decay.

The flux measurement

Three methods are used and cross checked to count the muons.

Accidentals with a known random gate. This method which has been described by R.P. MOUNT[5] takes into account the loss of muons from the requirement of a reconstruction of hodoscope tracks. Let N_r be the number of reconstructed beam tracks in the random gate of width G and frequency f, I the unknown flux of muons which can be reconstructed : N_r = IxfxG allows a measurement of the muon flux I.

Dead time correction. Scalers counting the full muon beam can be corrected for discriminator dead times once the instantaneous intensity around each event is known.

Geometrical correction. Each beam hodoscope cell is counted, with an intensity divided therefore by $\sim$ 64. The sum of all cells is corrected for double counting by analysing the timing distributions.

The different methods agree to within 2%, which corresponds to an overall normalisation uncertainly of 3% when taking into account the additional contributions of beam phase space and acceptance correction to the error.

The acceptance domain

The muon experiments cover a Q^2 domain from 3 to $\sim$ 200 GeV^2, about 10 times higher than the ep, eD data from the SLAC MIT collaboration[6]. We see on Fig. 3 that these experiments are the only ones which stay safely away from a hadronic mass of w_H = 3 GeV, characteristic for instance of charm threshold, or resonance production.

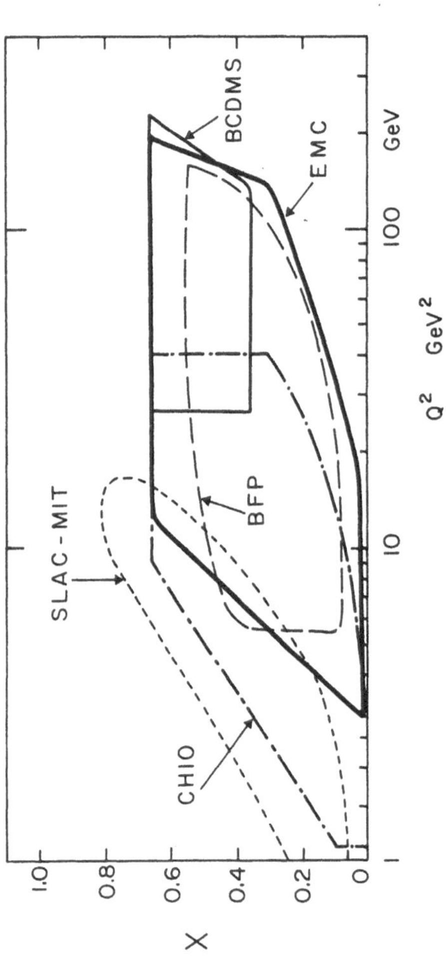

Fig. 3. Kinematical domains of deep inelastic muon experiments.

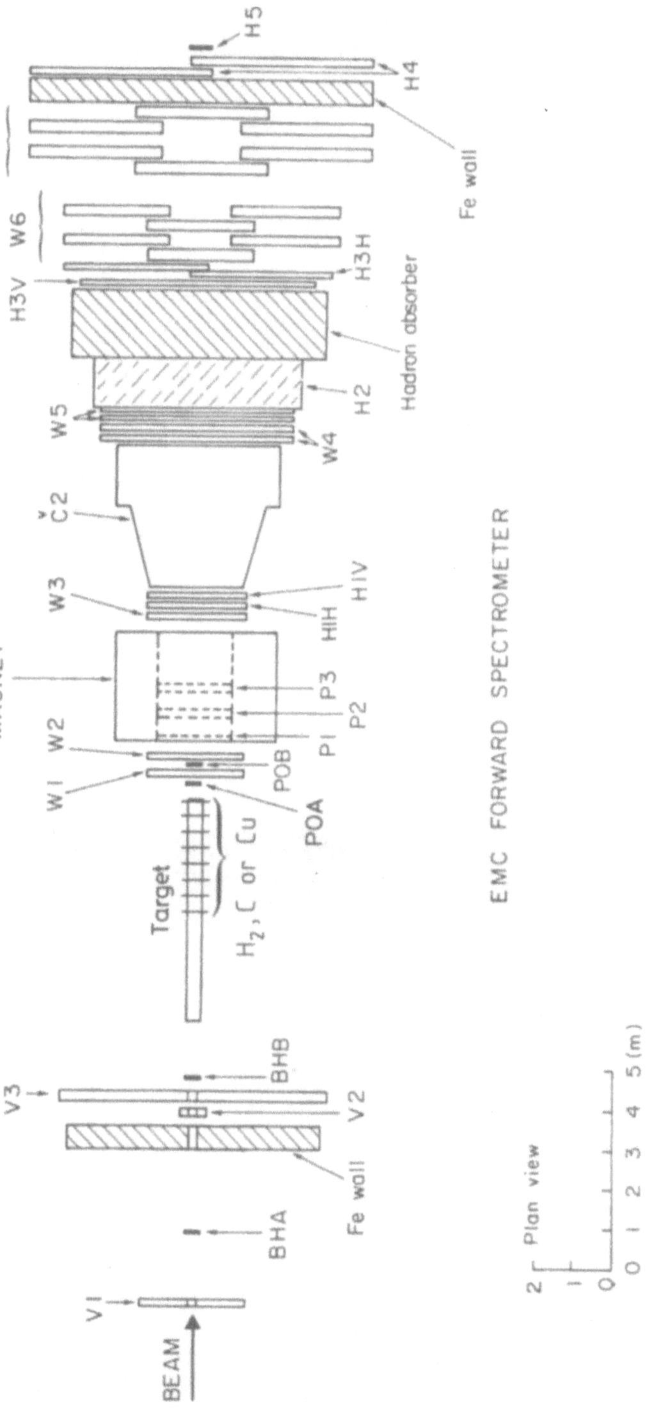

Fig. 4a. The EMC spectrometer.

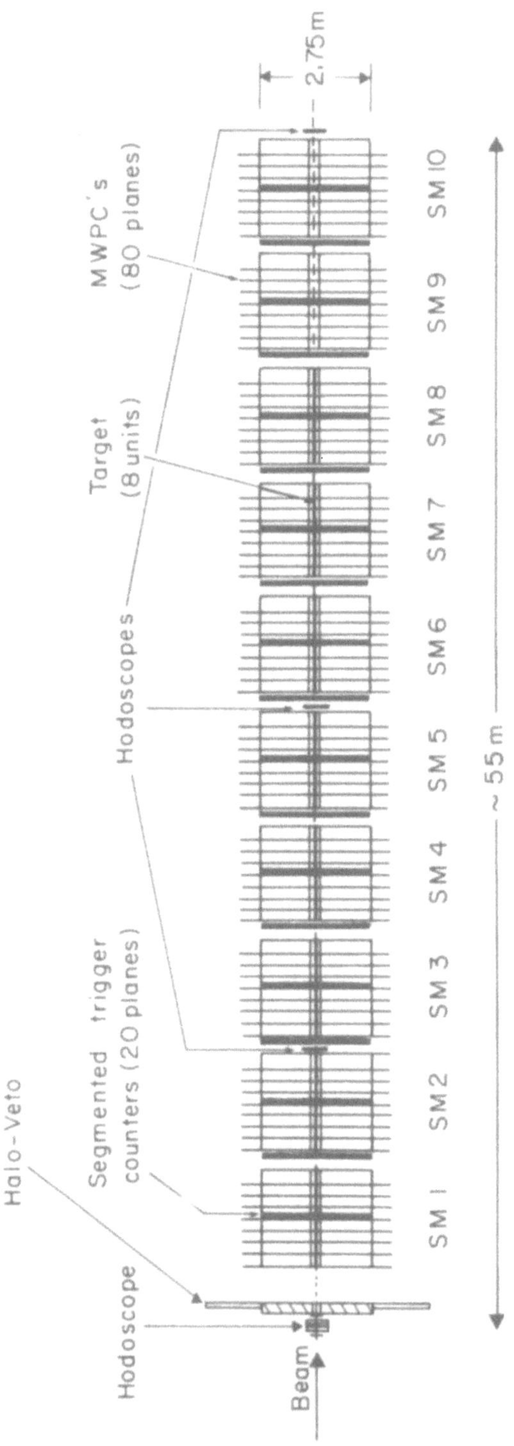

Fig. 4b. The BCDMS spectrometer.

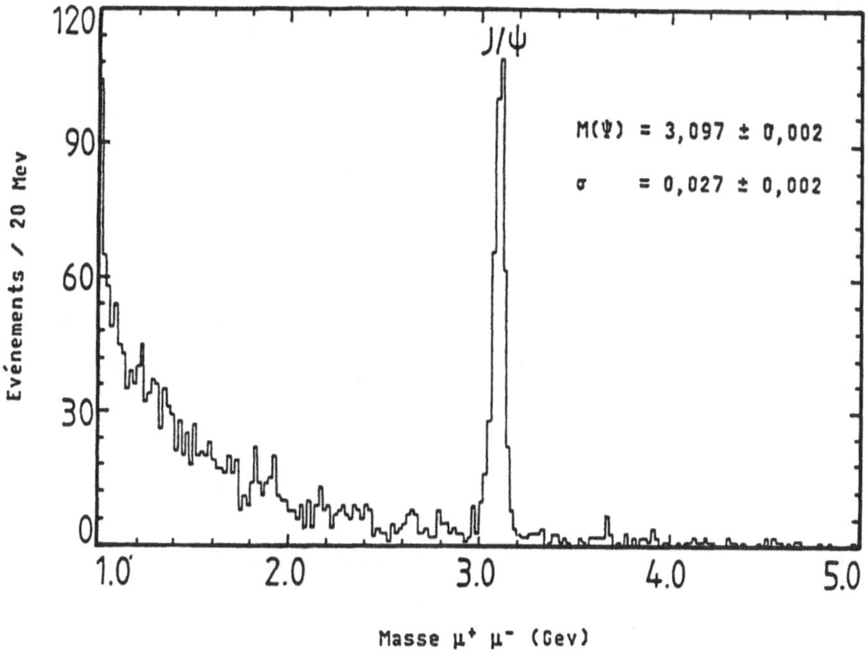

Fig. 5a. The Ψ mass in μD scattering.

The experimental lay out and resolutions

The EMC collaboration uses an air gap magnet with a field which can reach 20 kg. The spectrometer shown on Fig. 4a. has a wide acceptance domain and is characterised by an accuracy $\Delta p/p \sim 10^{-5}p$ (GeV), with a systematical calibration uncertainty of 3 10^{-3}.

These possibilities are illustrated on Fig. 5a. by the Ψ mass spectrum from $\mu D \rightarrow \Psi X$. When an iron target calorimeter (STAC) is used, the resolution is degraded by multiple scattering and radiative losses, as seen on Fig. 5b.

The same effects spoil the resolution of the iron toroids of the BCDMS collaboration, in the set up shown on Fig. 4b. The 40 m long Carbon target allows a high luminosity, but the momentum resolution is limited to $\Delta p/p = 7\%$ by multiple scattering. The resolution is further degraded by radiative fluctuations. Their presence is manifest on Fig. 6, which shows the energy spectrum of a 190 GeV muon after 16 m of C (or 2 m of Fe).

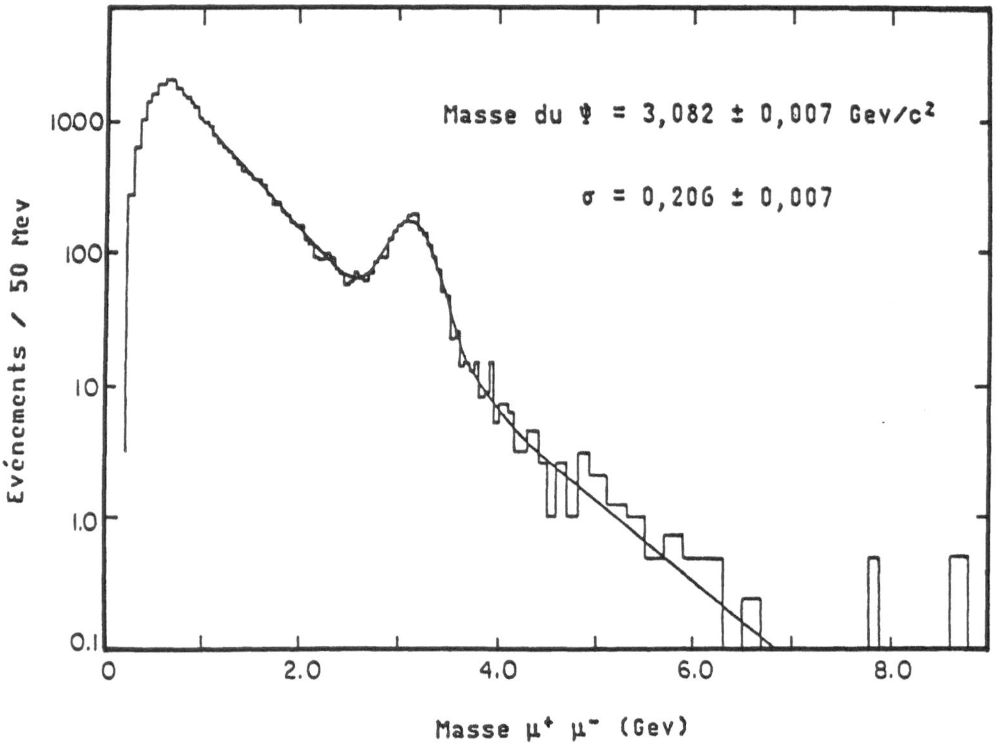

Fig. 5b. The Ψ mass in μFe scattering.

RADIATIVE CORRECTIONS

Radiative corrections must be applied to the observed cross section to take into account two effects : contribution of the elastic tail in the scattering of muons on heavy nuclei, and internal Bremstrahlung from the muon line in deep inelastic μN interactions. The corrections are evaluated within QED. The elastic tail contributes only when $x < .1$. On a carbon target, it reaches 50% at $x = .05$ for $Q^2 = 20\ \text{GeV}^2$ with a beam of 250 GeV. Bremstrahlung corrections amount to less than 6% for $x > .2$ but reach 20% at $x = .05$.

Three programs have been used to compute the corrections written using the formalism of L. MO and TSAI[7,8] by D. BARDIN[9,10] according to the techniques C. CHAHINE[11]. They all agree as shown by Fig. 7 where D. BARDIN's and DREES's computations are compared.

Determination for $R = \sigma_L/\sigma_T$

Notation for R, F_1, F_2. We briefly recall the relation between R, F_1, F_2 and its derivation :

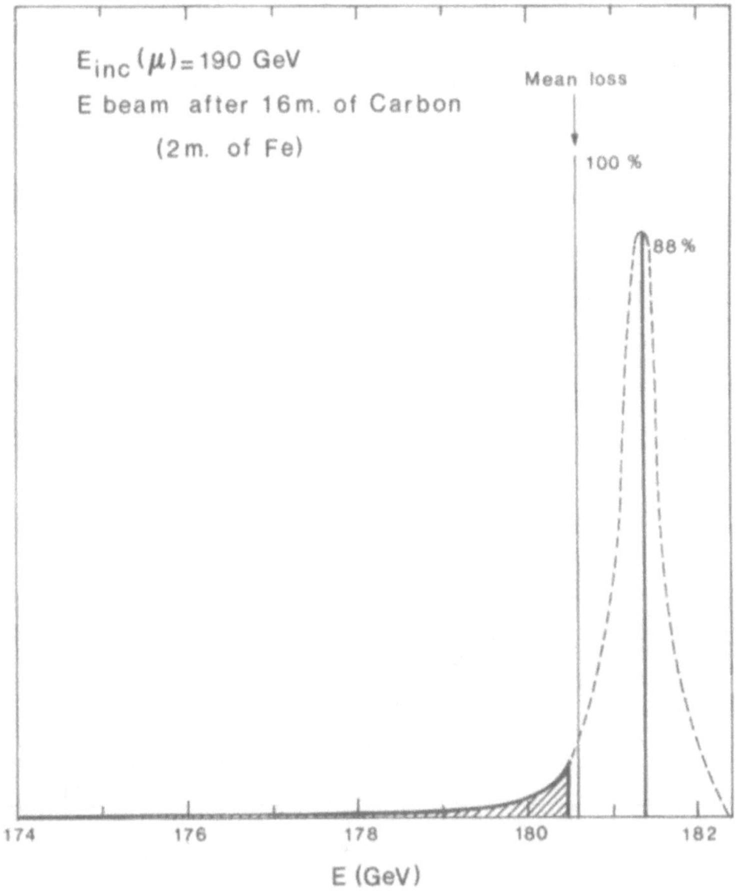

Fig. 6. Radiative tails in the energy loss.

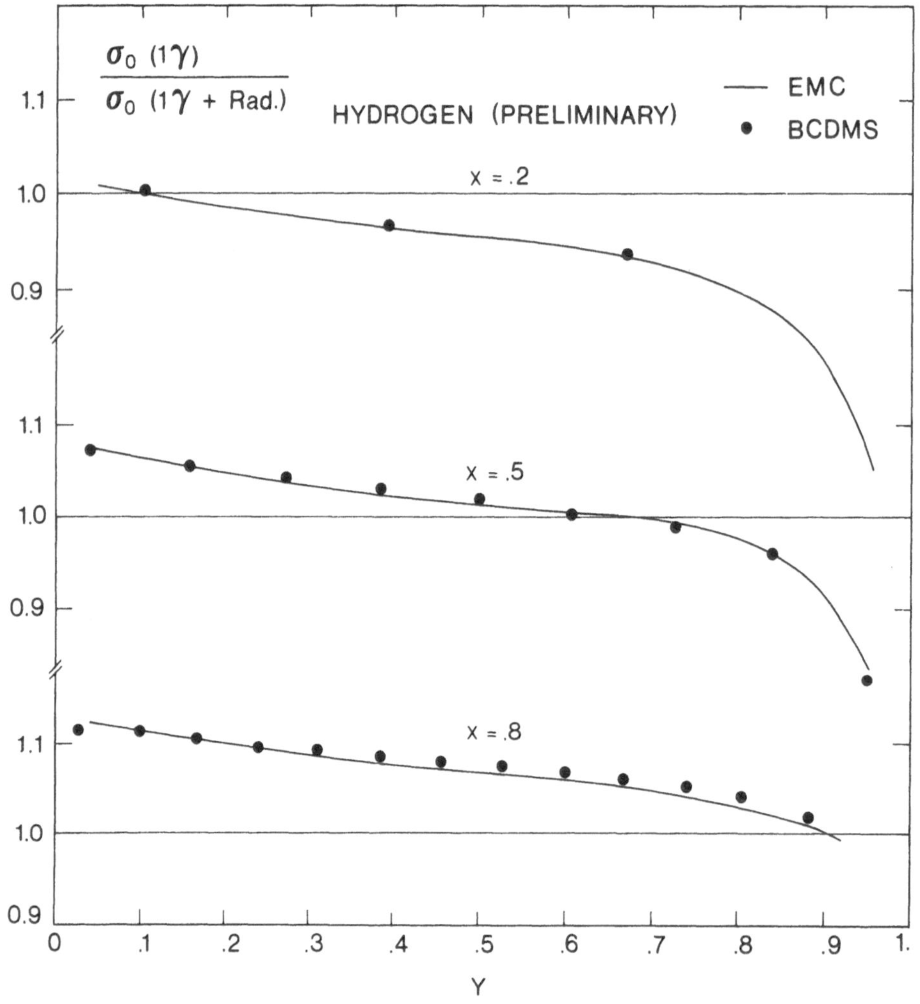

Fig. 7. Comparison of EMC[7] and BCDMS[9] radiative corrections.

$$d\sigma = \frac{1}{4\sqrt{(p_1 \cdot p_2)^2 + M^2Q^2}} \frac{d^3k}{(2\pi)^3 2E} W_{\mu\nu} \cdot L^{\mu\nu} ,$$

$$W_{\mu\nu} = (-g_{\mu\nu} + \frac{q_\mu q_\nu}{q^2}) W_1 + (p_\mu - \frac{p \cdot q}{q^2} q_\mu)(p_\nu - \frac{p \cdot q}{q^2} q_\nu) W_2 .$$

σ_L, σ_T : the longitudinal and transverse cross sections are defined as suggested by HAND[12] by selecting appropriate polarisation vectors.

a) Transverse cross section

$$A_T^2 = -1 \ ; \ q.A_T = 0 \ , \ \vec{q}.\vec{A}_T = 0 \ \rightarrow \ A_T^\circ = 0 \ ,$$

$$\sigma_T \ \alpha \ W_{\mu\nu} A_T^\mu A_T^\nu = \frac{4\pi^2\alpha}{\sqrt{Q^2 + \nu^2}} W_1 = \frac{4\pi^2\alpha}{\sqrt{Q^2 + \nu^2}} \frac{F_1}{M} .$$

b) Longitudinal cross section

$A_T^2 = +1$, $q.A_L = 0$, $\vec{A}_L // \vec{q}$ in the laboratory frame. A_L cannot therefore be obtained from A_T by a rotation.

$$\sigma_L \ \alpha \ W_{\mu\nu} A_L^\mu A_L^\nu = \frac{4\pi^2\alpha}{\sqrt{Q^2 + \nu^2}} \left[(1 + \frac{\nu^2}{Q^2}) W_2 - W_1 \right] ,$$

$$R = \frac{\sigma_L}{\sigma_T} = (1 + \frac{\nu^2}{Q^2}) \frac{W_2}{W_1} - 1 ,$$

with 2 limits as $Q^2 \rightarrow 0$:

if ν is fixed,

$$\left. \begin{array}{l} \sigma_L \rightarrow 0 \\ \sigma_T \rightarrow \sigma_\gamma \end{array} \right\} R \rightarrow 0 ,$$

if x is fixed,

$$\left. \begin{array}{l} \sigma_L \sim 1/\sqrt{Q^2} \\ \sigma_T \sim \sqrt{Q^2} \end{array} \right\} R \rightarrow \infty .$$

The latter limit has clearly no physical meaning.

Interest into R. In the parton model, a naive kinematical argument by FEYNMAN[13] suggests

$$R \sim \frac{4 \langle k_T^2 \rangle}{Q^2}$$

so that at low Q^2, R gives some information on the "transverse momentum of quarks in the nucleon". At large Q^2, on the other hand, ALTARELLI and MARTINELLI[14] have derived a contribution to R which is linear in α_S :

$$F_L^{\mu} = \frac{\alpha_S}{2\pi} x^2 \int_x^1 \frac{dz}{z^3} [\frac{8}{3} F_2(z) + \frac{40}{9} (1 - \frac{x}{z}) zG(z)] .$$

→ R should not be zero, but the large contribution from the gluon distribution G(z) at small z cannot be determined easily in muon experiments.

Measurement of R

The differential cross section $d^2\sigma/dQ^2d\nu$ can be reexpressed as a function of σ_T, σ_L :

$$\frac{d^2\sigma}{dQ^2d\nu} \propto \{\cos^2\frac{\theta}{2} W_2 + 2 \sin^2\frac{\theta}{2} W_1\}$$

$$\propto \{\frac{\cos^2\frac{\theta}{2}}{1 + \nu^2/Q^2} (\sigma_T + \sigma_L) + 2 \sin^2\frac{\theta}{2} \sigma_T\}$$

$$\propto \sigma_T + \varepsilon\sigma_L \quad \text{with} \quad \varepsilon = \frac{1 - y - Q^2/4E^2}{1 - y + \frac{y^2}{2} + Q^2/4E^2}$$

$$\varepsilon \sim \frac{1 - y}{1 - y + .5y^2} .$$

At a given (x,Q^2) point, the determination of $R = \sigma_L/\sigma_T$ requires the measurement of the differential cross section at two different energies :

$$R = \frac{\sigma(y_2) - \sigma(y_1)}{\varepsilon(y_2)\sigma_1 - \varepsilon(y_1)\sigma_2} .$$

The measurement error is about

$$\Delta R \sim \sqrt{2}\, \Delta\sigma/(\sigma\Delta\varepsilon) .$$

In the accessible domain, $\Delta\varepsilon$ is about .3, so

$$\Delta R < .05 , \frac{\Delta\sigma}{\sigma} < 1.0\ 10^{-2} ,$$

i.e. an accurate measurement of R requires a 1% accuracy on the absolute normalisation. One might try to readjust the relative normalisation of the two data sets in a region insensitive to R, but in existing experiments, it turns out that the statistical uncertainty in such a procedure is again ∿ 2%.
The EMC experiment has published a determination of R in H_2 scattering in a domain schematically represented in Fig. 8. A typical slope measurement in (Q^2,ν) bins where 3 energies are available is given in Fig. 9. R remains compatible with zero over the whole measurement

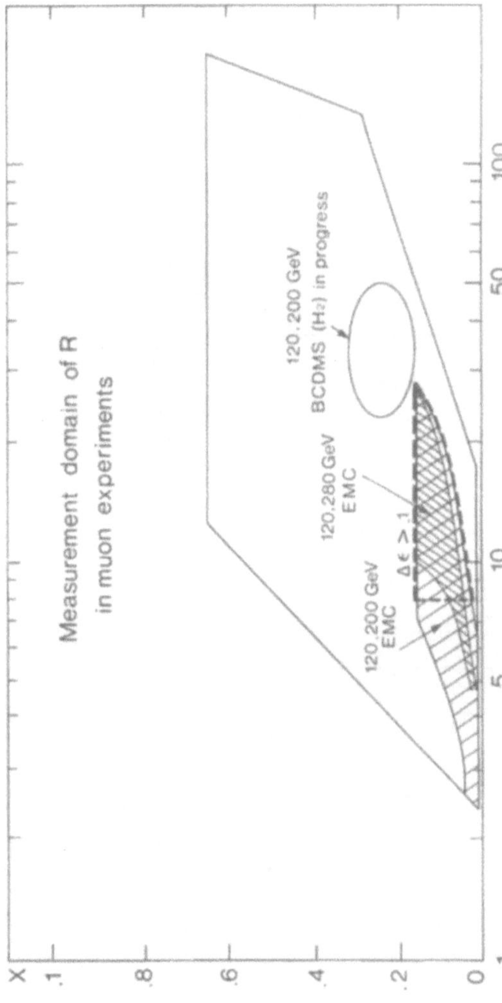

Fig. 8. Domain for the determination of R in μH_2 scattering.

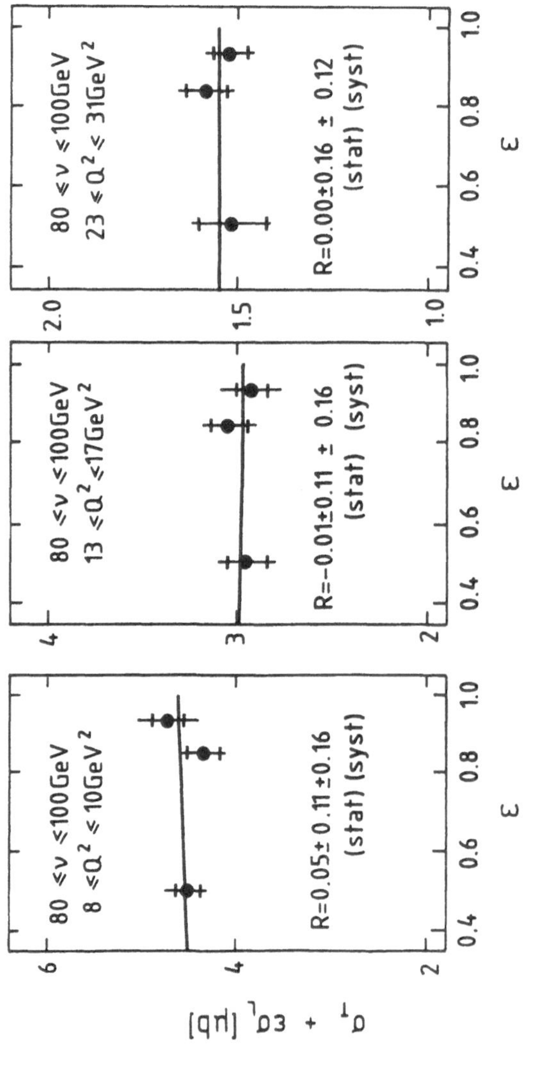

Fig. 9. Dependence of the total cross section upon Σ in μH_2 scattering.

range, as seen in Fig. 10. The mean values for μH_2 by EMC[15], and for μ Fe by P. PAYRE[16] are :

$$R_{H_2} = 0 \pm .1 \qquad \begin{cases} .05 < x < .2 \\ 8 < Q^2 < 40\ GeV^2 , \end{cases}$$

$$R_{Fe} = -.05 \pm .02 \pm .1 \qquad \begin{cases} .05 < x < .4 \\ 5 < Q^2 < 160\ GeV^2. \end{cases}$$

This should be constrated with the result obtained by BODEK et al. at a lower Q^2 : R = .2 ± .1, compared on Fig. 11 with the EMC data : it is not yet clear whether the disagreement reflects systematic uncertainties or a physical dependence upon Q^2, with $k_T^2 \sim .5\ GeV^2$.

MEASUREMENT OF F_2

H_2 and D

The structure function F_2 has been measured on hydrogen by the EMC collaboration at 120 and 280 GeV[17]. The range of these measurements extends from x = 0.3 to x = .65, with a low statistical significance beyond x = .45. It should be noted that the two sets of measurements at 120 and 280 GeV on Fig. 12 are quite compatible, which implies the absence of significant systematical errors.

The deuterium data collected by EMC can then be used to compare the distributions of u and d quarks in the nucleons :

$$F_2^n / F_2^p = \frac{1 + 4\ d/u}{4 + d/u} , \qquad x > .3.$$

The results on Fig. 13 confirm the trend observed by the SLAC MIT Collaboration[6] at lower Q^2 : $F_2{}^n / F_2{}^p \leq .25$ as $x \to 1$, which suggests $d/u \to 0$ as pointed out by Feynman and Fields[18].
The difference

$$\Delta(x) = F_2^n - F_2^p = \frac{1}{3}(u + \bar{u} - d - \bar{d})$$
$$= \frac{1}{3}(u - \bar{u} - (d - \bar{d}) + 2(\bar{u} - \bar{d})),$$

is not a pure valence distribution if $\bar{u}(x) \neq \bar{d}(x)$, as was suggested by the CFS collaboration[19] in an analysis of DRELL-Yan scattering. $\Delta(x)$ is compared on Fig. 14 with the neutrino data at $Q^2 = 7\ GeV^2$ to indicate that the contribution of $\bar{u} - \bar{d}$ is small. Another estimate is obtained when the former relation is integrated :

$$\int (F_2{}^p - F_2{}^n) \frac{dx}{x} = \frac{1}{3} + \frac{2}{3} \int dx(\bar{u} - \bar{d}).$$

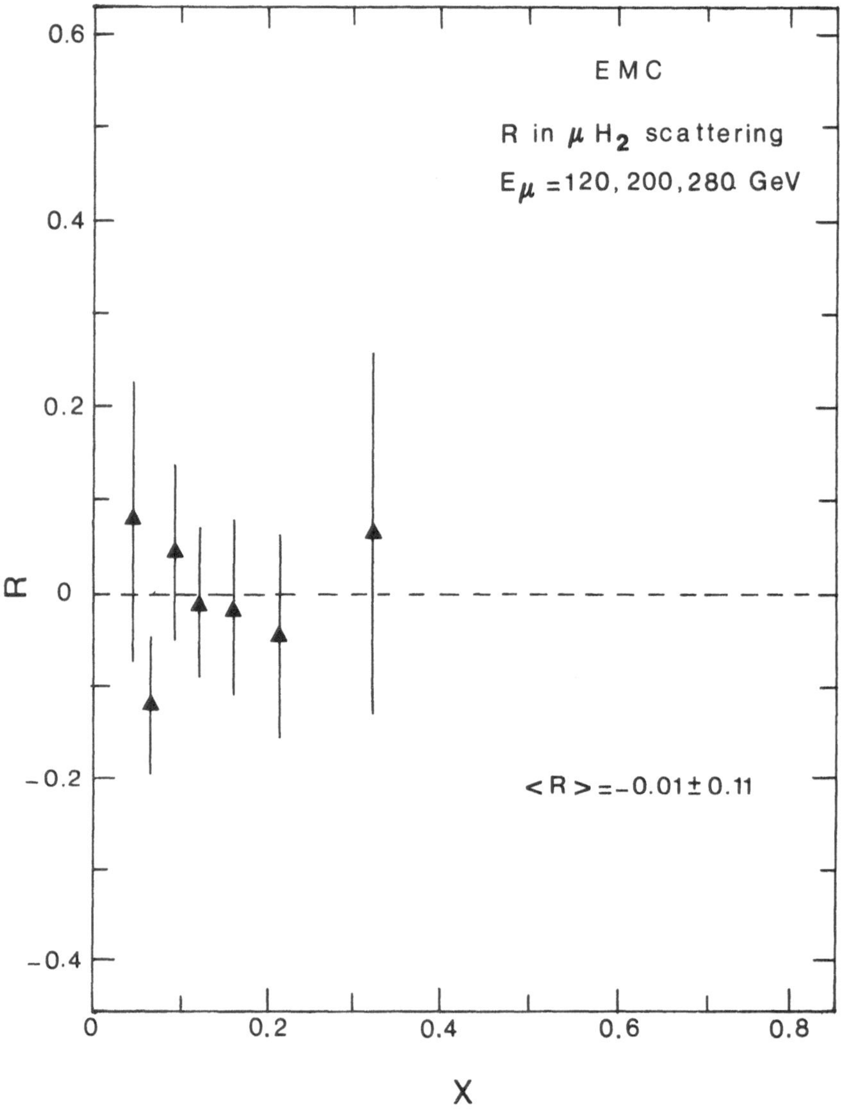

Fig. 10. R as a function of x.
(a) in μH_2 scattering.

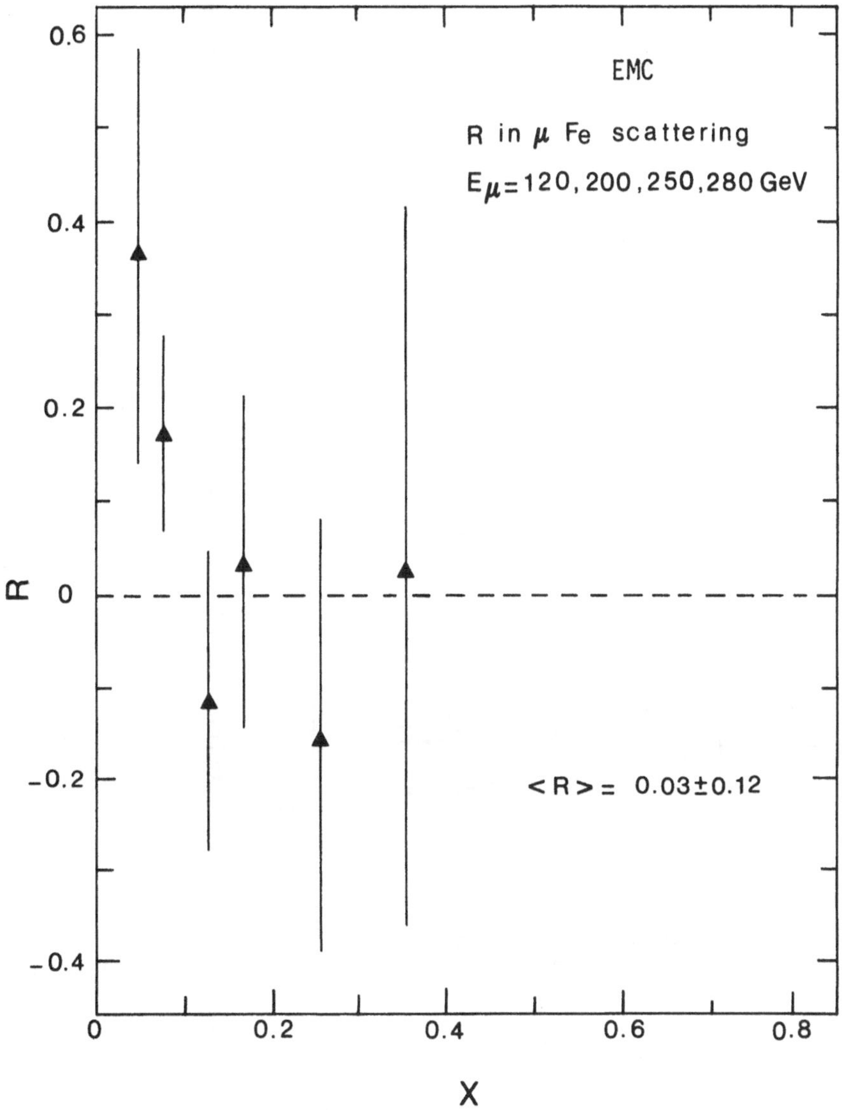

Fig. 10. R as a function of x.
(b) in μFe scattering.

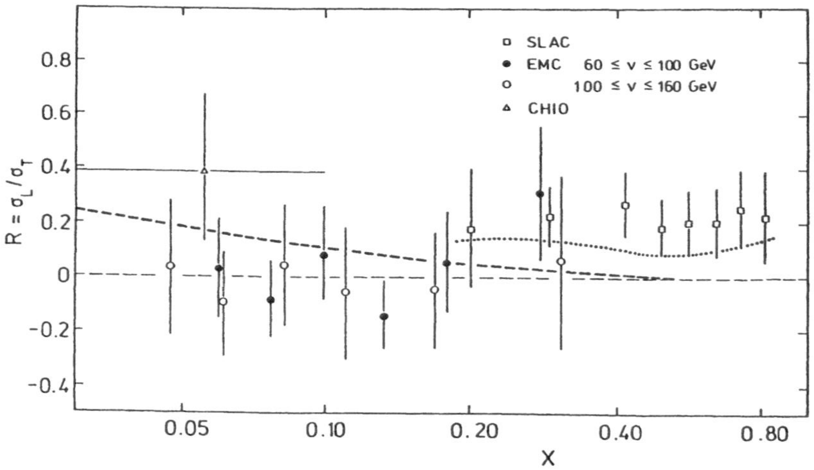

Fig. 11. Comparison of EMC determination of R ($Q^2 > 6\ GeV^2$) and SLAC-MIT measurement[6].

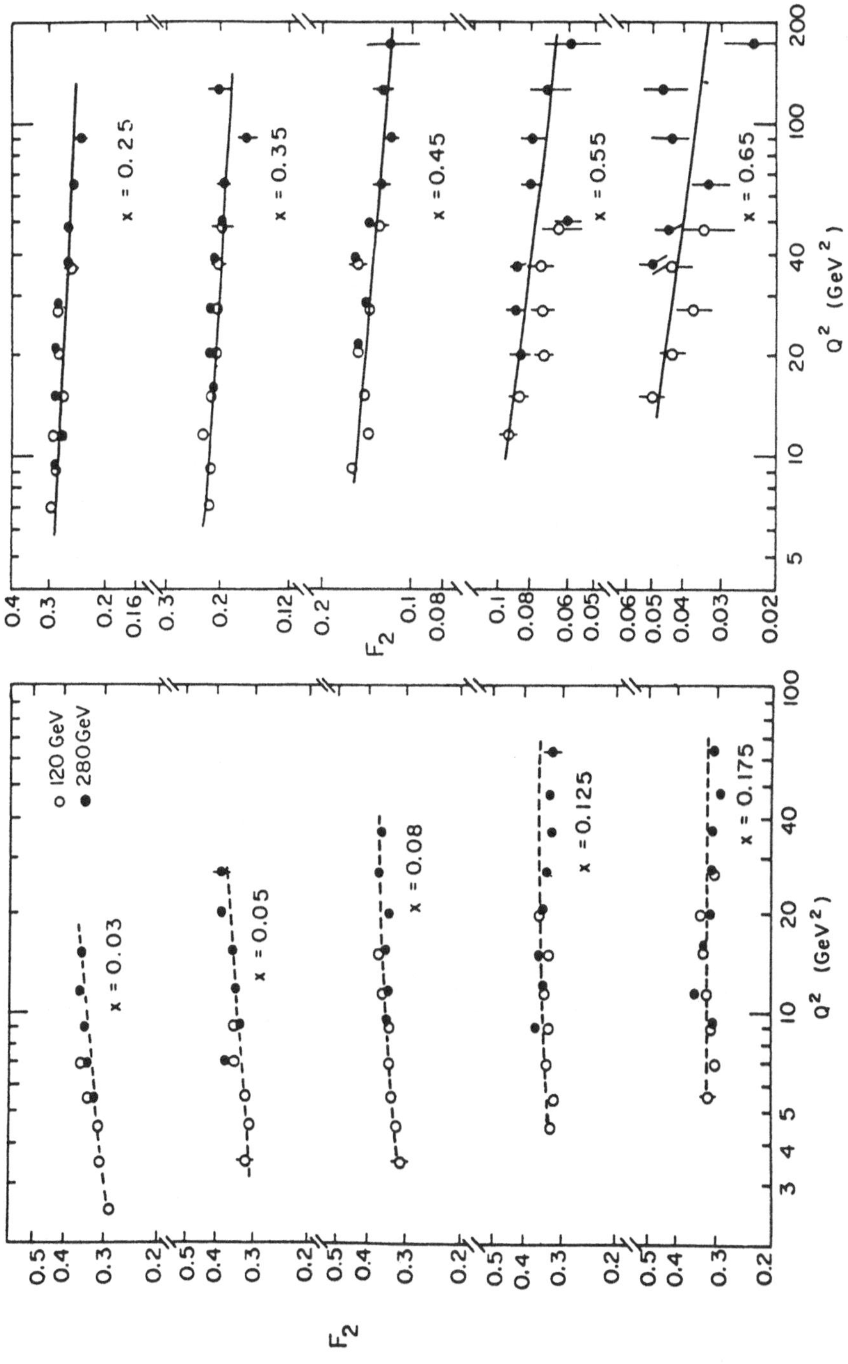

Fig. 12. F_2 function in μH_2 scattering (EMC).

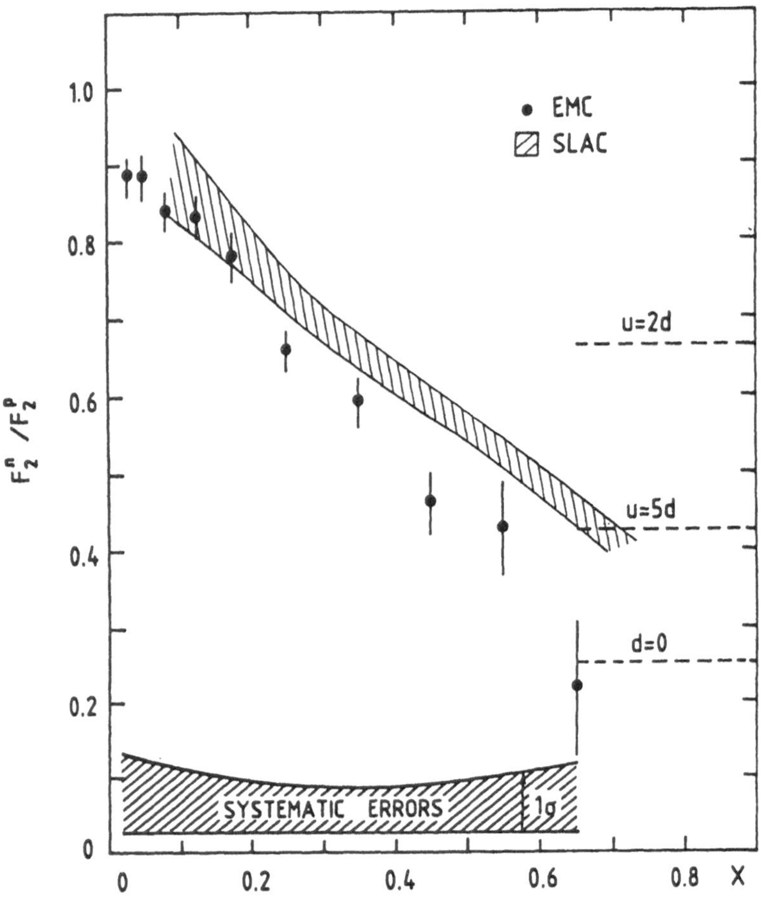

Fig. 13. The ratio of structure functions on n and p as a function of x.

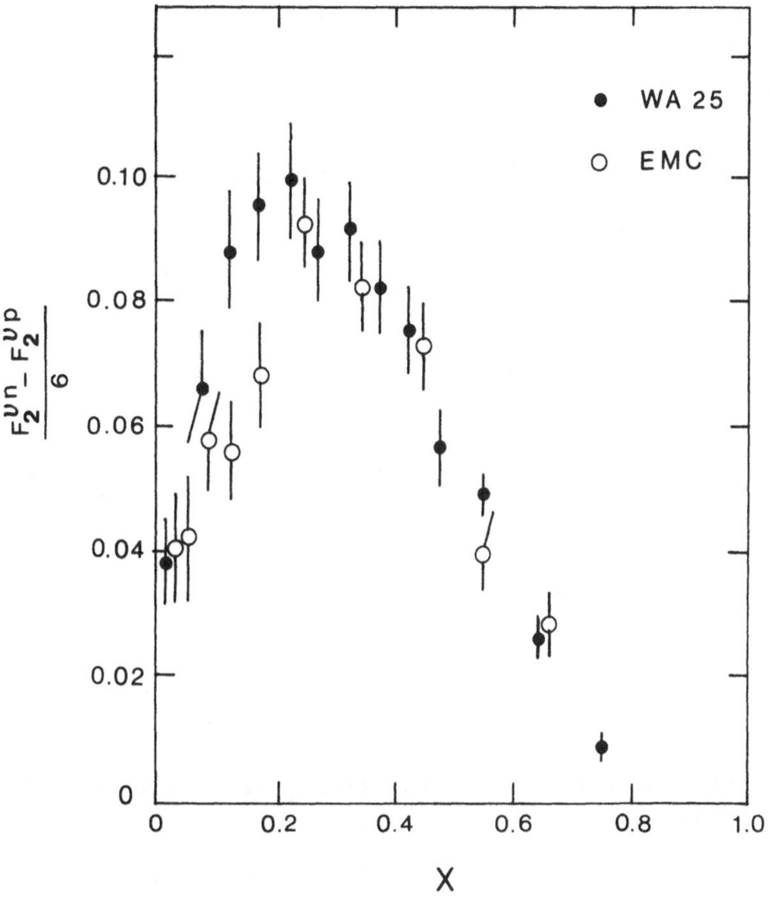

Fig. 14. Comparison of $\Delta(x) = F_2^n - F_2^p$ with $\frac{1}{6}(F_2^{\nu n} - F_2^{\nu p})$.

The experimental data of EMC[20] gives

$$\int_{.03}^{.65} \frac{dx}{x} (F_2^p - F_2^n) = .18 \pm .01 \pm .07 \text{ syst.}$$

$$\int_{0}^{1} \frac{dx}{x} (F_2^p - F_2^n) = .24 \pm .02 \pm .13 \text{ syst.}$$

by extending the parametrisation to the full range x = (0,1).

The large systematical uncertainty does not allow a determination of $\int(\bar{u} - \bar{d})$. A recent measurement in ν D_2 suggests $\bar{u} = \bar{d}$ with a better accuracy.

F_2 on Iron and Carbon

The heavy targets allow a better statistical accuracy at large x, as seen from the STAC measurement of EMC on iron[22]. Some discrepancies on Fig. 15 between the data at 120 and 250 GeV imply the presence of substantial systematical errors. The effect of a typical calibration error of 4 10^{-3} on F_2 is shown on Fig. 16. As x reaches .75, the shift of F_2 caused by the calibration offset exceeds 20% below Q^2 = 100 GeV^2. It is therefore no surprise that the data from different energies should not match well in high statistics experiments at large x.

Typical uncertainties arise form :

- beam calibration	: $\Delta E/E$	$\sim$	2 10^{-3}	
- spectrometer calibration	: $\Delta p/p$	$\sim$	4 10^{-3}	
- acceptance	: $\Delta A/A$	$\sim$	10^{-2}	
- muon counting	: $\Delta\phi/\phi$	$\sim$	2 10^{-2}	
- backgrounds	:	$<$	10^{-2}	(EMC)
		$<$	2 10^{-3}	(BCDMS).

Ratio of structure functions on Iron and Deuterium

The iron data is a combination of results at four energies : 120, 200, 250 and 280 GeV, while the deuterium measurement is performed at 280 GeV.

The EMC collaboration was able to compare the two data sets in the same apparatus[23], in a Q^2 range which varies with x :

$$9 < Q^2 < 27 \text{ GeV}^2 \quad x = .05 ,$$

$$36 < Q^2 < 170 \text{ GeV}^2 \quad x = .65 .$$

The ratio of the two structure functions is shown on Fig. 17 as a function of x. It is obtained by averaging over Q^2 a point by point determination of R = F_2 (Fe)/F_2 (D). The systematical uncertainties previously described appear as a shaded area on Fig. 17. In the available x range, R may be parametrised as R = a + b x with

$$b = -.52 \pm .04 \pm .21 \text{ (syst.)}.$$

Discussion of the result on F_2 (Fe)/F_2 (D)

The systematic contribution Δb = .21 should not be understood

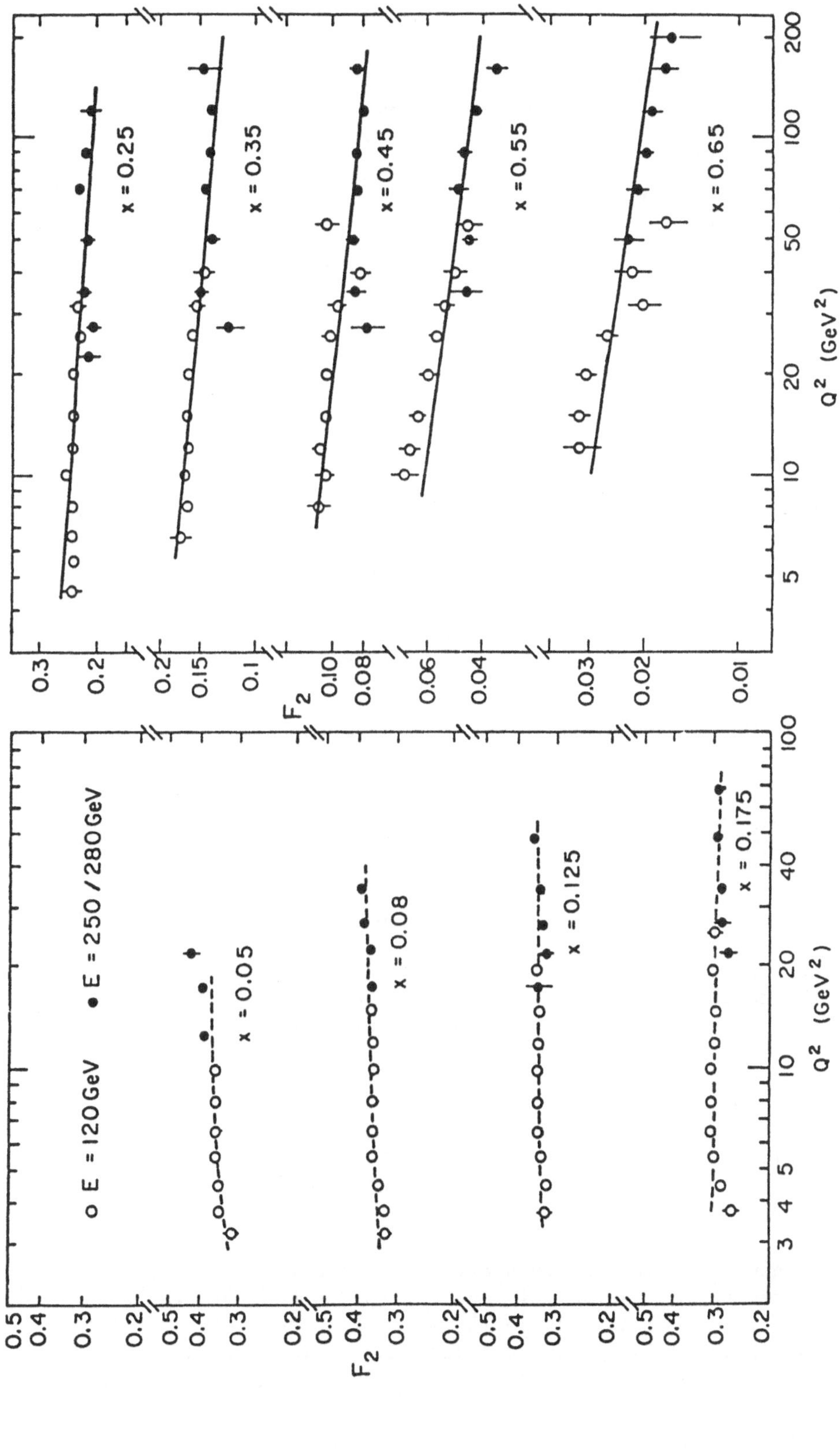

Fig. 15. F_2 function in μFe scattering.

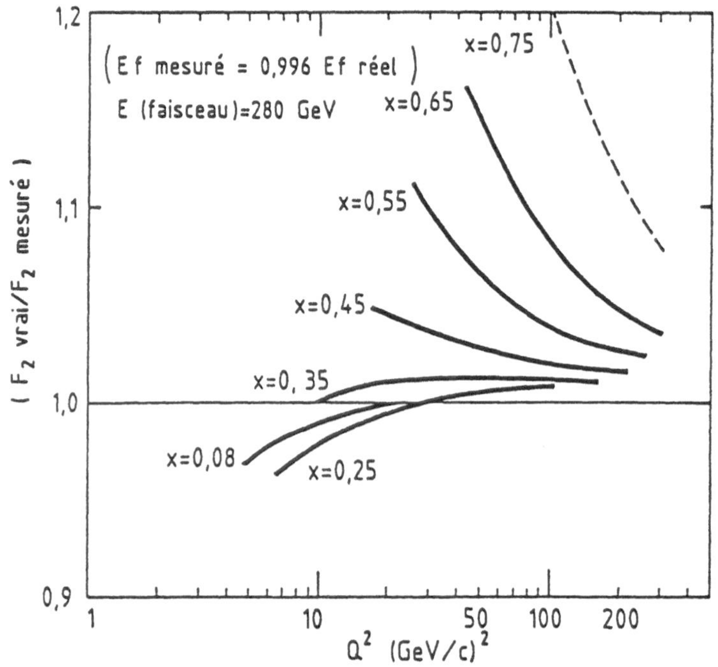

Fig. 16. Effect of a calibration error of 4 10^{-3} on F_2.

as 1 standard deviation. It is closer to a worst case evaluation.

The iron cross section is increased for all sources of errors while the deuterium structure function is decreased. All individual sources of systematics are then added quadratically.

There is additional evidence supporting the dependence upon x of the iron to deuterium cross sections : the carbon data of BCDMS can be substituted to the iron data as done on Fig. 18.(a) borrowed from K. RITH[24]. The x distribution of μD scattering by EMC[20] can also be replaced by the measurement of CHIO[25]. In all cases, the trend of Fig. 17. is confirmed.

The neutrino data on νFe scattering indicates directly the size of the systematics involved, if one interprets the 10% rise with x of F_2^{ν} (Fe)/F_2^{μ} (Fe) on Fig. 18.(b) as a measurement error. The ratio F_2^{ν} (Fe)/F_2^{μ} (D) is nevertheless decreasing with x as $x \to 1$.

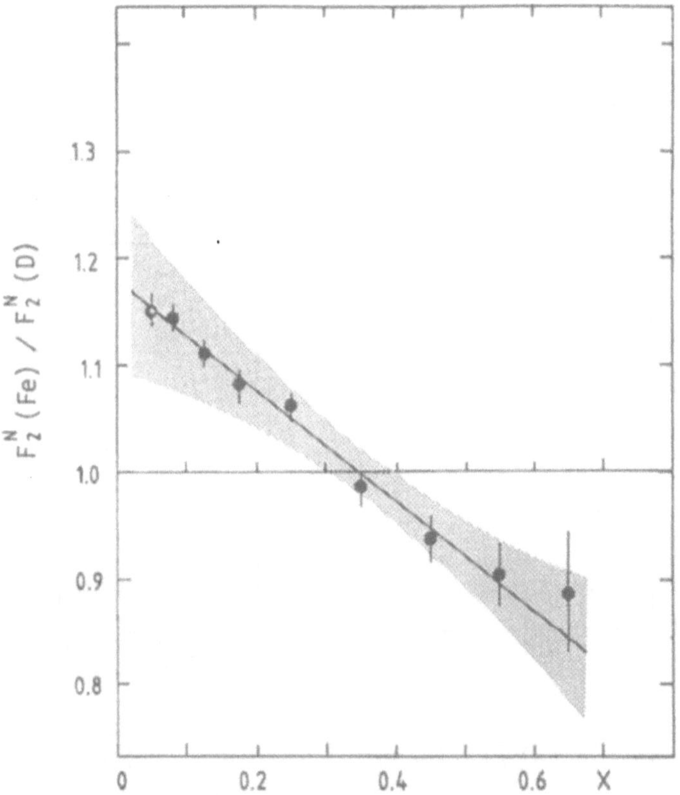

Fig. 17. Ratio of the structure functions of iron and deuterium as a function of x for $Q^2 \geqq 10$ GeV^2.

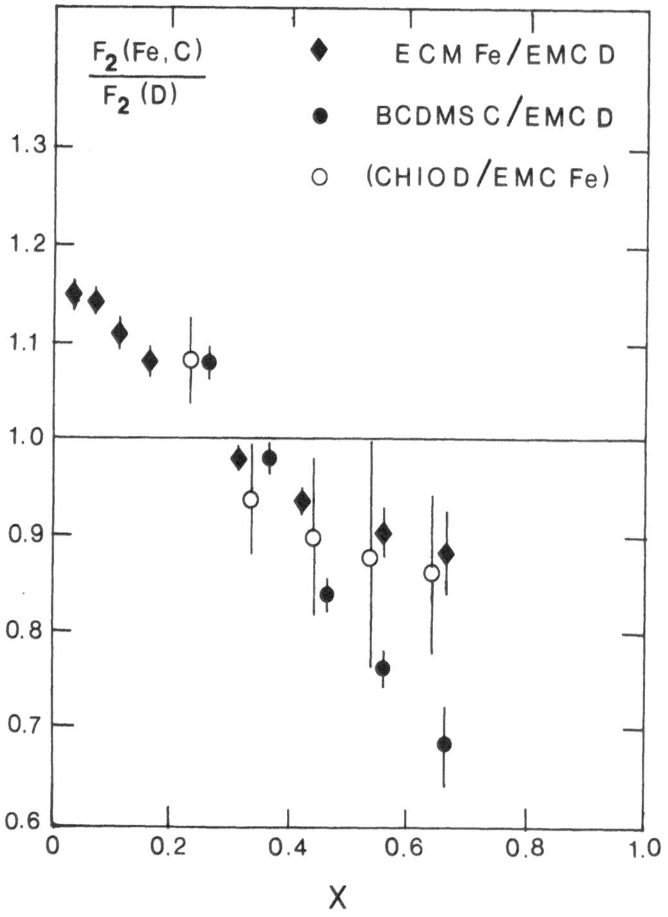

Fig. 18.(a) Ratios of structure functions
. F_2^{carbon} (BCDMS)/F_2^{D} (EMC) ;
∘ F_2^{Fe} (EMC)/F_2^{D} (CHIO).

The empty target data of the SLAC MIT experiments have been reanalysed[26] and confirm the results of EMC[23]. It is however observed on Fig. 19. that when $Q^2 < 1$ GeV^2, an additional "shadowing" occurs at $x < .1$, depleting the inclusive cross section.

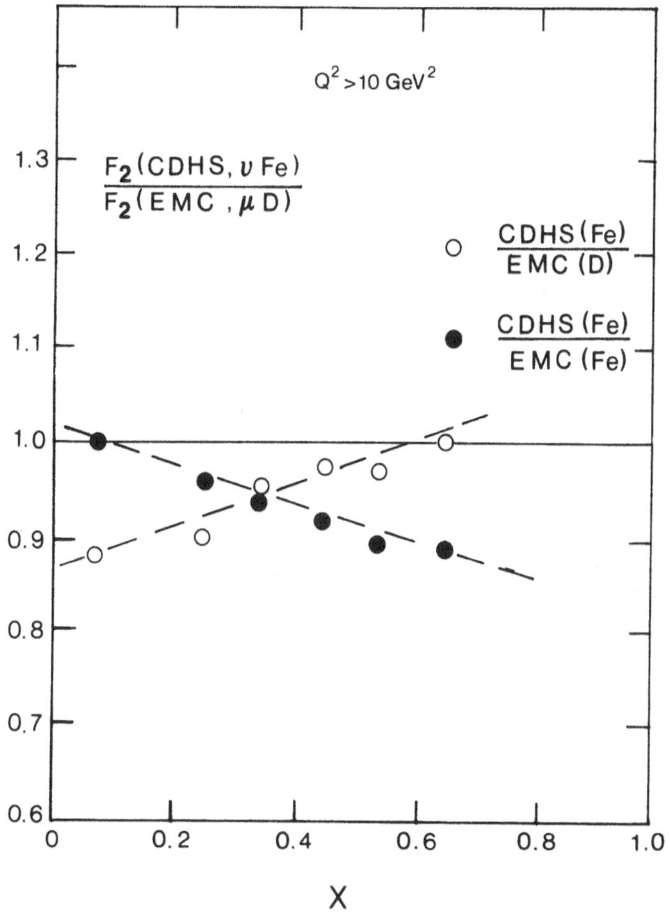

Fig. 18.(b) F_2^{Fe} (CDMS-ν)/F_2^{D} (EMC-μ).

The previous results should be contrasted with the natural expectations which prevailed before the measurement. Fermi motion effects in the target nucleus had been computed by several authors[27] and give in all cases contributions rising with x by ∿ 25%, as summarised on Fig. 20. It should be stressed that in the traditional understanding of Fermi motion corrections, even a flat ratio was unexpected.

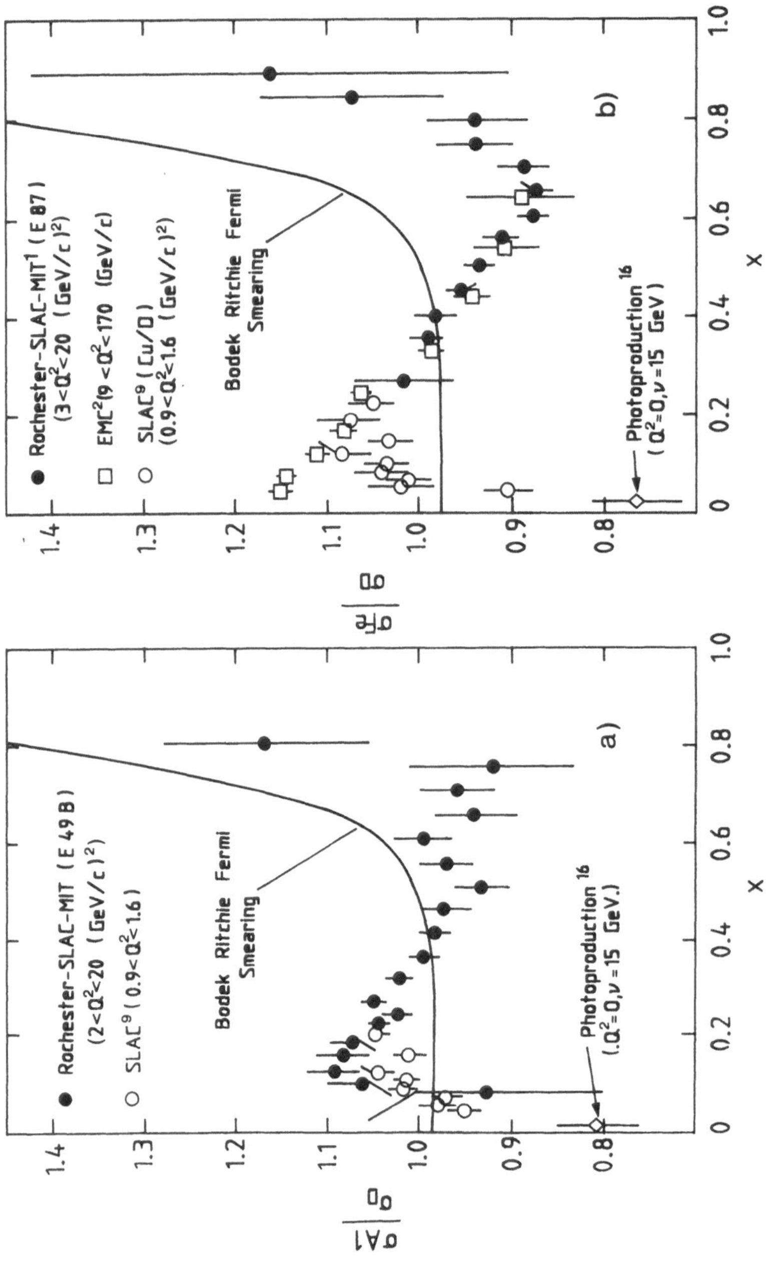

Fig. 19. SLAC MIT Rochester : ratio of F_2 on nuclear targets and D.

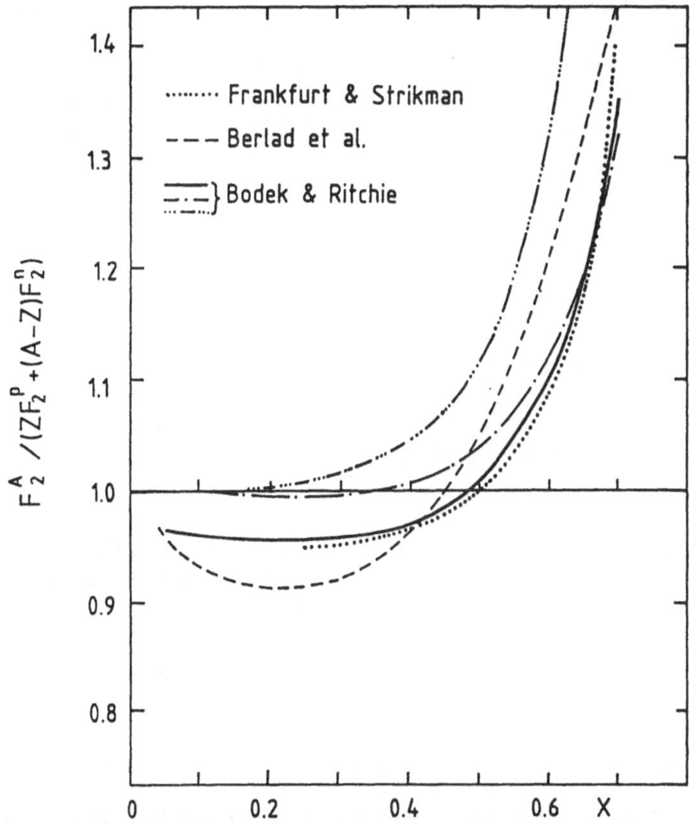

Fig. 20. Fermi momentum corrections according to reference[27].

Interpretation of the nuclear effects

Many tentative explanations of the EMC observation have been recently suggested, but none that actually covers the full x range in a compelling way. They may be classified into a few (non contradictory) groups :

- "bag model" computations of[28] can reproduce the observation if the radius of the nucleon is changed inside the nuclear matter.
- C.H. LLEWELLYN SMITH, M. ERICSSON and A.W. THOMAS[29] explain the enhancement at low x by an increased pion contribution from nuclear forces. One virtual pion contributes an amount :

$$\delta F_2^N = \int_x^1 f(y)\, F_2^\pi(x/y)\, dy \ ,$$

$f(y)$ = pion momentum distribution of virtual pions, related to the form factor $F_{\pi NN}$ (t) by

$$f(y) = \frac{3g^2}{16\pi} \int_{t_{min}}^{\infty} dt\ t\ \frac{|F(t)|^2}{t + m_\pi^2} \ ,$$

F_2^π (x) is the pion structure functions from DRELL-YAN process.[30]

- generalised scaling is used by O. NACHTMAN and H.J. PIRNER[31] to relate scaling violations in Q^2 to nuclear effects :

$$\frac{1}{A}\ F(x,Q^2,A) = F\ (x,R_A^{\ 2}Q^2),$$

$$R^2 \ \alpha\ A^{2/3}\ .$$

Such a relation generates the correct trend but the x value at which the ratio F_2 (Fe)/F_2(D) equals unity is x = .35, while the scaling violations vanish at x = .2.

- Multiquark states were introduced long ago by KRZYWICKI and others.[32] A microscopic description of valence, sea and gluon partons can be described as Π (Val x Sea x Gluons) $\delta(1-\Sigma x_i)$ and may be approximated by a thermal like behaviour x^d exp(λx) where λ is a "longitudinal" temperature. At large x the DRELL-YAN-WEST counting rules[33] will lead to

$(1-x)^3$ for 3 quarks (the usual prescription)

$(1-\frac{x}{2})^9$ for 6 quarks,

and the two functional forms may be combined with an appropriate weighting.

CONSEQUENCES

The nuclear effects do not rule out the parton model, but it is unnatural at first sight to retain the additivity of quarks when the additivity of nucleons in the nucleus is dropped.

Once one uses $q^A\ (x, Q^2)$ more accurate checks of the relation between μ and ν scattering are needed : is the quark charge relation :

$$F_2^{\pi}(x)\Big|_{x>.3} = \frac{5}{18} F_2^{\nu}(x)$$

really true on heavy isolar targets?
The momentum sum rule

$$\int_0^1 (F_2 + xG)\ dx = \frac{5}{18}$$

relates changes in F_2 to changes in G which is therefore modified. G(x) in the nucleon should then differ from $G^A(x)$.

The measurement of the DRELL-YAN "normalisation" factor , K, is also clearly sensitive to the choice of the structure functions but the effect is small when compared to the present accuracy of such measurements.

ANALYSIS OF SCALING VIOLATIONS

The evolution equations

All recent analyses of scaling violations use directly the evolution equations of ALTARELLI and PARISI[34] rather than moments. If

$$t = \text{Log}\ Q^2/\Lambda^2 \quad :$$

$$\frac{dF_2(x,t)}{dt} = \frac{\alpha_s(t)}{2\pi}\ x \int_x^1 \frac{dy}{y^2} \left[F_2(y,t)P_{qq}\left(\frac{x}{y}\right) + \sum_1^{2f} e_i^2\ y\ G(y,t)\ P_{qG}\left(\frac{x}{y}\right)\right]$$

$$\frac{dG(x,t)}{dt} = \frac{\alpha_s(t)}{2\pi} \int_x^1 \frac{dy}{y} \left[\frac{1}{\Sigma e_i^2}\ \frac{F_2(y,t)}{y}\ P_{Gq}\left(\frac{x}{y}\right) + G(y,t)\ P_{GG}\left(\frac{x}{y}\right)\right]$$

at firt order in α_s, the dependence of F_2 and G upon t on the right hand side of the equations should be ignored.

$$\alpha_s(t) = 12\pi/((33 - 2f)\text{Log } Q^2/\Lambda^2)$$

at first order.

$$P_{qq} = \frac{4}{3}\frac{1+z^2}{1-z}, \qquad P_{Gq} = \frac{4}{3}\frac{1+(1-z)^2}{z},$$

$$P_{qG} = \frac{1}{2}(z^2+(1-z)^2), \qquad P_{GG} = 6[\frac{1-z}{z} + \frac{z}{1-z} + z(1-z)].$$

These equations have been extended to the second order in α_s by E.G. FLORATOS[35] and other authors.
Several limitations occur in the practical application of evolution equations :
- The integration should extend to A rather than 1 for heavy targets when x is still defined as $Q^2/2M\nu$.
- On the other hand, there is no measurement beyond x = .7, while the unknown large x region is emphasized by the factor 1/1-z in P_{qq}.
- There is some arbitrariness in the choice of the number of flavours f in $\alpha_s(t)$, or in the right hand side of equations (1,2) : charm production contributes to the cross section for x < .1, so f = 4 is justified. Threshold effects are however still present in charm production, and should not (?) be included in an evolution analysis : one may then substract the charm contribution and use f = 3.
- The gluon distribution G(x) is not well known. The determination of α_s is performed by selecting a domain x > .3 with G(x) = 0, else by borrowing the gluon distribution G(x) from CDHS[36] :

$$G = 2.5\ (1+3.5x)\ (1-x)^{5.9}.$$

- A Q^2 cut should be applied to suppress the "soft" contributions varying like μ^2/Q^2.

A set of values of Λ summarising the results of fits to leading and next to leading order is given in Table 1. In practice, $Q^2 > 4$ GeV^2 (EMC) and $Q^2 > 20\ \text{GeV}^2$ in the BCDMS data sets.

Sensitivity of Λ to systematics and to the method of analysis

The large systematic errors on Λ reflect the sensitivity to absolute calibrations : an uncertainty of $5\ 10^{-3}$ on the beam energy (present error $\sim 3\ 10^{-3}$) causes a change of 50 MeV on Λ. The presence of systematic uncertainties is reflected in the high value of χ^2 typically 2 to 3 per degree of freedom:

Table 1.

Target	Analysis	Λ (MeV)	α_s (Q^2 = 100 GeV^2)
H_2 (EMC)	x ≧ .25, leading order	$110^{+58}_{-46}\,^{+124}_{-69}$	.167 ± .030
	Next to leading order (MS)	$139^{+68}_{-56}\,^{+156}_{-87}$	.144 ± .026
	$x \to \xi = \dfrac{2x}{1 + \left(1 + \dfrac{4m^2x^2}{Q^2}\right)^{\frac{1}{2}}}$	$154^{+70}_{-56}\,^{+173}_{-87}$	.147 ± .025
	NLO ($\overline{\text{MS}}$)		
	Singlet, non singlet All x, lowest order	$81^{+36}_{-30}\,^{+44}_{-32}$	.156 ± .020
Fe (EMC)	x ≧ .25 L. O.	$107^{+7}_{-7}\,^{+100}_{-60}$	.166 ± .028
	Singlet + non singlet Leading order	$170^{+25}_{-25}\,^{+105}_{-70}$	.185 ± .022
	x ≧ .25 NLO ($\overline{\text{MS}}$)	$150^{+26}_{-26}\,^{+130}_{-91}$	.146 ± .035
C (BCDMS)	x ≧ .25 LO	$85^{+60}_{-40}\,^{+90}_{-70}$	.158 ± .035

χ^2/Nd	Singlet (all x)	Non singlet x > .25
H_2	198/112	97/166
Fe	265/133	200/100

The presence of systematics weakens the quality of the "QCD" test performed on Fig. 21 where the logarithmic slopes

$$\frac{d \log F_2}{d \log Q^2} ,$$

measured in iron for $Q^2 > 20$ GeV^2 are shown as a function of x : the data points do not follow the curves of the evolution equations for Λ = 100, 200, 300, but are certainly compatible with these curves once the systematic uncertainties (shaded areas) are taken into account.

Λ is also sensitive to ill defined phenomenological or experimental inputs in the evolution equations :

	$\Delta\Lambda$ (MeV)
$R = 0 \rightarrow R = .1$	- 40
$x \rightarrow \xi = \dfrac{2x}{1 + \left(1 + \dfrac{4m^2x^2}{Q^2}\right)^{1/2}}$ (target mass effects)	+ 15
$x \geqq .2 \rightarrow x \geqq .3$ (Fe analysis)	- 40
Charm substracted 3 flavours	- 30
$x\,G(x) : (1-x)^5 \rightarrow (1-x)^4$ gluon distribution	+ 50
Change x distribution according to nuclear effects keep Q^2 slopes	+ 15 .

PRODUCTION OF OPEN CHARM IN MUON INTERACTIONS

The production of charmed particles is detected thanks to their semi-leptonic decay modes, so that they appear as dimuons or trimuons in the final state. Two experiments have publised extensive analyses of these channels : the BFP (Berkeley-Fermilab-Princeton)[37] and the EMC Collaborations[38]. The BFP apparatus, a target calorimeter , is particularly well adapted to the measurement of multimuons, as it can be observed on Fig. 22 that it has a full forward acceptance. The EMC collaboration has presented more recent results, with complementary analyses at higher energies.

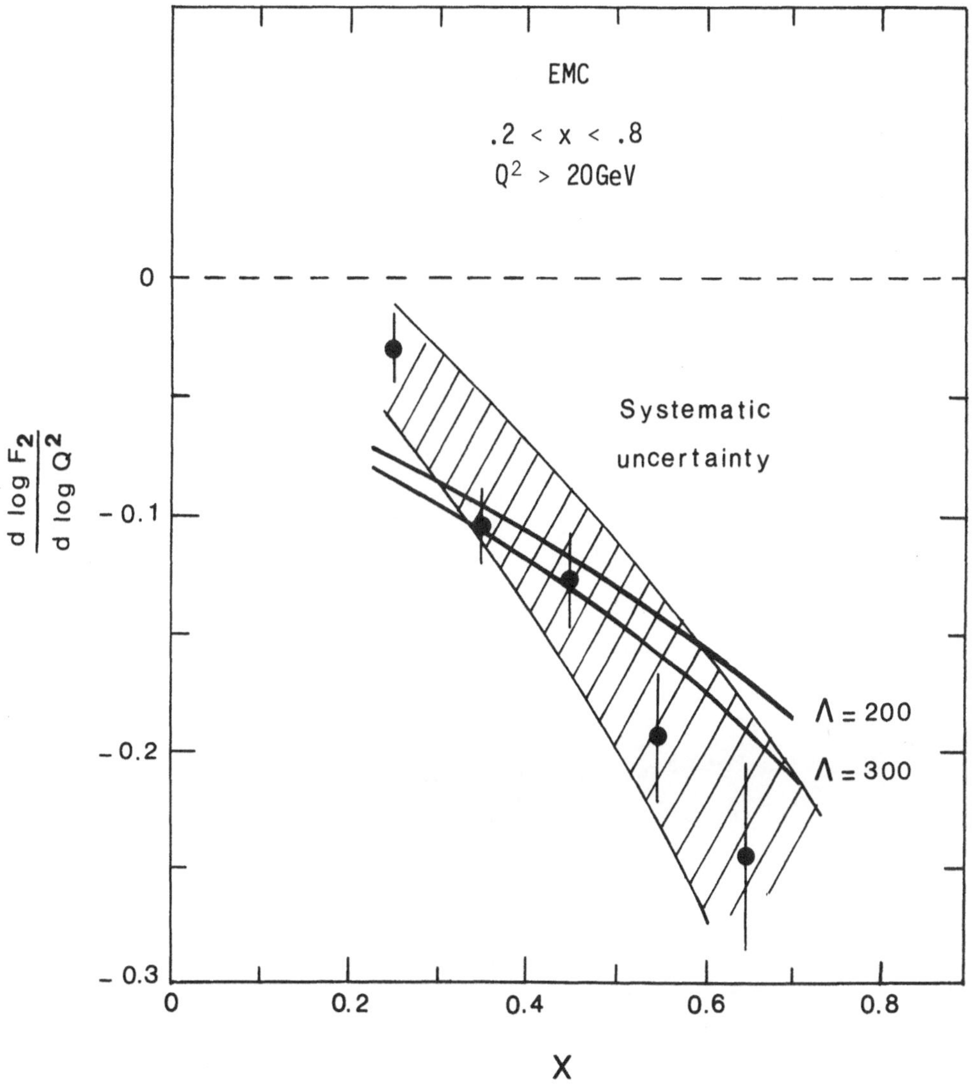

Fig. 21. $\frac{d\text{Log } F_2}{d\text{Log } Q^2}$ for $Q^2 > 20$ GeV^2/c compared with expected values for Λ = 100, 200, 300 MeV.

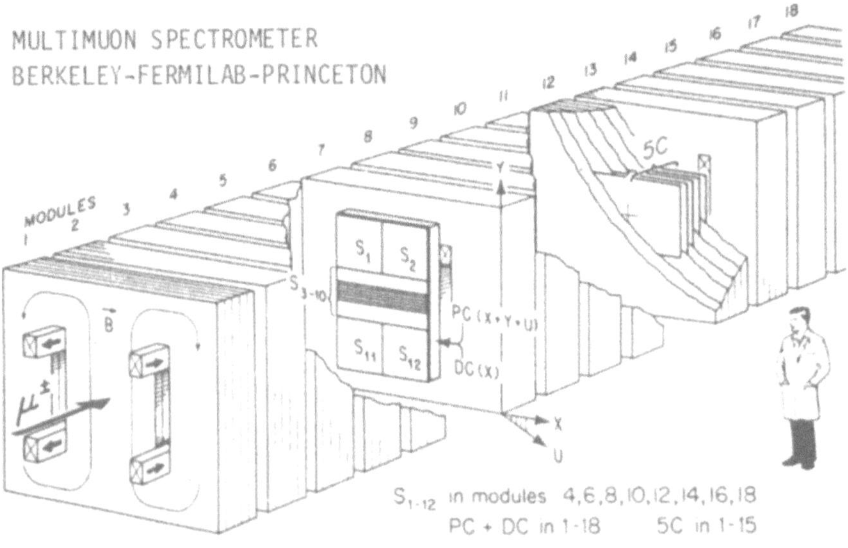

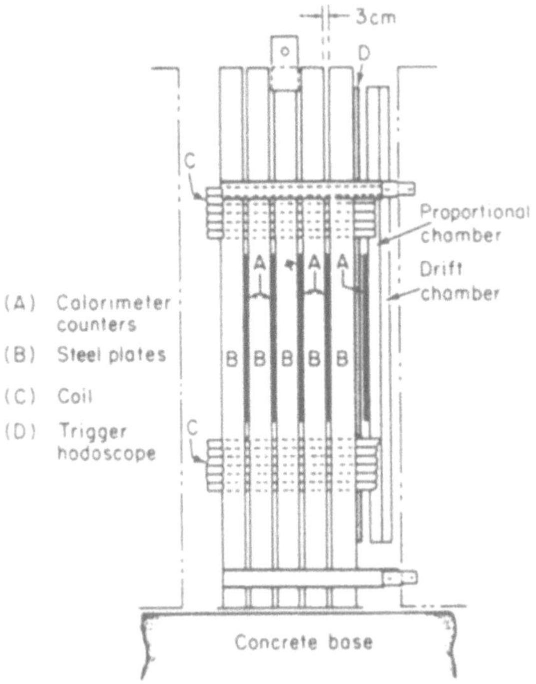

One module in the muon spectrometer.

Fig. 22. The apparatus of the BFP collaboration at Fermilab.

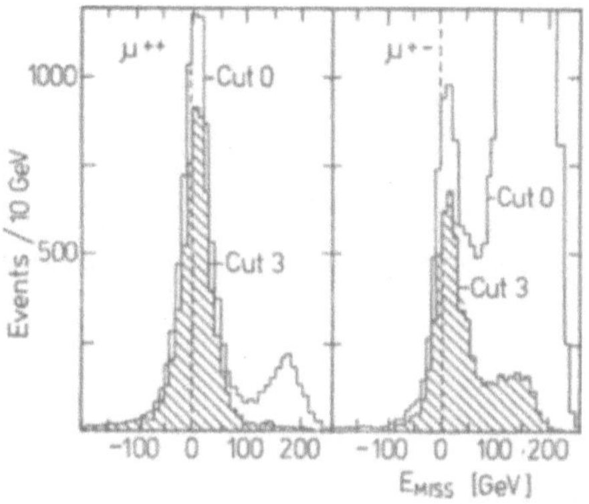

Fig. 23. Missing energy spectra of the selected dimuons.

Channels and backgrounds

The reaction considered is $\mu N \to \mu C\bar{C}X$ followed by the decay $C \to \mu^+ S\nu$, and/or $\bar{C} \to \mu^- \bar{S}\bar{\nu}$. The main backgrounds consist of :
- electromagnetic tridents, which appear as di/or trimuons, depending on the acceptance.

Their contribution is suppressed by the requirements of :
- missing energy $E_m < 90$ GeV (dimuons),
- a genuine hadronic shower $E_H > 30$ GeV.

The effect of these cuts is seen on Fig. 23.
- pion and kaon decays in dimuons.

After applying the cut $p_\mu > 15$ GeV, $p_\mu^{scat} > 20$ GeV, the remaining contamination is estimated to be 15 ± 7%. The total sample of dimuons is 20000 events in the BFP experiment, 1500 events in the EMC data.

Fig. 24. Interaction of a virtual photon with (a) a charmed quark of the sea; (b) a pair of charmed quarks from gluon fragmentation.

Models for charm production

Models are needed to correct the data for the acceptance losses and cuts, and present it in a form free of experimental biases. They represent also a first attempt to check QCD in a deep inelastic semi-inclusive process, although there is no reliable prescription for the treatment of final state masses.

Intrinsic charm. The interaction of the virtual photon with a charmed sea-quark as in Fig. 24a must take place at some level. BRODSKY, HOYER and PETERSON[39] have derived a very specific form for the momentum distribution of heavy flavour using

$$\xi = \frac{Q^2 + M_c^2}{2M\nu} \qquad M_c \sim 1.5 \text{ GeV} .$$

BRODSKY et al. claim :

$$C(\xi) = 1800\ \lambda\ \xi^2\ [\tfrac{1}{3}(1-\xi)\ (1+10\xi+\xi^2) + 2\ \xi\ (1-\xi)\ \text{Log}\ \xi] ,$$

$$\int C(\xi)\ d\xi = \lambda ,$$

the probability of finding the quark.

Previous results from the ISR concerning diffractive charm production[40] could be explained with $\lambda \sim 1\%$. The EMC collaboration has shown that $\lambda < .28\%$ with 90% C.L. and $\lambda < .6\%$ if QCD evolution is allowed to shrink the ξ distribution[41]. The function $C(\xi)$ does not describe the x distributions of dimuons on Fig. 25; no "hard" charmed quark is observed in the nucleon at present Q^2 contributing more than $\sim .5\%$.

Photon gluon fusion. (BETHE-HEITLER charm production) There are many unknown contributions to this symbolic diagram of Fig. 24b, which we shall discuss later on. It has nevertheless the merit of

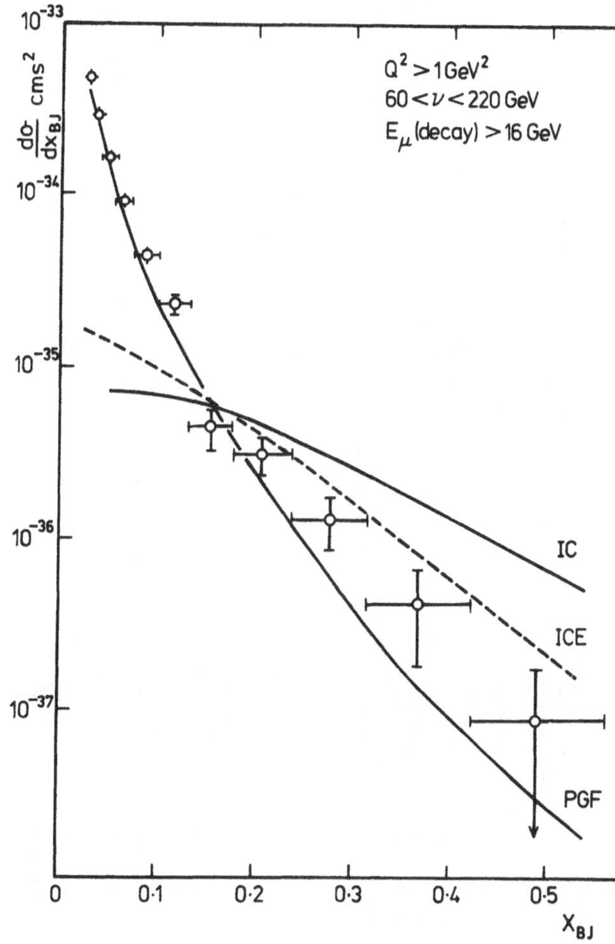

Fig. 25. Expected x distributions for photon gluon fusion and intrinsic charm.

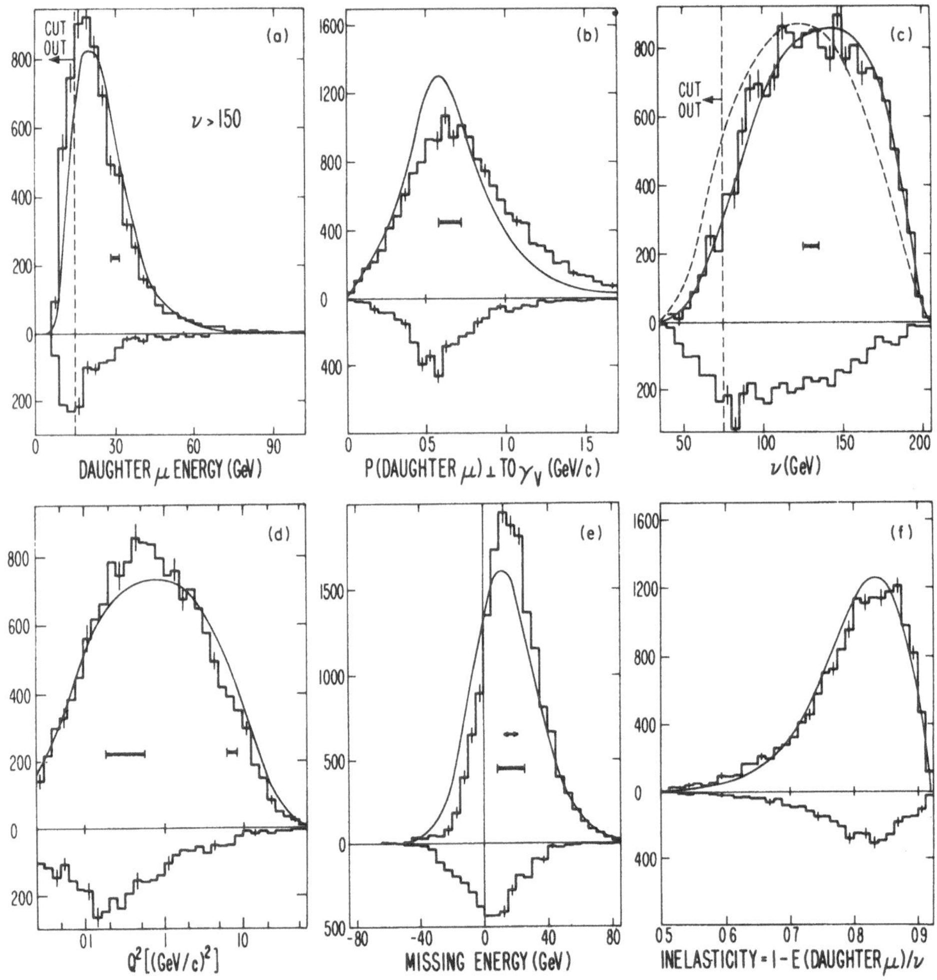

Fig. 26. Comparison of expectations of the photon gluon fusion with experimental distributions.

describing all the experimental distributions of Fig. 26, borrowed from BFP[37], rather well, in particular :
- the missing energy spectrum, E_m a signature of the presence of ν in the final state, is reproduced : its mean value E_m for data and

Monte Carlo compares as follows :

Missing energy $\langle E_m \rangle$ (GeV)	data	$= 18.2 \pm 2$ GeV
	γGF, Monte Carlo	$= 14.6 \pm 2$ GeV
	decays Monte Carlo	; MC $= 4.5 \pm .5$ GeV .

- The virtual photon energy distribution ν is shifted to high values, as expected from phase space effects associated to the production of heavy particles.

The photon gluon fusion model can therefore be used to compute acceptance corrections.

The charm structure function. This function is defined by

$$\frac{d^2\sigma\ (\mu N \to \mu c\bar{c}X)}{dQ^2 d\nu} = \frac{4\pi\alpha^2}{(Q^2)^2\nu}\ (1 - y + \frac{y^2}{2})\ F_2^{c\bar{c}}$$

which is comparable to the definition of F_2 in inclusive deep inelastic scattering if $\sigma_L = 0$. Below $x = .1$, $F_2^{c\bar{c}}$ reaches 5% of F_2 at large Q^2, which implies the possibility of a simple charm trigger in muon scattering. The scaling violations of $F_2^{c\bar{c}}$ on Fig. 27 are much larger than in F_2. Part of it can be understood as a consequence of the evolution equations (1,2) with a dominant gluon contribution : the term $F_2 \otimes P_{qq}$, which gives a negative slope, is negligible in $F_2^{c\bar{c}}$.

We shall see nevertheless that although the fusion model reproduces the evolution of $F_2^{c\bar{c}}$, it does not accomodate at present the same Λ as in deep inelastic, casting doubts on this picture.

Ingredients of photon gluon fusion. Several additional assumptions are needed to make use of the photon gluon fusion diagram of Fig. 24b.

- The gluon distribution G(x) is not constrained by muon deep inelastic data. It can be borrowed from neutrino experiments[36] or adjusted to reproduce the observed distributions, assuming a functional form $(1-x)^m$.
- The momentum transfer Q^2 is not large when compared to the hadronic mass $M^2_{c\bar{c}}$: $\langle Q^2 \rangle \sim 5$ GeV^2 so that a perturbative analysis may be questioned. R.J.N. PHILLIPS[42] has suggested the use of $Q^2 + M^2_{cc}$ as an argument of the running coupling constant

$$\alpha_s = \frac{12\pi}{25}\ \text{Log}\ \frac{Q^2 + M^2_{cc}}{\Lambda^2} \quad .$$

The high value of $\Lambda = .67$ GeV found in the fit discussed later on shows that a problem remains in this description, which is not

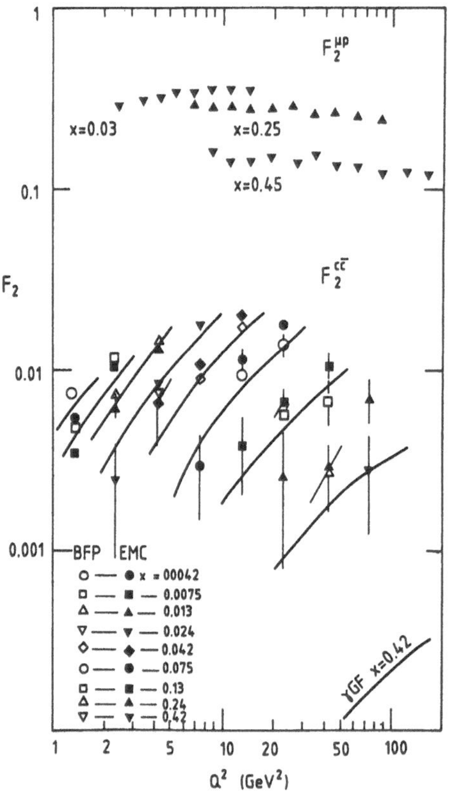

Fig. 27. Charm structure function $F_2^{c\bar{c}}$.

Table 2.

$$G(x) = (1-x)^{5.2 \pm .5}$$

$$M_c = 1.51 \pm .06 \text{ GeV}$$

$$\Lambda = .67 \pm .2 \text{ GeV}$$

Charm fragmentation : $\exp(A\, Z_D)$, $A = 1.6 \pm .6$

$$F = \left(\frac{C\bar{C} \rightarrow \text{Charmonium}}{C\bar{C} \rightarrow \text{open charm}}\right) = .2^{+.1}_{-0.05}$$

consistent with the rest of deep inelastic data.
- A "quark mass" M_c has to be chosen to define the effective threshold for open charm production in the diagram 22b.
- The charm fragmentation function is adjusted to reproduce the energy distribution of the muon pair : $\exp(AZ_D)$ for EMC, $(1-Z_D)^m$ for BFP. It is not in disagreement with experimental data from other sources[43] which point to a hard fragmentation.
- The branching ratio of semi-leptonic D decays : $D \rightarrow K\mu\nu + K\mu\nu$ is assumed to be 8.2% from the P.D.G. tables[44].

The ratio $\frac{\text{trimuons}}{\text{dimuons}}$ of 7% is compatible with this partial width.

Numerical results. We shall quote in Table 2 the results of a common fit to dimuon and Ψ production by EMC[38].

The quality of the χ^2 : 203/134 degrees of freedom reflects the general qualitative agreement of the fit with the experimental data, with persisting discrepancies already apparent on Fig. 24.

In fact, the large value of α_s about .5, associated with such fits casts doubts on a perturbative treatment.

The energy spectrum of the muon in the charm decay also allows to set an upper limit of 2% on the two body decay $D \rightarrow \mu\nu$.

TRANSVERSE MOMENTUM DISTRIBUTIONS

Fragmentation processes are complex, and no clear theoretical picture has emerged at present, as stressed by H.D. POLITZER[45]. One of the most striking effects, to be discussed here, is the high transverse momentum tails of final state hadrons in deep inelastic scattering. In particular, we shall try to summarise the experimental situation with respect to the intrinsic momentum of quarks : k_T, which is related to :

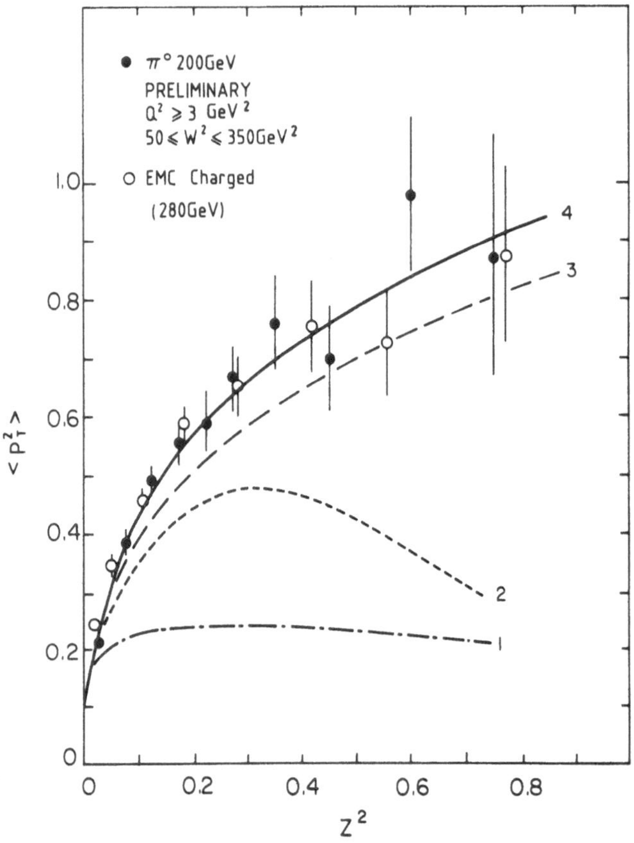

Fig. 28. Variation of the mean p_T^2 of hadrons. (a) as a function of z.

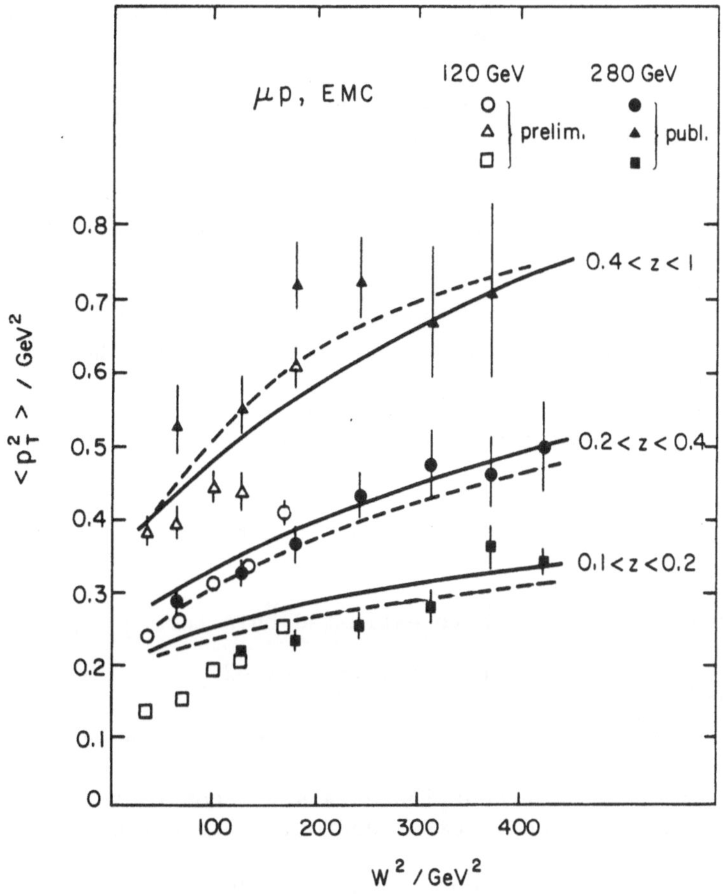

Fig. 28. Variation of the mean p_T^2 of hadrons. (b) as a function of the hadronic mass W^2.

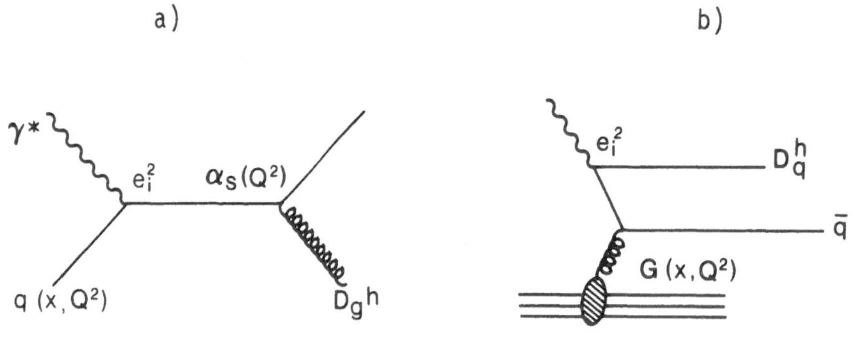

Fig. 29. Contribution of gluons to the p_T distribution. (a) gluon emission. (b) pair production.

($R = \sigma_L/\sigma_T$; $R = 4 \langle k_T^2\rangle / Q^2$), and to $1/Q^2$ contributions into F_2.

Contributions to p_T

The traditional description includes three contributions to the transverse momentum of hadrons :
- the intrinsic transverse momentum k_T of quarks.
Let k be the quark momentum $p_h = z_k \rightarrow \langle p_h^{T2}\rangle = z^2\langle k_T^2\rangle$: the quadratic dependence upon Z observed on Fig. 28. (a) is assigned to the $\langle k_T\rangle^2$ coefficient.
- The "QCD motivated" contribution of gluon radiation.
It is associated to the diagrams of Fig. 29. (a), (b) and shown quantitatively on Fig. 29. (c).
ALTARELLI and MARTINELLI[14] have estimated this term as approximately

$$\langle p_T^2\rangle_{QCD} \sim \alpha_s\,(Q^2)\,Q^2\,f(x,y) \sim \alpha_s\,(Q^2)\,W^2\ ,$$

where W is the hadronic mass. Such W^2 a contribution is observed on Fig. 28. (b).
- The quark fragmentation contributes a small constant amount σ_q = .31 GeV, as observed in usual hadronic interactions.

The observed distributions can be reproduced by a combination of the three previous terms, but the results are not compatible whether one works with inclusive p_T distribution or the total transverse momentum .

In the first case[46], k_T^2 = .65 GeV^2 ,

Λ = . 5 GeV .

In the latter[47] k_T^2 = . 4 GeV^2 .

The value of Λ needed to account for the high p_T tails is high :

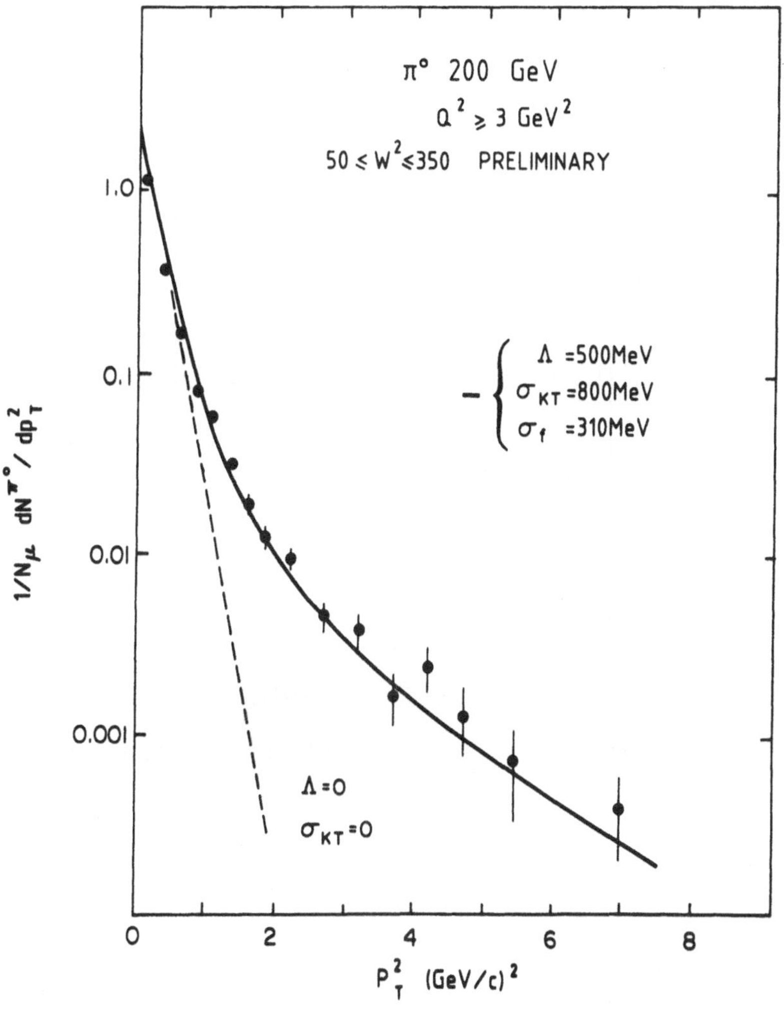

Fig. 29. (c) Contribution of gluons to the p_T distribution.

one needs a larger α_s than is observed in deep inelastic scattering. Furthermore the k_T parameter itself is larger than one would have guessed, and it is not really consistent from one spectrum to the other. It would be essential to work in a region with $Q^2 \sim 10$ GeV^2 rather than $Q^2 \sim 4$ GeV^2 and to concentrate on large p_T tails : this will only be possible at higher energies.

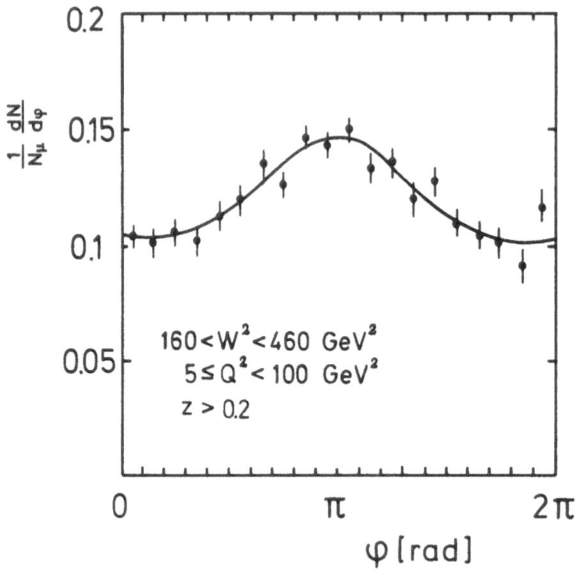

Fig. 30. Azimuthal distribution of final state hadrons.

<u>Azimuthal distributions around the virtual photon</u>

The inclusive azimuthal distribution of final state hadrons around the direction of the virtual photon is strongly anisotropic, as observed on Fig. 30 where the lepton scattering plane is used as azimuthal origin. It was pointed out by GEORGI and H.D. POLITZER[48], and by R. CAHN[49] that this effect can be understood as a consequence of quark transverse momentum, whether intrinsic or from gluon radiation.

$$M = \bar{u}_p' \gamma_\mu u_p \bar{u}_k' \gamma_\mu u_k,$$

$$|M|^2 \alpha 4 [(p+k)^2 + (p^1-k)^2].$$

Then,

$$\frac{1}{N}\frac{dN}{d\phi} = A + B f_1(y) \cos \phi + C f_2(y) \cos 2\phi + D P f_3 \sin \phi.$$

P is the polarisation of the virtual photon and $y = \nu/E$ the usual deep inelastic variable.

$$f_1(y) = \frac{(2-y)\sqrt{1-y}}{1 + (1-y)^2}, \qquad f_2(y) = \frac{1-y}{1 - (1-y)^2},$$

$$f_3 = \frac{y \sqrt{1-y}}{1 + (1-y)^2} .$$

If k_T is the quark transverse momentum

$$\langle\cos \phi\rangle = -\frac{2 \langle k_T\rangle}{Q^2} f_1(y)$$

$$\langle\cos 2\phi\rangle = 2 \frac{\langle k_T\rangle^2}{Q^2} f_2(y)$$

The variation of $\langle\cos \phi\rangle$, $\langle\cos 2\phi\rangle$ are given on Fig. 31 as a function of Z, and the trend is qualitatively consistent with the curves drawn for $\langle k_T^2\rangle = .7$ GeV^2 (a larger k_T would be necessary to reproduce the data).

Rapidity balance of transverse momentum

Although different values of k_T are found in different variables reflecting obvious inadequacies of the phenomenological description, all tend to be rather high, between .4 and 1 GeV^2. A correlation test by EMC[50] sheds some light on the sources of transverse momentum.

The highest p_T of the event is selected : $\vec{p}_T^{\,M}$. One then searches for the dependence of $\langle\vec{p}_h^{\,T}\ \vec{p}_T^{\,M}\rangle$ as a function of the rapidity y^*. The transverse momentum should be balanced in the backward hemisphere (target fragmentation) if the intrinsic quark momentum k_T were large. A large fraction of the transverse momentum compensation on Fig. 32 occurs for $y^* > 0$, in contrast with the expected effect for $k_T = .7$ GeV (dotted curve). The Lund model with soft gluon emission[51] happens to agree with the data, suggesting that a more refined theoretical analysis might improve the consistency of various parametrisations of transverse momentum.

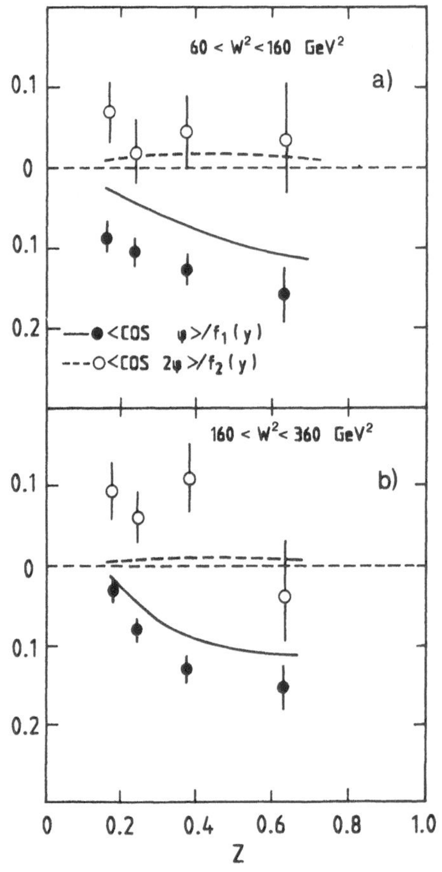

Fig. 31. Dependence of the moments of the azimuthal distributions of hadrons upon Z.

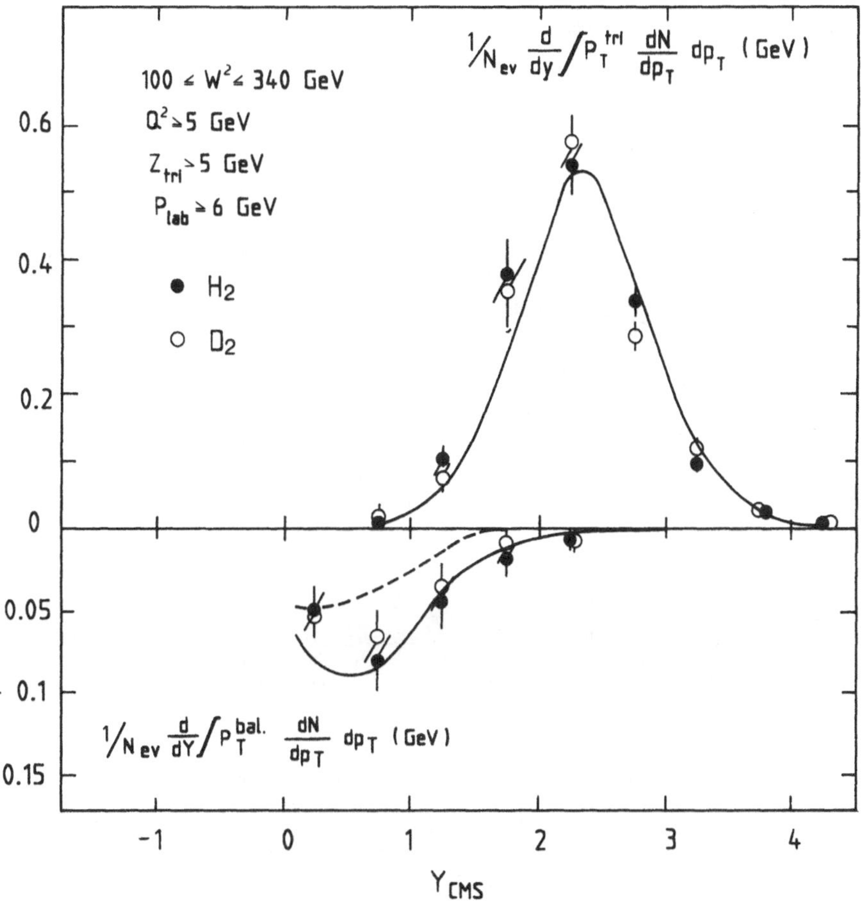

Fig. 32. Transverse momentum balance as a function of y^*.

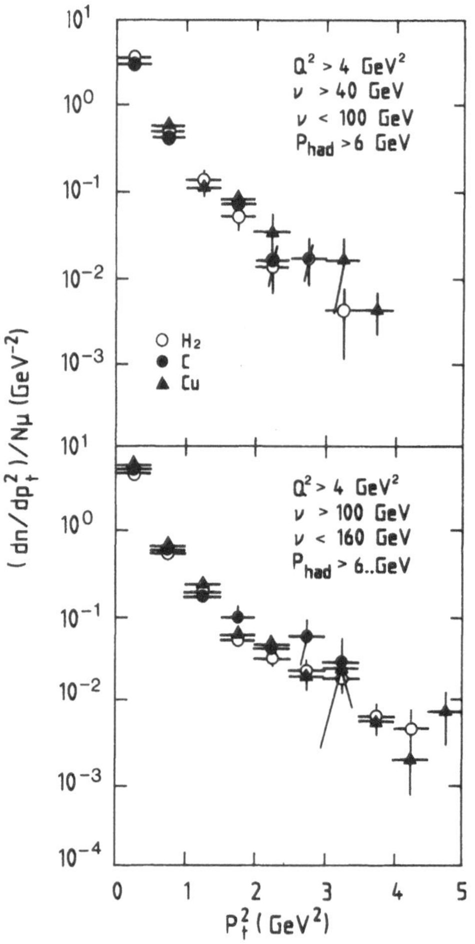

Fig. 33. Transverse momentum distributions on H_2, carbon and copper.

Nuclear effects

If quark structure functions are modified inside nuclei, as observed previously, one would expect their intrinsic transverse momentum to be also affected. This could be observed in inclusive transverse momentum distributions.

No such effect is apparent on Fig. 33. The momentum distributions are the same in Hydrogen, Carbon and Copper at the 20% level.

REFERENCES

1. EMC : European Muon Collaboration : CERN - DESY (Hamburg) - Freiburg - Kiel - Lancaster - LAPP (Annecy) - Liverpool - Oxford - Rutherford - Sheffield - Torino - Wuppertal.
2. BCDMS : Bologna - CERN - Dubna - Munich - Saclay.
3. BFP : Berkeley - Fermilab - Princeton.
4. CHIO : Chicago - Harvard - Illinois - Oxford.
5. R.P. Mount, Nucl. Instr. Meth. 187 (1981) 401.
6. A. Bodek et al., Phys. Rev. D20 (1979) 1471.
7. Y.S. Tsai, SLAC Pub 848 (1971).
8. L.M. Mo and Y.S. Tsai, Rev. Mod. Phys. 41 (1969) 205.
9. D.Y. Bardin and N.M. Shumeiko, Nucl. Phys. B127 (1977) 242.
10. D.Y. Bardin and N.M. Shumeiko, Yad. Fiz. 29 (1979) 969.
11. C. Chahine, Phys. Rev. D22 (1980) 2727, D22 (1980) 1062.
12. L.N. Hand, Phys. Rev. 129 (1963) 1834.
13. R.P. Feynman, Photon hadron interactions. (W.A. Benjamin, New York 1972).
14. G. Altarelli and G. Martinelli, Phys. Lett. 76B (1978) 89 E. Reya, Phys. Rep. 69 (1981) 195.
15. EMC : J.J. Aubert et al., Phys. Lett. 121B (1983) 87.
16. EMC : P. Payre, Thèse d'Etat, Grenoble (1983).
17. EMC : J.J. Aubert et al., Phys. Lett. 105B (1981) 315.
18. R.D. Field and R.P. Feynman, Phys. Rev. D15 (1977) 2590.
19. CFS : Caltech - Fermilab - Stony Brook; A.S. Ito et al., Phys. Rev. D23 (1981) 604.
20. EMC : J.J. Aubert et al., Phys. Lett. 123B (1983) 123.
21. WA25 Collaboration : D. Allasia et al., Contribution to the Brighton Conference (1983).
22. EMC (Fe) : J.J. Aubert et al., Phys. Lett. 105B (1981) 322. BCDMS (C) : D. Bollini et al., Phys. Lett. 104B (1981) 403.
23. EMC : J.J. Aubert et al., Phys. Lett. 123B (1983) 275.
24. K. Rith, Nuclear effects in muon scattering, presented at XVIII Rencontre de Moriond, La Plagne, March 1983.
25. CHIO : B.A. Gordon et al., Phys. Rev. D20 (1979) 2645.
26. Rochester - MIT - SLAC : A. Bodek et al., Phys. Rev. Lett. 51 (1983) 534.
27. P. Hogaasen, P. Sorba, and R. Viollier, Zeit. Phys. C4 (1980) 131. L. Bergstrom, S. Frederiksson, Rev. Mod. Phys. 52 (1980) 675.

H.J. Pirner and J.P. Vary, Phys. Rev. Lett. 46 (1981) 1376.
M. Namiti, K. Okano, and N. Oshimo, Phys. Rev. D25 (1982) 120.
L.L. Frankfurt and M.I. Strikman, Phys. Rep. 76C (1981) 215.
28. R.L. Jaffe, Phys. Rev. Lett. 50 (1983) 228.
J. Szwed, Phys. Lett. 128B (1983) 245.
29. C.H. Llewellyn-Smith, Oxford preprint 18-83 submitted to Physics Letters.
M. Ericsson and A.W. Thomas, CERN preprint TH-3553 submitted to Physics Letters.
30. NA3 : Badier et al., Phys. Lett. 93B (1980) 354.
31. O. Nachtmann and H.J. Pirner, Heidelberg University preprint, HD THEP 83-8.
32. A. Krzywicki, Phys. Rev. D14 (1976) 152.
33. S. Drell and T.M. Yan, Phys. Rev. Lett. 24 (1970) 181.
S.J. Brodsky and G. Farrar, Phys. Lett. 31 (1973) 1153.
V.A. Matveev, R.M. Muradyan, and A.N. Tavkhelidze, Lett. Nuovo Cimento 7 (1973) 719.
34. G. Altarelli and G. Parisi, Nucl. Phys. B126 (1977) 298.
35. E.G. Floratos , D.A. Ross, and C.T. Sachrajda, Nucl. Phys. B129 (1977) 66, and Erratum B139 (1978) 545.
A. Gonzales Arroyo, C. Lopez, and F.J. Yndurain, Nucl. Phys. B153 (1979) 161.
G. Curci, W. Furmanski, and R. Petronzio, Nucl. Phys. B175 (1980) 27.
E.G. Floratos, R. Lacaze, and C. Kounas, Phys. Lett. 98B (1981) 89.
36. CDHS : H. Abramowicz et al., Zeit. Phys. C12 (1982) 289.
37. BFP : G.D. Gollin et al., Phys. Rev. D24 (1981) 55.
38. EMC : J.J. Aubert et al., Production of charmed particles in 250 GeV μ Iron Interactions, Nucl. Phys. 213 (1983) 31.
39. S.J. Brodsky et al., Phys. Lett. 93B (1980) 451; Phys. Rev. D23 (1981) 2745.
40. For a review, see L.J. Koester, 20th International Conference of High Energy Physics, Madison, Wisconsin, 1981.
41. EMC : J.J. Aubert et al., Phys. Lett. 110B (1982) 73.
42. R.J.N. Phillips, Proc. 20th International Conference on High Energy Physics, Madison, Wisconsin, 1980.
43. TASSO : M. Althoff et al., Phys. Lett. 126B (1983) 493.
CDHS : H. Abramowicz et al., Z. Phys. C15 (1982) 19.
J.M. Yelton et al., Phys. Rev. Lett. 49 (1982) 430.
44. Particle Data Table, Rev. Mod. Phys. 52 (1980).
45. H.D. Politzer, 21st International Conference on High Energy Physics, Paris (1982).
46. H. Montgomery, CERN preprint EP/82-164, Lectures at the 1982 Arctic School of Physics.
47. EMC : J.J. Aubert et al., Phys. Lett. 100B (1981) 433.
48. H. Georgi and H.D. Politzer, Phys. Rev. Lett. 40 (1978) 3.
49. R.N. Cahn, Phys. Lett. 78B (1978) 269.
50. EMC : J.J. Aubert et al., Phys. Lett. 119B (1982) 233.

51. B. Andersson et al., Zeit. Phys. C9 (1981) 233; Zeit. Phys. C12 (1982) 49; Lund preprint LU TP 81-8 (1981).

and the general form "current-current" of the Lagrangian is assumed

$$\mathcal{L}_{eff} = - \frac{G}{\sqrt{2}} \{ J^{(\ell)\lambda} J_{\lambda}^{(H)} + h.c. \} .$$

The differential cross section is proportional to the product of the leptonic tensor $\ell^{(\nu)\lambda\rho}$ and the hadronic tensor $W_{\rho\lambda}^{(\nu)}$

$$\frac{d\sigma}{dE' \, d\Omega} \alpha \, \ell^{(\nu)\lambda\rho} \, . \, W_{\rho\lambda}^{(\nu)} .$$

i) The leptonic current $J^{(\ell)} = \bar{\ell}_{\mu} \gamma_{\lambda}(1+\gamma_5)\nu_{\mu} + \bar{\ell}_{\rho}\gamma_{\lambda}(1+\gamma_5)\nu_{\rho}$ is perfectly calculable via the V-A theory and has described up to now all the weak decay phenomena quite well. One should notice that at high energy neutrino interactions the open question was to know if the V-A theory was still valid. Two results have confirmed this validity :

a) The inverse μ decay $\nu_{\mu} + e^{-} \rightarrow \mu^{-} + \nu_{e}$ observed by GGM[2] in good agreement with the V-A theory : $\sigma(\text{inv. } \mu \text{ decay})/\sigma(\text{V-A}) = 0.9 \pm 0.2$.

b) The polarization of the positive muons in high energy antineutrino interaction by the CHARM collaboration[3]. A possible S,P,T (Scalar, Pseudoscalar, Tensor) mixture interaction can mimic a V-A spectrum distribution[4] but not the μ polarization. V and A coupling preserve the helicity while S, P and T couplings flip the helicity. An upper limit on S, P, T couplings is given $\sigma(S,P,T)/\sigma_{TOT} < 18\ \%$ with 95 % confidence level.

We see that the V-A form of the leptonic current is accepted as good but a 20 % possibility of "trouble" is still possible. More recent results on V + A current limits in μ decay[5] are more restrictive.

ii) The hadronic current $J^{(H)}$ is more complicated to calculate. The hadron tensor $W_{\rho\lambda}$ cannot be calculated from strong interaction principles. Basic principles or symmetry properties are used to establish the form of $W\rho\lambda$.

The Lorentz invariance is first used to establish the properties of $W_{\rho\lambda}$. The Lorentz tensor $W_{\rho\lambda}$ is constructed from the 2 four-vectors p and q and is a function of the two invariants

$$\nu = q.p/M \ , \qquad Q^2 = - q^2 \ ,$$

M is the mass of the nucleon.
The tensor $W_{\rho\lambda}$ is the sum of 6 scalars W_i

$$W_{\rho\lambda} = - g_{\rho\lambda} W_1 + \frac{p_\rho p_\lambda}{M^2} W_2 - \frac{i}{2M^2} \varepsilon_{\rho\lambda\alpha\beta}\, p^{\alpha} q^{\beta} W_3 + \frac{q_\rho q_\lambda}{M^2} W_4 + \ldots$$

$$+\frac{1}{M^2}(p_\rho q_\lambda + q_\rho p_\lambda)W_5 + \frac{i}{M^2}(p_\rho q_\lambda - q_\rho p_\lambda)W_6 .$$

These W_i's are called "structure functions". If we study the ν, $\bar{\nu}$ scattering on proton and neutron, we start a priori with 24 of these structure functions.
We will see that symmetry properties decrease the number of these structure functions.

If T invariance holds: $W_6 = 0$
W_4 and W_5 are proportional to the mass of the scattered lepton μ or e and are neglected. (This should not be correct with ν_τ interaction)

$$W_4 = W_5 \cong 0 , \quad \text{in neutrino interactions,}$$

$$W_1 = - W_4 \quad \text{in e, } \mu \text{ scattering,}$$

$$W_2 = - W_5$$

These properties decrease to 12 the number of structure functions describing the ν, $\bar{\nu}$, n, p scattering.

Charge symmetry is used for $\Delta S = 0$ weak current and neglecting the $\Delta S = 1$ and $\Delta C = 1$ current one can write

$$W_i^{(\nu,p)} = W_i^{(\bar{\nu},n)} ,$$

$$W_i^{(\bar{\nu},p)} = W_i^{(\nu,n)} .$$

On an isoscalar target ($N_n = N_p$) one defines

$$W_i^{(\nu,N)} = \frac{1}{2}(W_i^{(\nu,p)} + W_i^{(\nu,n)}) ,$$

$$W_i^{(\bar{\nu},N)} = \frac{1}{2}(W_i^{(\bar{\nu},p)} + W_i^{(\bar{\nu},n)}) .$$

Assuming charge symmetry :

$$W_i^{(\nu,N)} = \frac{1}{2}[W_i^{(\nu,p)} + W_i^{(\nu,n)}] = \frac{1}{2}[W_i^{(\bar{\nu},n)} + W_i^{(\bar{\nu},p)}] = W_i^{(\bar{\nu},N)}.$$

"The scattering ν, $\bar{\nu}$ on an isoscalar target is then described by 3 structure functions"

$$W_1(\nu,Q^2),\ W_2(\nu,Q^2),\ W_3(\nu,Q^2) , \quad \nu = q.p/M , \quad Q^2 = - q^2.$$

Björken scaling limits. Keeping the ratio $X = Q^2/2M\nu$ fixed and $Q^2 \to \infty$, $\nu \to \infty$ Björken had shown that the structure functions are function of X only :

$$2M\,W_1(\nu,Q^2) \to F_1(X)\ ,$$

$$\nu\,W_2(\nu,Q^2) \to F_2(X)\ ,$$

$$\nu\,W_3(\nu,Q^2) \to F_3(X)\ .$$

- This scaling behaviour works already at very low $Q^2 \geqq 2\ \mathrm{GeV}^2$ or $W^2 = M^2 + 2M\nu - Q^2 \geqq 4\ \mathrm{GeV}^2$.
- On the other hand a big part of the game is now to study the scaling violation ...

"En résumé", assuming : V-A, Lorentz invariance, T invariance, charge symmetry ($\Delta S = 0$), Björken scaling limits.

Dropping the propagator term : $1/(1 + Q^2/M_W^2)^2$ the scattering ν, $\bar{\nu}$ on an isoscalar target is formulated :

$$\frac{d\sigma^{\overset{(-)}{\nu}}}{dXdY} = \frac{G^2ME}{\pi}\left[XY^2F_1(X) + \left(1 - Y - \frac{MXY}{2E}\right) F_2(X) \pm Y\left(1 - \frac{Y}{2}\right) XF_3(X)\right]. \tag{1}$$

By analogy with the virtual γ exchange in the charged lepton scattering, the cross section can be separated as absorption cross section for virtual W bosons of right (σ_+) left (σ_-) and longitudinal (σ_L) polarization, neglecting the possible scalar contribution. Because of the V-A interference in the neutrino case $\sigma_- \neq \sigma_+$. Then :

$$W_1^{(\nu,N)} = \frac{K}{\pi} \times \frac{1}{2} \times (\sigma_- + \sigma_+)^{(\nu,N)}\ ,$$

$$W_2^{(\nu,N)} = \frac{K}{\pi}\,\frac{Q^2}{\nu^2 + Q^2}\left[\frac{1}{2}(\sigma_- + \sigma_+) + \sigma_L\right]^{(\nu,N)}\ ,$$

$$W_3^{(\nu,N)} = \frac{K}{\pi}\,\frac{M}{\sqrt{\nu^2 + Q^2}}\,(\sigma_+ - \sigma_-)^{(\nu,N)}\ ,$$

$$K = \frac{W^2 - M^2}{2M}\ ,$$

and one defines

$$R = \frac{\sigma_L}{\frac{1}{2}(\sigma_+ + \sigma_-)} = \frac{\sigma_L}{\sigma_T}\ ,$$

as in the μ, e scattering. Expressed in term of the F_i structure functions the exact formula of R is

$$R = \frac{\sigma_L}{\sigma_T} = \frac{(1 + \nu^2/Q^2)\, F_2 - 2XF_1}{2XF_1} \quad . \qquad (2)$$

Quark parton model. This simple model allows the identification of the F_i's with the distributions of the quarks and antiquarks in the nucleon. The discrepancies from the naîve QPM appeared the most interesting and the most difficult to solve and explain experimentally. We will point out two.

Assuming that the quark travels the same direction as the nucleon, sharing its momentum $X = Q^2/2M\nu$. In the Breit frame

A massless quark leads to the equality $\xi = X$.

1. The cross section is the coherent sum of quark cross sections weighted by the quark probability to be inside the nucleon. The discrepancy from this clear statement comes from the fact that maybe not only free quarks exist. Diquarks can be in the nucleon and this lead to the higher twist problem.
2. If the scattered particles have spin 1/2 then the Callan-Gross relation tells us that $R = \sigma_L/\sigma_T = 0$ or if one defines $F_2 = 2XF_1+F_L$, $F_L = 0$ i.e. $2XF_1 = F_2$.
   Notice that in this circumstance we are left with only two structure functions F_2 and XF_3.
   Here again it is the deviation from these formulas which may complicate the formalism of the F_i's. The determination of R, to which we will devote a chapter, is a very difficult experiment.

Cross section in the QPM. One can briefly give the following steps.
1 - The $\overset{(-)}{\nu}$, e^- scattering can be calculated completely; it involves only the $\ell_{\rho\lambda}$ tensor. One finds :

$$\frac{d\sigma}{dY}(\nu,e) = \frac{G^2}{\pi} \times s \; ,$$
$$s = 2.m_e.E \; , \qquad (3)$$
$$\frac{d\sigma}{dY}(\bar{\nu},e) = \frac{G^2}{\pi} \times s \times (1-Y)^2 \; .$$

2 - By analogy with the νe scattering one can calculate the simple ν-quark scattering

$$\frac{d\sigma}{dY}(\nu,q) = \frac{G^2}{\pi} \, . \, 2 \, . \, M \, . \, E \, . \, X \, ,$$
$$\frac{d\sigma}{dY}(\bar{\nu},q) = \frac{G^2}{\pi} \, . \, 2 \, . \, M \, . \, E \, . \, X \, . \, (1-Y)^2 \, . \qquad (4)$$

2 M E X is the s-center of mass energy of the νq system.

3 - Scattering on proton or neutron.

A simple arithmetic calculation leads to the ν and $\bar{\nu}$ reactions on quarks (forgetting about the Cabibbo angle)

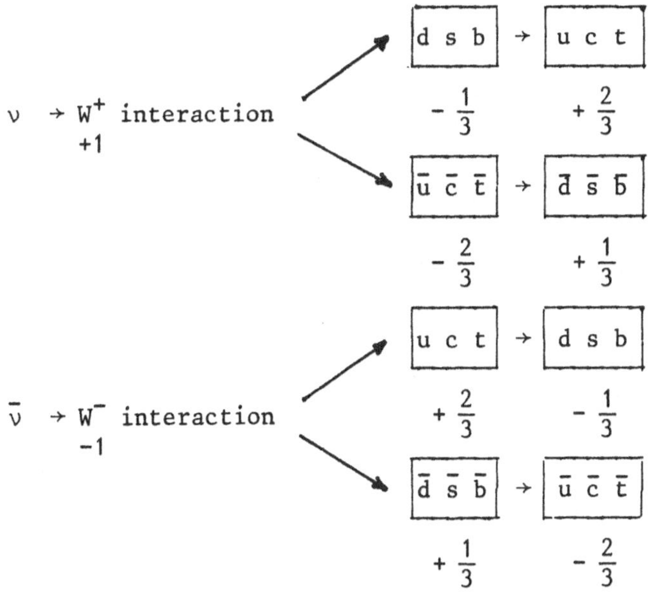

If we consider the case of neutrino interaction on proton and neutron to simplify with valence quark only we will get :

$$\frac{d\sigma}{dX}(\nu,p) = \frac{G^2}{\pi} \, . \, M \, . \, E \, . \, 2 \, . \, Xd(X) \qquad \text{u. .u} \; \overset{\circ}{\text{d}}$$

$$\frac{d\sigma}{dX}(\nu,n) = \frac{G^2}{\pi} \, . \, M \, . \, E \, . \, 2 \, . \, X(d_1(X)+d_2(X)) \qquad \text{d}_\circ \; {}_\circ\text{d} \; \overset{\bullet}{\text{u}}$$

$d(X)$ is the distribution of quark d in the proton,
$d_1(X)$ and $d_2(X)$ is the distribution of the d_1 and d_2 quark in the neutron.

Here we assume the isospin symmetry, it means that

$$d_1(X) + d_2(X) = d(X)_{neutron} = u(X)_{proton} \, .$$

This assumption which is very fundamental that proton and neutron are in the same doublet in the isospin space is in this case not obvious and one should test it seriously experimentally. Only bubble chamber experiments on deuterium can do it. The separation of neutrino antineutrino interactions on proton and neutron is not free of bias and statistics are poor. A good test of this important hypothesis is still not done. The $\bar{u}$ and $\bar{d}$ momentum distribution have already shown discrepancies from the naïve QPM as results from hydrogen have shown[6]. Nevertheless in what follows we assume

$$d(X)_{\text{neutron}} = u(X)_{\text{proton}} \quad .$$

It is why in all the formalism on neutrino interactions one says that we are using the quark density in the proton and one writes :

$$\frac{d\sigma}{dX}(\nu,p) = \frac{G^2}{\pi} \,.\, M \,.\, E \,.\, 2 \,.\, d(X) \,,$$

$$\frac{d\sigma}{dX}(\nu,n) = \frac{G^2}{\pi} \,.\, M \,.\, E \,.\, 2 \,.\, u(X) \,.$$

4. Complete calculation.

On an isoscalar target the cross section of neutrino and antineutrino an a nucleon is : (one neglects b and t contributions)

$$\frac{\nu p + \nu n}{2} = \frac{G^2.M.E}{\pi} \,.\, X(u+d+2s) + X(\bar{u}+\bar{d}+2\bar{c})(1-Y)^2 \quad ,$$

$$\frac{\bar{\nu} p + \bar{\nu} n}{2} = \frac{G^2.M.E}{\pi} \,.\, X(u+d+2c)(1-Y)^2 + X(\bar{u}+\bar{d}+2\bar{s}) \quad .$$

One defines q : sum of the quark momentum distribution = $(u+d+s+c)X$

$\bar{q}$: sum of the antiquark momentum distribution = $(\bar{u}+\bar{d}+\bar{s}+\bar{c})X$

$$\frac{d\sigma^{\nu}}{dXdY} = \frac{G^2.M.E}{\pi}\left[q + X(s-c) + (1-Y)^2\left[\bar{q} - X(\bar{s}-\bar{c})\right]\right] ,$$

$$\frac{d\sigma^{\bar{\nu}}}{dXdY} = \frac{G^2.M.E}{\pi}\left[\bar{q} + X(\bar{s}-\bar{c}) + (1-Y)^2\left[q - X(s-c)\right]\right] . \tag{5}$$

Identification of the F_i's and the quark, antiquark momentum distribution. From the general formula (1) established independently of the QPM one can write :

$$\frac{d\sigma^{\nu}}{dXdY} = \frac{G^2.M.E}{\pi}\left[\frac{Y^2}{2} \times 2XF_1^{\nu} + \left[1-Y-\frac{MXY}{2E}\right]F_2^{\nu} + \left[Y-\frac{Y^2}{2}\right]XF_3^{\nu}\right] ,$$

$$\frac{d\sigma^{\bar{\nu}}}{dXdY} = \frac{G^2.M.E}{\pi}\left[\frac{Y^2}{2} \times 2XF_1^{\bar{\nu}} + \left[1-Y-\frac{MXY}{2E}\right]F_2^{\bar{\nu}} - \left[Y-\frac{Y^2}{2}\right]XF_3^{\bar{\nu}}\right] . \tag{6}$$

For the sake of simplicity one drops the term $MXY/2E \propto Q^2/\nu^2$. One has by identification of formulas (5) and (6) the important result that due to the V-A structure of the weak interaction, one has the possibility to separate the valence and sea quark momentum distribution :

$$2XF_1^{\nu} = 2XF_1^{\bar{\nu}} = q + \bar{q} ,$$

$$F_2^{\nu,\bar{\nu}} = 2XF_1 + F_L , \qquad R = \frac{F_L\,(1 + Q^2/\nu^2)}{2XF_1} + \frac{Q^2}{\nu^2} ,$$

$$XF_3^{\nu,\bar{\nu}} = q - \bar{q} \pm 2X(s-c) .$$

One defines

$$XF_3 = \frac{1}{2}\,(XF_3^{\nu} + XF_3^{\bar{\nu}}) = q - \bar{q} .$$

Limit on right handed weak coupling. New results on the general form of the weak current have been given recently[7]. In the QPM the V-A interaction has the form

$$\frac{d^2\sigma^{\nu}}{dXdY} \propto q(X) + (1 - Y)^2\bar{q}(X) ,$$

$$\frac{d^2\sigma^{\bar{\nu}}}{dXdY} \propto \bar{q}(X) + (1 - Y)^2 q(X) .$$

as it can be obtained from equations (5) neglecting s and c quarks.

If right handed quark currents are present and if $\rho = |C_R/C_L|$ such as C_R and C_L are the coefficient of the right handed and left handed amplitude in the Lagrangian, the differential cross section is then written :

$$\frac{d^2\sigma(\nu)}{dXdY} \propto q(X) + \rho^2\bar{q}(X) + (1 - Y)^2\,[\bar{q}(X) + \rho^2 q(X)]$$

$$= q_L + (1 - Y)^2 q_R ,$$

$$\frac{d^2\sigma(\bar{\nu})}{dXdY} \propto (1 - Y)^2\,[q(X) + \rho^2\bar{q}(X)] + \bar{q}(X) + \rho^2 q(X)$$

$$= (1 - Y)^2\, q_L + q_R .$$

At large Y,

$$q_R(X) \propto \frac{d^2\sigma^{(\bar{\nu})}}{dXdY} - (1 - Y)^2\,\frac{d^2\sigma^{(\nu)}}{dXdY} .$$

Experimentally at large X one can see that $q_R(X) \ll q_L(X)$. The upper limit on ρ^2 is the upper limit on the ratio

$$\left[\frac{d^2\sigma^{\bar{\nu}}}{dXdY} - (1 - Y)^2 \frac{d^2\sigma^{\nu}}{dXdY}\right] \Big/ \left[\frac{d^2\sigma^{\nu}}{dXdY} - (1 - Y)^2 \frac{d^2\sigma^{\bar{\nu}}}{dXdY}\right]$$

for X and Y both large. On the base of 175 000 $\bar{\nu}$ events and 90 000 ν events the upper limit found by CDHS[7] is $|\rho|^2 < 0.009$ at 90 % confidence level.

This result limits very much the possible right handed quark current contribution and the formalism of the V-A theory can still be formulated.

Correction for non isoscalar target. All what has been written is for isoscalar target i.e. number of proton (Z) equal to the number of neutron (N) if A = Z + N. In most experiments the target is not an isoscalar one. We will evaluate the correction in the case of iron (Z = 26, N = 30) as an example.

$$d\sigma(Fe) = Z\, d\sigma(p) + N\, d\sigma(N).$$

At this level of correction one can still write this equality but the E.M.C. effect[8] seems to show that maybe this is not exactly correct. Nevertheless :

$$d\sigma(Fe) = \frac{A}{2}\, d\sigma(p) + \frac{A}{2}\, d\sigma(n) - \left(\frac{A}{2} - Z\right) d\sigma(p) - \left(\frac{A}{2} - N\right) d\sigma(n)$$

$$= A\, d\sigma(\text{Nucleon}) - \frac{N - Z}{2}\,[d\sigma(p) - d\sigma(n)].$$

From equation (5) if we assume $\bar{u} = \bar{d}$ and neglect s and c

$$[d\sigma(p) - d\sigma(n)]_{\nu} = d_v - u_v ,$$

$$[d\sigma(p) - d\sigma(n)]_{\bar{\nu}} = (d_v - u_v)(1 - Y)^2 .$$

d_v, u_v for u, d valence quark.
And the exact formulas are

$$\left.\frac{d^2\sigma^{\nu N}}{dXdY}\right|_{IRON} = \left.\frac{d^2\sigma^{\nu N}}{dXdY}\right|_{I=0} + \frac{N - Z}{N + Z}\, X(u_v - d_v)\, \frac{G^2.M.E}{\pi} ,$$

$$\left.\frac{d^2\sigma^{\bar{\nu} N}}{dXdY}\right|_{IRON} = \left.\frac{d^2\sigma^{\bar{\nu} N}}{dXdY}\right|_{I=0} - \frac{N - Z}{N + Z}\, X(u_v - d_v)(1 - Y)^2\, \frac{G^2.M.E}{\pi} .$$

The correction is applied using

$$X(u_v - d_v) \equiv \frac{u_v - d_v}{u_v + d_v}\, XF_3 = \frac{1 - d/u}{1 + d/u}\, XF_3 .$$

One can use the theoretical value of d/u of the QPM : d/u = 0.5 or for more refined analyses use the results found in ν, $\bar{\nu}$, hydrogen scattering[6] d/u = 0.57 (1 - X). This correction is small δ=0.023 XF_3 and has been calculated in the case of iron to determine the structure functions. In what follows the formulas are formulas for an isoscalar target.

Formulas used to extract structure functions. One can then easily see that :

- The difference of the neutrino and antineutrino differential cross section give XF_3

$$\frac{\pi}{G^2.M.E}\frac{d^2(\sigma^\nu - \sigma^{\bar{\nu}})}{dXdY} = XF_3\ [1 - (1 - Y)^2]\ , \tag{7}$$

where XF_3 is defined as

$$XF_3 = \frac{1}{2}\ (XF_3^\nu + XF_3^{\bar{\nu}}) = q - \bar{q}\ .$$

- The sum of the neutrino and antineutrino differential cross section gives :

$$\frac{\pi}{G^2M_NE_\nu}\frac{d^2(\sigma^\nu + \sigma^{\bar{\nu}})}{dXdY} = F_2[1 + (1-Y)^2(\frac{R}{1+R}(1+\frac{Q^2}{\nu^2}) - \frac{Q^2}{2\nu^2})]$$
$$+ 2X(s-c)\ [1 - (1-Y)^2]\ , \tag{8}$$

where :

$$R = \frac{\sigma_L}{\sigma_T} = \frac{(1+\nu^2/Q^2)F_2 - 2XF_1}{2XF_1}$$

will have to be evaluated and 2X(s-c) brings in the problems of the s → c charm threshold and the value of the strange sea. This part has been evaluated in "dimuon physics" and will be discussed later.

- One other very important formula is the antiquark distribution $\bar{q}$ given by :

$$\frac{\pi}{G^2M_NE_\nu}\left[\frac{d^2\sigma^{\bar{\nu}}}{dXdY} - \frac{d^2\sigma^\nu}{dXdY}(1 - Y)^2\right] = \overbrace{\frac{1}{2}[2XF_1 - XF_3 + 2X(s-c)]}^{\bar{q}}\ [1-(1-Y)^4]$$
$$+(F_2 - 2XF_1)\ [(1-Y) - (1-Y)^3]$$
$$- \frac{Q^2}{4\nu^2}\ F_2(2Y^3 - Y^4) \tag{9}$$
$$- 2X(s-c)[(1-Y)^2 - (1-Y)^4]\ ,$$

where $\bar{q}$ is defined as :

$$\frac{1}{2}\,[2XF_1 - XF_3 + 2X(s-c)] = \bar{q}^{\bar{\nu}} + \frac{Q^2}{4\nu^2}\,XF_3 \ .$$

This ensures $\bar{q}^{\bar{\nu}}$ always larger or equal to zero because of the D.J. Gross, Ch. Llewellyn inequality

$$2XF_1 \geqq \sqrt{1 + Q^2/\nu^2}\ XF_3 \ .$$

We shall see that to obtain $\bar{q}$ experimentally has given much more strength to the QCD analysis of the structure functions.

How can we get events ?

If the neutrino is the ideal tool for the weak interaction, one has a price to pay i.e. the very small total cross section $\sigma_T = 10^{-38}\ cm^2\ GeV^{-1}\ E_\nu$ (GeV). It is the reason why neutrino physics has been for so long time a physics where the neutrino was as a spectator or a missing particle ... mainly the physics of decay particles. It was necessary to gain as well on number of incident particles as in the mass of the target to have a large number of neutrino interactions. So the development of the "active" neutrino physics has followed the development of accelerators and massive neutrino detectors.

From the following table one can see the evolution of number of events with the accelerator possibilities.

From reactors to accelerator of 7, 30, 400 GeV.

Intensity from 10^{11} to 3.10^{13} .

Detectors from 1 ton to 1000 tons.

Number of events from ≅ 10 to ≅ 100.000.

These factors have allowed one to go from experiments leading to qualitative statements to very precise experiments. This big jump has been the most important after 1972 with the accelerators of 400 GeV.

Neutrino beam.

Principle. The neutrinos and antineutrinos are produced via the decay of π and K. The most important of these decays are :

$$\pi \rightarrow \mu + \nu \ ,$$

$$K \rightarrow \mu + \nu \ .$$

. One can see immediately that the production of ν_μ is more copious than ν_e in the ratio of 100 to 5. As a consequence the neutrino physics is mainly coming from ν_μ interaction and the ν_e physics is still not very well developed .

. These π and K are long-lived particle and the mean decay length for

a 100 GeV π is 5.7 km

a 100 GeV K is 0.7 km.

. $\pi \rightarrow \mu + \nu \qquad P^*_{cm} = 29.9$ MeV

$K \rightarrow \mu + \nu \qquad P^*_{cm} = 235.5$ MeV

. A flat distribution in $\cos\theta^*$ gives a flat energy spectrum

$0 \leqq E_\nu/E_\pi \leqq 0.43$,

$0 \leqq E_\nu/E_K \leqq 0.95$.

. The K's contribute to the high ν energy because of the Q value of the decay. So the high energy part of the spectrum will not be as intense as the low energy part due to π decay because K's are less produced than π's.

In principle if we know E_π or E_K and the angle θ, E_ν is determined

$$E_\nu = \frac{2\,\gamma\,P^*_{cm}}{1 + \gamma^2\theta^2} ,$$

and it should be possible to obtain monoenergetic neutrino beams.

Such beams exist and are called Narrow Band Beam (NBB). Since π's and K's are not separated a Narrow band beam will have two energies due to π and K decay. If one does not select monoenergetic π or K, but tries to use all π and K produced one obtains a so-called Wide Band Beam (WBB).

Narrow band, Wide band beams. The two types of beam differ essentially by the magnetic optics.

. In a narrow band beam the π and K produced are selected and analysed by classical beam transport quadrupoles and magnets. These momentum analysed particles are then decaying in a decay zone which is followed by the shielding.
One must let these particles decay after their production in a decay zone which should not be too small.

. From these decays μ's are also produced at the same rate as neutrinos. These muons if they reach the detectors have very bad conse-

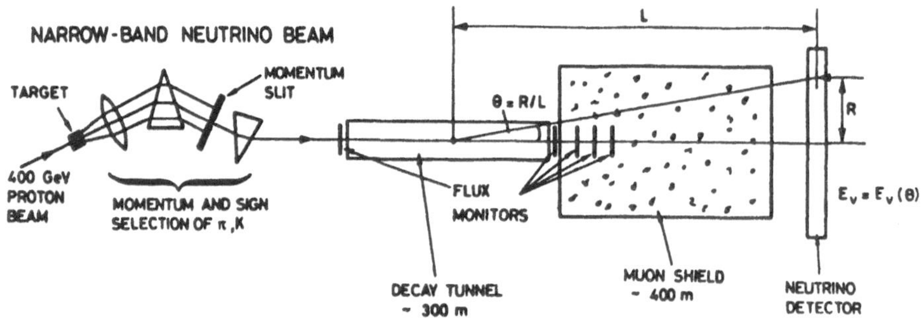

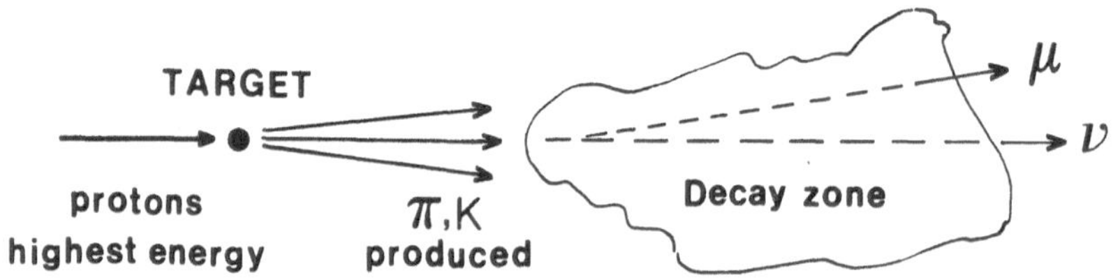

quences. For bubble chambers they increase the number of tracks on the picture and will hide the real interaction; in counter experiments they will increase the dead time of the electronics by wrong triggers. So protection from these muons is needed. If a μ loses roughly 1 GeV in one meter of iron one can see that to protect experimental set up from muons produced at a 400 GeV accelerator one needs about several hundred meters of shielding.

It is why at a 400 GeV accelerator the neutrino beams are of the order of 1 km long.

Neutrino energy. Two body decay for a given π, K energy gives the following kinematics :

$$E_\nu = \gamma(P^*_{cm} \cos\theta^* + \beta E_{cm}) ,$$

$$E_\nu = \gamma P^*_{cm}(\cos\theta^* + 1) ,$$

$$\gamma = E_{\pi,K}/m_{\pi,K} .$$

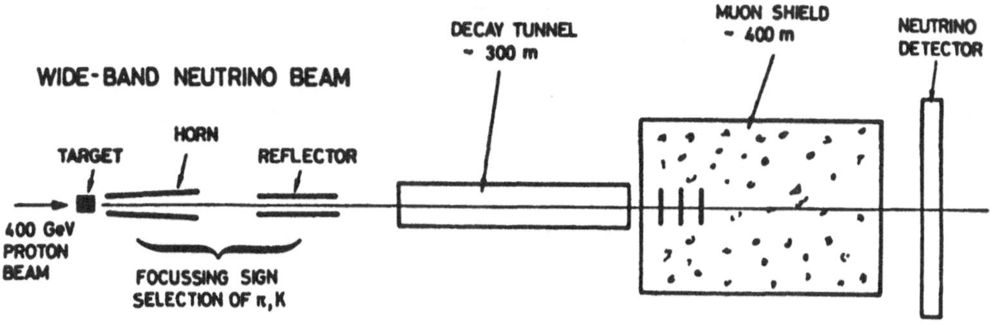

. In a wide band beam the focusing element is a horn which gives a toroidal magnetic field produced by strong pulsed currents. This magnetic field is almost zero in the center of the conductor. At zero degree positive and negative particles will not be separated i.e. neutrino and antineutrino will be produced. Sometimes the focussing is improved by a second horn called reflector.

The energy spectrum given by these two types of beam is presented in the following figure.

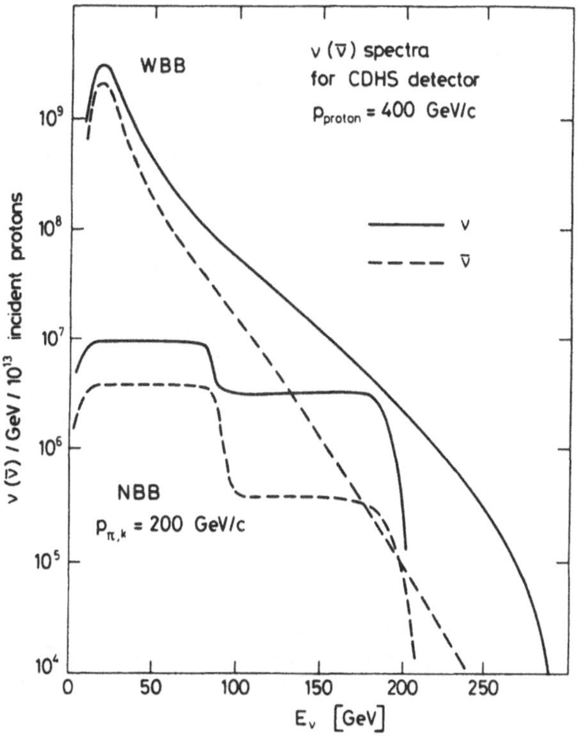

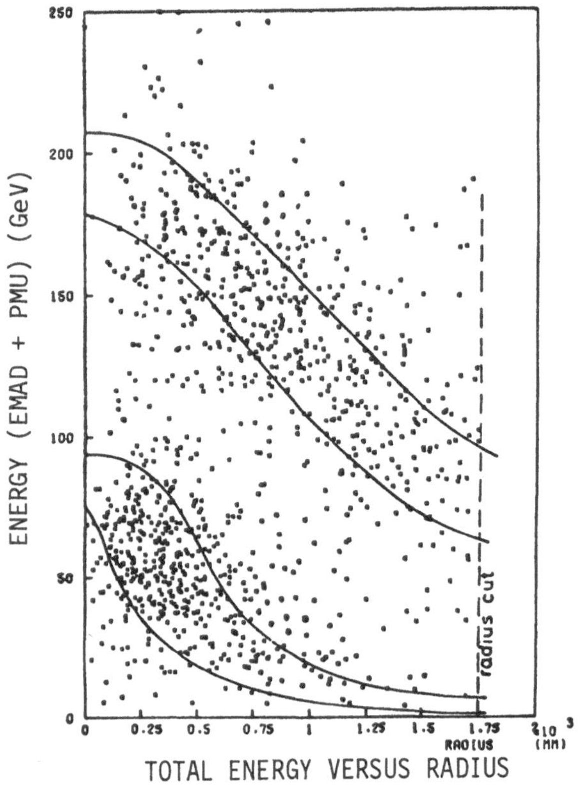

Fig. 1. Example of correlation energy versus radius in a NBB.

A Narrow band beam as in fact a resolution which is due to

- $\Delta P/P$ of the parent momentum (5 to 10 %),
- Divergence of parent beam (σ = 0.2 mrad),
- Uncertainty of the decay point along the tunnel (σ = 87 m means $\Delta\theta \cong 0.2$ mrad at 1 meter radius for CERN beam),
- Resolution in radial distance measurement where the vertex of the interaction is known at ± 20 cm in some electronic detector. This vertex is very well known in bubble chambers.

An example of resolution in E versus the radius is given in the figure 1. The overall resolution is

$$\sigma(E_\nu) \cong 30\ \%\ \text{for}\ \nu_\pi\ ,$$

$$\cong 15\ \%\ \text{for}\ \nu_K\ .$$

As we can see some events are outside the one σ resolution lines. These events are due to $K_{\mu 3}$ and K_{e3} decays which are 3 body decays and so do not follow the simple kinematic correlation E_ν versus θ of the two body decay, or the WBB background i.e. decays of π or K produced before the $\Delta P/P$ selection and which contribute to the NBB as neutrinos of all small energies.

One can summarize the comparison between NBB and WBB.

	NBB	WBB
ν energy	Known 10 to 30 %	Unknown
	In both cases one will assume $E_\nu = E_H + E_\mu$ for some type of events called charged current events.	
ν spectrum	Flat	Big slope after the maximum reducable with "quadrupole beam"
	Limited at $\cong \frac{E_{proton}}{2}$	Some events at high energy
Flux	$10^6/GeV/M^2/10^{-3}$	$5.10^8/GeV/M^2/10^{13}p$
$\nu/\bar{\nu}$ sign selected	Excellent 10^{-3}	Very bad at high energy 1 to 1 $\bar{\nu}/\nu$ at 100 GeV
Monitoring	Should be good	Difficult, rely on flux calculation

Follwing the type of experiment one wants to do a NBB or a WBB will be more or less suitable. It depends also on the detector which is used. One can take two examples. In the case of total cross section or structure functions when the calibration and the knowledge of the beam is important one will choose a NBB. But for the study of charm production via dimuon events, or the sea via antineutrino events, it is better to have a large sample of events : a WBB is better. It is also clear that the upper limit for charm changing neutral current for which one looks at wrong sign muon in a normal charge current sample cannot be determined with a WBB but in a NBB since one needs for this experiment a very good sign-selected ratio $\bar{\nu}/\nu$.

Beam monitoring. To know the number of incident neutrinos is essential for some experiments and is very useful also to understand the apparatus. The technics are different for two types of beams.

WBB - One cannotcount the number of π and K produced immediately after the target. One has to calculate the π and K production, sometimes to check these productions by an experiment and deduce the neutrino flux by a Monte Carlo spectrum calculation. The absolute flux which one obtains is known with a 15 % error. The experimental check of this Monte Carlo calculation is achieved by the determination of muon flux after the decay zone at different depths in the iron shielding. The muons are detected by solid state counters and one checks the radial shape distribution of the muons and their absolute rate. These muon detectors are also used in NBB and have been very useful to cross check some NBB calibrations.

NBB - Several detectors are installed in the Narrow band beam optics to count particles.

Beam current transformer. It is a coil which gives a signal proportional to the number of charged particles going through it. At the beginning of the decay zone one can then measure the number of π, K, proton or π and K in negative beam. The calibration of such a detector is known at ± 3 %. A problem which arises is the problem of the δ rays created in material surrounding the detector as the air. The correction of these δ rays are of opposite sign since :

$$BCT^{+} = (N_{\pi} + N_{K} + N_{p})(1-\delta),$$

$$BCT^{-} = (N_{\pi} + N_{K})(1 + \delta) .$$

The muon chambers can in principle help to determine the δ rays. This δ ray contamination although quite small ($\cong$ 6 %) is one of the problems which increase the error on the NBB monitoring. It is not right now completely understood. Tests are going on now in the very precise total cross sections measurement to understand these δ ray productions.

Ionisation chambers. In principle these detectors are used as crosscheck with the BCT when the flux is less than 10^8 particles per burst.

Cerenkov counter. They are used to count separately the π, the K and the protons. The number of particle being $\cong 10^8$ one can either reduce the intensity of the accelerator to make a single particle counting or one integrates all the light from one burst. The reduction of the intensity of the machine of a factor 10^3 to 10^4 needs the assumption that the targetting is the same at small intensity and so that the optics of the accelerator is reproducible. The pressure of the gas, the Cerenkov angle and the momentum are correlated by the formula

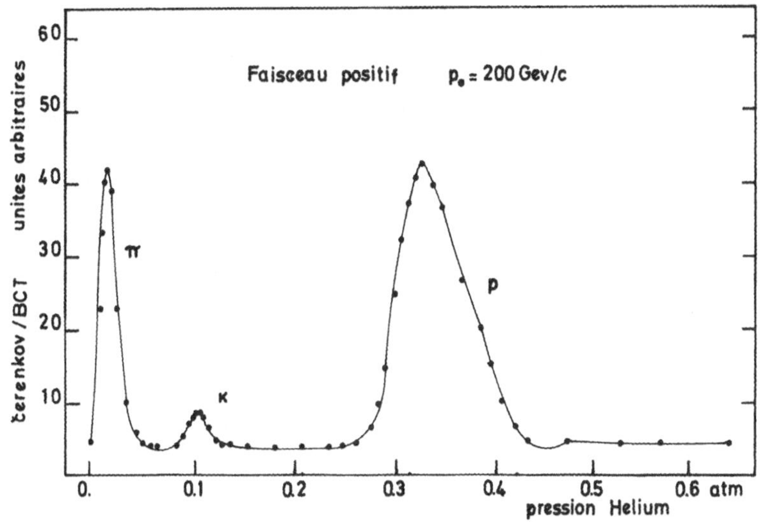

Fig. 2. Example of Cerenkov pressure curve obtained in a 200 GeV NBB at CERN.

$$P_r = \frac{1}{2n_\circ}\left(\theta^2 + \frac{m^2}{P^2}\right) ,$$

θ emission angle,
P momentum of the particle,
m mass of the particle,

$$n^2 - 1 = n_\circ \frac{P_r}{2}$$, $n_\circ$ is the index of the gas.

An example of a Cerenkov pressure is given in figure 2. The ratios π/K/p are very important in the monitoring of the beam and the use of Cerenkov counters appeared quite delicate. The use of two diaphragms in the optics of a Cerenkov counter helped the CDHS group to understand some fluctuations in the ratio π/K. This ratio has been compared to those obtained by a Monte Carlo programm. We give some ratios in a NBB of 200 GeV and the errors.

$$K^+/\pi^+ = 0.147 \pm 0.004 ,$$

$$p/\pi = 4.3 \pm 0.10 ,$$

$$K^-/\pi^- = 0.049 \pm 0.002 .$$

The following table gives the errors on the beam measurement in

$N_{\pi^-} = \dfrac{N(BCT^-)}{1 + (K^-/\pi^-)}\,(1 - \delta)$	$\dfrac{\Delta N_{\pi^-}}{N_{\pi^-}} = 6\ \%$
$N_{\pi^+} = \dfrac{N(BCT^+)}{1 + (K/\pi) + (p/\pi)}\,(1 + \delta)$	$\dfrac{\Delta N_{\pi^+}}{N_{\pi^+}} = 6\ \%$
	$\dfrac{\Delta N_{K^+}}{N_{K^+}} = 6.7\ \%$
$N_K = N_\pi * \dfrac{K}{\pi}$	$\dfrac{\Delta N_{K^-}}{N_{K^-}} = 7.2\ \%$
$R = \dfrac{N_{\pi^-}}{N_{\pi^+}} = \dfrac{1 + (K/\pi)^+ + (p/\pi)^+ * BCT^- * (1 - \delta)}{1 + (K/\pi)^- * BCT^+ * (1 + \delta)}$	$\dfrac{\Delta R}{R} = 5.4\ \%$

a narrow band beam of 200 GeV at CERN. These numbers are important because they fix the minimum errors one can expect in the determination of the total cross section or the structure functions.

Detectors. One can outline three important neutrino interactions

(1) $\nu_\mu + N \to \mu^- + X$

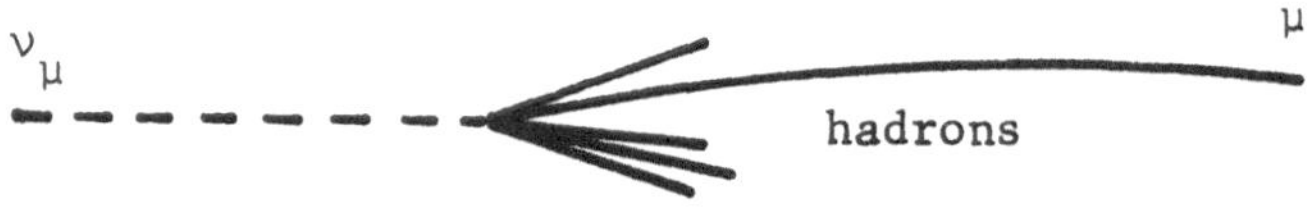

(2) $\nu_e + N \to e^- + X$

(3) $\nu + N \to \nu + X$

Obviously there is no unique and good detector to detect reaction

(1) (2) (3) ...

One can imagine one like the Track sensitive target (TST) which has a good vertex localisation, detects all charged particles, neutral particles, converted in the Ne and the muon is identified outside the bubble chamber. But such a detector has a small fiducial volume and so collects few events.

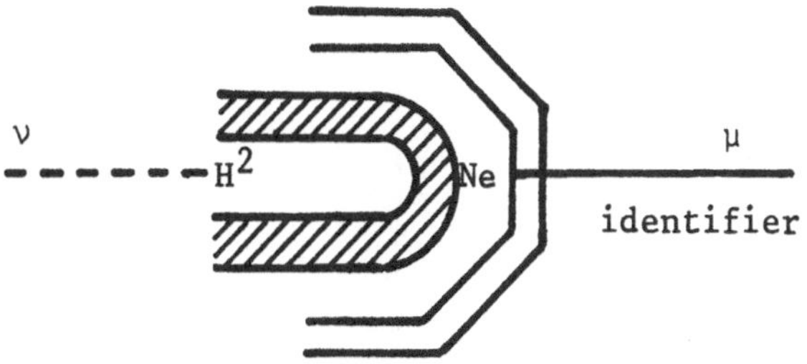

Different types of detectors will be suitable for different types of physics.

Inclusive physics. One measures the angle and the momentum of the lepton and the total energy of the hadron shower. [Reaction (1)]. For reaction (2) the difficulty is to separate the electron from the shower.

Exclusive physics. It is the type of reactions (1) (2) (3) where one can know what is inside the hadron shower, for instance strange particle production, charm production via hadronic decay ... Only the bubble chamber can expect to answer these questions and even with difficulties : identification of particles, conversion of γ, for instance ...

Neutral current physics. This study of reaction (3) is even more specific since the outgoing lepton is not detected. The total cross section of this reaction is easy to determine but anything else needs a good knowledge on the hadron shower : either detection of all hadrons or at least detection of the angle of the shower and its energy.

So neutrino event detection is a conflict or a compromise between

High density detector		Low density detector
↓	and	↓
Electronic detectors High statistics		Bubble chamber See everything, low statistics

$\mu^+\mu^-$-event

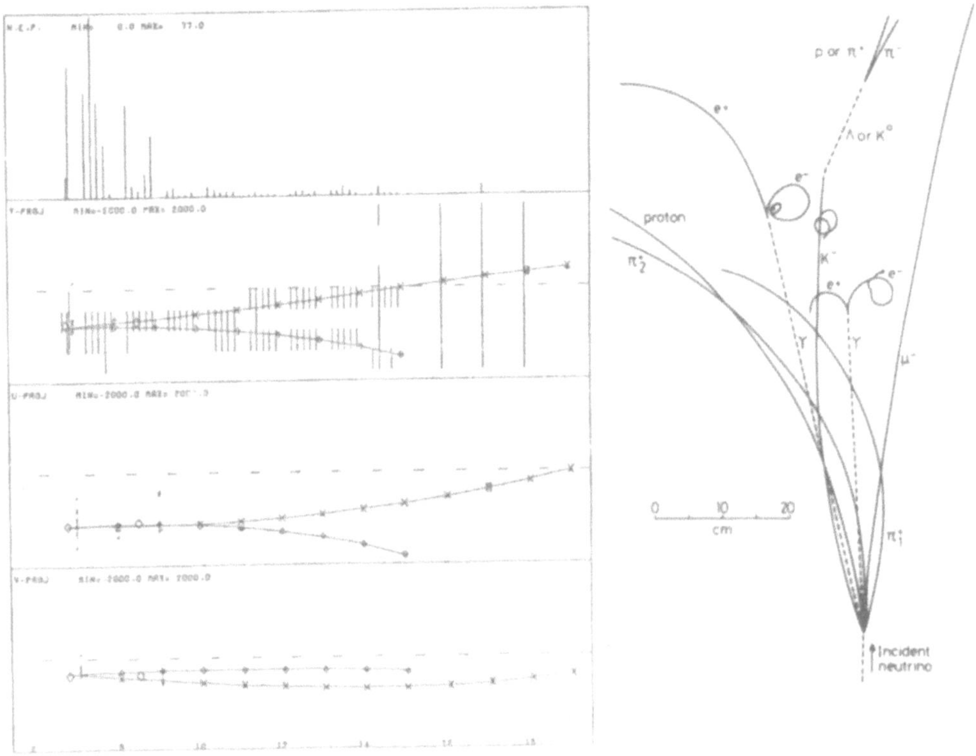

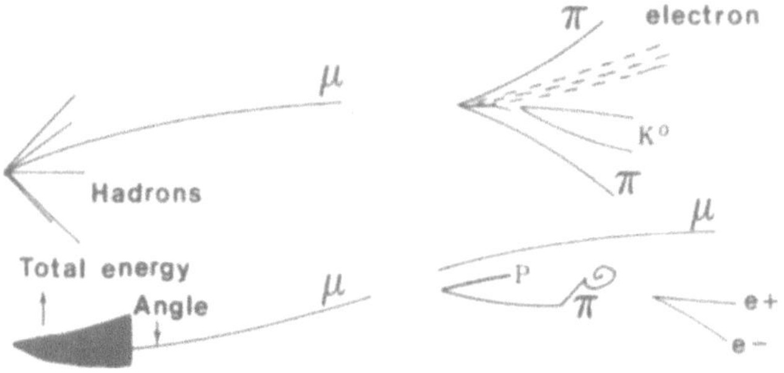

There exist now around FNAL (Fermilab at Batavia) and CERN different detectors of each type which are listed with their characteristics in the following table. This very old fight between the feasibility of physics in bubble chambers and electronics experiments comes to an end. This has been very clear when the bubble chamber group has proposed to put an electro-magnetic calorimeter in BEBC. Then around a very good vertex detector would have been all the other "electronics" gadgets : calorimeter, muon identifier. This proposal

	ELECTRONIC DETECTORS		BUBBLE CHAMBERS
	High density	Low density	
Target mass	1000 T	100 T	1 T (H^2) 20 T (N_e)
Muon identification	Use iron of target or after the target Excellent	Good	Need (EMI) around B.C. Good
Electron ident.	Not possible	Longitudinal devel. of shower (X_0, L_0)	Excellent
Energy hadronic shower	Sampling of the target with scintillators or proportional tubes Good	Good	Measure all particles Missing energy, π, γ, K_0 (20%)
Exclusive physics	Blind ... Except dimuons	Determination of angle shower θ_H neutral current	Good K_0, $K^{\pm}$(?), π^0 $\pi^{\pm}$, p
Some experiments	CERN : CDHS FNAL : CFRR E616	CERN : CHARM FNAL : E 594	CERN : BEBC (Gargamelle) FNAL : 15 Feet

has been refused for "scientific political" reasons but it is clear that after 10 years of running time the optimization of the detector for neutrino physics would have been of this type.

Resolution of detector. It is interesting to know now with what precision one can measure an event with the different types of detectors.

Muon momentum.

. In bubble chambers, BEBC for instance the measurement of a track is very good : $\Delta P/P = 4\ \%$ at 100 GeV/c for a 2 meters track length in a magnetic field of 35 kGauss.

. In electronic detectors in principle the measurement in an air core magnet will be as good but such a magnet will provide a very bad acceptance for muons and it is not possible to avoid some material from the target before this measurement. The measurement of muon momentum is usually made in toroidal magnets and the precision is then limited by the multiple scattering in iron. One obtains

$\Delta P/P \sim 0.2 \sqrt{L}$, 18 to 8 % following the length of the track L.

$\Delta\theta_\mu \sim 0.5\ L/P$.

The contribution of the error on position of the track becomes important only at very high energy :

$$(\Delta P/P)^2 = 0.027 \frac{P^2}{L^2} \Delta m^2 ,$$

Δm is the error on position. This error is still 1/4 of the error due to multiple scattering at 200 GeV/c.

In conclusion the error on the muon cannot be decreased very much by improvement of apparatus.

Hadron energy.

. In a bubble chamber one has to measure all the charged particles. The problem is the neutral particle γ or neutrons or K_L° which can be missing. Corrections can be applied but this gives a $\Delta E/E \cong 25\ \%$ and of course this question will not improve at high energy.

. In counter experiments it is by calorimetry that the hadron energy is measured.
In principle the average ionisation is measured and is proportional to the total energy dissipated by the hadronic shower.

The fluctuations of this measurement are due to

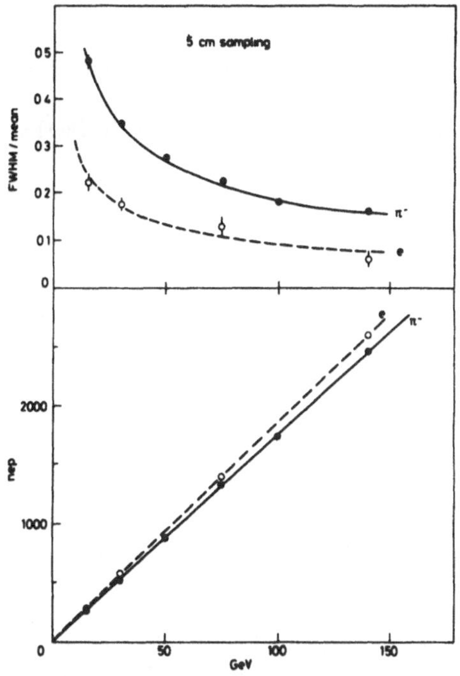

Fig. 3. Example of calibration of calorimeter.

- number of particles traversing iron,
- nucleon excitation not detected,
- neutral particle fluctuation.

One calibrates these calorimeters by measuring the response to well momentum analyzed hadrons (π usually). The typical responses are shown in figure 3.

An average resolution is :

$\Delta E/E_H$ = 1 to $0.6/\sqrt{E}$ hadrons,

$\Delta E/E_H$ = $0.1/\sqrt{E}$ electrons.

Contrary to bubble chambers the resolution in calorimeters will improve with higher energy.

Final test of calibration and resolution.

Because - as we have seen - we do not know very well the energy of the incident neutrino and also because it is difficult - sometimes impossible in counter experiments- to reconstruct some well known masses of particles, the calibration of the entire apparatus is not possible. It is one important question in neutrino physics.

One way to understand the apparatus is to compare the energy of the neutrino from a NBB per bin of radius.

. One measure the energy of the neutrino $E_\nu = E_\mu + E_H$, this involves the experimental errors on μ and hadrons energy,

. and one compares to a Monte Carlo which implies : beam optics
acceptance
resolution
→ smearing function.

To see the relative influence of the error on muon measurements or hadron energy one can do this test at small y → large P_μ or large y → small P_μ.

An example of such test is presented in the following figure 4.

This figure is very interesting for different reasons :

- For all radius bins one has a good agreement between the experimental distribution and the Monte Carlo. This implies that the calibration is correct - no shift in energy - and that the widths of the curve agreed - this tests that the smearing function on the variable P_μ and E_H are correctly performed. This will be important for the analysis of data later on.

- One sees clearly on this figure the evolution of the contribution of π and K neutrinos events with the radius of the apparatus : at small radius the π's contribute very much, disappear at a radius of 1 meter and only the K's stay. This clear separation could be also very useful in the analysis - for instance study of only ν_K or ν_π events if one thinks that the bias can be different in each case.

Extraction of structure functions.

Method. The formulas (7), (8), (9) show that to obtain XF_1, XF_2, XF_3 we need to measure the number of ν and $\bar{\nu}$ events in a ΔX, ΔY space in bins of Q^2 or ν. This implies flux monitoring of neutrino and antineutrino, choice of the variables, knowledge of the resolution on variables ...

Since the detection of neutrino is not 100 % efficient one has to know the acceptance of the detector via a "Monte Carlo" calculation. In most detectors the geometrical acceptance is quite good. More important are only the cut-off effects on the momentum of the μ and a cut-off on the hadronic energy E_H = 2 to 5 GeV (typically on electronic detectors). One can see as an example (Fig. 5) the effect in the X, Y plane of a cut-off on P_μ of 7 GeV at an energy E_ν = 70 GeV. The dashed area represents the area where 50 % or more of the events are dropped when flat X and Y distributions are generated.

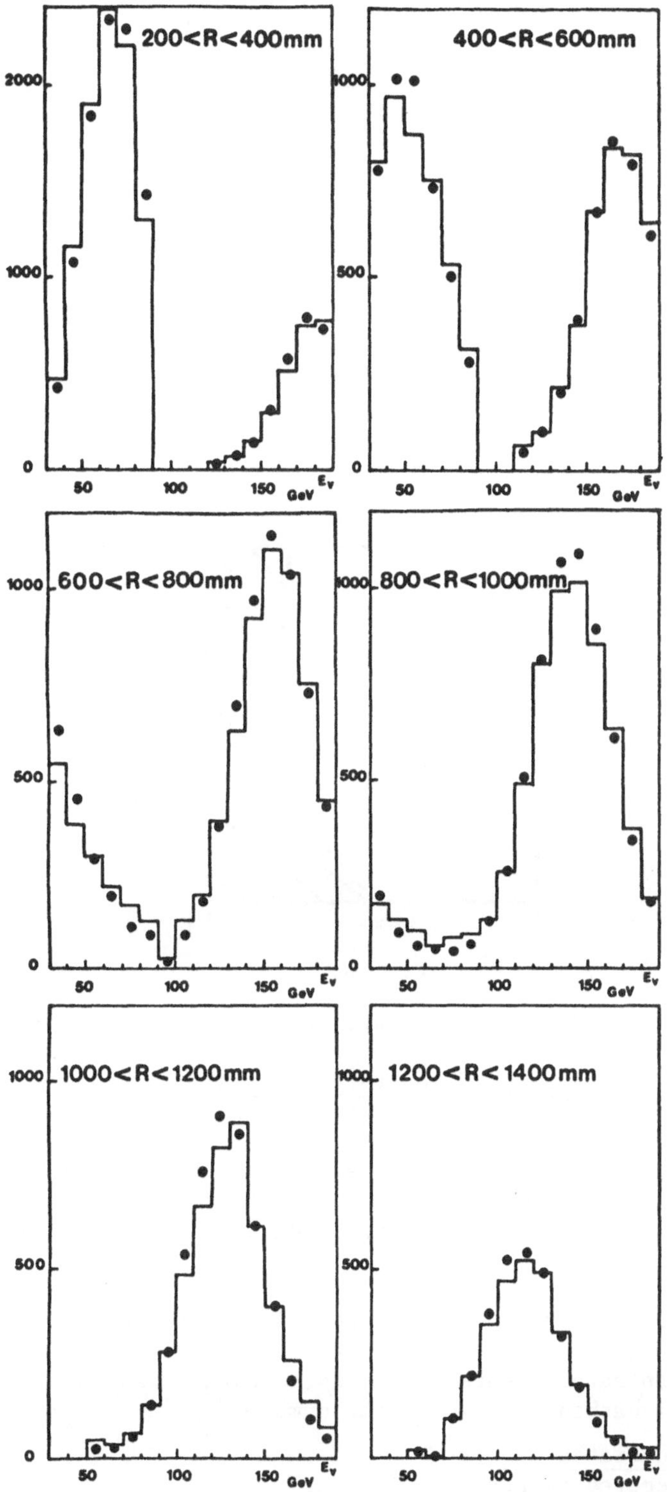

Fig. 4. Energy of neutrino events versus radius in a NBB.

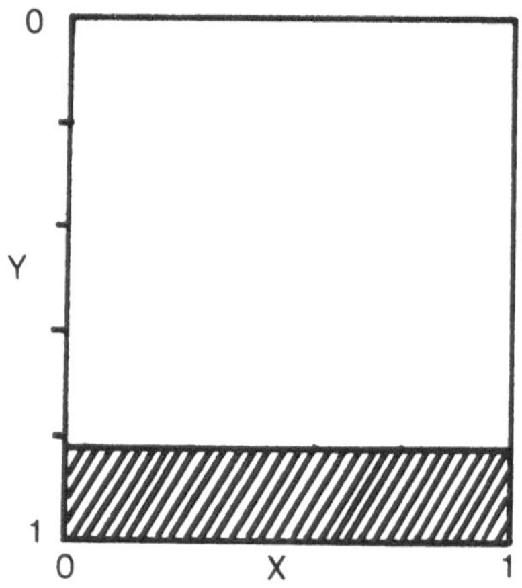

Fig. 5.

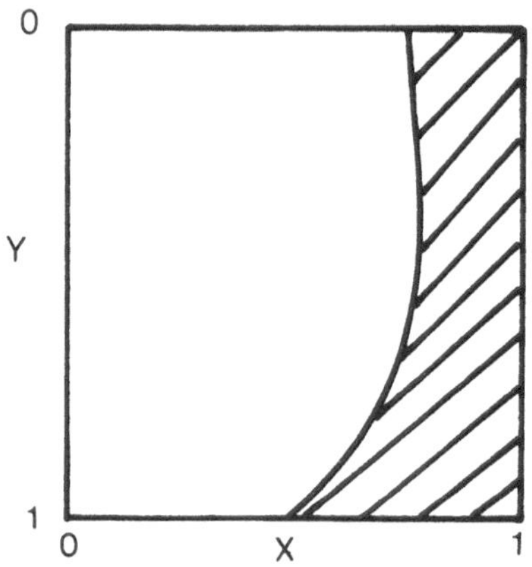

Fig. 6.

The number of "true" events are then calculated as :

$$N_{ev}\ \mathrm{TRUE} = N_{ev}\ \mathrm{OBS} \times \frac{\mathrm{MC\ generated\ events}}{\mathrm{MC\ accepted\ events}}\ .$$

The Monte Carlo generation of events not only needs the variable resolution but also some physics input. It is an iterative process which stops when the agreement between data and Monte Carlo is satisfactory.

. In the CDHS analysis to avoid too big a discrepancy between MC and data one has dropped regions where the difference of number of Monte Carlo generated events and accepted events was too large. Figure 6 shows the dashed area where the number

$$k = \frac{\mathrm{MC\ generated\ events}}{\mathrm{MC\ acceted\ events}}$$

differs by 40 %. The X distribution has a "typical" shape of the F_2 structure function. We see that this affects mainly large X regions. We then limit the analysis at $X \leq 0.7$, this does not affect too much the study of structure functions because there are few events at large X.

The antiquark distribution $\bar{q}$ is extracted from WBB because of obvious statistical reasons (155 000 $\bar{\nu}$ events and 30 000 ν events are used for $\bar{q}$ analysis with $E_\nu > 20$ GeV and $Y > 0.55$). The main problem is the knowledge of the energy spectrum of the data and their normalization by the total cross section. It is why a very precise total cross section is needed if one wants to use WBB data. A discrepancy on the last measurement of total cross section at FNAL in 1981[9] has "re-opened" this question and precise measurements are made again at FNAL and CERN. In the data presented on $\bar{q}$ by CDHS the cross section used for antineutrino is a linearly rising cross section with the slope $\sigma^{\bar{\nu}}/E = 0.30$ (10^{-38} cm^2/GeV) and for neutrino $\sigma^\nu/E = 0.62$ (10^{-38} cm^2/GeV) for $E_\nu > 70$ GeV and a rise of 11 % down from 70 GeV to 20 GeV[10]. This point is important because neutrino physics is the unique way to obtain $\bar{q}$ and this data can come only from WBB exposure to reach a good statistics sample.

In an exposure of 300 GeV on neutrino only the functions F_2 and XF_3 cannot be obtained by lack of antineutrino data. One defines then a function

$$F_+ \equiv \frac{1}{2}\,[2XF_1 + XF_3] = X\,(u+d+2S)\ ,$$

$$F_+ = \left(\frac{\pi}{G^2 M E_\nu}\,\frac{d^2\sigma^\nu}{dXdY} - X(\bar{u}+\bar{d}+2\bar{c})\,[(1-Y)^2 + R(1-Y)]\right) / \,[1 + R(1-Y)].$$

This function is used at large X when $2XF_1 = XF_3 = F_X$ and completes the large X, Q^2 region.

<u>Choice of variables.</u> Theory wants information on F_1, F_2, F_3 functions of X and Q^2 and mainly the scaling violation is studied by looking at X - distributions for different Q^2. For different reasons, not all obvious, but correlated to experimental problems, it is better to study the data in X, E_H bins. We would like to explain these points. Let's have a look to the following figures (7) which represent the kinematic lines of constant X in the Q^2, ν plane.

. If one takes a constant Q_0^2 bin one sees two limitations in the X distribution :
  - for large Q^2, the low X region is missing because the maximum available ν energy is limited. For a given Q_0^2 one will increase the low X region by increasing ν, i.e. the energy of neutrino, i.e. the energy of the incident protons.
  - For low Q_0^2 the large X region can be missing if one limits the value for experimental reasons- mainly when the error for $\nu \cong E_H$

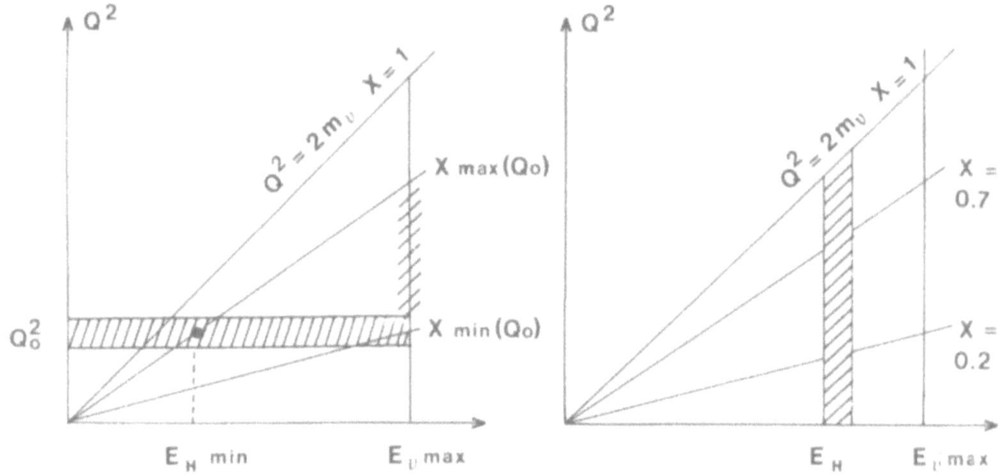

Fig. 7.

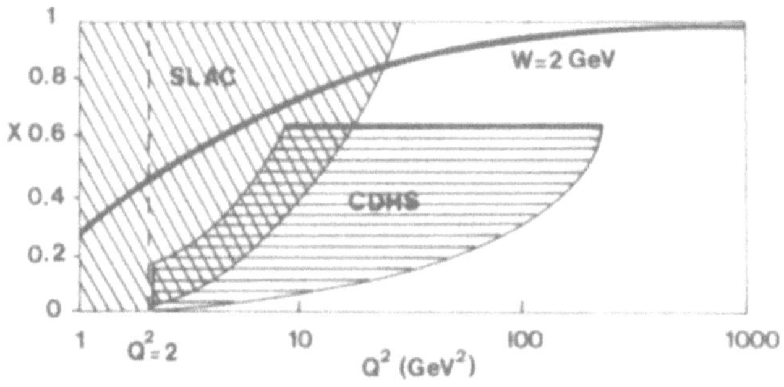

Fig. 8. Kinematic limits in X, Q^2 plan due to experimental cut-offs.

is too large and that one wants to limit these errors.
One sees that for a given Q_o^2 it is impossible to obtain from one experiment all the X distribution. These two intrinsic limitations on X distribution are in all experiments. If one wants to study the largest range in X one has to take results of two different experiments. It was the case of CGM and BEBC and SLAC and CDHS. The figure 8 illustrates the complementarity of the two results

SLAC-CDHS in the Q^2, X plane. The limitation of the CDHS data at X = 0.7 is due to a cut-off on data because the smearing factors are large for large X as we have discussed in the above chapter. If we consider in the Q^2, ν plane a bin in E_H, we see in the figure 7 that all the X distribution is seen. Then the study of the evolution of the X distribution with E_H was in this case much easier. It was the reason for the choice of E_H as variable to see the scaling violation for F_2 and XF_3 (See figures 9 and 10). But E_H as variable has also another advantage.

- For fixed Q^2, the neutrino energy which contributes to small values X and large values of X is very different and therefore the K/π ratio must be well known. See figure 10.
- For fixed E_H one sees on the same figure that the same energy of neutrino contributes for different X and so the errors on the flux and the K/π ratio do not affect the shape of the X distribution. This was very important when we were looking in the beginning at the scaling violation - i.e. at X distributions at different energies -and it is why E_H has been used as a variable to treat the data. Then one can express the data in Q^2, X bins.

Resolution in X, Y, Q^2. Since one studies distribution functions of X, Y, Q^2 or ν one must know very well the experimental errors on these variables. Sometimes the errors on a variable are so large that it becomes nonsense to use it any longer.

Error on Y or ν : $Y = E_H/E_\nu$, $\nu \cong E_H$.

One measures E_H and the error on ν is $\Delta E_H/E_H \cong 0.8/\sqrt{E_H}$ which is well known by calibration of the apparatus. One should notice that when this error is too large at small E_H (small ν) energy one does not use the data. A usual cut-off is done for $\Delta E_H/E_H > 25$ % because then this error is large compared to the average of all other errors. The error on Y is mainly the error on E_H and the error on the momentum of the muon :

$$Y = \frac{E_H}{E_H + E_\mu} .$$

Typical errors on Q^2 and ν in the Q^2, ν plane are shown in figure 8.

Error on X : $X = Q^2/2M\nu$.

Since the dependence of the structure functions is studied as a function of X and Q^2 the error on these two parameters must be very well known. Errors on X depend on errors on E_H, P_μ, θ_μ. The following figures show the relative errors of these variables for different values of Q^2.

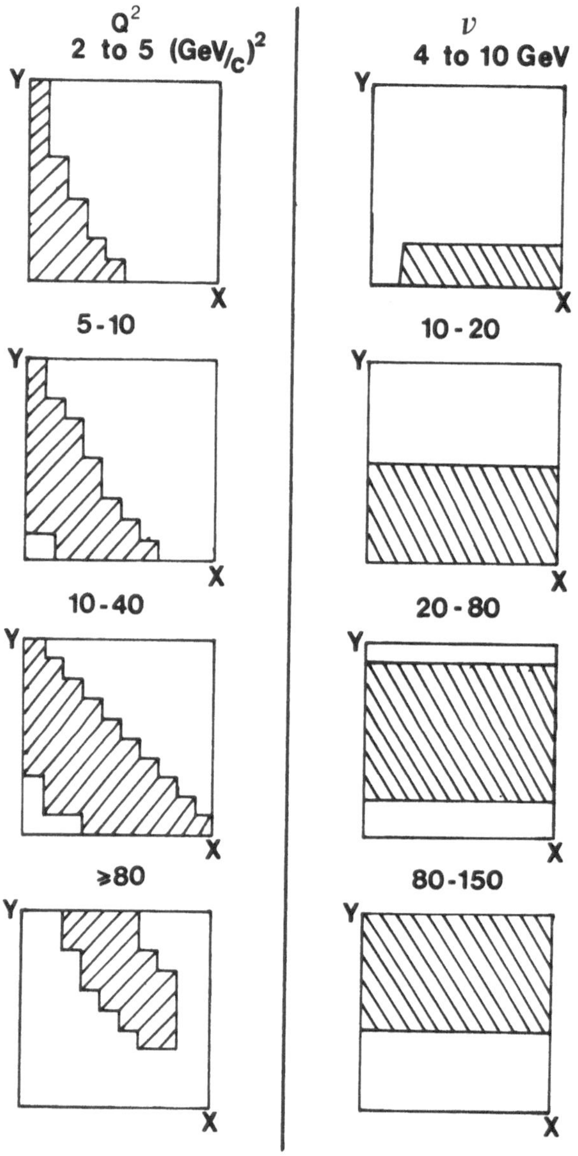

Fig. 9. Evolution of the X distribution seen in the variable Q^2 and ν.

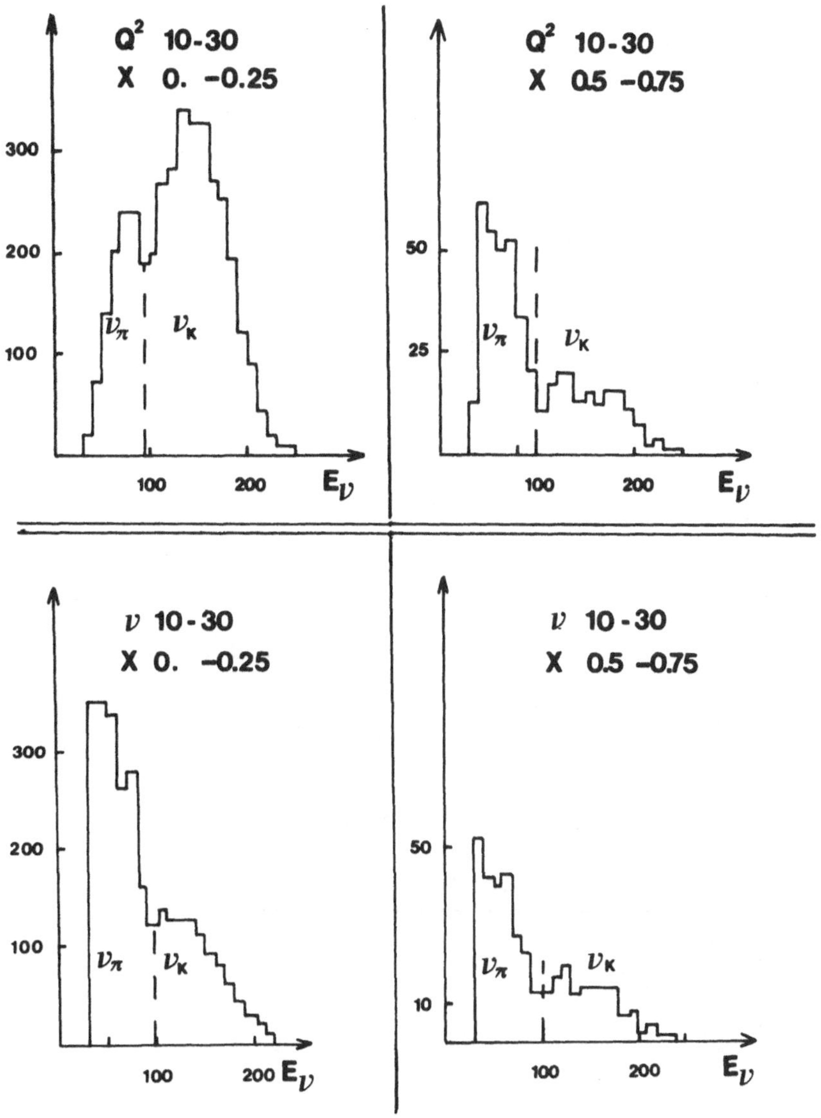

Fig. 10. Relative contribution of ν_π and ν_K for the two variables Q^2 and ν.

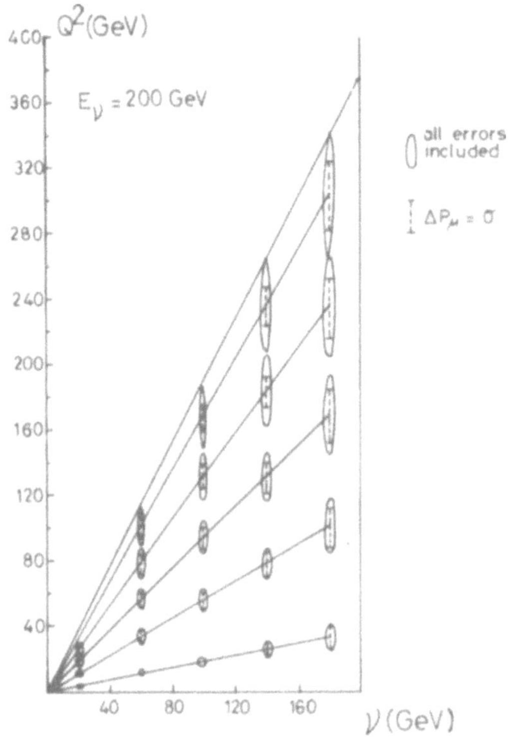

Fig. 11. Typical errors on Q^2 and ν.

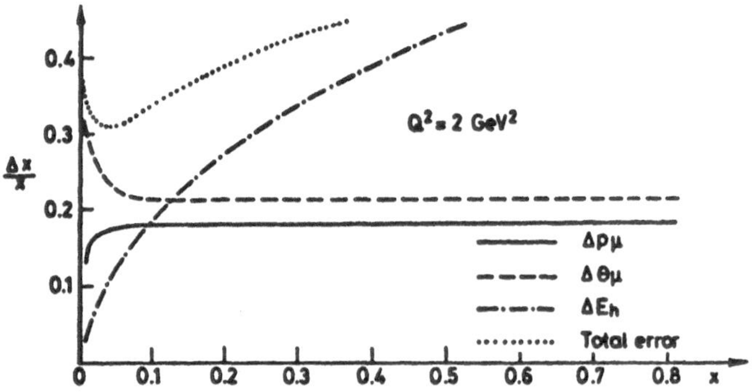

. Small X, small Q^2 errors on θ_μ dominate

$$\Delta\theta = 0.5\ L/P_\mu .$$

This error in iron targets is due to multiple scattering and cannot be very much decreased. It is much smaller in bubble chamber experiments.

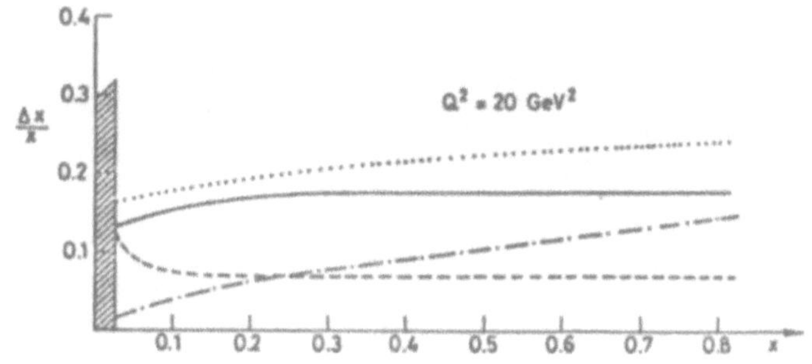

. Small Q^2, large X errors on E_H dominate

$$Q^2 = 2\,X \,.\, \nu \,.\, M$$

↓ ↓

small large small

↓

$$\frac{\Delta E}{E} = 0.8/\sqrt{E} \quad \text{large error.}$$

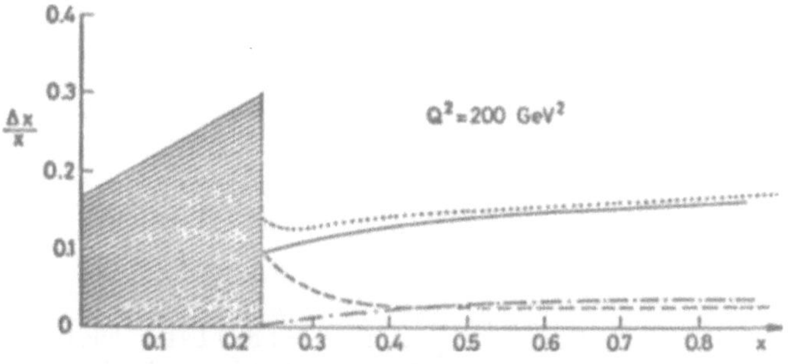

. For an average Q^2 of 20 GeV^2/c^2 the 3 errors on E_H, P_μ, θ_μ have similar amplitudes.
. Notice that ΔP_μ is constant over all Q^2 and dominates then for high Q^2.

We see a limitation at small X and small Q^2 due to experimental errors on E_H and θ_μ. This indicates that one detector does not cover the entire range of X and Q^2 even without kinematic limitations.

Study of systematic effects.

To be sure that a wrong resolution of variables does not mimic any scaling violation we have run a Monte Carlo programm in which we have analysed the structure function F_2' with an intentional error on resolution of E_H, P_μ and θ_μ. The results are presented in

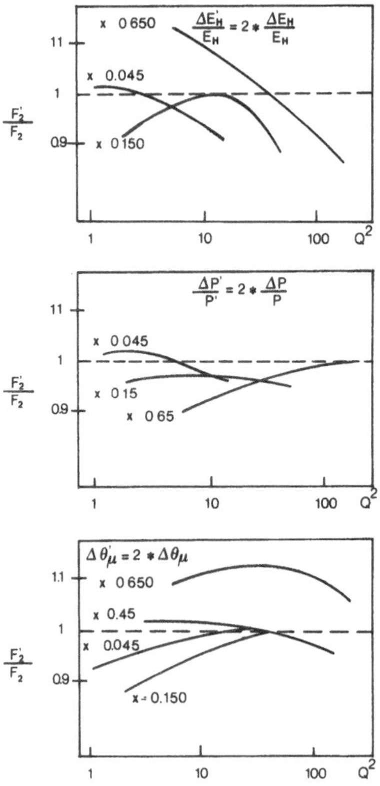

Fig. 12. Effect on the structure function F_2 of "wrong" resolution on E_H, P_μ, θ_μ.

figure 12 where the ratio F_2' on F_2 (generated flat) shows the effect of a large error (a factor two) on the resolution of these variables. This effect is quite small and cannot be responsable for scaling violations.

Corrections. We have mentioned already the correction for a not exactly isoscalar target I = 0. Two other corrections have to be applied to extract XF_3 and F_2.

1 - The radiative corrections - Radiative effects are corrected following the calculation of De Rújula et al.[11].

One writes

$$\frac{d\sigma}{dXdY}_{OBS} = \frac{d\sigma}{dXdY}_{WEAK} * \text{COR.}$$

The correction term COR is presented in the following table for E_ν = 90 GeV.

1	1.02	0.98	0.96	0.95	0.92
	1.025	0.99	0.975	0.95	0.92
Y	1.02	0.985	0.96	0.945	0.915
	1.01	0.98	0.96	0.94	0.91
0	0.99	0.96	0.95	0.93	0.91
	0		X		1

These corrections increase $d\sigma/dXdY$ (weak), at large X and one sees that the effect is generally less than 10 %. The correction factor is much larger at very small X. Another parametrization for radiative corrections has been calculated by Barlow and Wolfram[12]. This calculation differs for small values of X from the De Rújula's one. Nevertheless no "average" has been made and the correction is done following reference 11.

2 - In the expression of F_2 or $\bar{q}$ the correction due to the strange sea appears. We will present in more detail in a later chapter the extraction of the strange sea from dimuon physics. Since the results are far from a SU_3 symmetric sea, the next table shows the effect on F_2 for two values of $2(s-c)/\bar{u}+\bar{d}$: 0.4 and 0.2.

The correction is small and is sensitive mainly to small X. This small effect can be emphasized by the fact that in neutrino interactions the strange quark s leads to charm quark c which is kinematically suppressed by threshold mass effects. In some cases of analysis this effect has been evaluated using the slow rescaling model[13].

No correction for Fermi-motion has been made. These corrections affect the shape of the structure function at large X but have small effect on the Q^2 dependence. (Remember that events larger than $x = 0.7$ are not used).

The correction due to the value of R will be discussed in a special chapter.

Q^2	0.015	0.045	0.08	0.15	0.25	0.35	X
1.4	0.95	0.98	0.99	1	1		
3.5	0.93	0.96	0.98	0.99	1		
9.0		0.95	0.96	0.97	0.99	1	
18			0.95	0.97	0.99	1	
36				0.97	0.99	1	
72					0.99	1	

Results. The results are presented in figure 13, $F_2(x,q^2)$, figure 14, $XF_3(x,q^2)$, figure 15, $\bar{q}^{\bar{\nu}}(x,q^2)$ and figure 16, $F_+(x,q^2)$. Tables of values for these structure functions have been published and can be used for any analysis.

In figure 17 the x dependences of $2XF_1$, XF_3, F_+ and $\bar{q}^{\bar{\nu}}$ are shown at a fixed $Q^2 (4.5 < Q^2 < 10\ GeV^2/c^2)$.

The empirical fits to the data show that for $x > 0.4$, $2XF_1 \cong XF_3$ and the antiquark distribution $\bar{q}^{\bar{\nu}}$ disappears.

Comparison with charged lepton structure functions. This comparison is very interesting because the experimental method to obtain these structure functions is completely different and then the systematics errors are expected to be also different. The quark parton model relates the structure function F_2^{eN} observed in muon or electron scattering and the structure function F_2 observed in neutrino scattering :

$$F_2^{\nu N}(x,q^2) = \frac{18}{5} F_2^{eN}(x,Q^2).$$

This formula is strictely valid outside the sea region. The agreement of the two measurements is presented in figure 18. The experimental points compared with the CDHS results are these from EMC collaboration[14] and SLAC-MIT[15]. The structure functions $F_2^{\mu N}$ and $F_2^{\nu N}$ agree very well and the normalization factor 18/5 is correct. For the SLAC data this agreement is not as good ($F_2^{\nu N}/F_2^{ed} = 1.46 \pm 0.12$ for

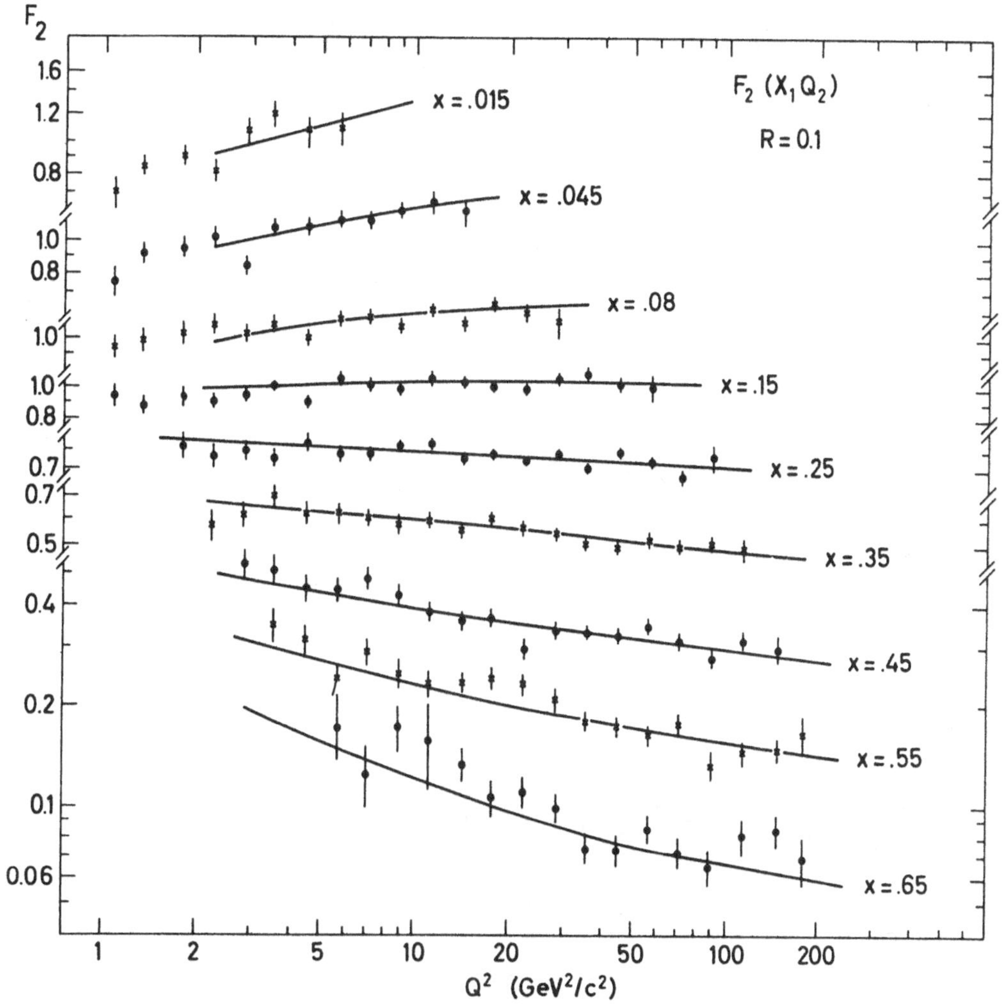

Fig. 13. Structure function F_2 versus Q^2 for different bins in X. The solid lines are the result of a leading-order QCD fit to F_2 and $\bar{q}^{\bar{\nu}}$.

x > 0.4 averaging over the whole Q^2 range, compare to 1.8 expected). The discrepancy may be due to the value of R as we will see in the next chapter.

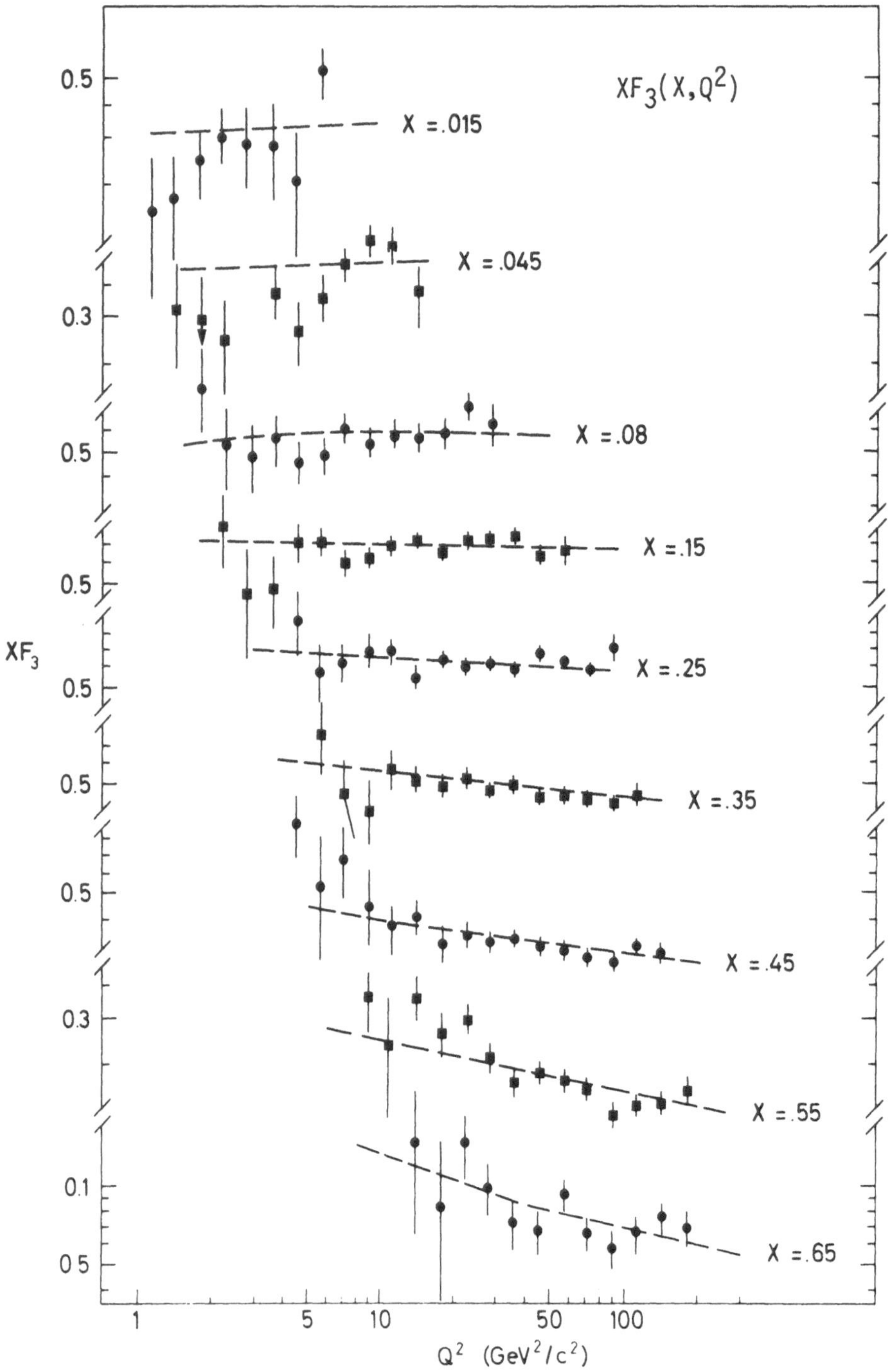

Fig. 14. The structure function XF_3 versus Q^2 for different bins in X. The dashed lines are the result of a leading-order QCD fit to the data.

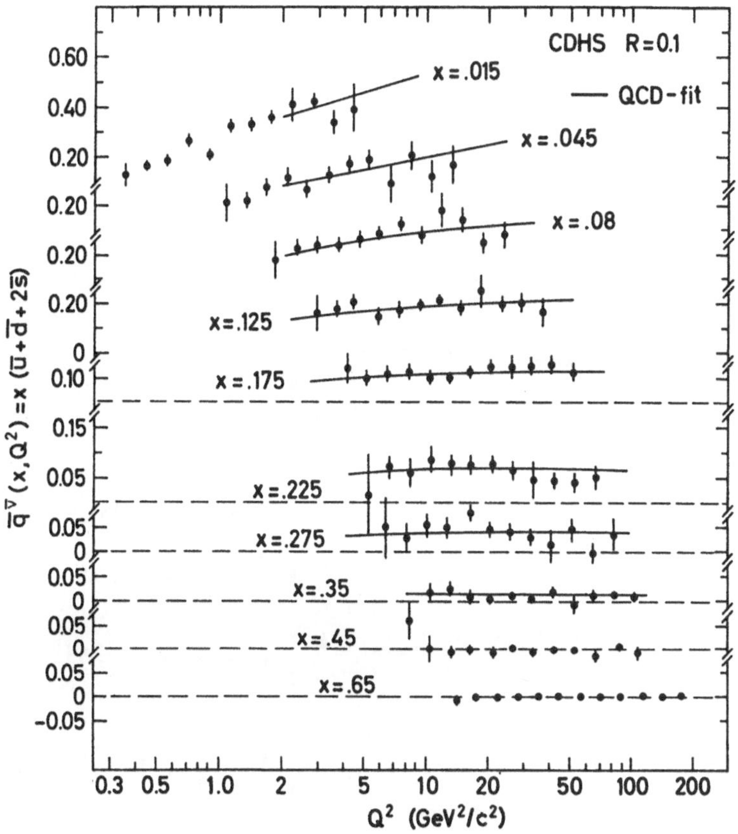

Fig. 15. The structure function $\bar{q}^{\bar{\nu}}$ versus Q^2 for different bins in X. The solid lines are the result of a QCD fit F_2 and $\bar{q}^{\bar{\nu}}$.

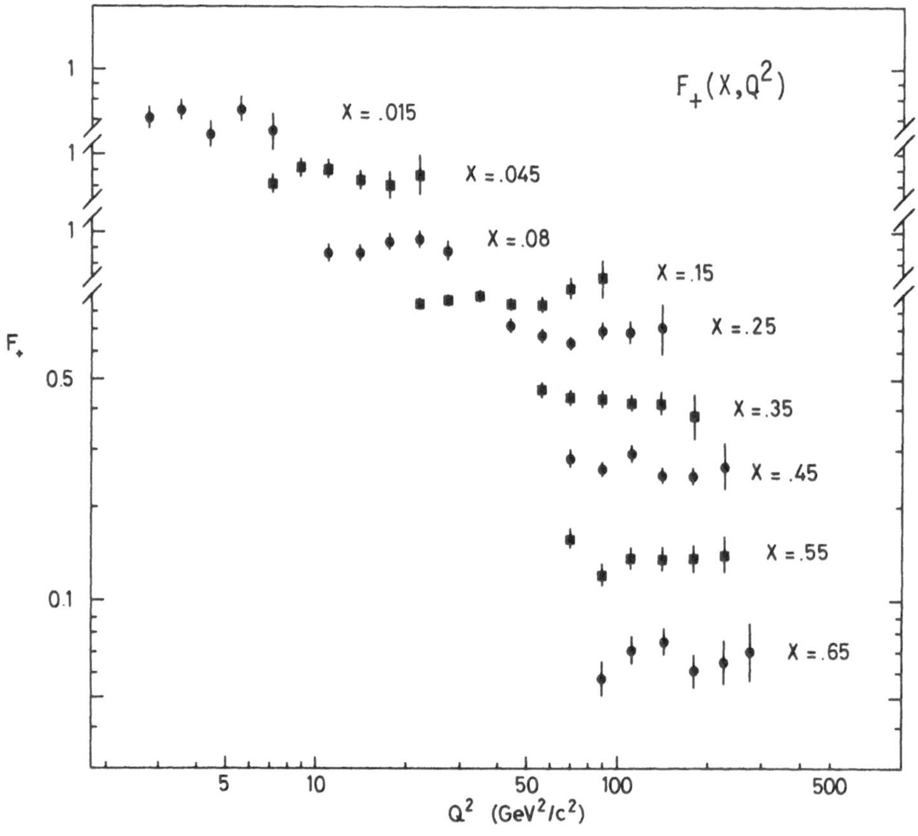

Fig. 16. The structure function F_+ versus Q^2 for different bins in X.

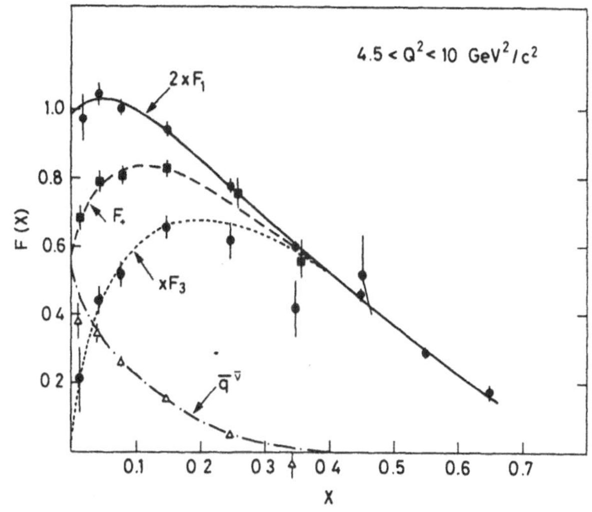

Fig. 17. Comparison of the structure functions $2XF_1$, F_+, XF_3 and $\bar{q}^{\bar{\nu}}$ for fixed Q^2 as a function of X. The curves are empirical fits to the data which fulfil the quark parton model relations between these structure functions : $2XF_1 = q + \bar{q}$, $XF_3 = q - \bar{q}$, $F_+ = q + Xs$, $\bar{q}^{\nu} = \bar{q} + Xs$ with $2s/\bar{q} = 0.4$.

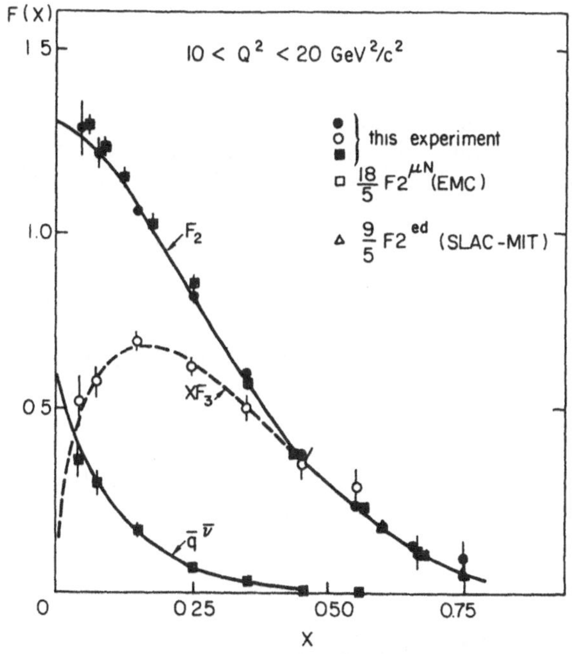

Fig. 18. Comparison of $F_2^{\mu N}$ by EMC[14] and F_2^{ed} by SLAC-MIT[15] multiplied by the quark parton model factors and CDHS structure functions[10].

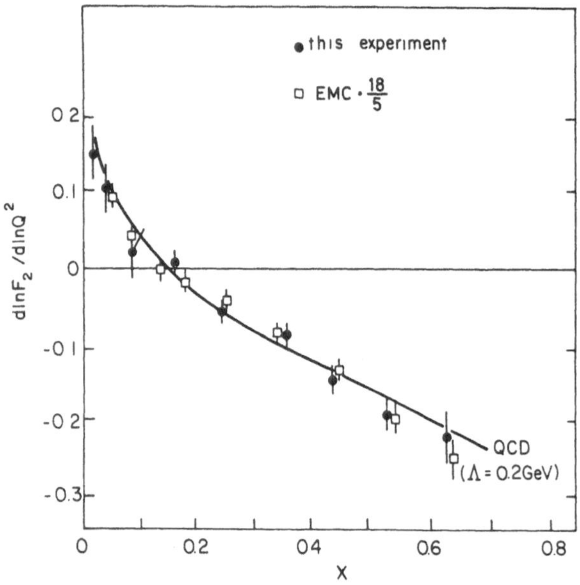

Fig. 19. The Q^2 dependence of the slope of the structure function F_2 versus X as measured by CDHS[10] compared with the results of EMC[14].

In figure 19 the variation of d ln F_2/d ln Q^2 of the structure function $F_2^{\nu N}$ and $F_2^{\mu N}$ is shown. These data show clear scaling violations and a very good agreement for both experiments. After several years of very hark work we can conclude that data on the F_2 structure function agree well in neutrino and muon scattering. The value of Λ, to the leading order in QCD should then be the same.

Longitudinal structure function : determination of $R = \sigma_L/\sigma_T$

We want to develop specially the discussion on the determination of R for two reasons :

- If R = 0 means $2XF_1 = F_2$, then the study of structure functions is reduced to two structure functions F_2 and XF_3. (We have seen that even if this relation R = 0 is not completely exact then R is used as a correction factor in the expression of F_2 and $\bar{q}^{\bar{\nu}}$).
- One can learn about the spin and the $P_\perp$ of the nucleon constituant.

In the quark parton model, R can be written as

$$R = 4\frac{(M_Q^2 + P_\perp^2)}{Q^2} ,$$

where M_Q^2 and $P_\perp$ are the mass and the transverse momentum of the quark. R=0 for massless spin 1/2 quarks and if $P_\perp = 0$.

The relation R = 0 or $2XF_1 = F_2$, called the Callan-Gross relation,

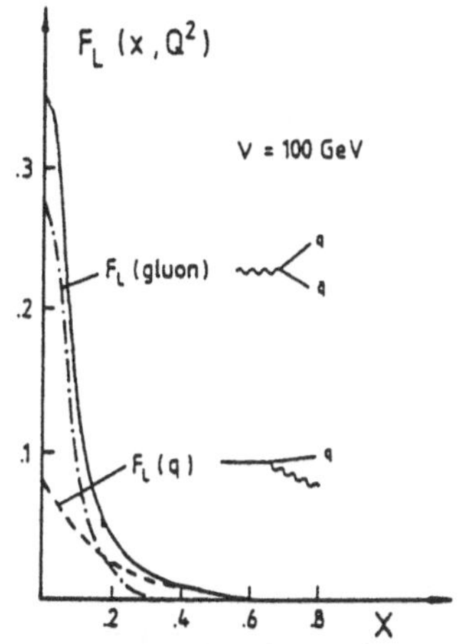

Fig. 20. QCD prediction for the longitudinal structure function F_L for ν = 100 GeV and Λ = 0.2 GeV. The contribution due to gluon emission and $g \to q\bar{q}$ are shown separately. (From F. Eisele[17]).

was derived in 1969[16] by Callan-Gross from current algebra considerations.

The longitudinal structure function $F_L(X,Q^2)$ allows an independent QCD test and therefore the study of $F_L(X,Q^2)$ can contribute to the scaling violation study.

The perturbative calculation of $F_L(X,Q^2)$ has the form

$$F^{QCD}(X,Q^2) = \frac{4\alpha_s(Q^2)}{3\pi} \int_X^1 \frac{dY}{Y} \left(\frac{X}{Y}\right)^2 F^2(Y,Q^2)$$

$$+ 6\left[\left(\frac{X}{Y}\right)^2 - \left(\frac{X}{Y}\right)^5\right] G(Y,Q^2) .$$

The different contributions from these two terms are shown on figure 20[17] and one sees that the one which involves the gluon distribution is the most important. A precise $F_L(X,Q^2)$ determination should then be very important to the determination of the gluon.

The problem is that the experimental distribution of R is very difficult to obtain ...

Analysis of the Y dependence. One recalls : $F_L = F_2 - 2XF_1$ and

$$R = \frac{\sigma_L}{\sigma_T} = \frac{F_2}{2XF_1}\left(1 + \frac{Q^2}{\nu^2}\right) + \frac{Q^2}{\nu^2} .$$

If one sums the neutrino and antineutrino differential cross sections the term XF_3 disappears :

$$\frac{d\sigma(\nu+\bar{\nu})}{dXdY} = \frac{G^2\, M\, E}{\pi}\,[(1 + (1-Y)^2)F_2 - Y^2F_L + 2X(s-c)(1 - (1-Y)^2)].$$

The term $2X(s-c)(1 - (1-Y)^2)$ being considered as a correction term for simplicity we will drop it and the Q^2/ν^2 term. Hence,

$$\frac{d\sigma(\nu+\bar{\nu})}{dY} = \frac{G^2\, M\, E}{\pi} \int_0^1 F_2(X,Q^2)\,[1 + (1-Y)^2 - RY^2]\, dX . \qquad (10)$$

QCD predictions for F_L or R are known in function of Q^2 and X. Usually analyses of R_ν were made from a general fit on Y distributions integrated over E_ν and Q^2. This mainly because of statistical problems. But since large Y corresponds to large Q^2 and similarly small Y to small Q^2 the result is affected by a priori unknown Q^2 dependence of the structure function. One could correct for this effect as measured in the structure function determination, but ones uses in this analysis the fact that $R_\nu = 0$ so this method is circular. To overcome the problem of the Q^2 variable one still has to make the choice of a variable other than Q^2. From equation (10) one sees that :

- One would like to see the Y distribution - on which one makes the fit - with a level arm in Y as complete as possible.
- One would like the expression $\int_0^1 F_2(X,Q^2)dX$ be independent of Y, to let the only Y dependence be

$$(1 - (1-Y)^2 - RY^2).$$

These two arguments cannot be fulfilled because of the correlation between the Q^2, X, Y variables which is obvious in figure 9. It is why the variable ν is more convenient :

- $\int_0^1 F_2(X,\nu)dX$ is Y independent.

- all the X distribution is seen but the Y distribution is not complete and the level arm varies with ν. (Figure 21a).

These arguments are clear enough to show the experimental difficulties to determine R. This is emphasized if on figure 21b one sees how small the differences are between two values of R or a scaling violation on the Y distribution. This depletion of events at large Y due to the experimental cut-off on momentum of the muon

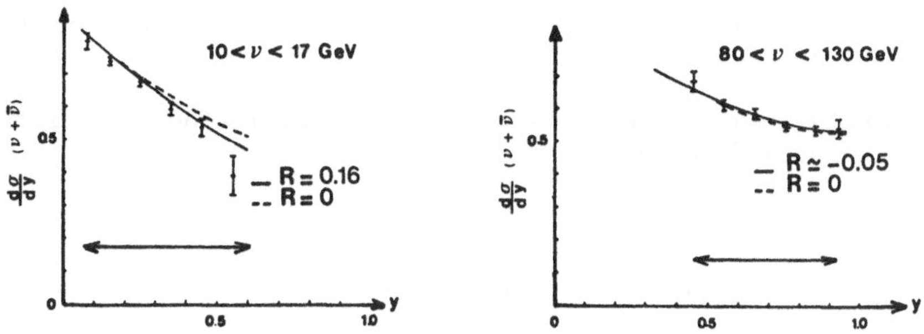

Fig. 21a. Fit on Y distribution to evaluate $R = \sigma_L/\sigma_T$ for two ranges of E_H.

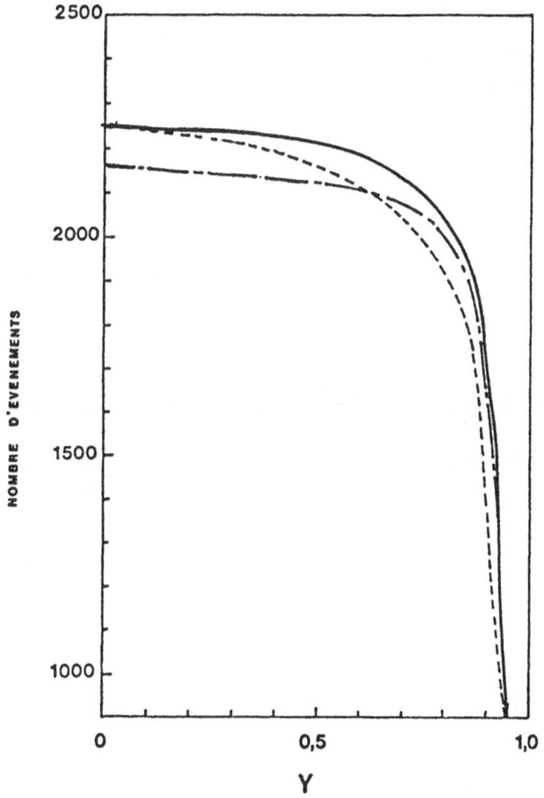

Fig. 21b. Y Monte Carlo distribution.
——— Monte Carlo without scaling violations and R = 0,
----- Monte Carlo without scaling violations with R = 0.2,
— - — Monte Carlo with scaling violations and R = 0.

(P_μ > 5 GeV) complicates the problem and contributes to the reduction of the level arm on Y.

Besides this unavoidable restriction a careful study of systematic errors is needed in such an analysis :

. strange sea $\frac{2s}{q - \bar{q}}$ = 0.04 or 0.065 $\Delta R = 0.02$

. radiative correction : without radiative correction $\Delta R = -0.05$

. other systematics error : ratio (K/π) - 3 % $\Delta R = +0.04$
calibration - 2 % $\Delta R = -0.03$

Still large systematic error will be present in the results.

As a first step the ν dependence of R after summing the data in X is determined and presented in figure 22[18]. In a second step the X dependence of R can be determinated but one has to assume a Q^2 independent value of R for every X bin (Fig. 23). Since no strong dependence on X or Q^2 seems present, the results are presented on an averaged value of R :

$$\langle R \rangle = 0.10 \pm \underset{\text{stat}}{0.025} \pm \underset{\text{syst}}{0.06} \cdot$$

<u>Upper limit on R at large X.</u> Using the exact formulas of neutrino and antineutrino scattering on isoscalar target one can derive the following expressions :

$$\frac{d^2\sigma(\bar{\nu})}{dXdY} - (1-Y)^2 \frac{d^2\sigma(\nu)}{dXdY} = \frac{G^2 M\, E_\nu}{\pi} \{\frac{1}{2}\, [1-(1-Y)^4][2XF_1 - XF_3 + 2X(s-c)]$$
$$+ q_L\, [(1-Y)-(1-Y)^3] - [(1-(1-Y)^2)]\; 2X(s-c)(1-Y)^2$$
$$+ (\frac{Y^4}{4} - \frac{Y^3}{2})\, \frac{Q^2}{\nu^2}\, F_2\},$$

$$\frac{d^2\sigma(\nu)}{dXdY} - (1-Y)^2 \frac{d^2\sigma(\bar{\nu})}{dXdY} = \frac{G^2 M\, E_\nu}{\pi} \{\frac{1}{2}\, [1-(1-Y)^4][2XF_1 + XF_3 + 2X(s-c)]$$
$$+ q_L\, [(1-Y)-(1-Y)^3] - [1-(1-Y)^2]\; 2X(s-c)(1-Y)^2$$
$$+ (\frac{Y^4}{4} - \frac{Y^3}{2})\, \frac{Q^2}{\nu^2}\, F_2\, \}.$$

The definitions :

$$\bar{q}^{\bar{\nu}} + \frac{Q^2}{4\nu^2}\, XF_3^{\bar{\nu}} = \frac{1}{2}\, [2XF_1 - XF_3 + 2X(s-c)]\ ,$$

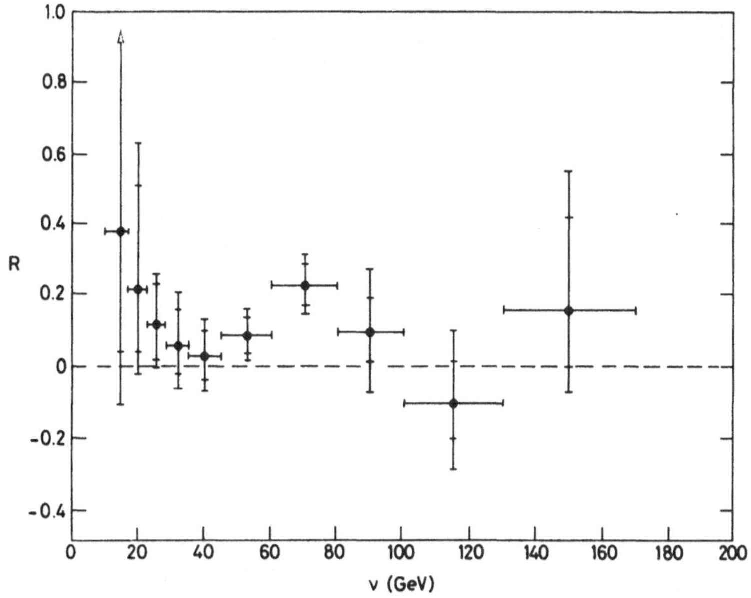

Fig. 22. R averaged over X as function of ν. Inner bars are statistical errors.

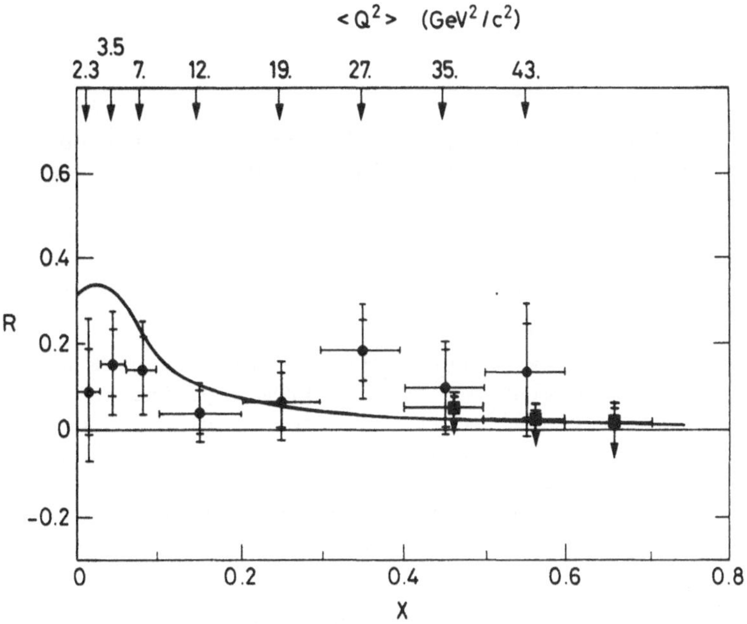

Fig. 23. R averaged over ν as a function of X. The curve shows a QCD prediction for a gluon distribution ≅ $(1-X)^5$ and calculated for ν=50 GeV[18]. The square data are limits on R from WBB data.

$$\bar{q}^{\bar{\nu}} + \frac{Q^2}{4\nu^2} XF_3^{\nu} = \frac{1}{2}\,[2XF_1 - XF_3 - 2X(s-c)]\ ,$$

$$2XF_1 \geqq XF_3 \left(1 + \frac{Q^2}{\nu^2}\right)^{1/2}\ ,$$

ensure $\bar{q}^{\bar{\nu}}$ and $q^{\bar{\nu}} \geqq 0$.
At large $Y \geqq 0.5$ and since experimentally for $X \geqq 0.4$ the term

$$\left(\frac{d^2\sigma^{\bar{\nu}}}{dXdY} - (1-Y)^2\,\frac{d^2\sigma^{\nu}}{dXdY}\right)$$

is compatible with 0 (see the determination of the right handed weak current) one can derive the expression :

$$\frac{1}{(1-Y)-(1-Y)^3}\left(\frac{\dfrac{d^2\bar{\sigma}}{dXdY} - (1-Y)^2\dfrac{d^2\sigma}{dXdY}}{\dfrac{d^2\sigma}{dXdY} - (1-Y)^2\dfrac{d^2\bar{\sigma}}{dXdY}}\right) \geqq \frac{R - \dfrac{Q^2}{2\nu^2}}{1 + \dfrac{Q^2}{2\nu^2}\left((1-Y) - (1-Y)^3\right)}\ ,$$

form which one obtains for $0.4 \leqq X \leqq 0.7$ the result

$$R \leqq 0.039 \pm \underset{\text{stat}}{0.014} \pm \underset{\text{syst}}{0.025}$$

for $\langle Q^2\rangle = 38\ \mathrm{GeV}^2/c^2$.

This method is extremely powerfull and one can see on figure 24 the improvement on errors on determination of R. Statistically this improvement is due to the use of narrow band and wide band beam data (maily for the $\bar{q}$ determination). The main systematic errors are due to the ratio $\sigma^{\bar{\nu}}/\sigma^{\nu}$. This method can be used only for large X where the antiquark contribution is small. But it is by far the best result on R in neutrino physics.

Comparison with other experiments. Several results are presented in figure 24. The SLAC-MIT results from e-p, e-d scattering measured by A. Bodek et al.[19] correspond to an $\langle\nu\rangle$ of 8 GeV and give an averaged value of $R = 0.21 \pm 0.10$. One should notice that the large value of R of the SLAC data are incompatible with the QCD expectation and the CDHS data at large X. This incompatibility can be explained if one accepts the proposed idea of diquark contributions at large X which will desapear rapidly with Q^2.

The other important results come from the μ-scattering. The determination of R in μ-scattering is very difficult and mainly necessitates different energy runs. Then problems of overlap of data normalisation of different runs induce large systematics

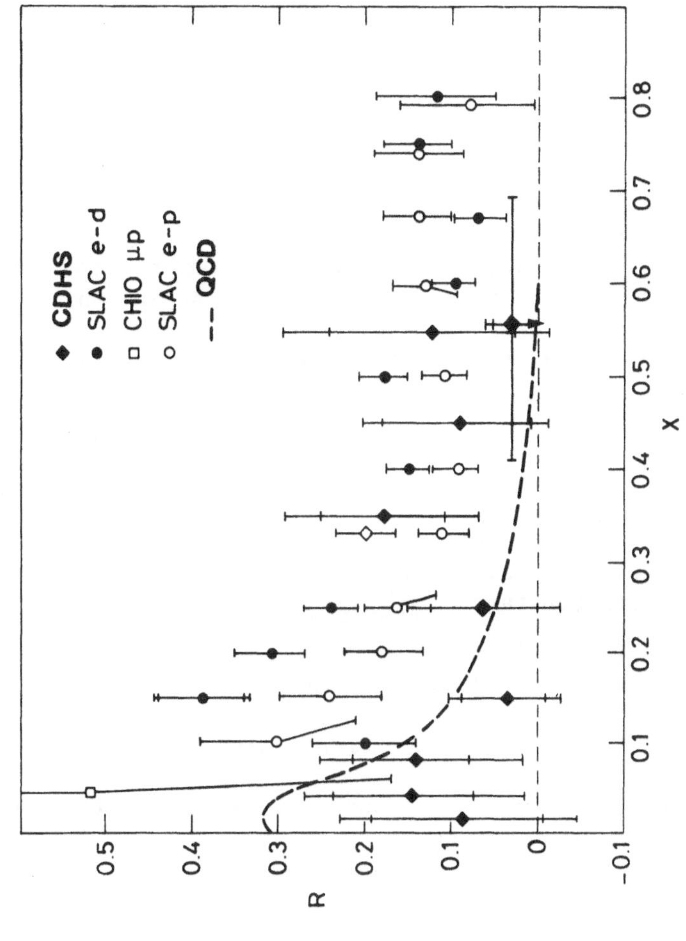

Fig. 24. R function of X for different experiments.

errors[20,21]. The X range of analysis is very limited. The old results of B.A. Gordon et al.[22] give a value of $R = 0.44 \pm 0.25$ at an averaged Q^2 of 12 GeV^2/c^2 for $0 < X < 0.10$. Recent results of the EMC group[20] for $0.05 < R < 0.4$ and $3 < Q^2 < 130$ GeV^2/c^2 in hydrogen and iron give the following value of R :

Hydrogen $\langle R\rangle = -\ 0.03 \pm 0.04 \pm 0.13$,

Iron $\langle R\rangle = -\ 0.05 \pm 0.03 \pm 0.10$,

compatible with the CDHS data.

Conclusion on R. The large systematics errors on R measurements as well in neutrino as e, μ scattering show the difficulty to measure R (this situation will not change for a while in muon scattering). The powerful method used to measure the upper limit on R in neutrino physics can be used for values of X larger than 0.35. But a good measurement of R at small X will not be obtained easily. The absolute need of R in the study of structure functions and the precise determination of these structure functions will be in conflict and the only possibility in analysis will be to see the effect on F_2 and $\bar{Q}$ of constant value of R as it is found now or the use of the QCD formula.

QCD analysis

The departure from scaling of the data are studied in the form of the QCD theory. In QCD the interaction is mediated by a massless vector gluon field. By analogy with the QED theory one can evaluate the "strong coupling constant"

$$\alpha_s(Q^2) = \alpha_s(\mu^2) \ [1 + \frac{\alpha_s(\mu^2)}{4\pi} C \ln Q^2/\mu + \ldots$$

where $C = 11 - 2f/3$, f being the number of quark flavours.

These series can be summed if one assumes the "leading log approximation" and the strong coupling constant has the form

$$\alpha_s(Q^2) = \frac{\alpha_s(\mu^2)}{[1 + b\alpha_s(\mu^2) \ln(Q^2/\mu^2)]} ,$$

$b = (33 - 2f)/12\pi$.

One usually defines $\alpha_s(Q^2)$ as

$$\alpha_s(Q^2) = \frac{12\pi}{33 - 2f} \frac{1}{(\ln Q^2/\Lambda^2)} ,$$

where $\Lambda^2 = \mu^2 \exp[1 - b\alpha_s(\mu^2)]$ is the parameter of the theory which has to be determined experimentally.

This parameter Λ has not a unique definition and depends on the normalization scheme choosen.

In most analyses one defines the parameter Λ evaluated in leading order (Λ_{LO}) and Λ_{MS} (minimum scheme) for second order correction[23]. Because of the difficulty to extract Λ from the Altarelli-Parisi equation most experimental groups using the same numerical programmes use this Λ_{MS}.

From this beautiful QCD theory one cannot extract at present an analytic form of the structure functions. The Altarelli-Parisi equations predict the Q^2 evolution of the slope $dF_i/d\ln Q^2$ of the structure functions.

$$\frac{dXF_3(X,Q^2)}{d\ln Q^2} = \frac{\alpha_s(Q^2)}{2\pi} \int_X^1 P_{qq}\left(\frac{X}{Z}\right) ZF_3(Z,Q^2) \frac{XdZ}{Z^2},$$

$$\frac{dF_2(X,Q^2)}{d\ln Q^2} = \frac{\alpha_s(Q^2)}{2\pi} \int_X^1 \left[P_{qq}\left(\frac{X}{Z}\right) F_2(Z,Q^2) + 2N_f P_{gq}\left(\frac{X}{Z}\right) G(Z,Q^2)\right] \frac{XdZ}{Z^2},$$

P_{ij} = "splitting" functions, calculated by QCD,
$G(X,Q^2)$ = gluon structure function,
N_f = number of active flavors which contribute to the color change.

Different combinations of quarks inside the nucleon lead to non-singlet or singlet structure functions. Following their transformation properties under flavour one can define :

$Q_j + \bar{Q}_j$ singlet flavour,

$Q_j - Q_i$ non singlet flavour ($i \neq j$).

Apart from the knowledge of the normalization scheme used in the analysis one must be careful to other aspects of definition or corrections :

- Target mass corrections[24] - for small values of Q^2 or W^2 the nucleon mass cannot be neglected.
- Number of flavours entering in the analysis : at a neutrino energy of 200 GeV far above charm threshold 4 flavours have to be considered.
- Cut-off on low invariant hadronic mass W^2. At too low W^2 one may expect non-perturbative contributions (we shall see in detail the higher twist effects). As a consequence, the cut-off on W^2 should be

be specified in the QCD analysis. One uses $W^2 > 11\ GeV^2$ and $Q^2 > 2\ GeV^2/c^2$ cut-offs in high energy sample data.

- Last, a parametrization of structure functions is needed to study their Q^2 evolution. In the CDHS[10] analysis, which we will consider in what follows, one has used :

$$XF_3(X,Q_\circ^2) = a_3(1 + b_3X)(1 - X)^{c_3} ,$$

$$F_2(X,Q_\circ^2) = a_2(1 + b_2X)(1 - X)^{c_2} ,$$

$$G(X,Q_\circ^2) = a_g(1 + b_gX)(1 - X)^{c_g} ,$$

$$\bar{Q}^{\bar{\nu}}(X,Q_\circ^2) = a_Q(1 - X)^{c_Q} ,$$

$$a_3 \rightarrow \int_0^1 F_3\ dX = 3\left(1 - \frac{\alpha_s(Q^2)}{\pi}\right) ,$$

$$a_g \rightarrow \int_0^1 G\ dX + \int_0^1 f_2\ dX = 1 .$$

We shall separate the QCD analysis of structure functions in three parts :

- non-singlet structure function XF_3, moment analysis, study of the slope, Λ from Altarelli-Parisi equation,
- F_2 and $\bar{Q}^{\bar{\nu}}$ - Λ from general Altarelli-Parisi equation, gluon distribution,
- what can we say on higher twist ?

Non-singlet structure function XF_3

The function XF_3, which is specific to the neutrino field and which can be obtained only in neutrino scattering, has several advantages :

- It is a non-singlet structure function,
- The gluon distribution is not needed in the study of the slope evolution,
- $R = \sigma_L/\sigma_T$ does not enter in the formula (7) used to extract XF_3 ,
- Very simple formulas exist for XF_3 in the moment analysis.

These good points are counter-balanced by the problem of poorer statistics since XF_3 is proportional to $d\sigma(\nu) - d\sigma(\bar{\nu})$ and that $\bar{\nu}$ samplings are smaller that ν ones.

i) Moment analysis of XF_3. Because of the simplicity of the QCD predictions of the Q^2 evolution on moments of the XF_3 non-singlet structure functions it was very fashionable and theoretically easy around 1977 to test the QCD theory by this method. One recalls the two "simple" predictions on XF_3 moments. One defines the moment N, the "Normal moment" or the CORNWALL-NORTON moments.

$$M_i(N,Q^2) = \int_0^1 X^{N-2} F_i(X,Q^2)\, dX ,$$

$$L.O. \rightarrow M_3(N,Q^2) = M_3(N,Q_\circ)\, e^{-d_N S} ,$$

$$d_N = \text{ANOMALOUS DIMENSION} = \frac{4}{33 - 2f}\left[1 - \frac{2}{N(N+1)} + 4 \sum_2^N \frac{1}{J}\right],$$

f : number of flavors,

$$S = \text{Ln}\,[\text{Ln}\, Q^2/\Lambda^2/\text{Ln}\, Q_\circ^2/\Lambda^2] .$$

1/ Plot of log M_i versus log M_j. If one considers the logarithms of two moment i and j we can write :

$$\frac{\log M_i(Q) - \log M_i(Q_\circ)}{\log M_j(Q) - \log M_j(Q_\circ)} = \frac{d_i}{d_j}$$

This ratio is dependent of N and m and is calculated by QCD theory. It depends on the scalar or vector character of the gluon.

2/ Variation of moments with Q^2 : determination of Λ. The second test of QCD on XF_3 is simply understood if one considers the formula

$$M_i^{-1/d_i} = \text{const}(\ln Q^2 - \ln \Lambda^2) .$$

- The linear dependence in ln Q^2 of this expression is well verified as shown in figure 26.
- The intercept of these lines with zero must be obtained for $Q^2 = \Lambda^2$ and this intersection allows one to measure Λ.

This extrapolation is very sensitive to mass corrections, the quark correction, the quark number and the second order correction.

These two simple predictions were in fact very quickly submitted to experimental as well as theoretical difficulties[26]. One would like to summarize them :

- What scaling variable ?
  To take care of the target mass effect : mass of the proton not negligible compared to the value of Q^2 considered, Nachtmann has proposed the variable

$$\xi = \frac{2X}{1 + \left(1 + \frac{4M^2X^2}{Q^2}\right)^{1/2}} ,$$

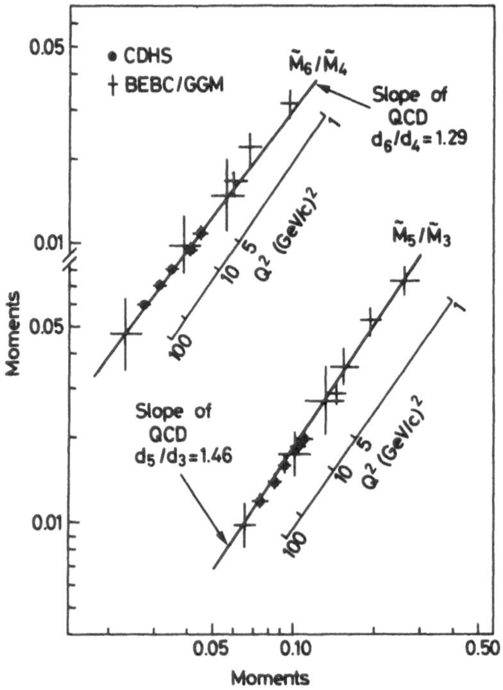

Fig. 25. Nachtmann moments from CDHS and BEBC/CGM data in log-log plot.

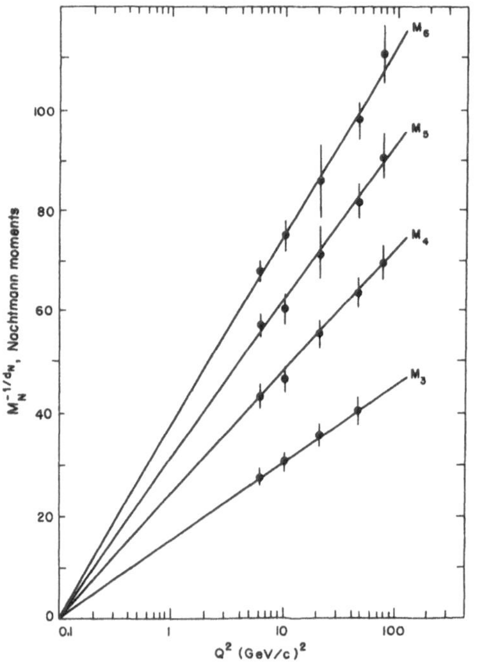

Fig. 26. Nachtmann moments as function of Q^2 for CDHS data.

and to calculate "Nachtmann moments" with this variable. One sees clearly that these two variables become identical for large Q^2 : $\xi \to X$. But QCD is "surprisingly" good at relatively small Q^2 and on the other hand this validity of Nachtman moments is sometimes discussed among theoreticians themselves. Cornwall-Norton or Nachtman moments ? The best for experimentalists is to analyse the data with both types of moments and to see the effects on the results.

- What data can be used, what cut-off on Q^2 ?
  This question was very important and was connected with the question of the cut-off on Q^2 in the QCD theory. Can we test QCD at Q^2=1 GeV^2/c^2 or less ?
  The characteristics of the data are specific to each experiment. For instance the SLAC ep scattering experiment detects elastic events and resonances and can reach a very small value of Q^2 and W. In the CDHS experiment for instance the cut-off which is applied $Q^2 > 6\ GeV^2/c^2$ and $X < 0.7$ excludes completely the region W > 2 GeV: i.e., all quasi-elastic and resonance region. This cut-off on Q^2 was not so important in the BEBC data and a fortiori in the GGM data. For any comparison of experiments, this experimental point must be very clear.
- What number of flavors to take in calculating α_s?
  In the calculation of the anomalous dimension it enters the number of flavors f. For a given energy two different experimental teams will not take the same value of f - is the charm fully developed or not, f = 3 or 4 - and then the comparison of data is not correct.
- What fraction of the moments is seen by each experiment?
  This figure 27 from R.M. Barnett[27] shows the part of the XF_3 distribution measured by CDHS. One has already mentioned the impossibility for one experiment to measure all X distribution for all Q^2. This is a good illustration.
  What can be done ?
  1/ Most of the analyses on moments are done with two sets of data from two different experiments: SLAC - CDHS
     GGM - BEBC
  2/ As can be seen in figure 23, this is not sufficient and one must parametrize XF_3 and make some extrapolation for $X \to 0$ and $X \to 1$ (Question of elastic events ...). This increases the errors on the moments. Mainly the extrapolation for $X \to 1$ is very sensitive for high moments and can reach a 25 % correction.
- Errors on moments
  We give two examples of errors on moments to help to compare this high error due to the extrapolation for $X \to 1$ to other errors that one usually has to deal with in this moment calculation.

Besides these errors one should stress the strong correlation which exists between moments. At large N this correlation is very important since the same measurements of the function XF_3 are used and only multiplied by a power of X. Then correlations should be

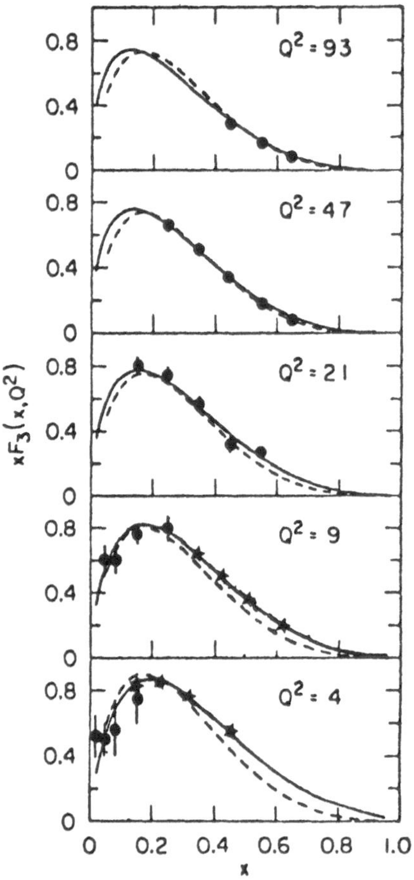

Fig. 27. $XF_3(X,Q^2)$ at different Q^2 values. The black-points are the CDHS data, the stars are the SLAC data. The curves are a fit from R.M. Barnett[27] for QCD, leading order and higher twist only.

N = 2	$Q^2 = 6.5$	N = 5	$Q^2 = 45$
NORTON	0.385	NORTON	0.0143
NACHTMANN	0.376	NACHTMANN	0.0140
STAT	± 0.017	STAT	± 0.0006
SLAC → CDHS	0.010	SLAC → CDHS	-
FERMI MOTION	-	FERMI MOTION	0.0008

correctly taken into account in the analysis.

All these restrictions on moment analysis have reduced the enthousiasm of experimentalists. Since the first moment analysis made around 1977[27] a new analysis of non-singlet moments has been done by A. Para[28] using new CDHS data[10] and old SLAC ed data[15,19]. Figure 28 shows the "slope plot" of the test number mentioned pre-

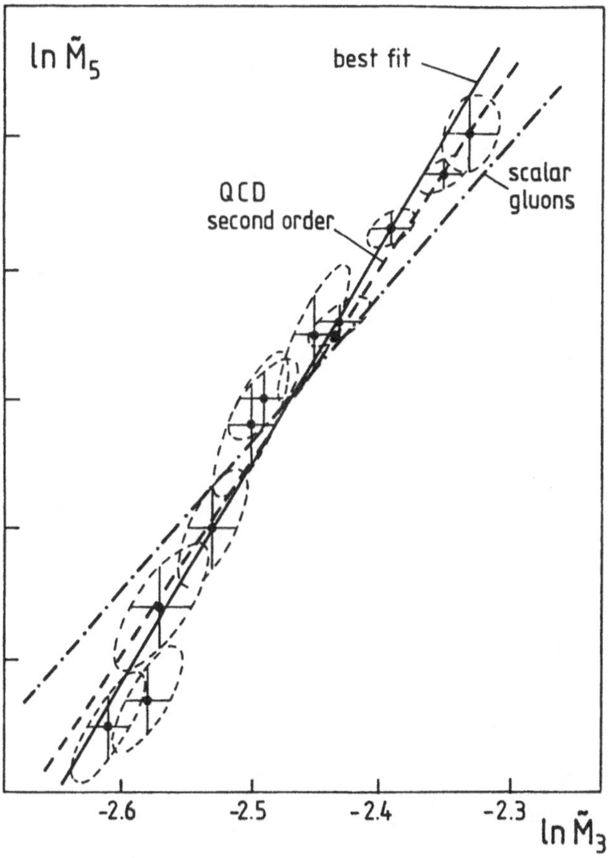

Fig. 28. Logarithm of Nachtmann moments M_3, M_5 of XF_3.

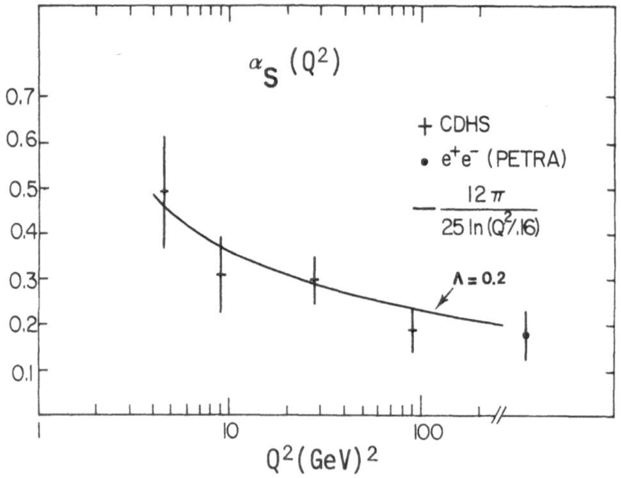

Fig. 29. $\alpha_s(Q^2)$ as obtained from the observed slopes $dXF_3/d\ln Q^2$ from CDHS compared to the QCD expectation for Λ = 0.2 GeV.

viously for the Nachtmann moments N_3 and N_5. The measured slope d_5/d_3 = 1.68 ± 0.11 agrees with the QCD second order prediction (vector gluons) and excludes scalar gluons.

A result on the second test can provide a value of Λ. Including all correlations a combined fit on M_3, M_4, M_5 gives the result :

$$\Lambda_{LO} = 0.3 \pm 0.1 \ ,$$

$$\Lambda_{\overline{MS}} = 0.25 \pm 0.08 \ .$$

These results are as a matter of fact quite good and the only point is to handle correctly this analysis and to know its limitation. Specialists claim that the moment analysis was suddenly non interesting. It is limited but these tests on moments of non singlet functions must still be calculated with care. A more complete analysis is done on the full X-dependence.

ii) Fits to XF_3 on full X-dependence.

a) Slope of XF_3. If one considers the Altarelli-Parisi equation for XF_3 :

$$\frac{dXF_3(X,Q_o^2)}{d\ln Q^2} = \frac{\alpha_s(Q_o^2)}{2\pi} \int_X^1 P_{qq}\left(\frac{X}{Z}\right) ZF_3(Z,Q_o) \frac{XdZ}{Z^2} ,$$

experimentally, one can measure the slope $dXF_3/d\ln Q^2$ and we measure XF_3. The term under the integral can be calculated, P_{qq} has been given by QCD. $\alpha_s(Q^2)$ then can be determined. Figure 29 shows $\alpha_s(Q^2)$ obtained by this method. The data are consistent with the QCD prediction, but a strong conclusion on the "running α constant" cannot be given. The determination of Λ from this type of analysis is still poor.

b) Fit on X. The Q^2 evolution of XF_3 can be studied from the Altarelli-Parisi equation. Numerical solutions of these equations have been worked out by different groups. The CDHS collaboration have used programmes developed by Abbott and Barnett and by Lopez and Yndurain for the second order calculations.

The result is

$$\Lambda_{\overline{MS}} = 0.2 \begin{array}{l} + 0.2 \\ - 0.1 \end{array} \text{GeV},$$

including the systematic uncertainties. Although this result is free from assumptions already mentioned it is still not very precise. The difference between $\Lambda_{\overline{MS}} = \Lambda_{LO}$ is indistinguishable.

To improve the statistical significance of these results one can analyse the structure function F_2 at large X where one can neglect the gluon contribution. This will be discussed in the next chapter. But the best solution will be to obtain better result on XF_3, it means more events mainly in antineutrino data. This can be accomplished using data from wide band beam antineutrinos. Systematic errors from such data will come from the relative normalization of neutrino and antineutrino runs. The necessity to know very precisely the cross section is there stressed again.

Study of F_2 and $\bar{Q}^{\nu}$: gluon distribution. The structure function F_2 is very precisly known statistically, but the analysis implies the knowledge of $R = \sigma_L/\sigma_T$, the sea contribution. To improve the statistical accuracy one can analyse the singlet structure function F_2 for large X as a non-singlet structure function.

i) Non-singlet analysis of F_2 at large X .

One can write :

$$F_N(X,Q^2) = F_2(X,Q^2) - 2\,[\bar{Q}^{\bar{\nu}} - X\,s(X,Q^2)]\ ,$$

and for $X > 0.3$ the correction is small. On the other hand R is well bounded at large W as we have seen. The effect of different assumptions on R is seen in the Table I[10].

We see from table I that the value of $R = 0.1$ which is usually accepted resulting from the R analysis gives a substantial different value of Λ, 0.21 instead of 0.28 for $R = 0$, or $R = R_{QCD}$ which is

Table I

Fit		Λ_{LO}	χ^2/DF
Standard fit	$R = R_{QCD}$	0.279 ± 0.09	48/49
	$R = 0.0$	0.295	48/49
	$R = 0.1$	0.21	53/49
	$R = 0.2$	0.10	58/49

almost 0 at large X. This difference (0.07) is comparable to the statistical error (0.09).

The result using the R_{QCD} value gives (statistical errors only)

$$\Lambda_{LO} = 0.275 \pm 0.08 \ ,$$

$$\Lambda_{\overline{MS}} = 0.30 \ \ \pm 0.08 \ .$$

These values are in agreement with the non-singlet function XF_3 analysis.

ii) Determination of the gluon distribution by analysis of F_2 and $\bar{q}^{\nu}$.

In principle the Q^2 evolution of the structure function F_2 gives the gluon distribution G(X) due to the QCD effect.

$$\frac{dF_2(X,Q^2)}{d\ \ln Q^2} = \frac{\alpha_s(Q^2)}{2\pi} \int_X^1 [P_{qq}(\tfrac{X}{2})F_2(Z,Q^2) + 2N_f\ P_{gq}(\tfrac{X}{Z})G(Z,Q^2)]\ \frac{XdZ}{Z^2}\ . \tag{11}$$

But the measurement of F_2 alone does not give a good determination of the gluon distribution and Λ. A strong correlation exists between the two. This can be seen in an analysis of F_2 in figure 30. We can also understand that by looking at the figure 31 which represents the slope of the structure function F_2. If α_s increases the contribution to the slope of $dF_2/d\ \ln Q^2$ of the term $P_{qq} \otimes F_2$ will become broader but this effect can be compensated by a broader positive contribution of the term $P_{gq} \otimes G(X)$ (gluon distribution).

An independent measurement when this effect is not so strongly correlated can decouple the gluon contribution to α_s (or Λ). This is the case of the structure function $\bar{q}^{\nu}$ (Fig. 31). Particularly

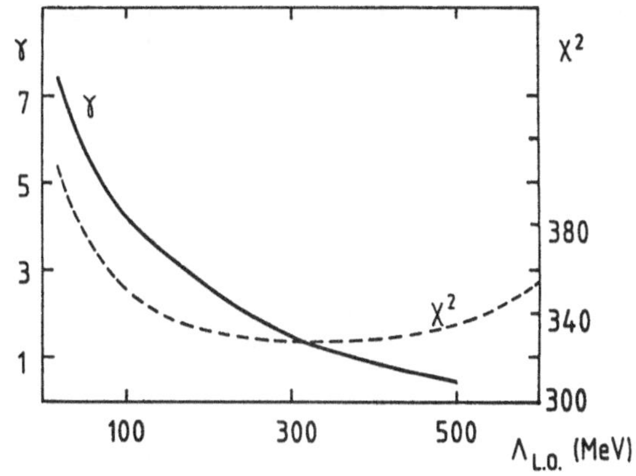

Fig. 30. Correlation between $\Lambda_{L\,O}$ and the gluon distribution G(X) parametrized as $G(X) = c/X(1-X)^{\gamma}$. From P. Payre (Thèse d'Etat 1983)

one sees that the correlation $P_{gq} \otimes G(X)$ is bound to be zero when no antiquarks are observed.

This analysis has been done by CDHS[29] by simultaneous numerical integration of equations (11), (12) and (13) :

$$\frac{d\bar{Q}(X,Q^2)}{d\ln Q^2} = \frac{\alpha_s(Q^2)}{2\pi}\int_X^1 [P_{qq}(\tfrac{X}{Z})\bar{Q}(Z,Q^2) + N_{\bar{q}}P_{gq}(\tfrac{X}{Z})G(Z,Q^2)]\,\frac{XdZ}{Z^2} \quad ,(12)$$

$$\frac{dG(X,Q^2)}{d\ln Q^2} = \frac{\alpha_s(Q^2)}{2\pi}\int_X^1 [F_2(Z,Q^2)P_{qg}(\tfrac{X}{Z}) + P_{gq}(\tfrac{X}{Z})G(Z,Q^2)]\,\frac{XdZ}{Z^2} \quad .(13)$$

The structure functions are parametrized for $Q^2 = Q_0^2$ as :

$$F_2(X,Q_0^2) = a_2\ (1 + b_2X)(1 - X)^{c_2} \ ,$$

$$G\ (X,Q_0^2) = a_g\ (1 + b_gX)(1 - X)^{c_g} \ ,$$

$$\bar{q}\ (X,Q_0^2) \doteq a_q\ (1 - X)^{c_q} \ .$$

The distribution $G(X,Q^2)$ cannot be compared with data and the parametrization choosen allows large variations in shape at small X.

Table II. Results of LO QCD-fits to different sets of structure functions. Target mass correction are included. $Q^2 > 2$ GeV^2/c^2, $W^2 > 11$ GeV^2.

Structure function	Assumptions about s(x), c(x)	Fit results	χ^2/DF
I) F_2, $\bar{q}$ all x	No threshold effect. $2(s-c)/(\bar{u}+\bar{d}) = 0.4$.	$\Lambda_{LO} = (0.18 \pm 0.02)$ GeV $G(x, Q_0^2 = 5) = 2.62(1+3.5x) \cdot (1-x)^{5.9\pm0.5}$	209/196
II) F_2' $\bar{q}$ for x > 0.3	$c(x) = 0$ for $Q^2 = 1$ GeV^2/c^2. $2s/(\bar{u}+\bar{d}) = 0.5$. Slow rescaling with $m_c = 1.5$ GeV.	$\Lambda_{LO} = (0.20 \pm 0.02)$ GeV $G(x, Q_0^2 = 5) = 2.86(1+2.8x) \cdot (1-x)^{6.3\pm0.8}$	166/137
III) F_2' $\bar{q}$ for x > 0.3	$c(x) = 0$ for $Q^2 = 1$ GeV^2/c^2. $2s/(\bar{u}+\bar{d}) = 1$. Slow rescaling with $m_c = 1.8$ GeV.	$\Lambda_{LO} = (0.206 \pm 0.02)$ GeV $G(x, Q_0^2 = 5) = 2.98(1+2.84) \cdot (1-x)^{6.65\pm0.9}$	173/137

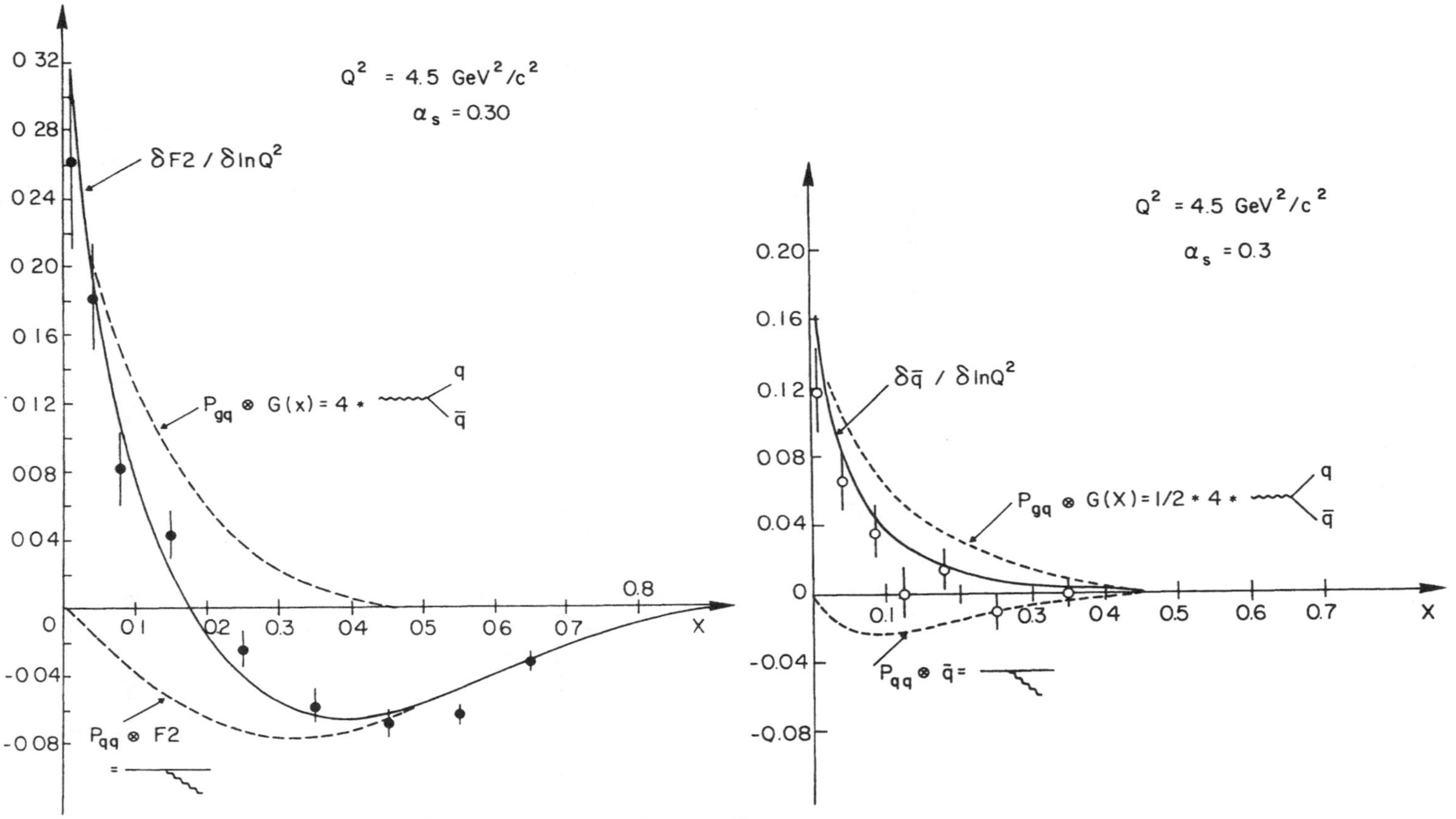

Fig. 31. Slope for the structure function $F_2^{\nu N}$ and $\bar{q}^{\bar{\nu}}$. The dashed lines show the contributions due to gluon bremsstrahlung and gluon pair production : the solid line is the sum of both contributions[29].

Table III

Structure function	Λ(GeV)	$F_i(X,Q_o^2)$	χ^2/DF
F_2 (X>0.03) $\bar{q}^{\bar{\nu}}$ (X>0.3) $R = R_{QCD}$ Corrected for slow rescaling	Λ_{LO}=0.29±0.03	$F_2=1/18(1+3.27X)(1-X)^{3.12}$ $\bar{q}^{\bar{\nu}}= 0.53(1-X)^{7.12}$ $G=1.75(1+8.9X)(1-X)^{6.03}$ $Q_o^2 = 5\ GeV^2/c^2$	136/130

The uncertainties on F_2 and $\bar{q}^{\bar{\nu}}$ at small X (i.e. on the gluon distribution) come from R, the strange sea, and the charm threshold effect. On figure 32 the effects on the slope of F_2 are shown. On table II the variations of these two last effects are shown assuming a constant value of R = 0.1.
After the new upper limits on R at large X have been obtained a second analysis was done and the results are presented in Table III.[10]

The gluon distribution from table III is broader compared to fit with R = 0.1 and Λ_{LO} increases to 0.29 GeV. One sees again the necessity of a good measurement on R for this analysis.

On figure 33 the shapes of F_2, $\bar{q}$ and G(X) are shown as they come out from this analysis at two values of Q_o^2.

A possible parametrization of the function $F(X,Q^2)$ although utilized with no specific theoretical basis can be the following :

$$G(X,Q^2) = a(1 + bX^c)(1-X)^d(Q^2)^e + fe^{-(gX+hX^3)},$$

with the values of the parameters :

$$a = 2.616 + 3.99\ s + 4.46\ s^2\ ,$$

$$b = 3.5 - 6.83\ s + 80\ s^2\ ,$$

$$c = 1 + 0.80\ s\ ,$$

$$d = 5.9 + 40\ s + 84.2\ s^2 - 64\ s^3\ ,$$

$$e = (-\ 0.033 - 0.28\ X + 59.1\ X^2)(s - s^2)\ ,$$

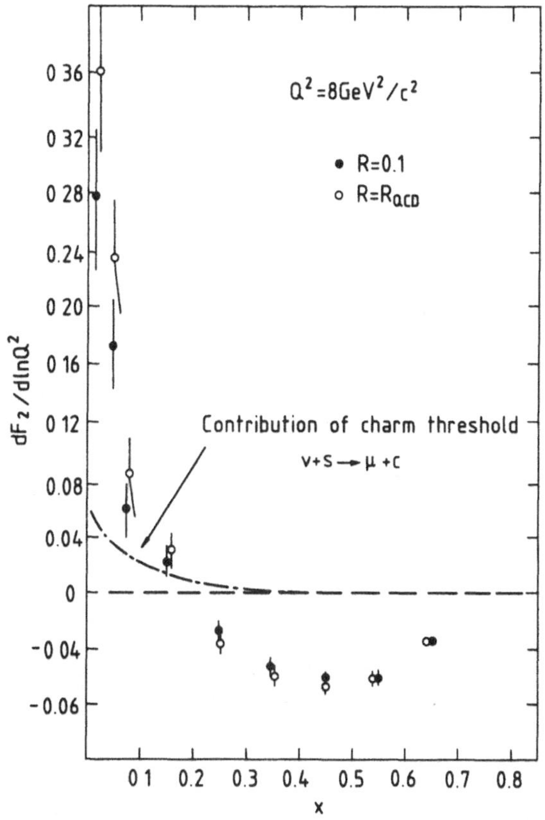

Fig. 32. Slopes $dF_2/d \ln Q^2$ for the structure function $F_2^{\nu N}$ as obtained from linear fits to the data in ln ln Q^2 at fixed Q^2 and two assumptions about $R = \sigma_L/\sigma_T$.
The contribution of charm threshold (solid curve) is calculated by the slow rescaling model.

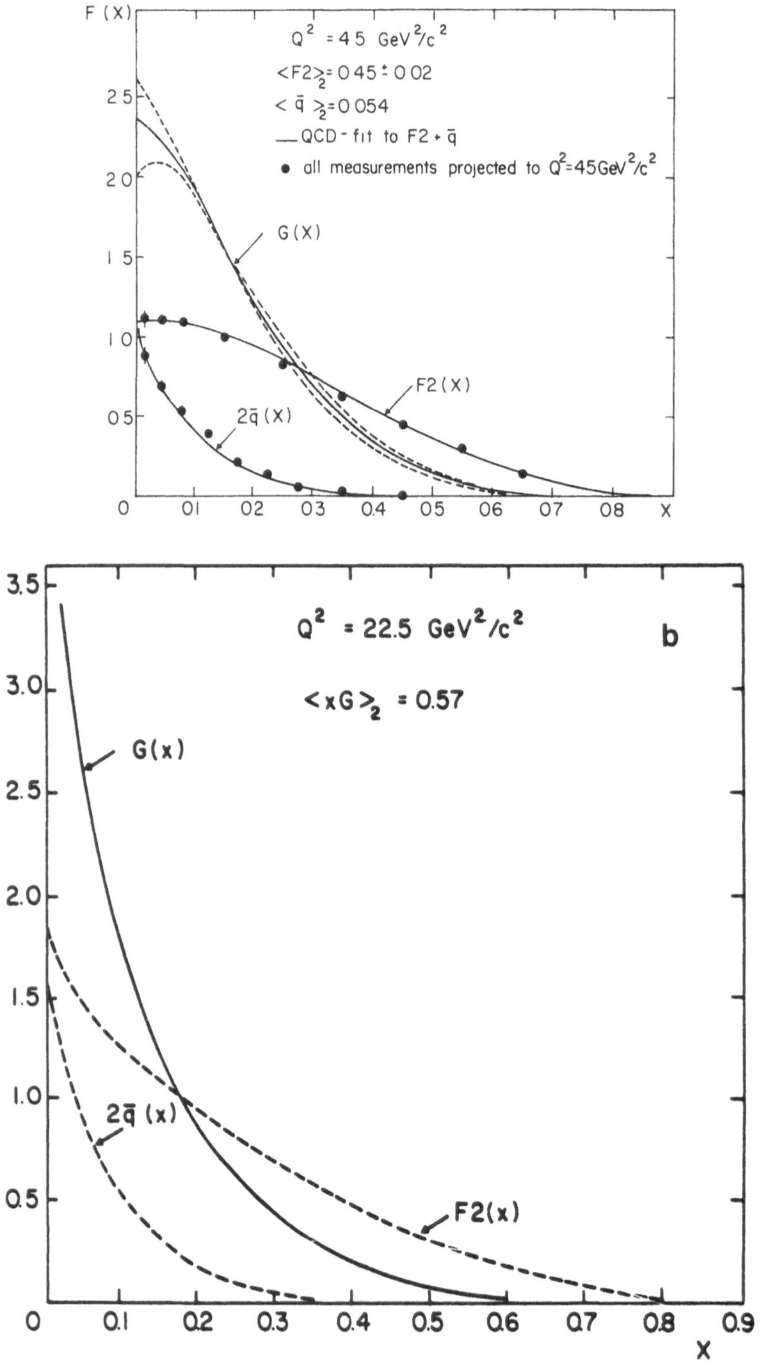

Fig. 33. Gluon distribution G(X), $F_2(X)$ and $2\bar{q}(X)$ for fixed Q^2 as obtained from leading-order QCD fit to F_2 and $\bar{q}^{\bar{\nu}}$,
a) $Q^2 = 4.5\ GeV^2/c^2 \pm 1\sigma$ bands for G(X),
b) $Q^2 = 22.5\ GeV^2/c^2$.

$$f = 5.16\ s - 0.955\ s^2 ,$$

$$g = - 0.48 + 12.2\ s + 0.38\ s^2 ,$$

$$h = 29.72 - 32.4\ s ,$$

with

$$s = \ln \frac{\ln(Q^2/0.04)}{\ln(5/0.04)} .$$

Although we should retain the uncertainties in the analysis the gluon distributions are reliable. It is a very good result from the stucture functions analysis in neutrino physics.

If one considers the overall fit on F_2, $\bar{q}^{\nu}$ which gives the gluon distribution, the value of Λ_{LO} obtained is $\Lambda_{LO} = 0.29 \pm 0.03$. This error is statistical only and the systematic errors are of the same order as in the previous analysis since the uncertainties have the same origin. One can then conclude $\Lambda_{LO} = 0.30 \pm 0.15$ in good agreement with the non-singlet F_2 study.

Higher twist contribution. In all the analyses that we have presented we have assumed that all the data were used for $W^2 > 11\ GeV^2$. This cut-off has its justification in the fact that non perturbative contributions to scaling violations may not be negligible. It is connected to the problem of higher twist.

What are these higher twists ? The Q^2 dependence of the scaling violation is assumed to be due to perturbative effects (term in $1/(\ln Q^2/\Lambda)$). Before a so clear statement can be confirmed it was pointed out at early experiments that non-perturbative terms ($1/Q^2$, $1/Q^4$, ...) could explain the data or at least a large part of it. These "higher twists" effects were expected. Besides the target mass effects of order $1/Q^2$ already discussed (Nachtmann moment) other effects like resonance formation, transverse momentum effects, diquark scattering were expected to contribute. The problem is that none of these non-perturbative contributions can be calculated and we are left with general statements on the general form of these contributions. Because these effects are expected to be large at small W and $W^2 = Q^2X/(1-X)$ one states that higher twist effects should contribute mainly at large X and small Q^2 and a form of higher twist, a/Q^2, b/Q^4, $cX/(1-X)Q^2$, $dX^{\alpha}/(1-X)Q^2$ with a, b, c, d unknown, is proposed to experimentalists.

To clarify and justify the cut-off on Q^2 and W^2 used in QCD analyses one would like to go through an analysis of "higher twist" effects studied by Franz Eisele in the Marella School in Spain[17].

2) A tentative to includes "higher twist" effects to fit SLAC data (per X bin separately) of the form $F(X,Q^2) = F^{QCD}(X,Q^2)+F^{H.T}(X,Q^2)$ with $F^{H.T}(X,Q^2) = \mu_4 X/Q^2$ or $\mu_3 X/Q^4$ leads to very good fits (continuous line in figure 34). The shape of the higher twist contribution agrees better with the form $F^{H.T}(X,Q^2) \sim X^2(1-X)/Q^2$ than the assumed form $F^{H.T} = F^{QCD}.X/(1-X)Q^2$.

3) A final fit to $F_2^{\nu N}$, $\bar{q}^{\nu}$ and SLAC data assuming a parametrization $F_2(X,Q^2) = F_2(X,Q^2)^{QCD}\ (1 + \mu[(XQ^\alpha)/((1-X)Q^2)])$ gives

$$\Lambda_{LO} = (0.19 \pm 0.02) \text{ GeV} ,$$

$$\mu = (1.5 \pm 0.2) \text{ GeV} ,$$

$$\alpha = 3.7 \pm 0.3 ,$$

$$\chi^2/DF = 210/206 .$$

The value of Λ_{LO} found is the same as a QCD fit with $W^2 > 11$ GeV excluding the SLAC data. We should notice that this small value of $\Lambda_{LO} = 0.19$ is obtained with $R = 0.1$. The point here is to stress the comprehension of the higher twist effect and not to determine Λ_{LO}.

4) A fit of data with higher twist terms only (Λ=0) can describe the data .

$$F(X,Q^2) = F(X)\ [1 + \frac{1.5X^{1.7}}{(1-X)Q^2} - 0.23\ \frac{X^{-0.26}}{(1-X)^2Q^4}\ ]$$

can describe the SLAC + CDHS data very well. Data are not precise enough to distinguish a $1/\ln Q^2$ and a sum $1/Q^2 + 1/Q^4$. Additional theoretical inputs on X dependence are needed ; a Q^2 dependence alone is not sufficient to separate perturbative QCD from higher twist effects.

The conclusion is the following[17] : "higher twist contributions are important in the SLAC range, they however are most likely small at high W and do not affect the QCD comparison unless the higher twist contributions mimic QCD in shape".

Conclusion on study of structure functions

It could appear peculiar to say anything on the EMC effect in a lecture on structure functions. We have assumed that up to now this effect is not Q^2 dependent and that all the work which has been done on the QCD analysis is still valid and should be persued. Nevertheless the knowledge of the X distribution of quark on hydrogen and deuterium stays the most important experimental result we should obtain.

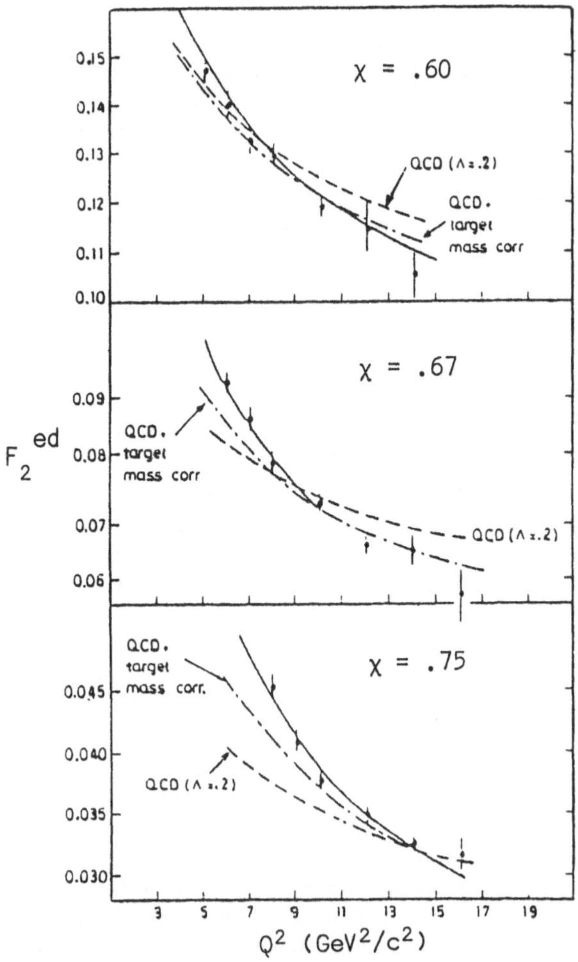

Fig. 34. The measurements of F_2^{ed} from SLAC-MIT compared to the results of the QCD fit to CDHS data with Λ = 0.2 GeV. From F. Eisele[17].

1) Study of SLAC data at $3 < W^2 < 12$ GeV^2

If one tries to analyse the SLAC data with the value of Λ = 0.2 GeV found in a CDHS analysis at $W^2 > 11$ one finds the results presented in figure 34. This value of Λ is not able to describe the SLAC data. Although the target mass correction - which has to be considered at this low mass invariant W - improves the fit, one still needs non-perturbative effects at low W. F. Eisele excludes the fact that this discrepancy with the two sets of data can be due to a relative normalisation because only the Q^2 shape dependence matters.

On the QCD study of the structure functions one now has an excellent sample of data as well in muon as in neutrino physics. The programme started in 1970 has been successfully accomplished. Statistics are not a problem anymore and the experimental teams have well understood their apparatus. Systematic errors are under control and studied with care. What will be done in the near future will be tedious work to have more events - but it cannot be improved by a factor 10 - and more precise data in the low X region. Some programmes have been accepted and will continue for the next two years. The step due to FNAL in this regard will be a step in energy ($\cong$ 1000 GeV) but not in statistics. So the overall improvement in neutrino physics due to the doubler may be minor except of course for new phenomena due to the energy threshold.

In the QCD analysis we have seen that the most unclear experimental result is the value of $R = \sigma_L/\sigma_T$. At each level the influence of R in the analysis is important. We do not see a possible neat improvement in this matter. The upper limit on R obtained in neutrino physics is by far the best experimental result but the value of R at small X stays badly known. The answer will not come from the muon experiment, because the systematic errors are very important and difficult to handle. The improvement from the present status can come from the new analysis on neutrino physics using large samples of wide band beam data. This question is a worrying one and can prevent reaching a better result on QCD parameters.

On the QCD analysis by itself, after restriction on some "higher twist" effects at small W^2, more and more precise data allows one to give evidence for the interpretation of the scaling violations by perturbative QCD. The vectorial nature of the gluon is clear, the value of Λ converges to $0.2 < \Lambda < 0.3$. From these results the gluon distribution is deduced and the new analysis of combined F_2 and $\bar{q}^{\bar{\nu}}$ structure functions is a success.

We have to face a fact. What was an exciting theory - "QCD" - ten years ago has become with time more and more difficult to "clearly" analyse. We have learned a lot, but we feel that one has not been completely successful in spite of tremendous efforts. I am not sure that definite answers will be given by neutrino physics in the matter. Here again should we expect an answer from high energy $p\bar{p}$ physics ?

DIMUON PHYSICS : AMOUNT OF STRANGE SEA

In the analysis of structure functions we have seen that the knowledge of the strange sea distribution enters at a small level but the effect is not negligible.

Results from the strange sea can be obtained by studying the production of charm in neutrino and antineutrino physics. The clear production of charm in neutrino physics has appeared as production of dimuons, the charm quark being tagged by an extra μ produced, this extra μ coming from the semi-leptonic decay

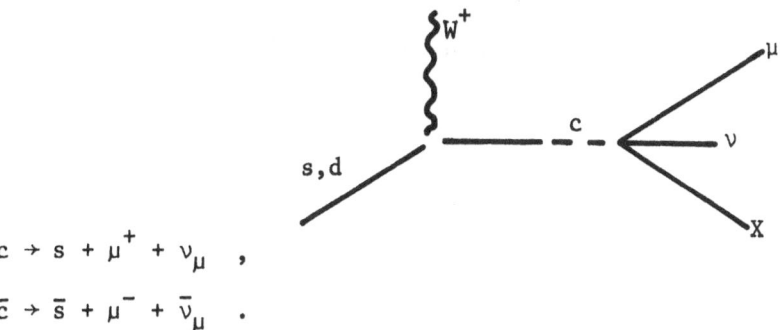

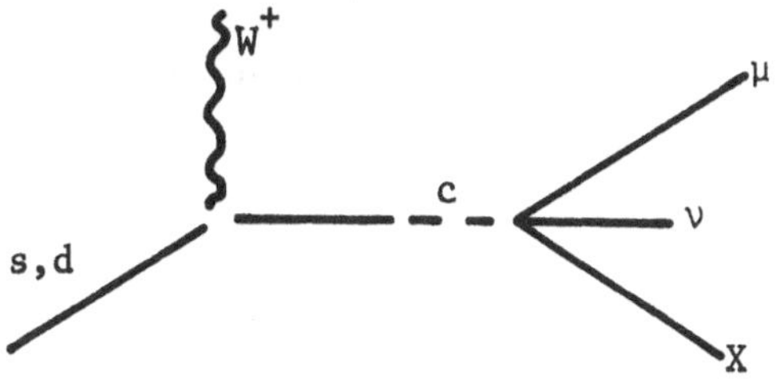

$$c \to s + \mu^+ + \nu_\mu \quad ,$$

$$\bar{c} \to \bar{s} + \mu^- + \bar{\nu}_\mu \quad .$$

The multilepton physics in neutrino interactions starts with the discovery in 1974 by the HPWF group of 2 events with 2 muons[30]. These events were clear and were interpreted as a weak production and weak decay of some new particle : the charm. For a while the "tool" multilepton was very exciting and was expected to be the key for new discoveries : b, t quark, heavy leptons... Trimuons were found, quadrimuons were found, but classical processes explained these productions and few little space was left for production of new objects.

A large increase of statistics permits furthermore a thorough study of charm production and charmed quarks with a sample of data of several thousands of dimuons[31]. The following properties were studied :

- charm cross section production,
- V-A structure of the charm producing current,
- charm fragmentation,
- strange sea structure function,
- Kobayashi-Maskawa angle θ_2.

Among these topics we will concentrate in this lecture on the amount of strange sea.

A/ Formalism

The model of charm-producing weak currents was proposed by Glashow, Iliopoulos and Maiani (GIM).

The GIM mechanism is expected to be the following :

$$\nu + d \to \mu^- + c \ ,$$

$$\nu + s \to \mu^- + c \ ,$$

$$\bar{\nu} + \bar{d} \to \mu^+ + \bar{c} \ ,$$

$$\bar{\nu} + \bar{s} \to \mu^+ + \bar{c} \ ,$$

and the current has the form :

$$J_\mu^{GIM} = (\bar{u},\bar{c})\gamma_\mu(1 + \gamma_5)\begin{pmatrix}\cos\theta_c & \sin\theta_c \\ -\sin\theta_c & \cos\theta_c\end{pmatrix}\begin{pmatrix}d \\ s\end{pmatrix} + h.c. \ ,$$

where θ_c is the Cabibbo angle.

The GIM mechanism is schematized in the following table :

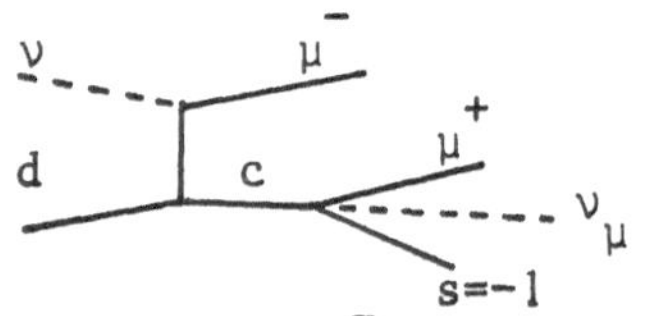

VALENCE QUARK $\sin^2\theta_c \cong 0.05$ Y FLAT

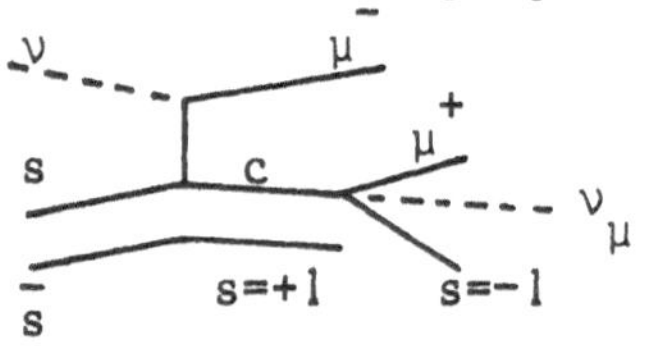

SEA QUARK $\xi_s \cos^2\theta_c \cong 0.04$ Y FLAT

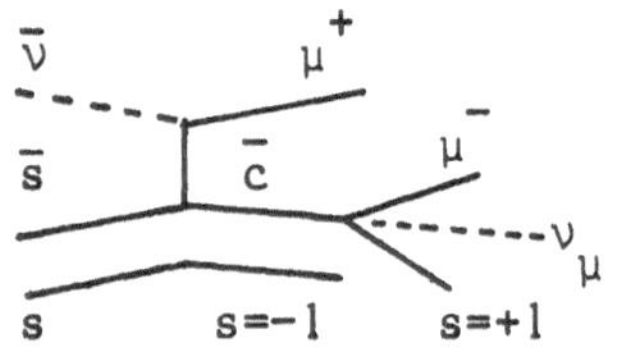

SEA ANTI-QUARK $\xi_s \cos^2\theta_c \cong 0.04$ Y FLAT

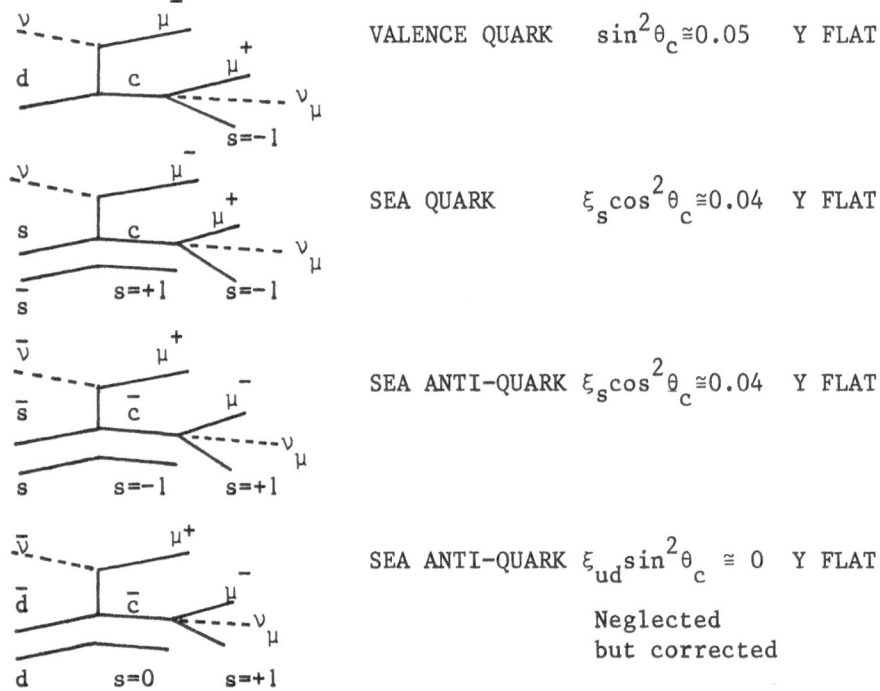

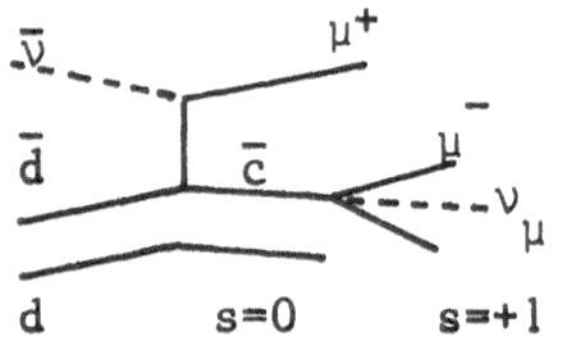

SEA ANTI-QUARK $\xi_{ud} \sin^2\theta_c \cong 0$ Y FLAT

Neglected
but corrected

Kobayashi and Maskawa[32] have extended to three flavour pairs the mixing angle procedure which involves three mixing angles and one phase. The K.M. matrix (U_{MK}) is the following 3 x 3 matrix :

$$u,c,t \left| \begin{matrix} C_1 & C_3 \, S_1 & S_1 \, S_3 \\ C_2 \, S_1 & C_1 \, C_2 \, C_3 - S_2 \, S_3 \, e^{i\delta} & C_1 \, C_2 \, S_3 + C_3 \, S_2 \, e^{i\delta} \\ S_1 \, S_2 & -C_1 \, C_3 \, S_2 - C_2 \, S_3 \, e^{i\delta} & -C_1 \, S_2 \, S_3 + C_2 \, C_3 \, e^{i\delta} \end{matrix} \right| \begin{matrix} d \\ s \\ b \end{matrix}$$

$C_j = \mathrm{Cos}\theta_j$, $S_j = \mathrm{Sin}\theta_j$, δ : CP violation phase.

The current has the form :

$$J^{KM} = (\bar{u},\bar{c},\bar{t})(1 + \gamma_5)\, U_{KM} \begin{pmatrix} d \\ s \\ b \end{pmatrix} + h.c.$$

One neglects the production of charm by b quark because of the high threshold and the small component of $b\bar{b}$ in the nucleon - for the same reason the conversion of charm quark to either d or s with production of its associated anticharm quark. Then the cross sections of charm production on isoscalar nuclei are :

$$\frac{d\sigma^{\nu}}{dXdY} = \frac{G^2 M E_{\nu} X}{\pi} \left[U_{cd}^2 \left[u(X) + d(X)\right] + |U_{cs}^2|\, 2s(X)\right],$$

$$\frac{d\sigma^{\bar{\nu}}}{dXdY} = \frac{G^2 M E_{\nu} X}{\pi} \left[U_{cd}^2 \left[\bar{u}(X) + \bar{d}(X)\right] + |U_{cs}^2|\, 2\bar{s}(X)\right].$$

In the GIM model $U_{cd} = \sin\theta_c$,

$U_{cs} = \cos\theta_c$,

θ_c Cabibbo angle[33]: $\sin\theta_c = 0.228 \pm 0.011$,

$\cos\theta_2 = 0.96 \begin{smallmatrix} +\ 0.04 \\ -\ 0.09 \end{smallmatrix}$.

In the KM notation $U_{cd} = \sin\theta_1 \cos\theta_2$,

$U_{cs} = \cos\theta_1 \cos\theta_2 \cos\theta_3 - \sin\theta_2 \sin\theta_3\, e^{i\delta}$.

Slow rescaling. When a heavy quark is produced, one cannot neglect the mass of the quark and we have to deal with threshold effects. This kinematical effect has been discussed by Brock[34]. If one considers the production of a c quark of mass M_c by a W of quadrimoment q hitting a quark of quadrimoment ξP one can write :

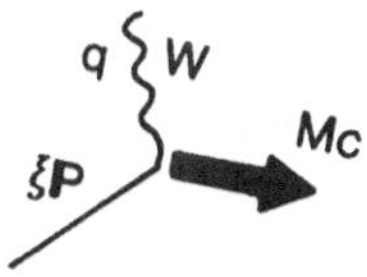

$$M_c^2 = (q + \xi P)^2 \cong q^2 + 2\,\xi\, P.q \quad ,$$

$$q.P = \nu M_N \;, \qquad \xi = X + \frac{M_c^2}{2M_N \nu} \quad .$$

One finds that for large Q^2 : $\xi \to X$, X being the definition of the quark distribution we have used up to now. This question of kinematical threshold is still important at neutrino energies of 100 GeV. It implies kinematical requirements on the hadronic mass W^2 ($W>M_c>W_{min}$) and reduces the X, Y space phase since

$$Y \geq \frac{W^2_{min} - M^2_N}{2\, M_N E} ,$$

$$X \leq 1 - \frac{W^2_{min} - M^2_N}{2\, M_N . E . Y} .$$

We see that the only parameter which is involved in the kinematical effect is the mass of the c quark M_c. In analyses on dimuons[31] this mass has been set to 1.5 GeV and we can rewrite the cross section[34]:

$$\frac{d^2\sigma^{\nu}}{dXdY} = \frac{G^2 M\, E}{\pi}\ \xi(1 - Y + \frac{XY}{\xi})\ [U^2_{cd}\ [d(\xi) + u(\xi)] + |U^2_{cs}|\ 2\ s(\xi)],$$

$$\frac{d^2\sigma^{\bar{\nu}}}{dXdY} = \frac{G^2 M\, E}{\pi}\ \xi(1 - Y + \frac{XY}{\xi})\ [U^2_{cd}\ [\bar{d}(\xi) + \bar{u}(\xi)] + |U^2_{cs}|\ 2\ \bar{s}(\xi)].$$

B/ Experimental sample.

Several experiments have studied dimuon production and results of about a hundred events were presented[35]. A big step in this respect has been made by CDHS[31] and results on 10 000 opposite sign dimuons from neutrino WBB and 3500 from antineutrino WBB have been analysed.

. In the case of "dimuons" the experimental samples are very clear, the only possible contamination being the π and K muonic decays from the hadronic shower. This contamination is very well calculated by Monte Carlo calculation and this background is about 13 % in neutrino and 6 % in antineutrino interactions. (Figure 35).

. The only price to pay for such a clear sample is the necessity of applying a cut-off on momentum of muons as large as 5 GeV to be able to select dimuons events and to measure these muons. This cut-off is necessary because of the need of several meters of iron to identify correctly a muon. In this respect the bubble chamber events "μe" are much more interesting since the electron detection and measurement is possible already around 500 MeV. But the statistics are poor (less than 100 events).

. A Monte Carlo is needed to know the acceptance of dimuons and specially the one which is the result of the charm decay. In this Monte Carlo the process of production of this muon is complicated

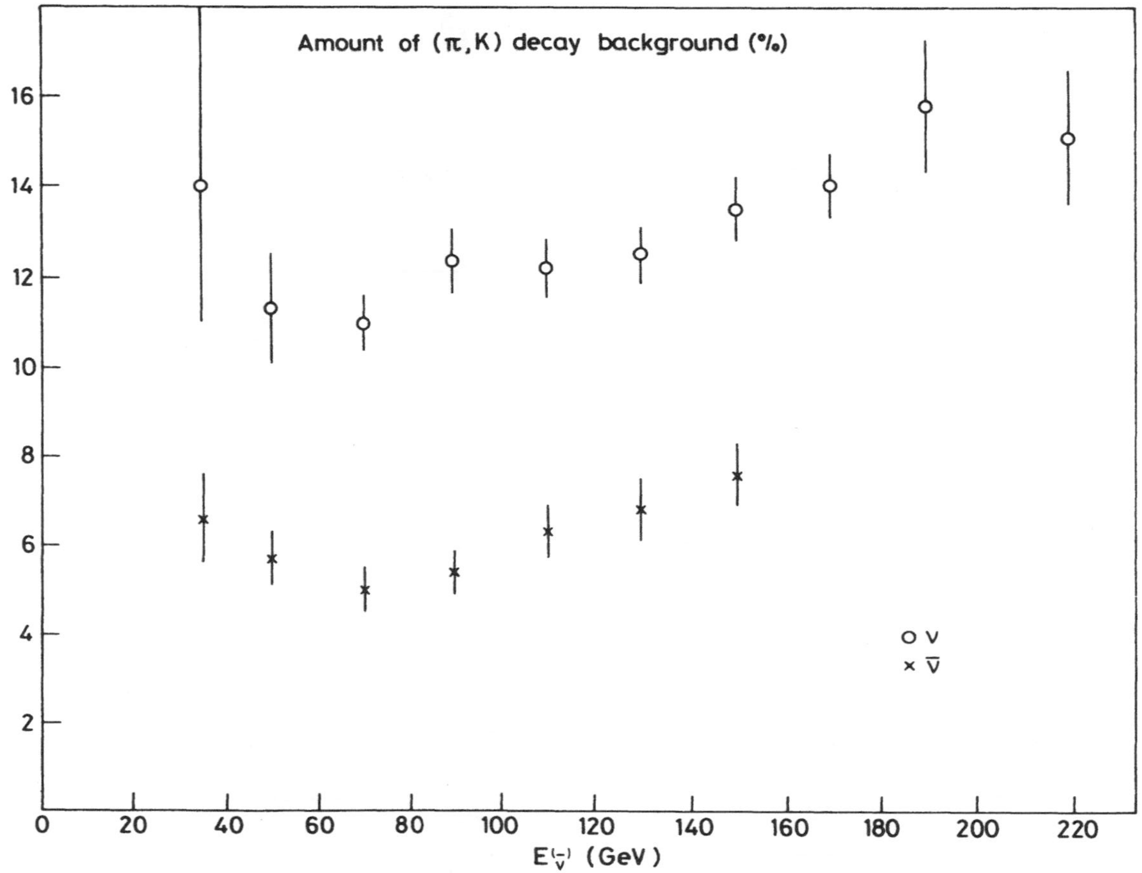

Fig. 35. Calculated background contribution from π and K decays for 350 GeV neutrino and antineutrino wide band beams.

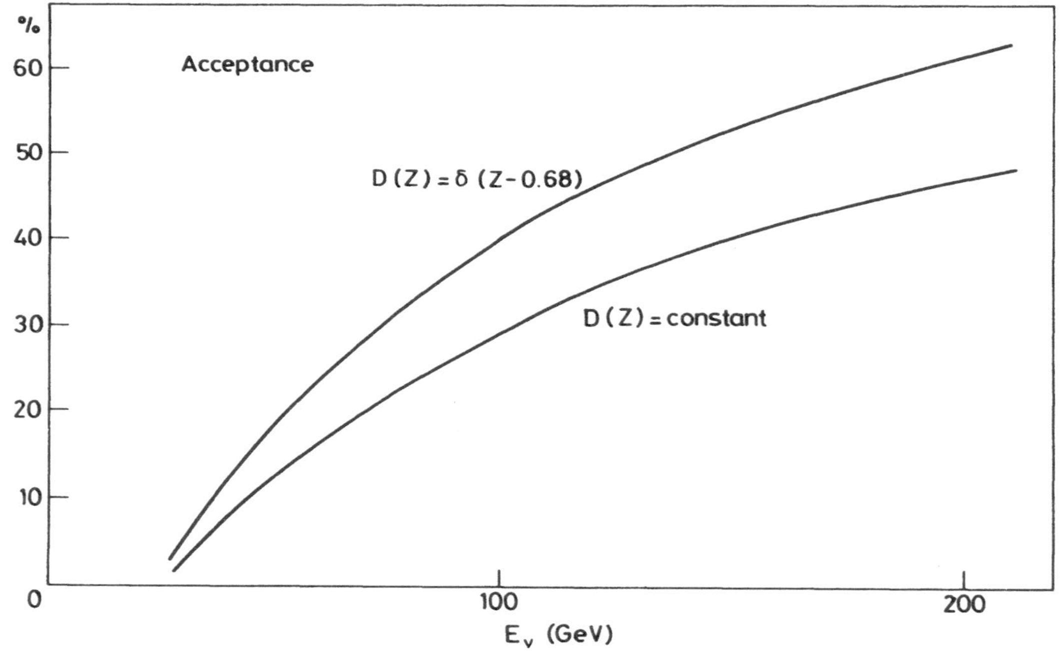

Fig. 36. Acceptance for neutrino dimuon events for two fragmentation functions $D_c(Z) = \delta(Z-0.68)$, $D_c(Z) = \text{const.}$

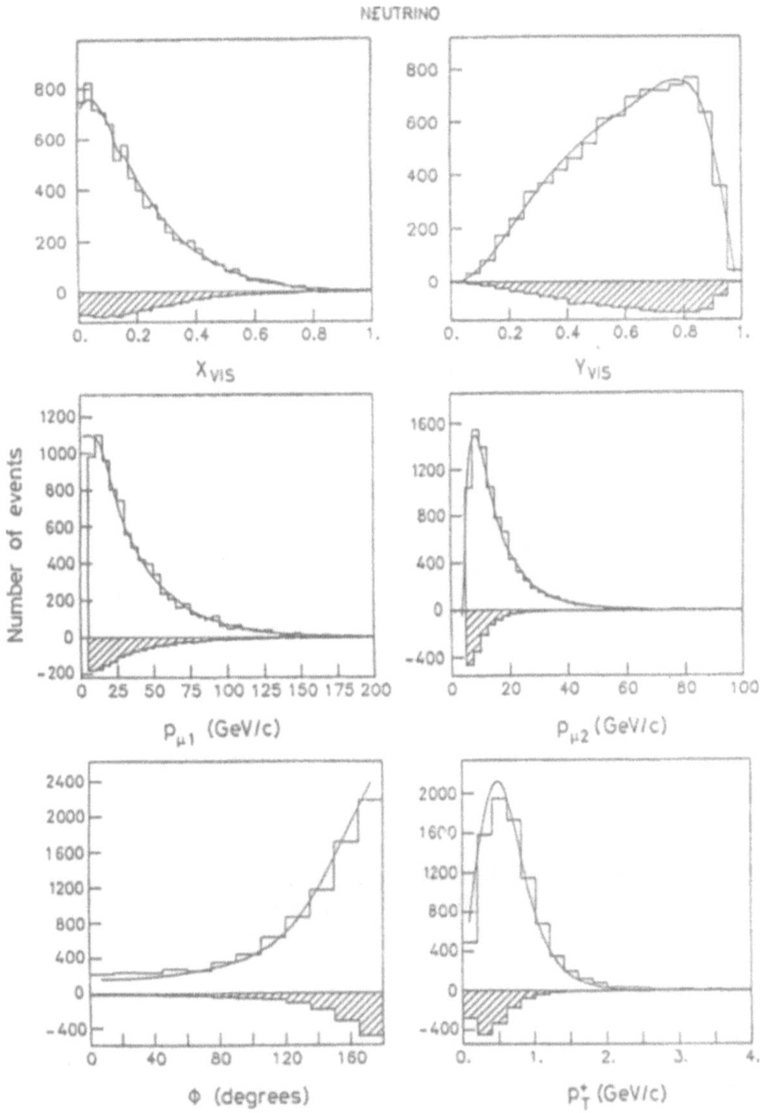

Fig. 37. Kinematical distribution for neutrino dimuon events. The dashed area represents the π and K backgrounds which have been subtracted from raw data. The errors are results of Monte Carlo simulations.

$X_{VIS} = Q^2/(2M_N.E_{HAD})$,

$Y_{VIS} = E_{HAD}/E_{VIS}$,

$P_{\mu 1}$ = momentum of the leading muon (lepton vertex),

$P_{\mu 2}$ = momentum of charm decaying muon,

ϕ= angle between the projection $\vec{P}_{\mu 1}$ and $\vec{P}_{\mu 2}$ on a plane perpendicular to the neutrino direction,

P_T = transverse momentum of muon relative to the hadronic shower direction.

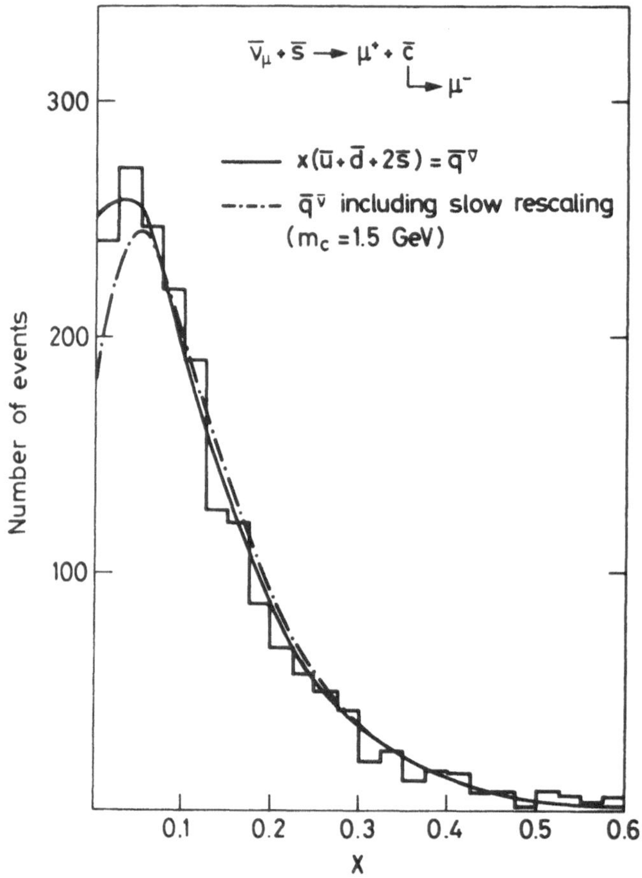

Fig. 38. X_{VIS} distribution for dimuon events. The solid curve represents the sea distribution obtained by CDHS in single-muon analysis[10], the dashed-dotted curve shows the effect of slow rescaling.

and several hypotheses have to be made :
- on the muon coming from the leptonic vertex : Y and X distributions;
- in the second muon : charm fragmentation, transverse momentum distribution, branching ratio for decay $K^* \rightarrow \mu\nu$, $K \rightarrow \mu\nu$ or $\pi \rightarrow \mu\nu$.

It is through this study that we have seen the importance of the charm fragmentation of the $\bar{c}$ quark. Figure 36 shows the difference in acceptance for different fragmentation functions. The charm fragmentation function has been determined[31] to be quite hard (peaked around Z = 0.68).

As an example the agreement with the Monte Carlo curves and the data is shown in figure 37 for neutrino events.

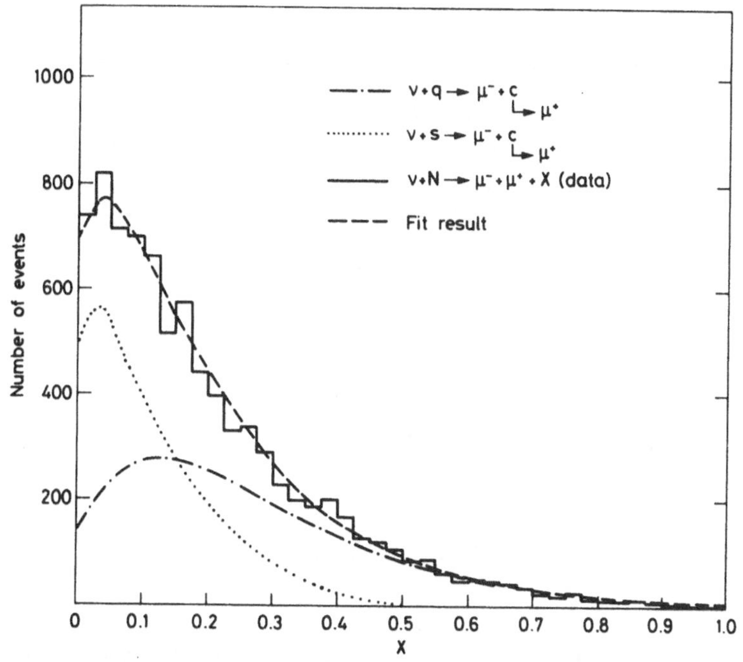

Fig. 39. Illustration of a fit to the neutrino induced dimuon event X distribution decomposed into 48 % of strange sea taken from antineutrino dimuon X distribution and 52 % of quark contribution. The dashed curve is the sum of both.

C/ Determination of the amount of strange sea

In the antineutrino dimuon production ,

$$\frac{d\sigma^{\bar{\nu}}}{dXdY} = \frac{G^2.E.M.X}{\pi} [U_{cd}^2[\bar{u}(X) + \bar{d}(X)] + |U_{cs}^2|2\bar{s}(X)] .$$

The term $|U_{cs}^2|2\bar{s}(X)$ is dominant : in the Cabibbo theory this term is 90 % of the cross section and this is also true if the KM mixing angles are small. In this condition one can compare the X distribution for antineutrino induced dimuon to the sea distribution $X[\bar{u}(X) + d(X) + 2s(X)]$ measured in single muon production. (figure 38). We see that these distributions are very similar. Now if we consider the shape of the neutrino X distribution of dimuons one can write it as a sum of contributions from the X(u + d) distribution and the strange sea. One can make a fit $F(X) = \alpha.sea(X)+(1-\alpha).X(u(X)+d(X))$ with :

- the sea distribution taken from the antineutrino dimuon X distribution;
- $(U(X) + D(X)) \cong 1/2\ [F_2(X) + XF_3(X) - 2Xs(X)]$ taken from the structure function analysis.

Table 4. The fraction of strange sea.

E_ν (GeV)	r_{sea}	r_q	$\frac{\lvert U^2_{cs}\rvert}{U^2_{cd}} \frac{2S}{U+D}$	$\frac{\lvert U_{cs}\rvert^2}{U^2_{cd}} \frac{2S}{\bar{U}+\bar{D}}$	$\frac{2S}{\bar{U}+\bar{D}}$ for $\frac{U^2_{cd}}{\lvert U_{cs}\rvert^2} = 0.056$
35-60	2.11	1.33	1.10±0.16	8.6±1.8	0.48±0.10
60-110	1.53	1.19	1.34±0.10	10.5±1.8	0.59±0.10
110-160	1.36	1.13	1.23±0.17	9.6±2.0	0.59±0.11
> 160	1.24	1.10	1.36±0.20	10.6±2.2	0.59±0.12
> 35	1.53	1.19	1.19±0.09	9.3±1.6	0.52±0.09

An example of such a fit is shown in figure 39. Results of fits are given on table 4.
r_{sea} and r_q are the slow rescaling factors for sea and quark distributions. The ratio $U^2_{cd}/|U_{cs}|^2 = 0.056$ is obtained by measuring the difference of ν and $\bar{\nu}$ total cross sections of dimuons[31].

What is interesting is the ratio of strange sea to non strange sea $2S/(\bar{U} + \bar{D})$. One can obtain this ratio from the results of table 4 utilizing from charged current analyses[10]:

- $\bar{U} + \bar{D} + 2\bar{S} = 0.070 \pm 0.005$ from $\bar{q}^{\bar{\nu}}$,
- $F_2(X)\, dX = 0.438 \pm 0.022$.

This gives $(\bar{U} + \bar{D})/(U + D) = 0.13 \pm 0.02$.
If one takes the averaged value of table 4 : $2S/(\bar{U} + \bar{D}) = 0.52 \pm 0.09$, one obtains

$$\frac{U^2_{cs}}{U^2_{cd}} \frac{2S}{\bar{U} + \bar{D}} = 9.5 \pm 1.6.$$

Results from the θ_2 KM mixing angle show that θ_2 is small and if we assume θ_3 also small one concludes

$$U^2_{cd}/|U_{cs}|^2 = \tan^2 \theta_c = 0.056 \pm 0.005 ,$$

$$2S/(\bar{U} + \bar{D}) = 0.52 \pm 0.09.$$

Slow rescaling effects and other possible bias have been studied. A variation of 5 % is possible. This result is very different from the ratio $2S/(\bar{U} + \bar{D}) = 1$ expected if the sea were flavour symmetric.

Except for difference due to S and U, D quark masses, this break-down of symmetry has not been explained.

We understand why in the structure function analysis the variation of the strange sea component was allowed within the limits

$$0.5 < 2S/(\bar{U}+\bar{D}) < 1.$$

This result $2S/(\bar{U}+\bar{D})$ cannot be improved so easily because the experimental sample is quite clear and the amount of events is large. One can expect shortly the Q^2 dependence of the strange sea and the correction to apply for structure functions will then be more precise.

ACKNOWLEDGEMENTS

I would like to thank the organizing committee of the Nato Advanced Study Institute : M. Lévy, J.L. Basdevant, M. Jacob, D. Speiser, J. Weyers, R. Gastmans, for their kind invitation and for the splendid stay in Cargèse. Special thanks to Marie-France Hanseler for her wonderful organization of the stay at the school.

I have been helped in the preparation of these lectures by F. Dydak and I had discussions on structure functions with F. Eisele and A. Para. I thank them very much.

I would like to thank Mrs. J. Thiolière and Mrs. S. Roussiez for their help on typewritting and figures of this text and J.P. Schuller for reading the manuscript.

REFERENCES

1. O. Nachtmann, "Interactions of neutrinos", T.E.P.P. Weak interactions, IN2P3, 11 rue Pierre et Marie Curie, 75231 Paris Cedex 05 (1977).
2. N. Armenise et al., Phys. Letters 84B (1979) 137.
3. M. Jonker et al., Z. Physik, Particles and Fields C17 (1983) 211.
4. M. Gourdin, Proc. of the International Neutrino Conference, Aachen 1976, p. 234.
5. J. Carr et al., "Search for right-handed currents in muon decay", LBL 16183 preprint, to be published in Physical Review Letters.
6. F. Eisele, 21st Intern. Conf. on High Energy Physics, Paris 1982, p. C3-337.
7. H. Abramowicz et al., Z. Physick, Particles and Fields, C12 (1982) 225.
8. J.J. Aubert, Phys. Letters B123 (1983) 275.
9. R. Blair et al., Contributed paper to the International Symposium on Lepton and Photon Interactions at High Energy, Cornell (1983).
10. H. Abramowicz et al., Z. Physik, Particles and Fields C17 (1983) 283.
11. A. de Rújula et al., Nuclear Phys. B154 (1979) 394.

12. R. Barlow and S. Wolfram, Phys. Rev. D20 (1979) 2198.
13. R.M. Barnett, Phys. Rev. D14 (1976) 70.
14. J.J. Aubert et al., Phys. Letters 105B (1981) 322.
15. A. Bodek et al., Phys. Rev. D20 (1979) 1471.
16. C.G. Callan jr. and D.J. Gross, Phys. Rev. Letters 22 (1969) 156.
17. F. Eisele, Xth International Winter Meeting on Fundamental Physics, Masella, Spain, 1982.
18. H. Abramowicz et al., Phys. Letters 107B (1981) 141.
19. A. Bodek et al., Phys. Rev. D20 (1979) 1471.
20. G. D'Agostini, Thèse de 3ème cycle, Nice (1982).
21. P. Payre, Thèse de Doctorat d'Etat, Université de Grenoble (1983).
22. B.A. Gordon et al., Phys. Rev. Letters 41 (1978) 615.
23. W.A. Bardeen and A. Buras, Phys. Rev. D20 (1979) 166.
24. R. Barbieri et al., Nuclear Phys. B117 (1976) 50.
25. J.G.H. de Groot et al., Phys. Letters 82B (1979) 292.
26. R. Turlay, "Weak Charged Currents", EPS Intern. Conf. on High Energy Physics, Geneva, 1979.
27. R.M. Barnett, Presented at SLAC Summer Institute - Particle Physics - Stanford, California, July 9-20 1979, Report SLAC-224, p. 416.
28. A. Para, Private communication ;
H. Abramowicz et al., Z. Physik, Particles and Fields C13 (1982) 199.
29. H. Abramowicz et al., Z. Physik, Particles and Fields C12 (1982) 289.
30. A. Benevenuti et al., Phys. Rev. Letters 34 (1975) 419.
31. H. Abramowicz et al., Z. Physik, Particles and Fields C15 (1982) 19.
32. M. Kobayashi and K. Maskawa, Progr. Theor. Phys. 49 (1973) 652.
33. R.E. Shrock and L.L. Wang, Phys. Rev. Letters 41 (1978) 1692 and 42 (1979) 1589.
34. R. Brock, Phys. Rev. Letters 44 (1980) 1027.
35. B.C. Barish et al., Phys. Rev. Letters 36 (1976) 939 ;
M. Holder et al., Phys. Letters 69B (1977) 377 ;
J. Blietschau et al., Phys. Letters 58B (1975) 361 ;
J. Von Krogh et al., Phys. Rev. Letters 36 (1976) 720 ;
C. Baltay et al., Phys. Rev. Letters 39 (1977) 62.

ACCELERATOR STUDIES OF NEUTRINO OSCILLATIONS:

FIRST RESULTS FROM A LOW-ENERGY EXPERIMENT AT CERN

C. Guyot

CERN

Geneva, Switzerland

ABSTRACT

After a review of the experimental situation, we present the first results of an experiment performed at CERN by the CDHSW* Collaboration, which compares the event rates at two distances. No evidence for ν_μ oscillations are found. At the 90% confidence level, δm^2 values between 0.3 and 90 eV^2 are excluded for maximal mixing.

INTRODUCTION

The problem of neutrino masses has become one of the key puzzles of high-energy physics and astrophysics. Grand Unified Theories of strong, weak, and electromagnetic interactions do not require exact lepton-number conservation or massless neutrinos. They may lead to a neutrino mass of $10^{1\pm1}$ eV, or even as low as 10^{-5} eV, as an alternative to masslessness[1].

Our knowledge of the ν_e mass, coming from the study of electron momentum spectrum in nuclear β decay is rather good ($m_{\nu_e} \lesssim 30$ eV). This is in contrast with the case of the muon neutrino, which has an experimental upper limit of ∿ 500 keV (from π-decay experiments), and the tau neutrino ($m_\tau \lesssim 300$ MeV from τ decays). These limits can be improved by a factor of 5 in the near future, but there is no hope of reaching the same accuracy as for ν_e mass measurements. So it appears that the only possibility to have access to the low-mass region ($\lesssim 1$ eV) is neutrino oscillations.

*CERN-Dortmund-Heidelberg-Saclay-Warsaw Collaboration.

After a brief reminder of the formalism, we review the situation in accelerators experiments. Then we present the first results of an experiment done at CERN by the CDHSW Collaboration.

BASIC FORMALISM

The theoretical framework of oscillation has been discussed first by Pontecorvo (1967)[2] after the experimental proof of the existence of at least two neutrino flavours (1964). For review articles, see Frampton and Vogel[1] and Bilenky and Pontecorvo[3].

Neutrino oscillations will occur if neutrinos are massive and the physical states (mass eigenstates) do not correspond to weak eigenstates. Let us consider the case of N physical states (ν_m, M_m). Neutrinos of definite flavours ν_f ($f = e,\mu,\tau$) are related to mass eigenstates by a unitary matrix U :

$$\nu_f = \sum_{m=1}^{N} U_{fm}\nu_m \ .$$

For N = 3, U can be parametrized in terms of three angles and one phase (like the Kobayashi-Maskawa matrix for quarks). Suppose at time t = 0, a pure weak eigenstate is born (from $\pi \to \mu\nu$) :

$$|\nu_f(0)> = \sum_m U_{fm}|\nu_m> \ .$$

The state evolves at time t into

$$|\nu_f(t)> = \sum_m U_{fm}e^{iE_m t}|\nu_m> \ .$$

The probability of finding $\nu_{f'}$ will be

$$P(\nu_f \to \nu_{f'}) = |<\nu_{f'}|\nu_f(t)>|^2 = |\sum_m U_{fm}U^{+}_{f'm}\ e^{iE_m t}|^2 \ .$$

For N = 3, $P(\nu_f \to \nu_{f'})$ is the sum of three oscillating terms of the form $A_{kj}[1 - \cos(E_k - E_j)t]$, A_{kj} being a function of the U parameters.

Usually, one makes the following approximations :
i) All ν_m are produced with the same momentum p_ν,
ii) $M_m << p_\nu$; $E_m \cong p_\nu + M_m^2/2p_\nu$.
For a complete description of the oscillation phenomena in terms of wave packets, see Kayser[4]. Then one can write :

$$P(\nu_f \to \nu_{f'}) \cong \sum 2A_{kj} \sin^2 \frac{m_k^2 - m_j^2}{4p_\nu}\ t \ .$$

Usually, data are analysed supposing that one term is leading (coming to the case N = 2):

$$P(\nu_a \rightarrow \nu_b) = \sin^2 2\theta \ \sin^2(1.27 \ \delta m^2 \ L/E) \ ,$$

where

θ is the mixing angle between ν_a and ν_b,
$\delta m^2 = |m_a^2 - m_b^2|$ (in eV^2),
L is the distance from the π decay to the detector (in m),
E is the neutrino energy (in MeV).

For a typical accelerator experiment :

$$E = 2 \text{ GeV}, \ L = 1000 \text{ m}, \ \delta m^2 = 1 \text{ eV}^2,$$

therefore

$$P(\nu_a \rightarrow \nu_b) = 0.35 \sin^2 2\theta.$$

In general, the mass range accessible to an experiment is given by $\delta m^2 \gtrsim E/L$, which can be considered as the figure of merit of an experiment. If $L/E \gg \delta m^2$, one observes the average of $\sin^2$, $P = 1/2 \sin^2 2\theta$, which could apply to solar neutrinos.

ACCELERATOR STUDIES OF NEUTRINO OSCILLATIONS

Neutrino oscillations can be detected by observing the appearance of a new type of neutrino, or by observing the disappearance of the formerly produced type of neutrino :

i) Appearance method (non-diagonal test); for example, $\nu_\mu \rightarrow \nu_e$. This method requires a good knowledge of beam contamination.
ii) Disappearance method (diagonal test) : for example $\nu_\mu \rightarrow \nu_\mu$, which is sensitive to the oscillation $\nu_\mu \rightarrow \nu_{anything}$. This needs the calculation of the expected event rate or two measurements at different distances.

Non-diagonal oscillation tests

The results, which give the best present limits on $\delta m^2/\sin^2 2\theta$ are given in table 1. Most of these results come from bubble chambers because of their good capability in low-energy electron identification. The corresponding limits in the plane (δm^2, $\sin^2 2\theta$) are shown in fig. 1.

Diagonal tests

This category includes solar neutrino experiments and reactor experiments. Up to the beginning of 1983, no dedicated accelerator experiment had been performed to look for disappearance of ν_μ. Nevertheless, experiments measuring neutrino total cross-section of $\overset{(-)}{\nu}_e$, $\overset{(-)}{\nu}_\mu$ yield from the decay of short-lived particles (charmed parti-

Table 1.

Experiment	Oscillation	L (m)	E (GeV)	δm^2 (eV^2)
GGM-PS (CERN)[5] (1978)	$\nu_\mu \to \nu_e$	100	3	1.5
LAMPF Electronic detector[6]	$\nu_\mu \to \nu_e$	9	0.035	2
FNAL-15 ft (Columbia-BNL)[7]	$\nu_\mu \to \nu_e$ $\nu_\mu \to \nu_\tau$ $\nu_e \to \nu_\tau$	1600	40	0.6 3 17
BEBC-PS (CERN) (June 1983)	$\nu_\mu \to \nu_e$	850	1.5	in progress

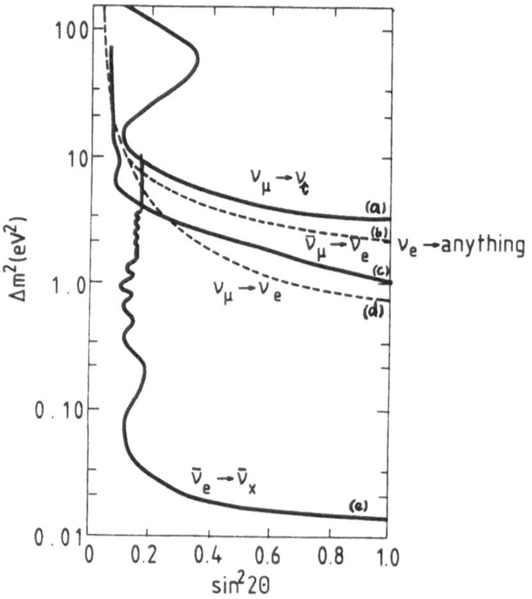

Fig. 1. Limits on the oscillation amplitude $\sin^2 2\theta$ and on the mass difference. Results are from : curves a and d - Ref. 7; curves b and c - Ref. 6; curve e - Goesgen.

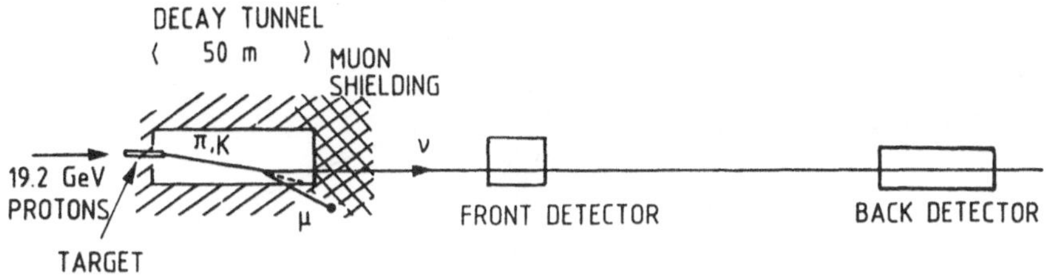

Fig. 2. Principle of the CDHSW experiment.

cles) produced in a beam-dump experiment can be analysed in terms of neutrino oscillations (see Wotschak[8]).

More recently, new results have been obtained at FNAL and CERN :

i) At FNAL, measurements of neutrino total cross-section have been performed at two distances (600 and 1000 m). No indications of oscillation have been found[9].

ii) At CERN, new beam-dump experiments (data-taking in 1982) with increased statistics and improved control of beam scraping gave[10] preliminary results on the ratio $R = \nu_e/\nu_\mu$. From CDHSW, $R = 0.84 \pm 0.13 \pm 0.15$, showing no indication for oscillation $(\nu_e \rightarrow \nu_\tau)$.

iii) At CERN, in 1983, two experiments [CDHSW, CHARM*] have taken data in a low-energy (1.5 GeV) PS beam using two detectors at two distances (130 m and 880 m). Here we report the first results of the CDHSW Collaboration.

FIRST RESULTS FROM THE CDHSW OSCILLATION EXPERIMENT

This experiment is looking for the disappearance of ν_μ and, so, is sensitive to oscillations $\nu_\mu \rightarrow$ anything.

Principles of the experiment

i) A low-energy ($E_\nu \sim 1.5$ GeV) bare target beam is used. (No magnetic elements : easy to simulate, ν flux $\sim 1/L^2$.) Neutrinos are produced in a new PS neutrino facility using 19.2 GeV protons interacting in a beryllium target. Figure 2 is a sketch of the experimental set-up; a general layout of the experimental area at CERN is given in fig. 3. The neutrino flux at both detectors has been calculated in a Monte-Carlo program (fig. 4). The departure from the $1/L^2$ law (due to π and K production angle spectra, absorption in the target, decay-tunnel wall effects, proton interactions in the air and in the shielding, etc.) stays below 5% and seems to be rather independent of the π and K spectra (fig.5).

*CERN-Hamburg-Amsterdam-Rome-Moscow Collaboration.

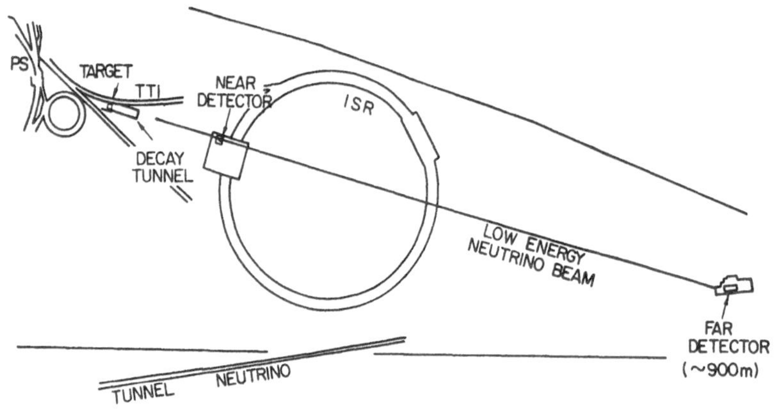

Fig. 3. Layout of the CERN PS oscillation facility.

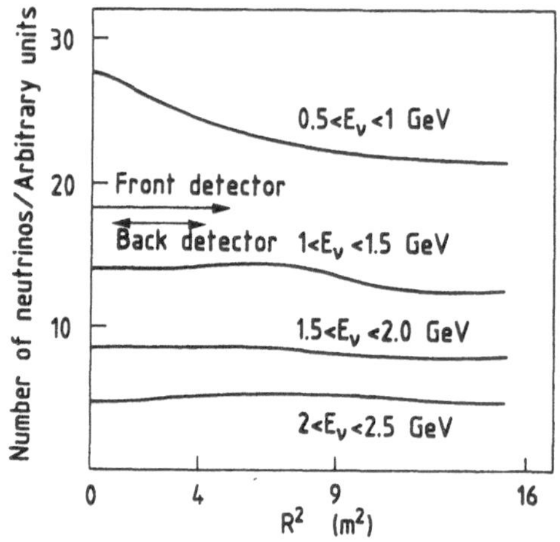

Fig. 4. Neutrino flux at 130 m from the target.

ii) Two detectors as identical as possible are used.

The far detector is the existing CDHSW detector. It is made of 21 Fe-scintillator modules. The iron sampling thickness is 12.5 cm for the first 10 modules, 5 cm for the next 5, and 15 cm for the last 6. The near detector consists of a reduced version of the far one, as shown in fig. 6.

For the purpose of this experiment, the scintillators are used for the pattern recognition of the events (see fig. 7).

iii) The rate of charged-current (CC) events [which appear as long events in contrast with events with no muons such as neutral-current (NC) events] in both detectors is compared :

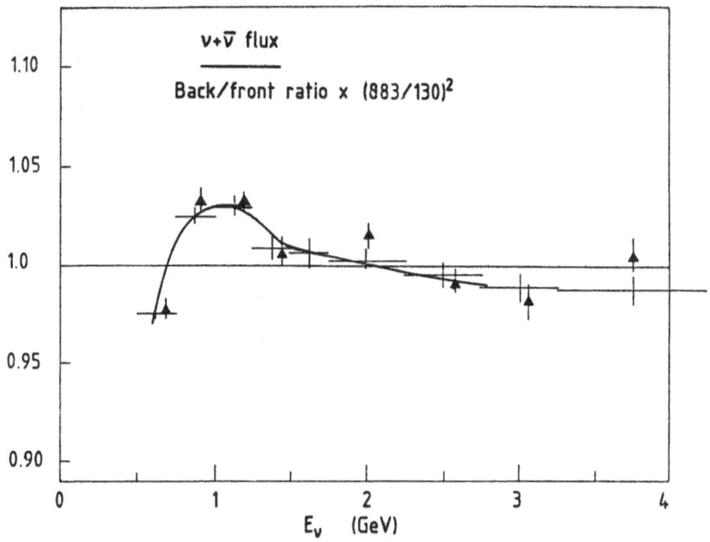

Fig. 5. Ratio of neutrino flux in the back detector to flux in the front detector, corrected for the different solid angles (Monte Carlo calculation).

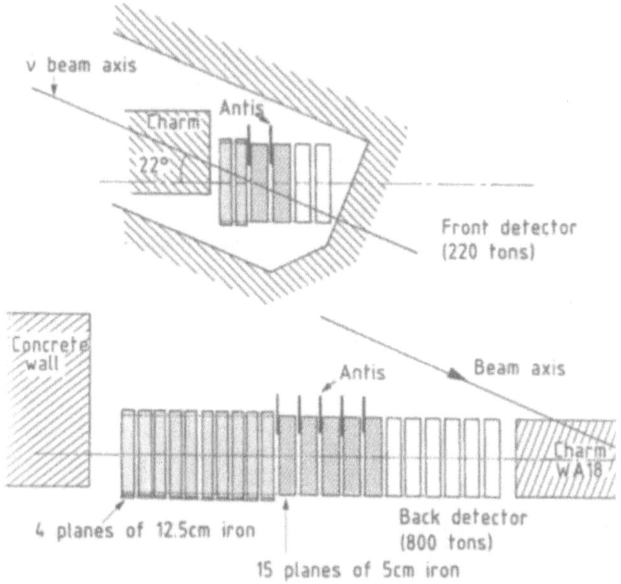

Fig. 6. Layout of the near and far detectors.

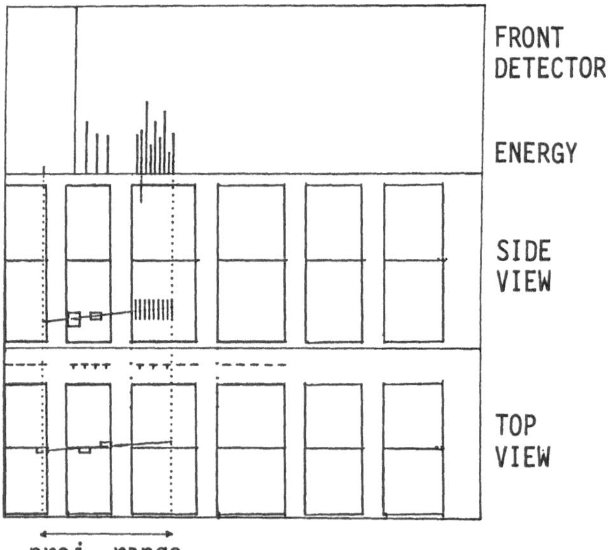

Fig. 7. Example of a typical event in the front detector, showing the scintillators hit.

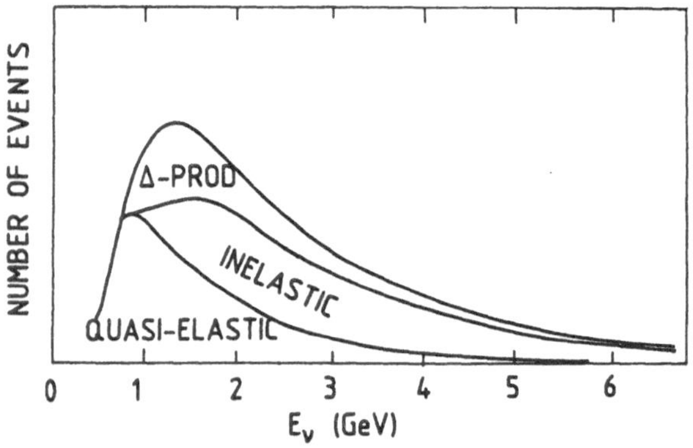

Fig. 8. Contributions of the different kinds of interactions to the observed charged current events (Monte Carlo calculation).

$\nu_\mu + N \rightarrow \mu + X$ (CC) "long" event (1 GeV $\mu \cong$ 80 cm Fe).

$\nu_\mu + N \rightarrow \nu_\mu + X$ (NC)

$\nu_e + N \rightarrow \nu_e + X$ (NC)

$\nu_e + N \rightarrow e + X$ (CC)

$\nu_\tau + N \rightarrow \nu_\tau + X$ (NC)

} "short" events (Event length $\leq$ 50 cm Fe)

E_ν being small, we are below the threshold for τ production. At such low energy, half of the CC interactions consists of quasi-elastic events (fig. 8). So one can use the correlation between event length (muon energy) and neutrino energy to compare the energy distributions in the two detectors. So, in principle, one has access to the two parameters :
- $\sin^2 2\theta$ is mainly correlated with the overall scale difference.
- δm^2 is given by the difference of the shape of the two distributions.

Data-taking and reconstruction

For triggering purposes, the detectors have been subdivided into "trigger planes" [one physical plane (12.5 cm Fe) in the first kind of module, and the sum of three physical planes (5 cm Fe) in the second kind of module]. The trigger condition asks for 3 out of 4 adjacent trigger planes (threshold on pulse height at 1/5 of the "minimum ionizing signal"), which ensures a detection efficiency $\geq$ 99% for tracks longer than 40 cm Fe.

The reconstruction of event candidates involves a pattern recognition by program, which determines the following quantities (see fig. 7)

- vertex,
- stop or exit point of the muon,
- angle θ_y,
- projected range in Fe.

The program classifies events according to the following criteria :

- good : if vertex inside detector,
- entering through front/side/top/bottom,
- leaving.

Data-taking took place in February-March 1983 using 10^{19} protons from Proton-Synchroton. Physics events were accepted inside a 3 μs spill gate. Cosmic background was estimated by taking data in an out-of-spill gate of 600 μs. In addition, these cosmic muons were used for calibration and monitoring of the apparatus. The numbers of physics triggers obtained are given in table 2.

Table 2.

	Total	Beam correlated (cosmic subtracted)
Front	150,000	145,000
Back	29,000	11,000

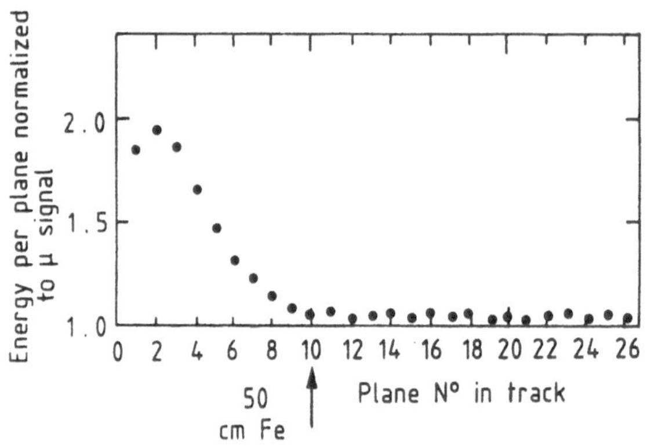

Fig. 9. Average energy profile of events.

Contamination backgrounds

i) Cosmic muons : contaminate mainly the far detector (see table 2). This background can be reduced by a cut on the angle θ_y (45°).

ii) Beam correlated muons : originate from neutrino interaction in surrounding materials. The muon environment was made as similar as possible by adding a concrete wall in front of the back detector. In addition, the following software vetoes have been applied :

- front plane of first module,
- radius cut (in the first part where we have x and y information),
- top and bottom scintillator (second part of the apparatus),
- anti-counter planes (second part of the apparatus : see fig. 6).

iii) Events with no muon : this contribution from events with a shower only can be estimated from the energy profile of events (fig. 9), which shows evidence for non-contamination above 50 cm projected range.

Two analyses have been performed : a global analysis integrating all events longer than a minimal projected range, and a projected range analysis. In fact, these analyses have been done separately for the two kinds of module. We present here the "average" of these analyses.

Global analysis

After all the cuts, we get the number of events, compared to a Monte Carlo (MC) calculation in table 3 (statistical errors only).

Systematic uncertainties in the Back/Front ratio are :

Table 3.

	Data (0.7×10^{19} p)	MC ($\delta m^2 = 0$)	Data/MC
Front	23,320	25,435	
Back	3,060	3,252	
Back/Front	0.131 ± 0.003		1.02 ± 0.03

- Overall scale shifts :
  - dead-time (Front) $\sim$ 1%
  - ν_μ induced muon background $\sim$ 1%
- Energy (range)-dependent effects :
  - π/K production spectra $\sim$ 1.0%
  - proton interactions downstream of the target $\sim$ 0.5%
  - reconstruction inefficiencies $\sim$ 2%

Total $\leqq$ 3%

The final result is

Back/Front = 1.02 ± 0.03 ± 0.03.

This result gives the limits on δm^2 and $\sin^2 2\theta$ shown in fig. 10.

Projected range analysis

The event length distribution in the front detector is given in fig. 11. The agreement between data and Monte Carlo calculation is good. Figure 12 shows the ratio of back distribution to front distribution corrected by a Monte Carlo calculation (correction smaller than 5%). The data show no evidence for oscillations if we compare the results with the expected effects also given in fig. 12. A fit to this distribution gives the limits in the plane (δm^2, $\sin^2 2\theta$) shown in fig. 13.

CONCLUSION

No evidence for the oscillation $\nu_\mu \to$ anything has been found in the range $0.3\ \text{eV}^2 < \delta m^2 < 100\ \text{eV}^2$ for mixing angles $\sin^2 2\theta \gtrsim 0.06$.

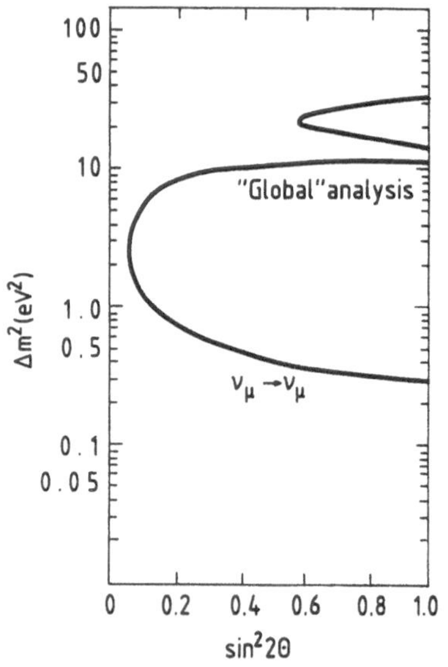

Fig. 10. Limits on oscillation parameters from the global analysis (90% C.L.).

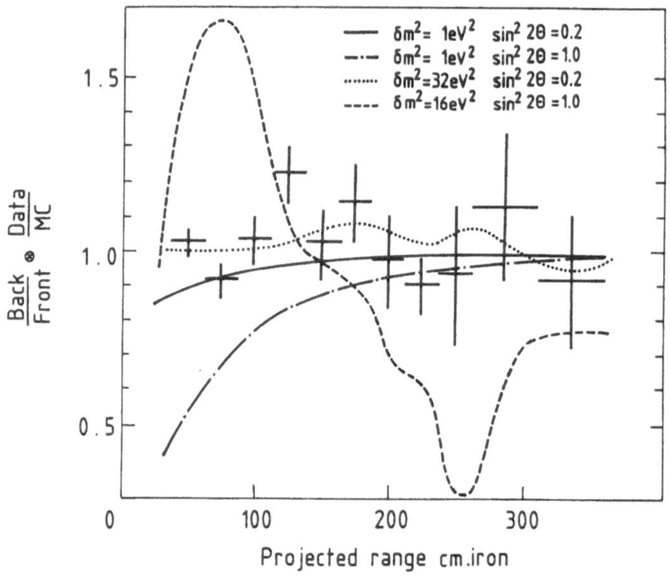

Fig. 12. Corrected ratio of back-to-front event-length distributions. The lines show the expected oscillation patterns for different values of $\sin^2 2\theta$ and δm^2.

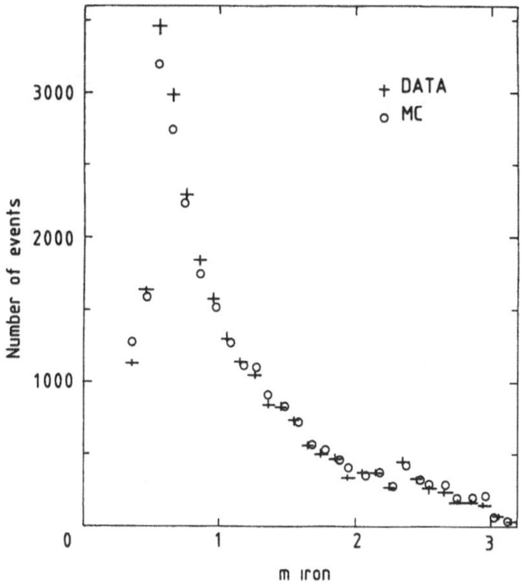

Fig. 11. Comparison of event-length distributions from data sample and Monte Carlo calculation.

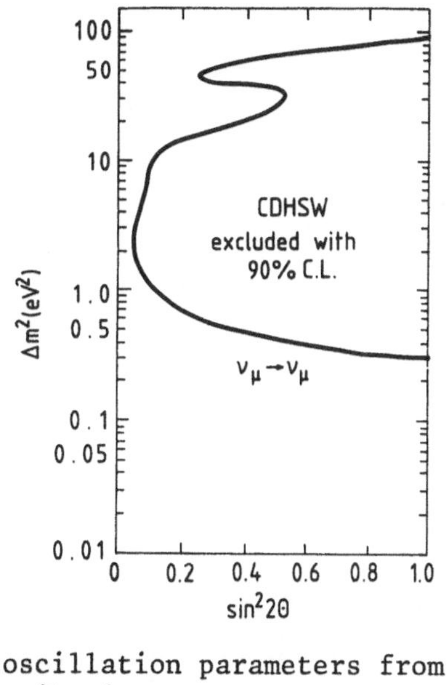

Fig. 13. Limits on oscillation parameters from the analysis of event-length distributions.

REFERENCES

1. P.H. Frampton and P. Vogel, Phys. Rep. 82 (1982) 342.
2. B. Pontecorvo, Sov. Phys. JETP 26 (1968) 984.
3. S.M. Bilenky and S. Pontecorvo, Phys. Rep. 41 (1978) 225.
4. B. Kayser, SLAC-PUB-2685 (1981).
5. J. Blietschau et al., Nucl. Phys. B133 (1978) 205.
6. P. Nemethy et al., Phys. Rev. Lett. 44 (1980) 522; Phys. Rev. 23 (1981) 262.
7. N.J. Baker et al., Phys. Rev. Lett. 47 (1981) 1576.
8. J. Wotschack, Proc. 4th Warsaw Symposium on Particle Physics, Kasimierz, 1981, p. 103.
9. W.H. Smith, presented at Brighton Conference (1983).
10. P. Perez, Preliminary results presented at Brighton Conference (1983).

SELECTED TOPICS IN NON-ACCELERATOR PHYSICS

G. Charpak

CERN

Geneva, Switzerland

INTRODUCTION

Recent theories in high-energy physics have given rise to speculative predictions; their proof relies on a great variety of experiments, most of which do not require high-energy accelerators.

The unified gauge theories permit the violation of previously "sacred" conservation rules and have led to the search for very rare decay modes. The predictions of highly exotic particles such as the magnetic monopole, with more specific properties than just carrying a magnetic charge, has led to an exciting and imaginative set of experiments.

The anomalously low flux of solar neutrinos observed has, among other interpretations, been attributed to oscillations between several neutrino flavours, and has encouraged scores of new experiments, both in nuclear physics and in particle physics.

Since these two lectures cover so many subjects, I will rarely refer the reader to the original contributions but rather to the review articles which I used when preparing my talks.

MASSIVE NEUTRINOS AND LEPTON CONSERVATION[1,2]

The idea of massive neutrinos goes back to the original hypothesis of the existence of the neutrino in the early 1930's. The existence of neutrino oscillations between different flavours was discussed as soon as a second neutrino flavour was discovered in 1962. The recent increase of interest in new measurements of the possible neutrino mass comes from several facts : i) Grand Unified

Theories (GUTs) unifying strong and electroweak forces may lead to a neutrino mass of 10 eV with an uncertainty of at least one order of magnitude, as an alternative to exactly zero mass for which there is no compelling argument; ii) astrophysical arguments show a heavy non-visible mass component in the universe which it is tempting to attribute to massive neutrinos; iii) the experimental claim, in 1980, of evidence for non-zero mass : this last experiment has sparked a great deal of activity aimed at the direct measurement of neutrino masses. Some tests of lepton-number conservation such as neutrino oscillations[3], or neutrinoless double β-decay[4], also have implications concerning the neutrino masses, and I wish here to underline some prospects for significative advances relative to the present situation.

The direct measurement of the neutrino mass in β-decay

By the end of 1982 our experimental knowledge on the mass of the neutrinos could be summarized by the following values :

ν_e : $14 \text{ eV} < m_{\nu_e} < 46 \text{ eV}$ (Lyubimov et al.) ,

ν_μ : $m_{\nu_\mu} < 500 \text{ keV}$ (Seiler et al.) ,

ν_τ : $m_{\nu_\tau} < 250 \text{ MeV}$ (DELCO).

The first affirmative measurement of a finite mass had led to a very sharp scrutiny of the sources of error in the experiment.

The experiment relies on the precise measurement of the end point of the electron spectrum in the decay of tritium,

$$^3H \rightarrow {}^3He + e^- + \bar{\nu}_e \, .$$

It is a three-body decay.

The energy distribution of the electron is given by the relation :

$$dN/dE = p_e^2 \, [(Q-E) \, \{(Q-E)^2 - m_\nu^2\}^{1/2}]F_c \, ,$$

where

Q is the energy balance $[M_i(\text{atom}) - M_f(\text{atom})]$,
F_c is a Coulomb attraction term ,
p_e is the electron momentum.

It is customary to represent the distribution by the Kurie plot (see Fig. 1 taken from Ref. 1) :

$$K = [(1/p_e F_c)(dN/dE_e)]^{1/2},$$

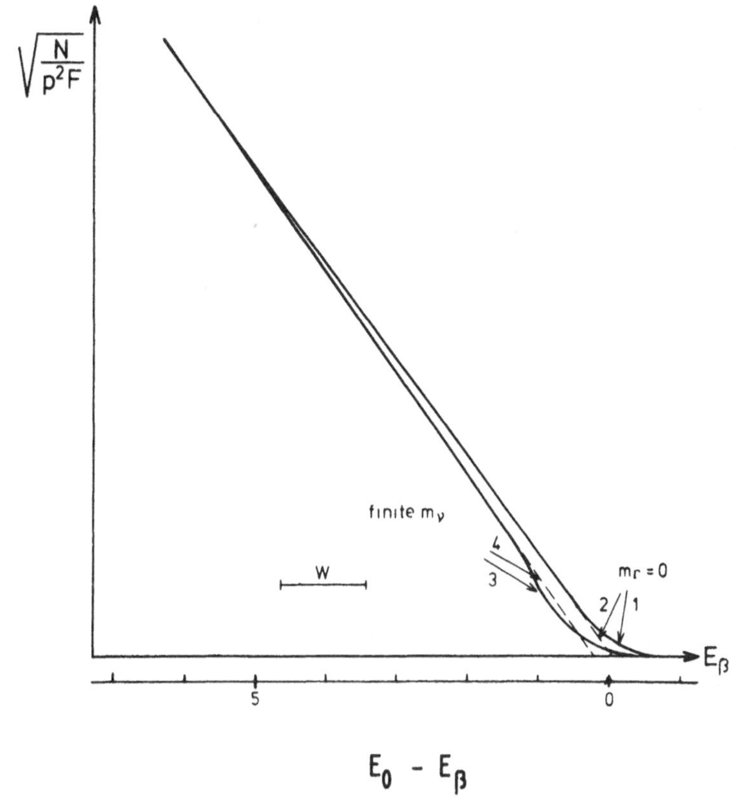

Fig. 1.

which is a linear function of E if $m_\nu = 0$.

The finite neutrino mass causes deviations near the end point. Theoretically the Kurie plot will have an infinite slope at the end point. In practice the slope will be decreased near the end point owing to the finite resolution of the spectrometer. The elements best suited to the study of the neutrino mass have the lowest end-point energy. It is easy to show that the figure of merit, defined as the ratio of the number of electrons in the Kurie plot between the extrapolated end point and the real end point, is proportional to $(m_\nu/Q)^3$. The element with the lowest balance is ^{187}Re with Q = 2.6 keV. Its lifetime of 4×10^{10} years makes it unsuitable for obtaining the large specific activities necessary for these experiments.

Tritium is particularly suitable for this measurement because of its low-energy end point, E_0 = 18.6 keV, and its convenient half-life, $T_{1/2}$ = 12.3 years. It has a long history of being used to give upper limits to the neutrino mass. As early as 1949 Hanna and Pontecorvo, using proportional counters, obtained a 500 eV limit.

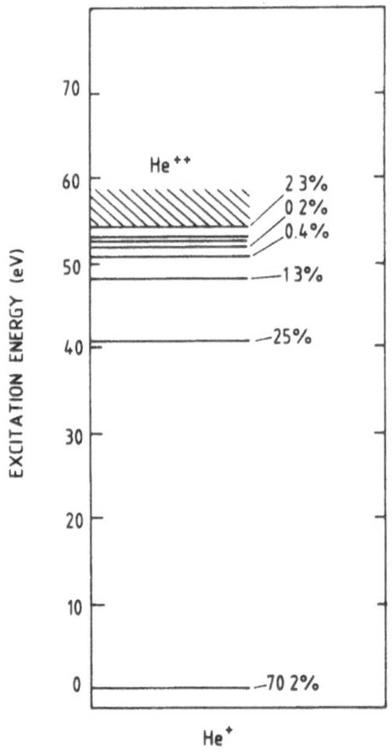

Fig. 2.

There have been numerous attempts since, but a breakthrough was achieved in 1972 by Bergkvist who increased the intensity by three orders of magnitude compared to previous studies. The tritium ions were implanted on Al backing, and the electron spectrum was measured by a combination of electrostatic and magnetic methods. The device had a resolution of 40 eV and the overall resolution, including energy losses in the source, was 55 eV. At this level of accuracy an additional difficulty arose : the initial state for a free tritium atom has one bound electron (E_o = 13.6 eV). During the β-decay nuclear charge suddenly changes by a factor of two. The initial bound spectator electron can be "shaken up" to the $2S_{1/2}$ state or to some higher excited state of the final ^3He atom. Bergkvist estimates that for a free atom the probability of the final $1S_2$ state is 70%, whilst that of the $2S_2$ state is 25%, and that the remaining 5% of the electrons probably go to the higher excited states or to the continuum (see Fig. 2). From the point of view of the neutrino-mass searches this leads to a decay having several end-point energies and leading to a modification of the line slope, essentially a broadening to $\sim$ 70 eV.

Figure 3 represents these atomic effects on the line shape.

The conclusion of Bergkvist is that there is no evidence for a neutrino mass above 55 eV (90% CL). The tritium end point is 18,610 ± 16 eV.

An experiment, with the aim of improving the accuracy, was undertaken at ITEP, in Moscow. The tritium was in the form of ^3H

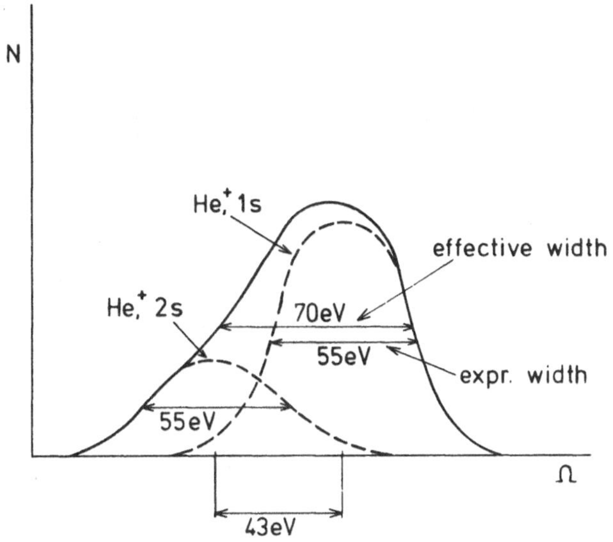

Fig. 3.

tagged valine ($C_5H_{11}NO_2$) with a much larger activity than in the work of Bergkvist. The background was reduced by a factor of up to 50. In 1976 the preliminary results gave $m_\nu \leq 33$ eV (90% CL). The end-point energy, 18,575 ± 13 eV, was 35 eV lower than the one determined by Bergkvist, which could be accounted for by the fact that the atomic surroundings of the ^3H atoms were different in the two experiments. The result published in 1980 finally gave positive results for the neutrino mass : the study of 16 independent samples gave evidence for a non-vanishing mass

$$14 \leq m_{\bar{\nu}} \leq 46 \text{ eV (99\% CL)}.$$

Assuming the structure of the final ^3He to be the same as in the case of Bergkvist one obtains narrower limits $26 \leq m_\nu \leq 46$ eV (90% CL). The limit of 14 eV is obtained by assuming the worst case, where only one state of ^3He is involved. The authors claim that this limit is independent of the atomic final-state effect.

It should be noted that these limits are narrower than the experimental resolution of the spectrometer, and the conclusions are based on an exhaustive discussion of all the sources of error and the energy calibration methods.

A recent remark by Simpson has cast doubts on the significance of this result. The spectrometer resolution function is derived from a calibration line, the M-shell internal conversion lines of the 20 keV transition in ^{169}Tm. Because of the finite width of about 10 eV of the electron M levels, it appears that the ^3H spectrum was deconvoluted with a resolution function wider than appropriate.

The effect on the end point would be that the evidence for a finite neutrino mass would be weaker or even disappear :

$$0 \leqq m_\nu \leqq 30 \text{ eV } (95\% \text{ CL}).$$

We may then conclude that the evidence from this experiment is weak. However, repetition of the experiment with stronger sources and taking the calibration line width into account yielded repeated finite values for the mass : m_ν = 35 eV. This experiment had the great merit of triggering half a dozen similar experiments aimed at avoiding some of the difficulties encountered in its interpretation. Experiments are under way using free tritium atoms or molecules. For instance, using an atomic beam source with known branching to $^3He^+$ states (no solid source effects), improved background rejection by electrostatic acceleration of the β's, and detailed study of instrumental response by the use of an electron gun, an unambiguous result at a mass below a limit of 10 eV is aimed at. [See, for example, Fackler et al.[5]]

Measurement of the neutrino mass in electron capture

Bennett et al.[6] and, independently, Raghavan[7] have conceived the idea that a finite neutrino mass alters the rates of electron capture from different electronic shells by an effect of the order of

$$1/2m_\nu^2/Q_x^2 ,$$

where Q_x is the electron capture decay energy excess over the binding energy of shell x. De Rújula[8,9] has shown that this effect, caused by the phase-space deficit due a finite ν-mass, is related to a characteristic distortion at the tails of X-ray or Auger lines, owing to a resonance effect between these transitions and the bremsstrahlung.

While electron capture is a two-body process, bremsstrahlung leads to a three-body process. The spectrum of the photon is governed as in tritium β-decay by the three-body phase space and has essentially the same shape. The determination of the photon distribution shape near the end point should lead to a measurement of the neutrino mass. In most cases the bremsstrahlung spectrum is very weak; however, when the photon energy near the end point is only slightly higher than the energy of an atomic transition it results in a considerable enhancement of the bremsstrahlung intensity. The case of ^{163}Ho illustrates the interest of this method. The Q value is 2.6 keV. With a factor of merit of $(m_\nu/Q)^3$ the advantage is clear. Another advantage of the method is that the perturbing effects of the sudden charge variation of the nucleus are smaller than in β-decay.

There remain the experimental problems, which are challenging and formidable. One has to measure the spectra of photons in the keV range, in coincidence with Auger electrons of very low energy, and be able to discriminate between electrons of about 50 and 500 eV. While this is reasonably feasible the accuracy required for a precise measurement of the neutrino mass is out of reach for the low intensity available. The best solid-state detectors have a resolution of the order of 100 eV, far worse than the magnetic spectrometers. The efforts undertaken to solve this problem can probably at best give the lowest upper limit of the ν mass.

De Rújula has also analysed the case when the third particle is an electron of the atomic shell ejected by the sudden change of charge of the electron-capturing nucleus. The intensity is higher than in the case of bremsstrahlung. In my opinion this method is experimentally more attractive since it is possible to accelerate the electrons by a fixed, accurate amount, to an energy much larger than their initial one which is below a few keV, and make use of magnetic analysis for an accurate measurement of the energy in the few electronvolt range.

However Raghavan[7] has found a new element to be a good "neutrino balance", offering intriguing possibilities; it is ^{158}Tb, which decays by an ultra-low energy K capture. It has a branch decaying to an excited state of 1187.3 keV in ^{158}Gd with an energy release of only 156 ± 17 eV above the K-shell threshold. The distortion in the shape of the K line should be a most sensitive test of the neutrino mass. A sensitivity of 25 eV is expected in a first generation of experiments using this approach.

THE DOUBLE β-DECAY[4]

One year after the publication of Fermi's theory, Goeppert-Mayer[9] computed the lifetimes of atoms decaying by double β-decay and found them to be of the order of 10^{25} years. In 1939 Furry[9] realized that Majorana neutrinos could be exchanged rather than emitted with a much smaller phase suppression than the 2ν process. However the belief in lepton conservation, in zero mass neutrinos, led to the prejudice of an interdiction of this process.

Double β-decay has been investigated for three decades. It is a possible process because the pairing energy in nuclei with an even number of protons and neutrons leads to tighter binding than the adjacent odd-odd nuclei. In many instances even-even nuclei are forbidden to decay to the neighbouring odd-odd nucleus, while it is energetically possible to decay to the next even-even nucleus via the second-order weak interactions.

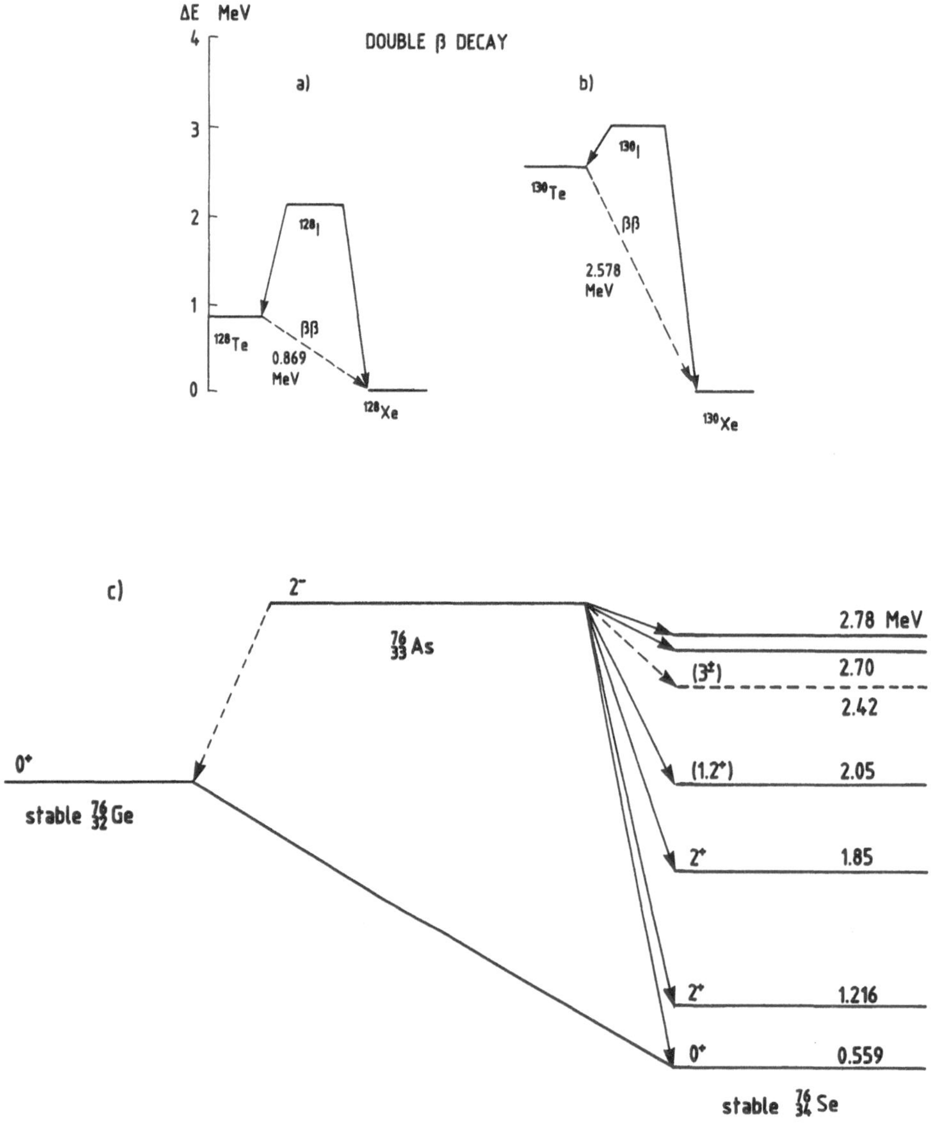

Fig. 4.

$$[A,Z] \rightarrow [A,Z+2] + 2e^- + 2\nu_e.$$

Three examples of such situations are given in Fig. 4. The first two isotopes are important because of the possibility of theoretically eliminating some sources of error by comparing their decay rates, while the third isotope has considerable possibilites since it is the constituent of the most accurate solid-state detector and, as we will see, provides favourable information from the branching ratio to two different levels.

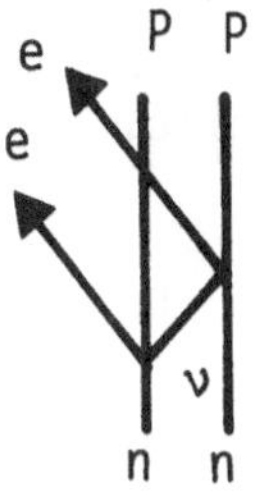

Fig. 5.

The lifetime of the second-order β-decay with 2ν emission is very long. It is well above 10^{20} years, which makes the experiments difficult. The point of interest is that if one relaxes lepton conservation laws, as is advocated in some GUTs, and if one puts forward the hypothesis of a finite neutrino mass, another process may enter into the game, a neutrino-less double β-decay :

$$[A,Z] \to [A,Z+2] + 2e^-.$$

This process violates lepton-number conservation (see Fig. 5). It requires that the virtual neutrino is identical to an antineutrino, or that the neutrino is a Majorana particle. This looks at first sight improbable since it is familiar to us that neutrino beams and antineutrino beams from high-energy accelerators produce different particles, for instance μ^- or μ^+. But this does not contradict the hypothesis of the neutrinos being Majorana particles. It is then the left-handed or right-handed helicity imposed by the interaction which is responsible for the selection of a given sign of the produced muon. The condition of $\nu \equiv \bar{\nu}$ is not sufficient. For the process to occur it requires also either the presence of right-handed currents (RHCs) in the Hamiltonian since the neutrino emitted at the first vertex has the wrong helicity to be reabsorbed, or the existence of a finite neutrino mass since the helicity can be inverted in another coordinate frame. These last two conditions are independent of each other and in the case of observation of the phenomena their relative importance can be measured. The effect of the finite neutrino mass is to eject two electrons with the same helicity in a relative S-wave, giving the selection rule $\Delta J^P = 0^+$, while the RHC ejects electrons of opposite helicity, in a relative P-wave, along $\Delta J^P = 0^+, 1^+, 2^+$.

Taking the examle of the germanium isotope, if the $0^+ \to 2^+$ transition is not observed but the $0^+ \to 0^+$ is, then the rate gives a measure of the neutrino mass. If both are measured, then a comparison of the rates gives the relative strength of the contributions of RHCs and finite neutrino mass. It should be pointed out that if the neutrinoless double β-decay is not forbidden it is strongly favoured with respect to the 2ν emission by large phase-space factors of the order of 10^6.

If there is an admixture of RHCs J_R, the total weak current can be written as

$$J = J_L + \eta J_R.$$

If $m_\nu = 0$ the rate $\Gamma(0\nu) \sim \eta^2 = (M_{WL}/M_{WR})^4$, where W_L is the usual intermediate vector boson, while W_R mediates the RHC. While for a transition mediated by the finite neutrino mass

$$\Gamma(0\nu) \sim (m_\nu/m_e)^2.$$

The present experiments give some impression of confusion, making it impossible to draw conclusions.

Two methods are available : the geochemical, where products of a double β-decay are looked for in ores whose age is known, containing the parent nucleus, and the direct measurement in the laboratory. Both methods have been used for ^{82}Se, which lead to ^{82}Kr by double β-decay from rocks about 10^9 years old. The measurements on rocks from four continents gave the same result within a factor of 3 to 4. They yield, however, a lifetime which is a factor of 28 larger than the one obtained directly in the laboratory by Moe and Lowenthal (See Ref. 4) : $(1.0 \pm 0.4) \times 10^{19}$ years. With such discrepancies there is not much room left for theoretical speculations.

Pontecorvo was the first to point out the advantage of looking for the relative decay rate of the two tellurium isotopes ^{128}Te and ^{130}Te. The energy release in ^{130}Te is almost three times larger than that in ^{128}Te; the ratio of their lifetimes is very sensitive to the number of leptons in the final state. The ratio is expected to be 5×10^3 for the 2ν decay and 25 for 0ν with no RHC and 300 with RHC.

What is observed in a first experiment is $(1.59 \pm 0.05) \times 10^3$, which means that 0ν exists as well as 2ν and, in a second experiment $(10^{+\infty}_{-5}) \times 10^3$, which is rather less compatible with 0ν. This has led to controversy concerning the relative values of the matrix elements in the two isotope decays, and in any case the calculated lifetimes differ drastically from the experimental ones, except for ^{82}Se where it agrees with the laboratory experiments.

If all these problems are ignored, and the most accurate experimental number is adopted, the calculation concerning the relative lifetime of $^{128}Te/^{130}Te$ gives $m_\nu \cong 10$ eV for no RHC for a calculation which treats more rigorously than others the relativistic phase and gives a result very different from previous ones, which range from 1 to 30 eV.

All this makes it understandable that a new generation of experiments is under way, aimed at measuring lifetimes of the order of 10^{24} years, at a new level of reliability, hoping to distinguish clearly between the different processes involved.

The difficulty is obvious : a lifetime of 10^{23} years requires 10^{2} grams of matter to give a few counts per year. The energy of the β-particles is a few MeV, i.e. of a small range. Two approaches seem promising :

i) Tracking devices, permitting to give a clear signature to the emission of two β-particles from a single point, thus reducing background by requiring a strict geometrical condition. While the first experiments along this line used spark chambers and cloud chambers, the present approach is based on time projection chambers (TPCs), which are a combination of drift and wire chambers. The fact that ^{136}Xe is a good filling for such chambers and a candidate for double β-decay is a favourable factor. Unfortunately, the abundance of ^{136}Xe in natural xenon is only 8.9%.

ii) The other approach makes use of the very favourable fact that Ge is both an ideal detector and presents a favourable decay scheme for checking the selection rules.

In principle the 0ν or 2ν decays should yield a peak and a continuous spectrum. Such an experiment was first performed by a Milan group, in the Mont Blanc Tunnel as early as 1972, with 70 cm^3 of Ge.

A set of new experiments is planned with Ge. A characteristic one aims at using 1400 cm^3 of Ge split into 8 crystals of 170 cm^3, i.e. 20 times more than the first Milan experiment. A thick heavy NaI shield serves many purposes : if a 3.6 MeV γ from ^{208}Pb, which is at the end of the thorium chain, gives a 2.045 MeV pulse in Ge, by a Compton scattering, the remaining γ can be detected either in the NaI or in another Ge crystal, and thus be rejected.

Also a transition $0^+2^+ \rightarrow 1.486$ MeV γ, in coincidence with the 0.559 line of 2^+0^+ transition, can be detected. If the lifetime is 2×10^{23} years the authors expect 1 count per month. Needless to say a formidable effort is being made to eliminate all sources of noise.

The best measurements so far done give some upper limits on the parameters relevant to the process (see Table 1).

If the process is not observed it will be compatible with any neutrino mass with lepton conservation barring the 0ν double β-decay. If the 0ν process is observed it may lead to information on several new aspects of weak interactions such as the mass of W_R (intermediate

Table 1

Source	$T_{1/2}(0\nu)$ (yr)	m_ν (eV)	η
^{48}Ca	$10^{21.3}$	< 30	
^{76}Ge	$10^{21.7}$	< 15	< $3x10^{-5}$
^{81}Se	$10^{21.5}$	< 12	< $1x10^{-5}$

vector boson responsible for RHC), and the mass of the neutrino in the range of 1 eV for no RHC. Several of the experiments in preparation hope to run very soon.

In 1955 Winter[10] estimated the probability of double electron capture, with two neutrinos and no neutrino emission. Bernabeu et al.[11] have shown that in some cases where the daughter nucleus is excited the neutrinoless decay may be enhanced by its proximity to a virtual resonance.

NEUTRINO OSCILLATIONS AND THE NEUTRINO MASS[1,3]

Grand Unified Theories lead to finite neutrino mass and lepton non-conservation. Let ν_e and ν_μ be the neutrinos produced in the weak interaction

$$\nu_\ell = \sum_{\sigma=1,2} U_{\ell\sigma} \nu_\sigma \qquad (\ell=e,\mu) ,$$

and ν_σ (σ = 1,2) is the state vector of the neutrino with mass m_σ, momentum p, and helicity equal to − 1 (we assume $p \gg m_\sigma$).

The matrix U has the form :

$$U = \begin{pmatrix} \cos\theta & \sin\theta \\ -\sin\theta & \cos\theta \end{pmatrix} .$$

What happens to a beam of neutrinos produced in some ordinary weak process? At the time t = 0, the beam is described by ν_e; at the time t the state vector of the beam is given by :

$$\nu_e(t) = e^{-iHt}\nu_e = \sum_{\sigma=1,2} U_{e\sigma} \; e^{-iE_\sigma t}\nu_\sigma .$$

Table 2. Theoretical expectations (in eV)

10^{-6}	10^{-4}	10^{-2}	1	10^{2}
Sun				
	Cosmic rays			
		Reactor		
			Low-energy accelerator	
				High-energy accelerator

This is not a stationary state since ν_ℓ is not an eigenstate of nature's Hamiltonian.

Calculations show that the probability $w_{\nu e';\nu e}(R)$ of finding $\nu_{e'}$ at a distance R from a source of ν_e, is :

$$W_{\nu e;\nu e}(R) = W_{\nu\mu;\nu\mu}(R) = 1 - 1/2 \sin^2 2\theta \,(1 - \cos 2\pi R/L), \qquad (1)$$

$$W_{\nu e;\nu\mu}(R) = W_{\nu\mu;\nu e}(R) = 1/2 \sin^2 2\theta \,(1 - \cos 2\pi R/L) \,, \qquad (2)$$

where $L = 4p/|m_1^2 - m_2^2|$ is the oscillation length.

Observing the oscillation consists in checking the existence of the cosine term in the neutrino intensity, or establishing the value of the constant terms 1 and 0 in the expressions (1) and (2). Because of the two independent parameters θ and $\Delta m^2 = |m_1^2 - m_2^2|$ the interpretation of the experiments is not straightforward.

Table 2 shows the neutrino masses accessible to oscillation experiments.

Since the solar neutrino puzzle led to the possible interpretation of it as due to the oscillations, great interest has been shown in checking the existence of oscillations in a laboratory experiment. Since the reactor energies cover a domain where the hypothetical neutrino mass of tritium lies, reactor experiments have been the subject of much attention. The neutrino fluxes are quite high :

$$F(\bar{\nu}_e/\mathrm{cm}^2.\mathrm{s}) \cong 1.5 \times 10^{12}\ P/L^2,$$

where P is the thermal power of the reactor in megawatts and L is the distance in metres. Only $\bar{\nu}$ are produced and the principle of the experiment consists in detecting the e^+ produced in the inverse reactions.

Two approaches are possible. One is to use a movable target and look for a change with distance of the total number and/or the spectrum of positrons. This is difficult because of the shielding. No result has so far been obtained with this method. The other method consists in an absolute measurement of the positron spectrum at a stable distance from the reactor. The observed spectrum is then compared with the expected spectrum.

There are two ways of determining the electron antineutrino spectrum. The most straightforward involves determination of the β branches of all fission fragments. The second relies on an experimentally determined electron spectrum associated with fission, which is then directly converted into the corresponding ν spectrum. One has to use the data from 700 fission fragments, of which 250 nuclei are well known but the others have a reduced weight. There is no sign of oscillation. The interpretation gives a rather complicated two-dimensional region of exclusion, which is of little interest in the absence of any theory for θ. Clearly while a positive result would have considerable implications, negative ones are of little value.

THE SOLAR NEUTRINO PUZZLE

Davis et al.[2] have measured the neutrino flux from the sun. They measured the production of ^{37}Ar from the reaction

$$\nu + {}^{37}\mathrm{Cl} \rightarrow {}^{37}\mathrm{A} + e^-$$

by irradiating 3.8 x 10^5 litres of C_2Cl_4 in a mine 1500 m deep. The difficulty of the experiment is illustrated by the observed rate : 0.42 ^{37}Ar atoms per day, for an estimated background of 0.08 atoms per day.

The ^{37}Ar is concentrated by flushing the liquid with He and measuring the ^{37}Ar in a proportional counter which has a background noise of only 1 count per month.

The solar neutrinos are produced in the thermonuclear reactions of the hydrogen cycle, the main reactions being tabulated in Table 3.

The threshold for neutrino capture in ^{37}Cl being 0.812 MeV, the reaction is unfortunately insensitive to the most abundant sources of neutrinos, in the above-listed reactions, and detects the neutrinos from ^{8}B.

Table 4 compares the neutrino fluxes with the rates shown by this reaction.

Table 3. The solar neutrino puzzle

$p + p \to d + e^+ + \nu$
$p + e^- + p \to d + \nu$
$p + d \to {}^3He + \nu$

86% ↓ 14%

${}^3He + {}^3He \to 2p + {}^4He$ | ${}^3He + {}^4He \to {}^7Be + \nu$

99.9% ↓ 0.1%

${}^7Be + e^- \to {}^7Li + \nu$
${}^7Li + p \to 2{}^4He$

$p + {}^7Be \to {}^8B + \gamma$
${}^8B \to {}^8Be^* + e^+ + \nu$
${}^8Be^* \to 2{}^4He$

Table 4

ν source	E_ν max (MeV)	Flux	^{37}Ar (SNU)[a]
$p + p \to d + e^+ + \nu$	0.42	6×10^{10}	0
$p + \bar{p} + p \to d + \nu_e$	1.44(line)	1.5×10^8	0.3
${}^7Be + e^- \to {}^7Li + \nu_e$	0.86,0.38	4.5×10^9	1
${}^8B \to {}^8Be + e^+ + \nu$	14	5.4×10^6	4.5

[a]An SNU (solar neutrino unit) is a rate of 1 event per 10^{36} detector atoms.

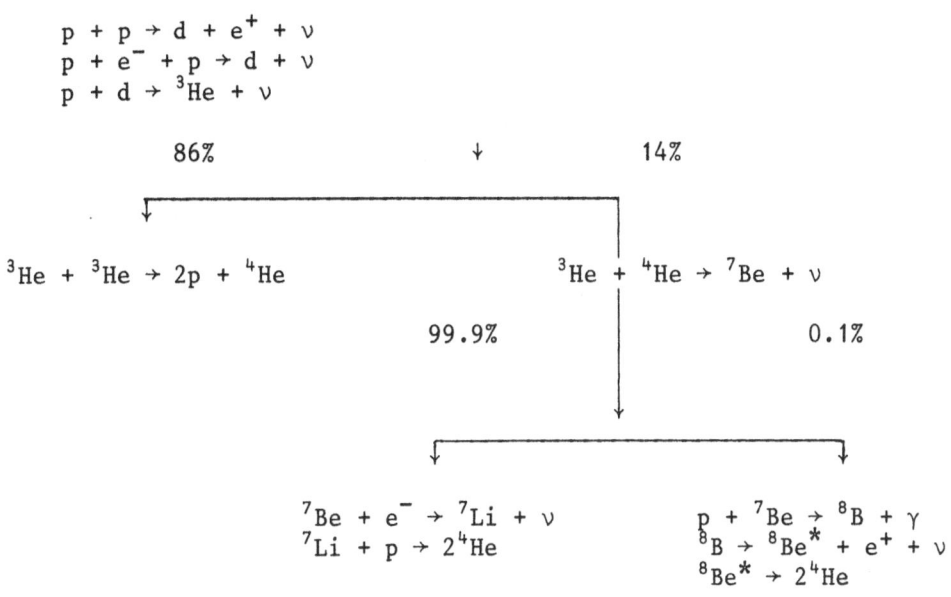

It is clear that the predicted ratio depends on specific solar models, but these were estimated to be reasonably well established until they came into contradiction with the observations of Davis. A factor of three is missing in the observed rate to be compatible with the most commonly admitted model. While this result led to a speculation on very different solar models and to a flurry of experiments on low-energy nuclear reactions playing a major rôle in the different steps of the thermonuclear cycle, more interesting for us is the speculation that the model is correct and that the lack of neutrinos is due to oscillations and the transformation of the electron neutrino into another flavour.

At the same time, new detectors were suggested for the detection of solar neutrinos from the most abundant nuclear reactions. Table 4 illustrates the very favourable position of the most promising one, ^{71}Ga; 65% of the counts would come from the pp reactions. The rate is insensitive to solar physics if the sun is static and the prototype works. The need for 50 tons of Ga makes the experiment expensive, at the level of a large high-energy physics experiment. The expected rate of about 100 SNU in the case of the validity of the static solar model and the absence of oscillations make this experiment a very sensitive test of many of the most exciting hypotheses in present-day physics.

NUCLEON DECAY[10]

Different models permit various possibilities of baryon-number violation, each of them characterized by different mass scales and selection rules.

One class predicts $M_X \cong 2.4 \times 10^{14}$ GeV and $\tau_p \cong 3 \times 10^{29\pm2}$ years. The correct predictions of $\sin^2\theta_W \cong 0.214$ gives strong support to this class of theories.

Other possibilities of baryon-number violation include n-$\bar{n}$ oscillation, and cosmic-ray antiprotons. There have been observations of cosmic-ray antiprotons with flux ($\bar{p}/p \geq 10^{-4}$), an order of magnitude larger than expected from secondary production in the interaction of cosmic-ray primaries. Especially surprising is the report of low-energy (hundreds of MeV) antiprotons. Possible explanations include n-$\bar{n}$ oscillations; secondary neutrons oscillate into $\bar{n}$'s which then β decay. The observed $\bar{p}$ flux would require an oscillation time of $\tau = 10^4$ s, which is apparently ruled out by direct measurements [$\tau_{nn} > 10^5$, Baldo-Ceolin[10]] and nuclear stability ($\tau_{nn} > 10^5 - 10^7$ s).

A great variety of nucleon-decay detectors have been constructed: water Cherenkov counters, fine-grained sampling calorimeters (with slabs of material separated by scintillation counters), planes of proportional tubes, planes of Geiger tubes, planes of drift chambers, and liquid scintillators. A difficult balance has to be found between cost, size, and characteristics. Far more ambitious detectors are considered for a possible second generation, such as high-density TPC or BaF_2 calorimeters.

Water, being the cheapest material, has permitted the construction of the largest type of detectors. However it is preferentially sensitive to one channel : $p \rightarrow \pi^0 + e^+$. If this channel is the most abundant one, as in the SU(5) minimal model, the water Cherenkov counter is the best one.

The largest of these detectors, with 8000 tons in total mass and 4000 tons in fiducial mass has now yielded no visible event, while seven were expected in the running period of 140 days. This is a significant number and points either to the non-validity of the model or to the validity of models favouring decay modes to which it is not or is less sensitive, such as the $K\mu$ decay.

The observation of one candidate in the NUSEX experiment[10], if confirmed, would point to a $K\mu$ decay. This would favour the contribution of fine-grained detectors with more detailed information permitting the identification of particles. Two candidates emerge, the high-density TPC and the BaF_2. Unfortunately the very high cost involved makes it difficult to excite enthusiasm for funding as long as no convincing sign of p decay is found with the existing detectors[12].

THE SEARCH FOR SUPER-HEAVY MAGNETIC MONOPOLES[13]

The possibility of isolating electric poles but not magnetic ones is a fundamental distinction between electricity and magnetism. The explanation is well known. The magnetism of an ordinary magnet arises from loops of electric currents.

If magnetic monopoles exist it would be easy to symmetrize the Maxwell equations and have the magnetic currents and charges playing the same role as their electric counterparts.

The main argument which makes the existence of monopoles plausible was introduced by Dirac in 1931. In classical non-relativistic mechanics the existence of a conserved total angular momentum for a system consisting of a magnetic pole and an electric charge leads to an interesting conclusion. The contribution of the fields to the angular momentum is a vector s directed from charge to pole.

$$s = \int d^3r\ [r \times (E \times B)/4\pi c].$$

If electric and magnetic charges are measured in the same units, then the magnitude of s is simply $\hat{r}(eg/c)$, where $\hat{r}$ is the vector unit. It is thus independent of the distance from pole to charge.

The quantization of angular momentum leads to :

$$ge/c = nh/2\ ,$$

where n is integer and

$$g = (n/2)(hc/e^2)e = (n/2\alpha)e.$$

The minimum monopole charge is thus

$$g = (1/2\alpha)e.$$

For a given value of g, if s is quantized, the electric charges are also quantized. The monopole g in e.m.u. is about 68.5 times the value of e in e.s.u. A Dirac monopole would feel a force of 2 MeV/m in a 1 G magnetic field.

All these theories would be disproved if a small difference were found between the magnitude of the electron and the proton charge because it would require an enormous increase in the unit of magnetic charge quantization. Since experiment limits that difference to less than $10^{-20} \times e$, any possible difference would correspond to magnetic charges experiencing forces of more than 3000 tons in a magnetic field of 1 G.

Charge to pole interactions are such that reflections in space or time show motions that may be possible only if all poles change sign. Thus pole strength must be treated as a pseudoscalar quantity in order for parity and time reversal symmetries to hold.

For many decades monopoles were searched for in several directions :

- In cosmic rays, where hope of finding relativistic monopoles with a 5000 times larger ionization than single-charge electric particles led to the misinterpretation of heavily ionizing events.
- In the running in of new accelerators, at every jump in the energy scale, here again without any hypothesis on the mass, it was a shot in the dark with no positive indication.
- In the possible storage of monopoles bound in various materials[14]. For instance, when a monopole is in the neighbourhood of a ferromagnetic crystal it is attracted by its image charge; one can show that the binding energy is greater than 30 eV by using classical laws for distances greater than 1000 Å. A monopole can escape from the magnetic trap if exposed to a strong magnetic field.

Eberhard et al.[13,17] at LBL have thus investigated lunar materials with negative results. The method consists in running samples in a superconducting loop and looking for induced currents with a sensitive magnetometer, the "SQUID"*. But the field of research changed drastically when 't Hooft and Polyakov showed that monopoles are required by SU(5) and that they also yield charge quantization, with a rather precise prediction on the mass, around 10^{16} GeV, i.e. 20 ng or the weight of an amoeba. This permits guesses as to the velocity of the monopoles on the earth surface and gives guidance as to the different detection methods to be used.

*SQUID = superconducting quantum interference device.

It is tempting to detect monopoles by the electromagnetic effects induced by a moving magnetic charge. The smallness of the effect has led to the use of supraconductive magnetometers, the SQUID. The flux in a closed loop of supraconductor is quantized in units of $2\phi_o$ where $\phi_o = hc/2e$. A Dirac monopole passing through the loop changes the flux by $4\pi g = hc/e = 2\phi_o$.

The first claim concerning the possible observation of a monopole, by Cabrera[15], was based on the signal obtained in a coil imbedded in a magnetic shield, giving a residual field of only 5×10^{-8}G.

The passage of the monopole induces a change of current

$$I(t) = \phi_o/L \{1 + [\gamma vt/\sqrt{(\gamma vt)^2 + b^2} \} ,$$

where b is the radius of the coil and L the inductance. This current is persistent. It can be measured with the SQUID, which permits a measurement of single $2\phi_o$ jumps with a high signal-to-noise rate.

The signal which Cabrera recorded on 14 February 1982 made a sensation because it indicated fluxes of the order of 0.5 $m^{-2}sr^{-1}d^{-1}$. The intense interest in this result has triggered many new experimental projects with coil diameters larger than the initial 20 cm^2 of Cabrera. Cabrera himself has improved the sensitivity x running time of his set-up by a factor $\sim$ 15, with no new candidate observed.

Rubbia[16] has considered using ordinary cooled coils of large surface, which are much less expensive. He comes to the conclusion that the signal-to-noise ratio can be favourable and thus permit larger surfaces than can be hoped for with supraconductors as well as multiple observation of tracks by hodoscoping methods.

In my lectures I will not devote more time to the magnetic detection of monopoles, mainly for the reason that a recent work by Drell et al.[17] shows that it is possible with gaseous detectors to cover the range of low velocities, which is the main interest of magnetic detection, with the advantage of much larger detecting surfaces.

Theoretical models on how supermassive monopoles enter and move in the galactic fields and in the solar system suggest that their velocities at the earth's surface would be of the order of the earth's orbital velocity about the sun ($\beta \sim 10^{-4}$) and in any event no less than the escape velocity from the earth ($\beta \sim 3 \times 10^{-5}$). This leads to less vague approaches to the problem of detection of the monopoles by their ionization.

For many years the estimate of the ionization of matter by monopoles was simply an extrapolation of the effects of slow moving charges where the theory is weakest. Figure 6 shows a rough repre-

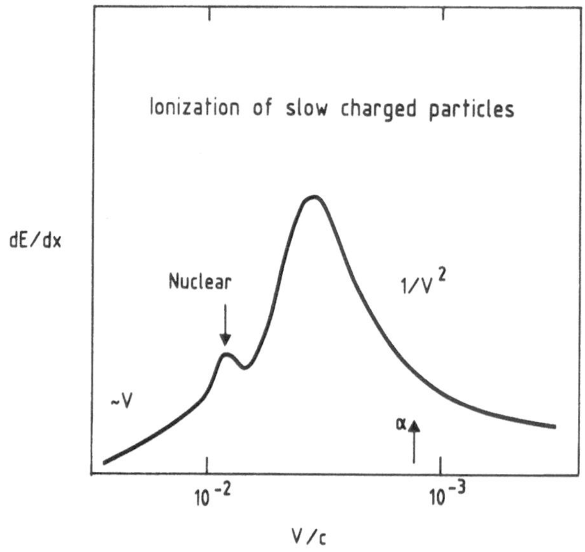

Fig. 6.

sentation of the ionization loss of charged particles in matter as a function of velocity. The high-energy physicist is familiar with the curve at the right of $\beta = 10^{-2}$, which corresponds to about the same velocity as an electron on the Bohr orbit of hydrogen. At velocities between 10^{-3} and 10^{-4} the energy loss varies like β and it is easy to extrapolate it to a massive monopole. The monopole being massive the relative velocity is the velocity of the orbital electron $\sim \alpha c$. The force exerted on an atomic electron by a field $\vec{E}$ is

$$F_E = e\vec{E} = eq(1/r^2). \tag{3}$$

The force exerted on a moving magnetic charge is :

$$F_m = e\vec{v} \times \vec{B} = evg(1/r^2) \; ,$$

$$v_{relative} = v_{electron} = \alpha c, \quad g = e/2\alpha \; ,$$

$$F_m = e^2/2r^2 \; . \tag{4}$$

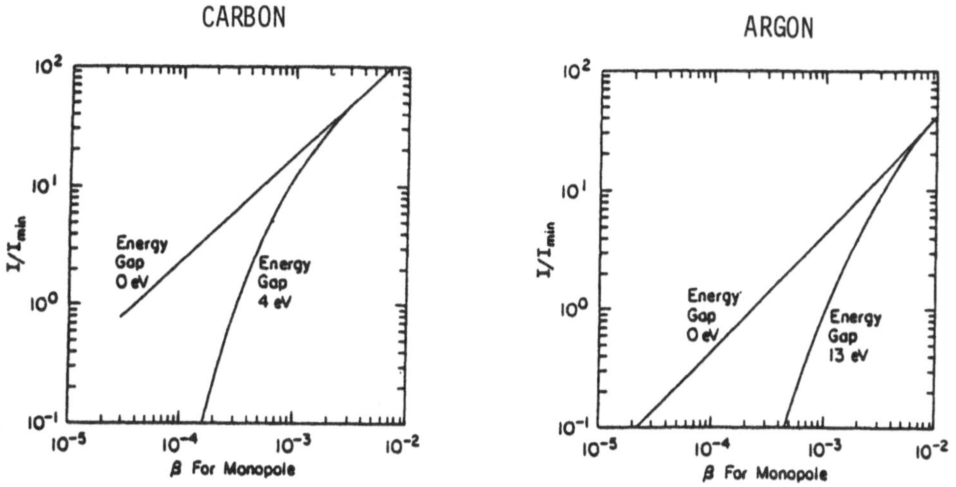

Fig. 7.

Comparing Eqs. (3) and (4), one may conclude that a monopole ionizes half as much as a proton of the same velocity, well above threshold.

The threshold for excitation corresponds to energy transfers of the order of $\Delta E = 1/2\ m_A \times v^2 > E^*$, which leads to β_{min} ranging from 1.4×10^{-4} for hydrogen to 0.11×10^{-4} for xenon. These naïve considerations have been the subject of more precise calculations, taking for the atoms a Fermi-Thomas model, without threshold and with threshold. Figure 7 shows the effect of threshold cutting sharply in the ionization at low velocities.

These calculations were basic in interpreting the experiments performed with existing large-surface detectors, usually made of the scintillator or gaseous detector planes used in nucleon-decay experiments. Moderate thresholds, of the order of 10^{-4}, were expected and Cabrera's results encouraged some physicists to investigate the possibility of using detectors with the lowest possible threshold, namely silicon detectors. However, the experimental perspectives have changed drastically with the recent results of Drell et al.[17], showing that naïve calculations do not describe properly the behaviour of slow monopoles in gases. In gases where rigorous calculations are possible, as in H and in He, they showed that the effect of the strong magnetic field is to mix levels, which leads to excited states which can easily be detected. Figure 8 shows the energy loss in H and He compared to the more conventional calculations. We see that at a β of 2×10^{-4} the energy loss by excitation is almost 2 orders of magnitude larger than what would have been expected from the ioniza-

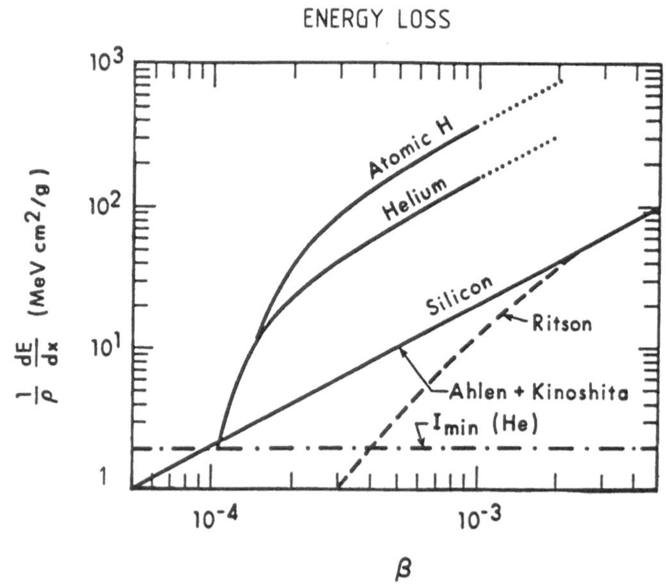

Fig. 8.

tion calculation. We also observe that with He there is a sharp kinematical threshold at about 10^{-4}.

With heavier noble gases the threshold is lower and reaches $\beta \sim 10^{-5}$ for Xe. It is quite easy to convert the excitation energy of these gases by addition of a proper vapour with an ionization potential lower than the excitation level.

This has given rise to new proposals which aim at reaching sensitivities 4 or 5 orders of magnitude larger than that obtained with the direct measurement of signals induced in coils by the moving monopole. The level at which future experiments aim is set by interesting debates on the possible limits compatible with astrophysical considerations.

The lower limit is set by considerations on the long-term effect of a flux of monopoles on the galactic fields. Accelerated monopoles gain energy at the cost of the energy stored in the circulating currents responsible for the magnetic fields. A uniform galactic field of a few microgauss can have survived only if the monopole flux

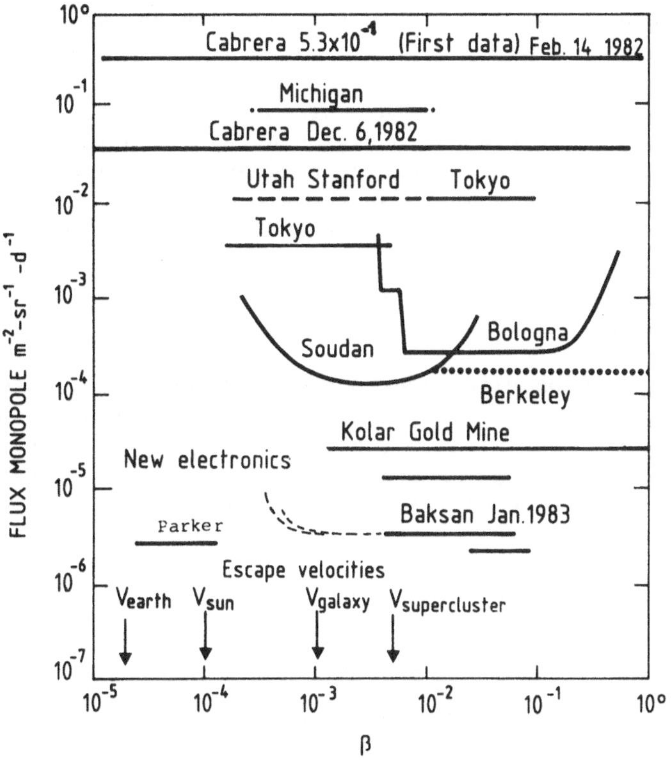

Fig. 9.

is lower than $10^{-5}/m^2$sr yr, which is five orders of magnitude smaller than what would be found if Cabrera's event was a monopole.

However this argument was considered as subject to objections on various grounds. Eberhard[18] pointed out that it was possible to conceive processes with non-stationary galactic magnetic fields, the fields and the monopoles' speed oscillating synchronously. Dimopoulos et al.[19] have shown that models leading to strongly enhanced local densities of monopoles are conceivable. In such a case the effect on the average galactic field is strongly reduced; the monopoles could be concentrated on orbits like meteorites in the solar system, and their velocity with respect to earth could range from 10 to 70 km/s. These considerations are rather encouraging for experiments, with a threshold of sensitivity compatible with the lowest conceivable monopole velocities.

Ritson and Loh, following the work of Drell et al., propose a detector of 1000 m^2, where the background rejection is obtained mainly from time-of-flight selection between successive layers of wire chambers. It is claimed that thresholds of $\beta = 10^{-4}$ will be reached with a proper choice of the gas fillings.

Figure 9 shows the present limits on the monopole flux set by the different experiments or theories.

It is easy to conceive detectors with thick drift spaces, with a detection threshold such that even ionization losses a hundred times smaller than from minimum ionizing particles are detectable, leading to a threshold of the order of $\beta = 10^{-5}$ with a high-Z noble gas filling. The cost would be certainly higher than the proposal of Ritson and Loh, but the sensitivity range would be practically the same as that of the magnetic loop methods, with the advantage of giving the velocity and direction of the monopole. The method is based on a comparison of the monopole velocity and the drift velocity of electrons in gases.

REFERENCES

1. P.H. Frampton and P. Vogel, Massive neutrinos, Phys. Rep. 82 (1982) 339.
2. S.M. Bilenky and B. Pontecorvo, Lepton mixing and neutrino oscillations, Phys. Rep. 404 (1978) 225 .
3. F. Boehm, Status of lepton conservation and neutrino mass, submitted to Surveys, 3 March 1983.
4. D.O. Caldwell, Proc. 1982 Summer Study on Elementary Particle Physics and Future Facilities, Snowmass, Colorado (Amer. Phys. Soc., Division of Particles and Fields, New York, 1983), p.600.
5. O. Fackler et al. (Rockfeller University-Fermilab-Lawrence Livermore Lab. Collaboration), An experiment to measure the ν_e mass, private communication from R. Cool.
6. C.L. Bennett et al., The X-ray spectrum following ^{163}Ho electron capture, Phys. Lett. 107B (1981) 19. This idea was independently conceived by Raghaven in 1980 and stimulated the spectroscopic program which culminated in the results on ^{158}Tb, to be published (see Ref. 7.).
7. R.S. Raghaven (Bell Telephone Lab.), Ultra-low energy K-capture decay of ^{158}Tb : a new neutrino "balance", to be published.

8. J.U. Andersen et al., A limit on the mass of the electron neutrino : the case of ^{163}Ho, Phys. Lett. 113B (1982) 72.
9. A. De Rújula and M. Lusignoli, Calorimetric measurements of ^{163}Ho decay as tools to determine the electron-neutrino mass, Phys. Lett. 118B (1982) 429.
10. Proc. 1982 Summer Workshop on Proton Decay Experiments, Argonne, Ill. (ANL-HEP-CP 82-24, 1982).
11. J. Bernabeu, A. De Rújula and C. Jarlskog, Neutrinoless double electron capture as a tool to measure the electron-neutrino mass, Nucl. Phys. B223 (1983) 15.
12. R.M. Bionta et al. (IMB Collaboration), A search for proton decay into $e^+\pi^\circ$, submitted to Phys. Rev. Lett., 1983.
13. S. Coleman, The magnetic monopole fifty years later, Harvard preprint HUTP 82/A032 (1982).
14. D.B. Cline, Binding of monopoles in matter and search in large quantities of old iron ore, Invited talk at Wingspread Monopole Workshop, Racine, Wisconsin, October 1982.
15. B. Cabrera, First results from a superconductive detector for moving magnetic monopoles, Phys. Rev. Lett. 48 (1982) 1378.
16. C. Rubbia, Hunting the supermassive monopole without superconductivity, Int. report CERN-EP/82-01 (1982).
17. S. Drell, N. Kroll, M. Mueller, S. Parker and M. Ruderman, Energy loss of slowly moving magnetic monopoles in matter, Phys. Rev. Lett. 50 (1983) 644.
18. P. Eberhard, Magnetic monopoles and the galactic magnetic field, LBL, Group A physics note No. 946, 23 July 1982.
M.S. Goodman et al., A sensitive search for neutron-antineutron transitions, Oak Ridge Proposal ORNL/Phys.-82/1 (1982).
19. S. Dimopoulos, S.L. Glashow, E.M. Purcell and F. Wilczek, Magnetic monopoles : A local source?, Harvard preprint HUTP-82/A016 (1982).

SELECTED TOPICS IN SPECTROSCOPY

A. Martin

CERN

Geneva

In this unique lecture, it is impossible for me to cover all spectroscopy, which is still, experimentally and theoretically, a flourishing field. I was in fact counting on the fact that J.-L. Basdevant would speak about what I would call light quark spectroscopy in which both radial and angular excitations are still being discovered and "explained". Unfortunately, Jean-Louis will not speak. I shall restrict myself mostly to heavy quark spectroscopy and make a timid incursion in the domain of baryons containing light quarks.

Initially, the quark model was largely a mnemonic trick to remember which states are allowed and which are forbidden. In the eyes of Murray Gell-Mann, it was a procedure to write currents with the right commutation relations. It is only in 1974, when S. Ting and B. Richter discovered the J/ψ and when people understood that this very narrow particle (60 keV for a mass of 3 GeV !) could not be something else than a heavy quark-antiquark bound state, that things changed. At that moment it was understood that quarks, or, at least heavy quarks like the c quark predicted by S. Glashow - J. Iliopoulos - L. Maiani in 1970, behaved very much like real particles, except for the fact that maybe they cannot be isolated. This was in fact from the very beginning the point of view of George Zweig.

It was natural to describe the $c\bar{c}$ system, of which two levels were known from the very beginning, the J/ψ (mass 3095 MeV) and the ψ' (mass 3685 MeV) as a system of two heavy objects interacting by a potential and obeying the Schrödinger equation. However, unlike what happens in the hydrogen atom, the quark-antiquark potential was not known. However, the theoretical prejudice is that one gluon exchange dominates at short distances while at large distances a linear rising confining potential seems to be the most natural thing to prevent quarks from escaping. So the initial potential used was

$$V = -\frac{4}{3}\frac{\alpha_s}{r} + Cr.$$

The potential model description was temendously successful in a qualitative and semi-quantitative way :

- the existence of P states between the ψ and ψ' was predicted. A triplet of P states was discovered. The non-observation of the singlet P state is due to the difficulty of producing it.
- A D state, with the same quantum number as the photon, was predicted with an energy slightly higher than the ψ' and was observed in e^+e^- collisions.
- Radial excitations of the S states were predicted and seen.
- Pseudoscalar states η_c and η_c' respectively associated with the vector states $J\psi$ and ψ' were predicted and, after a complicated history, observed experimentally.

In 1977 another heavy quark system was found, the Υ, made of a $b\bar{b}$ pair, as indicated by an indirect measurement of the absolute value of the charge of the constituent quarks. The Υ, with a mass 9459.9 ± 0.1 MeV (notice the fantastic precision !) is much heavier than the $J\psi$ and logically the potential description should work still better. Now, in fitting the energy levels of the Υ, we have the extra constraint that the potential should be approximately the same. This is "flavour independence" : because QCD is a non-Abelian gauge theory, the quark-quark-gluon coupling is related to the three gluon coupling and, therefore, cannot depend on the particular flavour of the quarks. It is not completely obvious that the resulting potential (if such a notion makes sense) will be exactly flavour independent, but the deviations from flavour independence should go to zero when the mass of the quarks goes to infinity. However, we can try to assume "precocious flavour independence" and try to impose that the $b\bar{b}$ and $c\bar{c}$ potentials be exactly the same.

It is at this point that we enter into the domain of quantitative spectroscopy. The question is what is the quark-antiquark potential. In principle, the quark-antiquark potential could be obtained from lattice gauge theory calculations, and this is on the way. However, lattice gauge theory would have a hard time getting the short distance behaviour, at distances less than the lattice spacing. However, the short distance behaviour can be obtained from asymptotic freedom. This was done first by H. Krasemann and S. Ono[1], J. Richardson[2] and W. Buchmüller, G. Grunberg and S. Tye[3]. One has at short distances :

$$V = -\frac{4}{3}\frac{\alpha_s(r)}{r} ,$$
$$\alpha_s(r) = \frac{12\,\pi}{25\log\left(\frac{1}{r^2\,\Lambda^2_{\overline{MS}}}\right)} . \qquad (1)$$

Notice that this attenuation of the Coulomb singularity is already not negligible for the Υ spectrum but becomes crucial for the $t\bar{t}$ system, since we know that the mass of the t quark is superior to 20 GeV.

At larger distances, while H. Krasemann and S. Ono just connect the potential to the linear confining potential with a logarithmic part, as was done before by G. Bhanot and S. Rudaz[4], J. Richardson and W. Buchmüller (whose more refined short range potential contains two loop effects) have a nice analytic interpolating formula in momentum space.

In contrast with these QCD inspired approaches, one can also take a purely phenomenological point of view and try to find the simplest potential fitting the $c\bar{c}$ and $b\bar{b}$ spectrum. Historically, the first attempts in this direction were made by Y. Tomozawa, C. Quigg and J.L. Rosner. The latter pointed out that if the spacings between all levels of the $c\bar{c}$ system were the same as the spacings between the levels of the $b\bar{b}$ system, this would be a proof that the potential is exactly $V = C\log(r/r_\circ^*)$. It is a simple exercise to prove, rescaling T, that if $V = \log r$, the distances between the levels do not depend on the mass of the quarks. The converse is more difficult to prove. In any case, as one can see from Table 1 (which is much more complete now !) this property is not very far from being satisfied by the $c\bar{c}$ and $b\bar{b}$ spectrum. We notice, however, a tendency of the $b\bar{b}$ levels to have a somewhat smaller spacing. To take this into account, we can consider the simplest generalization of the log r potential, $V = A + Br^\alpha$, with α slightly positive. This is what I tried[5], taking

$$V = -\,8.064 + 6.870\ r^{0.1} , \qquad (2)$$

where the units are GeV for energies and GeV^{-1} for distances.

In Table 1, we give the level spacings, from the 1^{--} ground state, to the various L = 0, L = 1 and L = 2 states (for L = 1, we give the centre of gravity of the triplet states), and the relative leptonic width of the L = 0; $\Gamma_{ee}(nS)/\Gamma_{ee}(1S)$ calculated "naively" using the formula $\Gamma_{ee} = C|\psi(0)|^2/M^2$. We give the experimental numbers[6], the theoretical predictions of the Buchmüller potential and the predictions of potential (2). The absolute masses of the various states are not significant since the ground state is adjusted by fixing the effective quark mass.

Table 1. Excitation energies and relative leptonic widths for the $c\bar{c}$ and $b\bar{b}$ systems compared with the theoretical predictions of Buchmüller et al. and Martin.

	$c\bar{c}$			$b\bar{b}$		
1^{--} ground state	3095 MeV			9460 MeV		
Quark mass	Exp.	Buchmüller 1480 MeV	Martin 1800 MeV	Exp.	Buchmüller 4870 MeV	Martin 5174 MeV
2S - 1S	589	600	592	561	560	560
$\Gamma_{ee}(2S)/\Gamma(1S)$	0.46 ±0.06	0.45	0.40	0.43	0.44	0.41
3S - 1S	935	1020	937	890	890	900
$\Gamma_{ee}(3S)/\Gamma(1S)$	0.16?	0.32	0.25	0.32	0.32	0.35
4S - 1S	1319?	1380	1185	1116	1160	1140
$\Gamma_{ee}(4S)/\Gamma(1S)$				0.21	0.26	0.27
1P - 1S	425	420	407	440	430	400
2P - 1S				796	790	780
1D	677	710	692			

Table 1 does not incorporate the data on the pseudoscalar states η_c and η_c' whose spacing to the corresponding vector states are

$$J\psi - \eta_c = 112 \text{ MeV}, \quad \psi' - \eta_c' = 93 \text{ MeV},$$

for which W. Buchmüller et al.[3], as well as previously D. Beavis et al.[7] have predictions of the right order of magnitude but, however, too low by a factor 0.8.

There are many other fits of the data but I have chosen these two as representatives of "theoretical" and "phenomenological" potentials. In both cases, the agreement with experiment is strikingly good. The flavour independence of the potentials is perfect (in fact too perfect maybe !). In the phenomenological potential the number of parameters is only three. If we look closer at the figures, we see that the phenomenological potential is somewhat better for higher L = 0 excitations. This may, however, not be terribly significant since open and closed coupled channel effects are pre-

Table 2.

		Exp.	Buchmüller	Eichten Feinberg	Beavis
$c\bar{c}$	$2^{++}-1^{++}$	44 MeV	57 MeV	70 MeV	69 MeV
	$1^{++}-0^{++}$	93 MeV	83 MeV	72 MeV	71 MeV
$b\bar{b}$	$2^{++}-1^{++}$	20 MeV	19 MeV	26 MeV	23 MeV
χ_b	$1^{++}-0^{++}$	21 MeV	28 MeV	25 MeV	24 MeV
	$2^{++}-1^{++}$	16 MeV	18 MeV	17 MeV	17 MeV
χ_b'	$1^{++}-0^{++}$	21 MeV	21 MeV	17 MeV	24 MeV

sent (remember that the $2m_D = 2 \times 1865$ MeV = 3730 MeV and $2m_B = 2 \times 5274 = 10548$ MeV). What matters more is that the QCD motivated potential reproduces much better the P states, especially the 1P state, than the phenomenological potential. This is presumably due to the fact that a non-singular potential overestimates the mass of the L = 0 states; this, however, is corrected by adjustment of the quark mass and leads to underestimating the mass of the L = 1 states.

In the most ambitious attemps, one tries also to reproduce the separation between the P states due to spin dependent forces. This is especially interesting now that we have two triplets of P states. Generally, the spin forces appear from a reduction of the Breit equation, but the nature of the interaction in the Breit equation is model dependent. We give in Table 2 some theoretical predictions by D. Beavis et al.[7], W. Buchmüller[8] and E. Eichten and F.L. Feinberg[9].

We see that orders of magnitude are reasonable but agreement is not perfect in any of the three calculations. For $c\bar{c}$, QCD sum rules also give reasonable numbers but we have no time to develop this approach.

We conclude that the potential model description of $b\bar{b}$ and $c\bar{c}$ works very well, that the potential is very well fixed in the interval 0.1 to 1 Fermi, where the two potential models coincide (after a shift due to quark mass differences), and that the potential which has the correct QCD behaviour at short distances is favoured. Also flavour independence is very well tested.

Now we want to move towards lighter quark masses, because, after all, the non-relativistic description of $c\bar{c}$ works very well as far as energies are concerned in spite of the fact that $\langle v^2/c^2 \rangle \cong 0.25$ (E1 or M1 transition rates are not as satisfactory but relativistic corrections can improve the situation). So if it works for $c\bar{c}$ why not try to fit $s\bar{s}$ with potential (2)? Naturally, one can reproduce the ϕ mass, 1020 MeV by adjusting the s quark effective mass to be 518 MeV. However, there are checks :

1) One predicts a ϕ', radial excitation of the ϕ, 615 MeV above the ϕ, while an experimental candidate from Orsay is 630 MeV above the ϕ.

2) Once the strange quark mass is fixed, one can predict the F and the F^* masses :

$$m_F = 1990 \text{ MeV},$$

$$m_{F^*} = 2110 \text{ MeV}.$$

Experiments give masses for the F ranging from 2020 MeV to 1970 MeV.[10]

This success has encouraged J.-M. Richard[11] to use potential (2), together with a phenomenological spin-spin force of the Fermi type adjusted to the $J\psi$-η_c mass difference to calculate the mass of the Ω^- which is a sss system. The recipe giving the quark-quark potential inside a baryon which is a colour singlet is

$$V_{QQ} = \frac{1}{2} V_{Q\bar{Q}} . \tag{3}$$

This is suggested by what happens for one gluon exchange or more generally colour octet exchange, and has the advantage that it is insensitive to the choice of quark masses : if one neglects changes in kinetic energies, a fit of the $Q\bar{Q}$ system with quark mass m_Q and potential $V_{Q\bar{Q}}$ will be stable in the change

$$\begin{aligned} m_Q &\to m_Q + \Delta , \\ V_{Q\bar{Q}} &\to V_{Q\bar{Q}} - 2\Delta. \end{aligned} \tag{4}$$

The prescription (3) is such that also the three-quark system mass will be left approximately invariant in the change (4). Anyway, J.-M. Richard has taken seriously this prescription and found

$$m_{\Omega^-} = 1665 \text{ MeV},$$

while experiment gives

$$m_{\Omega^-} = 1672 \text{ MeV}.$$

This kind of success may, of course, be fortuitous but it leads to reconsidering the systems containing even lighter quarks, i.e., up and down quarks, mesons and baryons, and trying to fit them in a non-relativistic way with an effective mass of about 300 MeV for the up and down quarks and a flavour independent potential which may or may not be constrained to satisfy property (3). In fact, in the past, this has been done to a certain extent by A. De Rújula, H. Georgi and S. Glashow[12] and they have been incredibly successful, justifying to lowest order in $m_s - m_{u,d}$ the Gell-Mann-Okubo mass formulae for the octet and the decuplet, showing that the spin-spin force derived from one gluon exchange has the right qualitative features :

$$m_{J=1} > m_{J=0}$$

for mesons,

$$m_{J=3/2} > m_{J=1/2}$$

for baryons, and leads, with $m_s/m_u = 0.6$, to a lot of interesting predictions like

$$\frac{M_{K^*} - M_K}{M_\rho - M_\pi} = 0.6 \qquad \text{(exp. 0.64)}$$

$$\frac{M_{D^*} - M_D}{M_{K^*} - M_K} = 0.29 \qquad \text{(exp. 0.35)} \qquad (5)$$

$$\frac{M_{\Xi^*} - M_\Xi}{M_{\Sigma^*} - M_\Sigma} = 1 \qquad \text{(exp. 1.12)}$$

$$\frac{1}{2}\,\frac{2M_{\Sigma^*} + M_\Sigma - 3M_\Lambda}{M_\Delta - M_N} = 1 \qquad \text{(exp. 1.04)} \qquad (6)$$

$$\frac{M_\Sigma - M_\Lambda}{M_\Delta - M_N} = \frac{2}{3}\left(1 - \frac{m_u}{m_s}\right) = 0.26 \qquad \text{(exp. 0.26)}$$

$$\frac{M_{\Sigma_c} - M_{\Lambda_c}}{M_\Delta - M_N} = 0.55 \qquad \text{(exp. 0.56),}$$

with $m_c = 1.8$, and naturally, the magnetic moment predictions of the quark model :

$$\mu_N/\mu_P = -\,0.67 \qquad \text{(exp. - 0.68)}$$

$$\mu_\Lambda/\mu_P = -\,0.2 \qquad \text{(exp. - 0.22)}$$

$\mu_{\Sigma^+}/\mu_p = 0.95$ (exp. 0.85±0.01)

$\mu_{\Sigma^-}/\mu_p = -0.38$ (exp. -0.50±0.09)

$\mu_{\Xi^-}/\mu_p = -0.16$ (exp. -0.25±0.01)

$\mu_{\Xi^\circ}/\mu_p = -0.49$ (exp. -0.45±0.05) .

What is missing in the list is μ_{Ω^-} ! It could be measured and this measurement would be very interesting !

However, A. De Rújula et al. did something very crude : they worked to lowest order in the symmetry breaking due to the u, d and s mass differences which meant that they could use a variational approach and disregard the distortions of the space wave function due to the breaking. This might be a very bad approximation, especially if heavy charmed or beautiful quarks are present. Calculations using a flavour independent potential for mesons and baryons, and the Schrödinger equation or a modified version, in which relativistic kinematics are used, have been made by A. Bhadury et al.[13] and by D.P. Stanley, D. Robson[14] first. However, it is not very easy to get a feeling from these calculations which end with long lists of numbers predicted but are sometimes criticized because of the many parameters introduced. Also, it is difficult to be certain that calculations are sufficiently accurate.

I proposed to present the results in a way in which the improvements, with respect to the "naive" A. De Rújula et al. treatment are manifest so as to measure the effects of the wave function distorsion by symmetry breaking by the quark effective masses. For instance, the expression in the Gell-Mann-Okubo mass formula is not exactly zero and one should try to reproduce this. Similarly, one can study the deviations of the right-hand side of Eqs. (5) and (6) from unity. Two groups have done this: S. Ono and F. Schöberl using variational methods with superpositions of Gaussians[15], J.-M. Richard and P. Taxil using an expansion in hyperspherical harmonics and solving a system of coupled differential equations[16] (a very efficient procedure for quark-quark potentials). Here, I report the J.-M. Richard-P. Taxil results (the others are very similar).

The Gell-Mann-Okubo mass formula for the octet gives :

$$2(M_N + M_\Xi) - M_\Sigma - 3M_\Lambda = -25 \pm 5 \text{ MeV}$$

experimentally, where the uncertainty is due to the spread due to electromagnetic effects.

A naïve estimate disregarding the distortion of the wave functions but only the spin-spin interaction of A. De Rújula et al. gives + 35 MeV ! However, J.-M. Richard and P. Taxil using a reasonable soft potential of the type (2) get - 22 MeV, in perfect agreement with experiment. Similarly, they get

$$\frac{M_{\Xi^*} - M_{\Xi}}{M_{\Sigma^*} - M_{\Sigma}} = 1.08 \text{ (exp. 1.12)} ,$$

$$\frac{1}{2}\,\frac{2M_{\Sigma^*} + M_{\Sigma} - 3M_{\Lambda}}{M_{\Delta} - M_N} = 1.07 \quad \text{(exp. 1.04)} .$$

It is encouraging to see that these results all go into the right direction.

Finally, I want to report another calculation of J.-M. Richard and P. Taxil to obtain the mass of the newly discovered hadron (csu) at CERN[17], with a mass of 2.46 GeV. A previous calculation by N. Isgur[18], using a harmonic oscillator potential gave 2.50 MeV for this state, a result already quite nice.

J.-M. Richard and P. Taxil[19] have taken a soft potential of the type (2) allowing for a change of the coefficients and fitting the nucleon, the Δ, the Ω^- and the Λ_c, and choosing $m_{u,d}$ = 300 MeV. They get

$$m_s = 600 \text{ MeV}, \ m_c = 1895 \text{ MeV} ,$$

$$V_{qq} = 1/2 \, [- 8.3377 + 6.9923 \, r^{0.1}]. \qquad (7)$$

Notice the constants in (7) are not very different from those in (2). Their results are listed in Table 3. The states A and S have spin-1/2 but correspond to different spin structures. In the limit $m_s = m_u$, A reduces to Λ_c while S reduces to Σ_c. S* has spin-3/2.

What do we conclude from these results ? A soft potential is better than a harmonic oscillator potential. However, J.-M. Richard and P. Taxil have also tried another exercise. They take three-body forces inspired by the string picture instead of two-body forces, with a potential which is the sum of the distances of the three quarks to a point minimized over the location of the point, plus a constant. They get results almost as good.

We see therefore that this kind of approach has a great predictivity but, on the other hand, it does not allow us to solve all problems concerning the nature of the interaction. We think that it should be pursued in spite of the fact that relativistic effects are disregarded. Of particular interest are, for instance, the corrections to the naïve quark model predictions for magnetic moments because at present we begin to have very precise measurements of the magnetic moments, for instance, in the case of Σ^+[20].

Table 3.

	theory	exp.		input	exp.
N	input	0.939 GeV	Ω^-	input	1.672 GeV
Δ	input	1.232 GeV	Λ_c	input	2.283 GeV
Λ_o	1.111 GeV	1.115 GeV	Σ_c	2.443 GeV	2.450 GeV
Σ	1.176 GeV	1.193 GeV	Σ_c*	2.542 GeV	
Ξ	1.304 GeV	1.318 GeV	A(csu)	2.457 GeV	2.460 GeV
Σ*	1.392 GeV	1.383 GeV	S(csu)	2.558 GeV	
Ξ*	1.538 GeV	1.533 GeV	S*(csu)	2.663 GeV	

REFERENCES

1. H. Krasemann and S. Ono, Nucl. Phys. B154 (1979) 283.
2. J.L. Richardson, Phys. Lett. 82B (1979) 272.
3. W. Buchmüller, G. Greenberg and S.H.H. Tye, Phys. Rev. Lett. 45 (1980) 103;
   W. Buchmüller and S.H.H. Tye, Phys. Rev. D24 (1981) 132.
4. G. Bhanot and S. Rudaz, Phys. Lett. 78B (1978) 119.
5. A. Martin, Phys. Lett. 100B (1981) 511.
6. The latest data on the Υ system can be found in the review by P.M. Tuts, Electron Photon Conference, Cornell (1983), to be published.
7. D. Beavis et al., Phys. Rev. D20 (1979) 743 and 2345.
8. W. Buchmüller, Phys. Lett. 112B (1982) 479.
9. E. Eichten and F.L. Feinberg, Phys. Rev. Lett. 43 (1979) 1205; Phys. Rev. D23 (1981) 2724.
10. S.L. Wu, Nato Advanced Study Institute on Particles and Fields, Cargèse 1983, to be published. The new values are lower.
11. J.-M. Richard, Phys. Lett. 100B (1981) 515.
12. A. De Rújula, H. Georgi and S. Glashow, Phys. Rev. D12 (1975) 147.
13. A. Bhadury et al., Phys. Rev. Lett. 44 (1980) 3180.
14. D.P. Stanley and D. Robson, Phys. Rev. D21 (1980) 3180 ;
    D.P. Stanley and D. Robson, Phys. Rev. Lett. 45 (1980) 235.
15. S. Ono and F. Schöberl, Phys. Lett. 118B (1982) 419.
16. J.-M. Richard and P. Taxil, Marseille CPT prerint (1983).
17. S.F. Biagi et al., Phys. Lett. 122B (1983) 455.
18. N. Isgur, New Flavours and Hadron Spectroscopy, Proc. XVIth Rencontre de Moriond, (Les Arcs, France, 1981), eds. J. Tran Tranh Van (Editions Frontières, Dreux, France, 1981).
19. J.-M. Richard and P. Taxil, Phys. Lett. 128B (1983) 453.
20. C. Ankelbrandt et al., Fermilab preprint 83/52 Exp, to appear in Phys. Rev. Lett.

REALISTIC KALUZA-KLEIN THEORIES ?*

C. Wetterich

Institute for Theoretical Physics
University of Bern
Sidlerstrasse 5, CH-3012 Bern, Switzerland

ABSTRACT

An introduction to Kaluza-Klein theories is given with emphasis on the question if a realistic model can be found. The theory works well in the bosonic sector, but has major difficulties so far in describing the observed spectrum of fermions.

Kaluza-Klein theories[1] are theories of gravity in more than four dimensions, where the supplementary D spacelike dimensions are somehow curled up to form a compact "internal" space with very small characteristic length scale. If this characteristic length scale is of the order of the Planck length, no direct experimental observation of these supplementary dimensions is possible. We are left with an effective four dimensional theory obtained by integrating out the extra coordinates in the action.

These theories unify gauge interactions and gravitation. Gauge symmetries are nothing else than isometries of the internal space. Imagine that the ground state of a d dimensional theory of gravity is $M^4 \times S^D$. The symmetries leaving this ground state invariant include of course the Poincaré transformations leaving four dimensional Minkowski space M^4 invariant. They also include a SO(D+1) group of rotations leaving the sphere S^D invariant. Furthermore, the SO(D+1) transformations can be applied independently at every space-time point in Minkowski space. Thus SO(D+1) is a local symmetry and

*Talk given at the First Capri Symposium "Supersymmetry and Frontiers in Elementary Particle Physics" May 1983, and at the 1983 Cargèse Summer Institute on "Particles and Fields" July 1983

after dimensional reduction we end indeed with a four dimensional theory with SO(D+1) gauge invariance.

The metric tensor $\hat{g}_{\hat{\mu}\hat{\nu}}$ of the d dimensional theory contains components $\hat{g}_{\mu\nu}$ transforming as second rank symmetric tensors with respect to four dimensional coordinate transformations. Other components $\hat{g}_{\mu\alpha}$ have only one Minkowski index μ and one internal index α. They transform as four dimensional spin one bosons and the gauge bosons are obtained from these components. Finally, the components $\hat{g}_{\alpha\beta}$ describe four dimensional scalars. Thus the higher dimensional theory of pure gravity describes a four dimensional theory of gravity, gauge interactions and scalar fields. Even if the extra dimensions cannot be observed directly, they reflect themselves in the effective four dimensional theory, predicting in principle many of the free parameters of the usual unified gauge theories. Four dimensional fermions can be obtained by adding a spinorial matter field to d dimensional gravity.

In this talk I will describe attempts to construct a realistic theory along these lines. Realistic, that means that a higher dimensional theory of gravity coupled to spinor fields should reproduce the known particles and their interactions and hopefully also give some new predictions related to this unification. We are still far from obtaining this goal, and the main problem today seems to be the difficulty to obtain the observed quantum numbers of fermions. This will be described in the second half of my talk. However, important progress has been made to describe the purely bosonic sector since the early days of Kaluza-Klein theories some sixty years ago.

Let me first give some necessary criteria for a realistic unification of gravity and gauge interactions.

1) The gauge group should at least contain SU(3)xSU(2)xU(1). The standard group is experimentally in a very good status and I am reluctant to give up part of it. In many respects our task would be easier if the gauge group includes a unification group like SU(5) or SO(10). The fermions belong then to only a few irreducible representations and this may help to reproduce their quantum numbers. Furthermore, relations between the low energy gauge couplings could be understood.

2) A ground state with the structure $M^4 \times K^D$, where K^D is some compact space with isometries, should follow from the dynamics of the d dimensional gravity theory. This ground state should correspond to a minimum of the effective four dimensional scalar potential, the effective four dimensional cosmological constant should vanish, four dimensional gravity should be governed by a term linear in the scalar curvature with positive Newton's constant and all kinetic terms should have the right sign.

3) In the limit of unbroken SU(3)xSU(2)xU(1) there should be massless fermions in the effective four dimensional theory with different hypercharges of the left-handed and the right-handed Weyl spinors (chiral fermions).

4) There are also some requirements of consistency. The d dimensional theory should be invariant under general coordinate transformations and Lorentz transformations in d dimensions. This symmetry should be maintained by quantum fluctuations, which necessarily are to be included since we know that gauge interactions are a quantum theory. No anomalies destroying this symmetry should occur.

I think that requirements of renormalizability and unitarity in perturbation theory are less stringent. First, this question is related to a particular expansion method. Second, the gravity theory to start with may well be not a good short distance theory (compared with the Planck length !), but rather an effective theory obtained from some more complete theory. I also leave completely open the option of supergravity. Of course, it would be beautiful to have also a unification of bosons and fermions. Nevertheless, the central idea of Kaluza-Klein theories is not related to supersymmetry and the fermion-boson unification may well occur at another level. I think it would be a great achievement to unify all bosonic interactions, even if we may have to wait to implement also the unification with fermions.

The bosonic sector in Kaluza-Klein theories looks nowadays already quite satisfactory. Starting from pure gravity in d dimensions one obtains a ground state of the form $M^4 \times K^D$ with an interesting gauge group (for example SO(10)). This symmetry can also be spontaneously broken. For an appropriate choice of parameters (fine tuning) the four dimensional cosmological constant is zero and one can even describe an interesting cosmology in these models.

Let me first recall how the gauge interactions come out of these models. The original Kaluza-Klein idea of a fifth dimension forming a circle S^1 with an abelian U(1) group has been generalized to the non abelian case where the internal space is a group space[2], a homogeneous space[3] or even an arbitrary compact space K^D admitting isometries[4]. If the ground state is $M^4 \times K^D$, we have to expand the components $\hat{g}_{\mu\nu}$, $\hat{g}_{\mu\alpha}$, $\hat{g}_{\alpha\beta}$ of the d dimensional metric tensor in harmonics on K^D :

$$\hat{g}_{\mu\nu}(x,y) = g^{i}_{\mu\nu}(x)Y^{i}(y)$$

$$\hat{g}_{\mu\alpha}(x,y) = A^{i}_{\mu}\ (x)\ X^{i}_{\alpha}(y)$$

$$\hat{g}_{\alpha\beta}(x,y) = \phi^{i}\ (x)\ Z^{i}_{\alpha\beta}(y)\ . \qquad (1)$$

Integration of the action over the internal coordinates y leaves us with an equivalent effective four dimensional theory for infinitely many fields $g^i_{\mu\nu}$, A^i_μ and ϕ^i depending only on the Minkowski coordinates x. Most of these fields, however, will be very massive and may be neglected in the low energy region. Nevertheless, there will be massless fields due to the symmetries of the ground state : the graviton and the gauge bosons. They are described, neglecting all massive excitations for a moment, by the following terms in the harmonic expansion :

$$\begin{aligned}
\hat{g}_{\mu\nu} &= g_{\mu\nu}(x) + g^2 A^z_\mu(x) A^y_\nu(x) K^\alpha_z(y) K^\beta_y(y) \mathring{g}_{\alpha\beta}(y) \\
\hat{g}_{\mu\alpha} &= g\, A^z_\mu(x) K^\beta_z(y) \mathring{g}_{\beta\alpha}(y) \\
\hat{g}_{\alpha\beta} &= \mathring{g}_{\alpha\beta}(y).
\end{aligned} \tag{2}$$

Here $\mathring{g}_{\alpha\beta}(y)$ is the metric of the internal space K^D. This metric is left invariant by the infinitesimal transformations

$$y^\alpha \rightarrow y^\alpha + \theta^z(x) K^\alpha_z(y), \tag{3}$$

forming the group of isometries of K^D with one Killing vector $K^\alpha_z(y)$ for every generator of the group. We have one gauge field $A^z_\mu(x)$ for every group generator and g is the gauge coupling. It has to be determined from the parameters of the d dimensional theory.

Calculating the d dimensional curvature scalar $\hat{R}$ for the ansatz (2), one finds

$$\hat{R} = R + \mathring{R} - \frac{1}{4} g^2 \mathring{g}_{\alpha\beta} K^\alpha_z K^\beta_y F^z_{\mu\nu} F^{y\mu\nu}. \tag{4}$$

Here R is the four dimensional curvature scalar, $\mathring{R}$ is the curvature scalar of the internal space and $F^z_{\mu\nu}$ is the usual non abelian field strength for gauge fields. Integration over the internal coordinates gives the usual kinetic term for gauge fields. The gauge coupling can be determined for an arbitrary space K^D[4,5] by requiring the correct normalization of the kinetic term. It is typically

$$g^2 = f \frac{16\pi}{L^2 M_p^2} , \tag{5}$$

with L the characteristic length of the internal space, M_p the Planck mass and f a model dependent factor. If the gauge coupling is not extremely small, L must roughly be of the order of the Planck length $L = M_p^{-1}$, but it may easily be a factor of ten or more larger[6].

Let me next discuss the problem of finding the ground state of a higher dimensional gravity theory. Clearly, the ground state has to fulfil the field equations derived from the effective action relevant at the characteristic length scale of spontaneous compactification. Since we unify gravity with a quantum theory of gauge interactions, the d dimensional gravity theory to start with is necessarily a quantum theory. The only relevant quantity for a discussion of the ground state or cosmology is therefore the effective action which includes all quantum fluctuations.

Unfortunately, there is no consistent renormalizable quantum theory of gravity at our disposal. Postponing the question how to find it, we can nevertheless discuss some general features of such an effective action. If the theory has no anomalies, quantum fluctuations should preserve the symmetry of general coordinate- and Lorentz-transformations in d dimensions, except fluctuations with wavelengths large compared with the characteristic scale of spontaneous compactification which "feel" that the ground state is not invariant under this full symmetry. Neglecting these symmetry breaking effects, the effective action at the scale of spontaneous compactification for the pure gravitational interactions has the form

$$S = - \int d^d x \sqrt{|\hat{g}|}\; f(\hat{g}_{\hat{\mu}\hat{\nu}}, \partial_{\hat{\rho}}\, \hat{g}_{\hat{\mu}\hat{\nu}} \ldots) \ , \tag{6}$$

with $\hat{g} = \det \hat{g}_{\hat{\mu}\hat{\nu}}$ and f a scalar function of the metric and its derivatives.

In general, the function f may be very complicated, even if we start with a simply theory. For example, quantum fluctuations generate typically all sorts of terms with more than two derivatives. As a toy model, let me demonstrate how an expansion of f up to four derivatives can already give a quite satisfactory description of spontaneous compactification[7]. Consider an effective action with

$$f = \alpha\hat{R}^2 + \beta\hat{R}_{\hat{\mu}\hat{\nu}}\hat{R}^{\hat{\mu}\hat{\nu}} + \gamma\hat{R}_{\hat{\mu}\hat{\nu}\hat{\sigma}\hat{\lambda}}\hat{R}^{\hat{\mu}\hat{\nu}\hat{\sigma}\hat{\lambda}} + \delta\hat{R} + \varepsilon \ , \tag{7}$$

with $\hat{R}_{\hat{\mu}\hat{\nu}\hat{\sigma}\hat{\lambda}}$, $\hat{R}_{\hat{\mu}\hat{\nu}}$ and $\hat{R}$ the d dimensional curvature tensor, Ricci tensor and curvature scalar respectively. For ε different from zero, d dimensional Minkowski space is not a solution of the field equations derived from eqs. (6) and (7). For

$$D(D-1)\alpha + (D-1)\beta + 2\gamma \equiv \zeta > 0$$

$$(D-2)\beta + 4\gamma > 0$$

$$\gamma > 0 \tag{8}$$

$$\delta > 0 \tag{9}$$

$$\varepsilon = \frac{1}{4}\,\delta^2\,\frac{D(D-1)}{\zeta} \ , \tag{10}$$

a solution of the field equations is $M^4 \times S^D$. The radius $L_\circ$ of S^D

is given by

$$L_o^{-2} = \frac{\delta}{2\zeta} \tag{11}$$

and the Planck mass is

$$M_p^2 = 16\pi\delta V_D \frac{(D-1)\beta + 2\gamma}{\zeta} \tag{12}$$

with V_D the volume of S^D. Note that equation (10) involves a fine tuning of the effective d dimensional cosmological constant.

This solution can be easily understood in terms of the effective four dimensional scalar potential. This effective potential is derived from the part of the action involving only scalars obtained from $\hat{g}_{\alpha\beta}$ and no Minkowski derivatives of these scalar fields. It is therefore given by

$$V = \int d^D y \, |g_D|^{1/2} f_D \tag{13}$$

with $g_D = \det \hat{g}_{\alpha\beta}$ and f_D the function f restricted to internal space. The inequalities (8) guarantee that f_d is bounded from below and the fine tuning (10) implies that f_D and therefore V is positive semidefinite with a minimum at $V = 0$. Eq. (8) also implies that the minimum must correspond to vanishing traceless parts of the internal curvature tensor and Ricci tensor and therefore to a maximally symmetric space. Finally, the inequality (9) insures that the ground state scalar curvature is negative and the solution therefore a sphere.

I would like to stress that $M^N \times S^{d-N}$ is not a solution of the system (8), (9), (10) unless $N = 4$. The three flat space dimensions are singled out by the particular fine tuning (10). The puzzling feature in this kind of models is not the spontaneous compactification to a space with very small length scales of some of the space dimensions. This is very natural since the effective d dimensional cosmological constant is not expected to be zero. But why have three space dimensions become so flat or, why holds the fine tuning of the cosmological constant with such an incredible accuracy ? The only speculation I have to offer is that quantum fluctuations regulate somehow themselves to produce this fine tuning. For example, eq. (10) may correspond to an infrared fixpoint of the relevant renormalization group equations. In this context it is interesting to remember that three flat space dimensions are the only possibility where long range fluctuations from gauge fields can contribute.

If eqs. (8), (9), (10) hold, $M^4 \times S^D$ corresponds to the absolute minimum of the scalar potential with respect to all the infinite number of scalar excitations contained in the harmonic expansion (1). However, a more complicated form of f can produce a scalar potential which is bounded from below without requiring that the traceless parts of the curvature and Ricci tensors vanish. Imagine that the potential is stabilised by terms with up to eight derivatives like $(\hat{R}_{\hat{\mu}\hat{\nu}\hat{\sigma}\hat{\lambda}}\hat{R}^{\hat{\mu}\hat{\nu}\hat{\sigma}\hat{\lambda}})^2$ etc. Then we can ask what happens if for example γ goes from positive to negative values. For $\gamma>0$, $(D-2)\beta + 4\gamma >0$, the maximally symmetric space S^D still corresponds to a minimum of the scalar potential (not necessarily the absolute minimum). However, for $\gamma<0$ it becomes a saddlepoint and there must be a minimum where the traceless part of the curvature tensor is different from zero. The symmetry SO(D+1) is broken at this minimum. This is typical for spontaneous symmetry breakdown of a unifiying gauge group and Kaluza-Klein theories can well acount for this phenomenon. If some phase transition characterising the spontaneous symmetry breakdown is second order, we even expect very light scalars if the parameters of the theory are near this transition.

Since our toy model has an acceptable ground state, we can start to ask questions about the cosmology of Kaluza-Klein models[8,9]. In its late history, the universe just does not feel anymore the supplementary dimensions directly. Fluctuations above the ground state lead to the standard Friedmann cosmology. However, going back in time, the Robertson-Walker scale factor becomes at some time comparable to the characteristic length scale of the internal space. It does not make much sense anymore to restrict solutions to a static behaviour of the internal space. The full dynamics of the d dimensional field equations become now relevant.

As a simple model I will discuss solutions where at any given time the spacelike dimensions form a space $S^3 \times S^D$ and the radii $R(t)$ and $L(t)$ of S^3 and S^D respectively are time dependent. ($R(t)$ corresponds to the Robertson-Walker scale factor for this case.) The field equations derived from the effective action (6), (7) involve up to four time derivatives of R and L. One can ask if there are solutions giving an inflationary scenario for the very early universe. Indeed, one finds solutions[9] where the internal dimensions form a sphere S^D with constant radius $L_H \neq L_o$ whereas the four remaining dimensions form a De Sitter space with positive Hubble constant $H = \dot{R}/R$. This solution corresponds to an extremum of the relevant scalar potential, but for the extremum at $L_H \neq L_o$ the effective four dimensional cosmological constant is positive, inducing an exponential expansion of three space dimensions.

Imagine the following scenario. At very early times both $R(t)$ and $L(t)$ were of the same order of magnitude. At some time some region of space came near to the described De Sitter solution. During the time t_c, when the solution remains near the De Sitter solution, three space dimensions expand exponentially whereas the other space dimensions remain small. The net three dimensional expansion is given by a factor $\exp(Ht_c)$, and for $Ht_c \geq 60$ this inflation can solve several outstanding cosmological problems, like the horizon problem or the flatness problem[10]. After the time t_c the solution goes away from the De Sitter solution and after some transition period L(t) approaches the ground state value L_o and we end with an effective four dimensional Friedman universe.

To be more quantitative, the De Sitter solutions are given by

$$Z = L_o^2/L_H^2 \tag{14}$$

$$H^2 = \frac{1}{24}\,\delta\,\frac{D(D-1)}{\zeta}(1-Z)^2(1+\sigma Z)^{-1} \tag{15}$$

$$a_1 Z^3 + a_2 Z^2 + a_3 Z + a_4 = 0 \tag{16}$$

$$a_1 = D(D-1) + D(D-1)\rho + \{D(D-1)-12\}\sigma - 12\sigma^2$$

$$a_2 = -3D(D-1) - 3D(D-1)\rho - \frac{3}{D}\{D^2(D-1)-4D + 16\}\sigma + \frac{12}{D}(D-4)\sigma^2$$

$$a_3 = \frac{3}{D}\{D^2(D-1)-4D-16\} + 3D(D-1)\rho + \frac{3}{D}\{D^2(D-1)-4D-16\}\sigma$$

$$a_4 = -D(D-1) + 12 - D(D-1)\rho - \{D(D-1) - 12\}\sigma \tag{17}$$

$$\sigma = -D(D-1)\alpha/\zeta$$

$$\rho = -(D-4)\beta/\zeta\ . \tag{18}$$

A linear expansion around the De Sitter solution gives for the critical quantity Ht_c

$$(Ht_c)^{-1} = \frac{4Z}{D}\frac{(1+\sigma Z)^2}{(1-Z)^2}(1+\sigma)\frac{\tilde{A}}{\tilde{B}} \tag{19}$$

with

$$\tilde{A} = (1+\sigma Z)(1-Z) - \frac{2}{D}\{1 + 2(1+\sigma)Z + \sigma Z^2\} \tag{20}$$

and $\tilde{B}$ a lengthy function of the parameters of the order of unity. For large values of D and small values of Z a value $Ht_c \geq 60$ is obtained rather easily. Kaluza-Klein cosmology looks very promising!

The purely bosonic sector of Kaluza-Klein theories looks very beautiful. Pure gravity in more than four dimensions can account for an interesting unifiying gauge group and a scalar potential which is bounded from below and which can have a sufficiently rich structure to account for spontaneous symmetry breaking. The four dimensional cosmological constant can be very small (this involves however an unnatural tuning of parameters). This opens the possibility of obtaining a realistic cosmology from these models and very early cosmology offers exciting prospects for solving some of the outstanding problems of the standard big bang model.

In the second half of my talk I will rather concentrate on the main problem of these theories : So far, they have failed to produce a realistic fermion spectrum. Introducing fermionic degrees of freedom in the Kaluza-Klein scheme requires to add matter fields in a spinor representation of the higher dimensional Lorentz group SO(1, d-1). This can be done in a supersymmetric way or not. In the first case, four dimensional fermions are typically obtained from dimensional reduction of a Rarita-Schwinger spinor in d dimensions and the maximal dimension for a consistent theory is generally believed to be eleven[11]. In the second case, we may consider Dirac, Weyl, Majorana or Majorana-Weyl spinors in d dimensions and the number of dimensions is a priori not limited.

In order to give a clear presentation of the problem, I will make two assumptions - which eventually have to be given up for the construction of a realistic Kaluza-Klein model. The first assumption is that the four dimensional gauge group is the group of isometries of the internal space. This is the original Kaluza-Klein idea and it implies that we do not consider extra gauge fields in higher dimensions (or that the four dimensional gauge group has nothing to do with extra gauge symmetries in d dimensions). We also do not consider the possibility that the four dimensional gauge group is realized on a composite level as has been proposed for the hidden SU(8) symmetry in d = 11 supergravity. The second assumption is that all observed particles are elementary (at the energy scales below the Planck mass considered in our context). This implies that all four dimensional particles have to be obtained from the harmonic expansion of the higher dimensional gravitational and spinor fields. Especially, composite models for the observed fermions are not discussed in this context. These assumptions imply immediately a lower bound on the number of dimensions. The minimal dimension for an internal space with SU(3)xSU(2)xU(1) isometry is seven, implying $d \geqq 11$. If we want to obtain a unified gauge group for the weak, electromagnetic and strong interactions like SU(5) or SO(10), the dimensions needed are $d \geqq 12$ or $d \geqq 13$ respectively, not consistent with the bound $d \leqq 11$ relevant for supergravity theories.

Let us first investigate what happens if we couple a Dirac spinor to higher dimensional gravity. (Many aspects of this discussion also apply to Rarita-Schwinger spinors.) A minimal covariant kinetic term for a Dirac spinor is given by

$$\mathcal{L}_\psi = i\bar{\psi}\,\gamma^{\hat{\mu}} D_{\hat{\mu}}\psi \tag{21}$$

with $D_{\hat{\mu}}$ the standard covariant derivative. The operator $\gamma^{\hat{\mu}}D_{\hat{\mu}}$ contains a four dimensional kinetic term $\gamma^\mu D_\mu$, a mass term $\mathring{\gamma}^\alpha D_\alpha$ which is the Dirac operator on the internal manifold characterizing the ground state, and interactions with gauge bosons, scalars etc.

An arbitrary $2^{[D/2]}$ component spinor can be expanded in harmonics on the D dimensional internal space K^D

$$\psi(y) = a_{nHMj}\ \psi_{nHMj}(y)\ , \tag{22}$$

where ψ_{nHMj} belongs to the representation H with respect to the group of isometries on K^D (this is the generalization of the different states of angular momentum obtained when one expands a spinor an a sphere). The index n denotes the component of the representation and for every representation one characterizes the independent terms in the expansion by their mass M

$$\mathring{\Gamma}^{\alpha}\ \mathring{D}_{\alpha}\ \psi_{nHMj} = M\ \psi_{nHMj} \tag{23}$$

and, if needed, by a supplementary index j. We then can expand the d dimensional spinor (with $2^{[D/2]}$ components) in the form

$$\psi(y,x) = \psi_{nHMj}(y)\ \phi_{nHMj}(x)\ , \tag{24}$$

where the coefficients of the expansion are now complex four component x-dependent fields transforming as Dirac spinors with respect to four dimensional Lorentz rotations. Every $\phi_{nHMj}(x)$ corresponds to an independent spinor with mass M after dimensional reduction to four dimensions.

What we need are four dimensional spinors which are chiral in the limit of unbroken SU(3)xSU(2)xU(1). This means that left-handed and right-handed four dimensional Weyl spinors belong to different (complex) representations of the gauge group. Chirality forbids a mass term, and the search for chiral fermions can concentrate on the zero modes of the internal Dirac operator

$$\mathring{\Gamma}^{\alpha}\mathring{D}_{\alpha}\ \psi_{nH0j}\ (y) = 0\ . \tag{25}$$

The harmonic expansion of a d dimensional Dirac spinor gives only four dimensional Dirac spinors and the left-handed and right-handed fermions have the same transformation properties with respect to the gauge group. No chiral fermions can be obtained from dimensional reduction of a d dimensional Dirac spinor and we need Majorana and/ or Weyl constraints in higher dimensions[12,13]. Majorana constraints still cannot do the job. Starting with a d dimensional Majorana spinor, one is left with four dimensional Majorana or Dirac spinors after dimensional reduction[13], still excluding chiral spinors. No chiral spinors can be obtained in odd dimensions.

A d dimensional Weyl constraint

$$\bar{\gamma}\ \psi(y,x) \equiv \psi(y,x) \tag{26}$$

could in principle induce a difference in the transformation properties of left-handed and right-handed spinors. The d dimensional generalization of the γ^5 matrix $\bar{\gamma}$ factorizes into the D dimensional "γ^5" matrix $\bar{\Gamma}_D$ and γ^5 :

$$\bar{\gamma} = \bar{\Gamma}_D \gamma^5 . \tag{27}$$

This implies for the harmonic expansion of a Weyl spinor

$$\bar{\Gamma}_D \psi_{nHMj}(y)\ \gamma^5 \phi_{nHMj}(x) \equiv \psi_{nHMj}(y)\ \phi_{nHMj}(x) . \tag{28}$$

Thus the + and - eigenstates of γ^5 are related to the + and - eigenstates of $\bar{\Gamma}_D$. Left-handed four dimensional spinors are obtained from the harmonic expansion of $\psi^+(y) = 1/2\ (1 + \bar{\Gamma}_D)\ \psi(y)$ whereas right-handed spinors derive from $\psi^-(y) = 1/2\ (1 - \bar{\Gamma}_D)\ \psi(y)$. If the harmonic expansions of ψ^+ and ψ^- contain different representations, we would be left with chiral fermions in four dimensions.

To be more explicit, let me define by n_c^+ the number of zero modes of the Dirac operator in ψ^+ transforming as a complex representation c of the gauge group. Defining similar numbers $n_{\bar{c}}^+$ for the complex conjugate representation $\bar{c}$ and n_c^-, $n_{\bar{c}}^-$ for the "internal" Weyl spinor ψ^-, the total number N_c of chiral fermions belonging to a representation c is given by

$$N_c = f_d(n_c^+ - n_c^- - n_{\bar{c}}^+ + n_{\bar{c}}^-) , \tag{29}$$

where $f_d = 1/2$ for Majorana-Weyl spinors in d = 2 mod 8 dimensions and $f_d = 1$ otherwise. For example, if c is the 16 dimensional spinor representation of a gauge group SO(10), N_c just counts the number of fermion generations in the four dimensional theory.

Chiral fermions require the dimension of the internal space to be D = 2 mod 4[13] since they can only be obtained if the spinor representations of the internal Lorentz-group SO(D) are complex. For D = 4 mod 4 the internal charge conjugation operator C_D commutes with $\bar{\Gamma}_D$. Since C_D transforms every representation in the harmonic expansion into a complex conjugate representation also contained in the expansion $[C_D, \bar{\Gamma}_D] = 0$ implies $n_c^+ = n_{\bar{c}}^+$ and $n_c^- = n_{\bar{c}}^-$ and therefore $N_c = 0$. In the contrary, for D = 2 mod 4 C_D and $\bar{\Gamma}_D$ anticommute implying $n_c^+ = n_{\bar{c}}^-$ and $n_c^- = n_{\bar{c}}^+$. Chiral fermions can only be obtained for d = 2 mod 4 !

I would like to emphasize that this result is just pure group theory. We only used the properties of the Clifford algebra and the general form of the harmonic expansion. It can therefore be extended immediately to Rarita-Schwinger fields and even higher spin representations. It is also independent of the concrete form of the mass operator. It still holds if we introduce dynamical torsion or other fields coupling to the spinor in d dimensions.

For d = 2 mod 4 we have to investigate more explicitly the relevant mass operator. This operator depends on the internal space K^D. A first important constraint is due to Lichnerowicz[14]. Squaring the Dirac operator yields

$$(\overset{\circ}{\Gamma}{}^{\alpha}\overset{\circ}{D}_{\alpha})^2 = \overset{\circ}{D}{}^{\alpha}\overset{\circ}{D}_{\alpha} - \frac{1}{4}\overset{\circ}{R} \quad . \tag{30}$$

The operator $\overset{\circ}{D}{}^{\alpha}\overset{\circ}{D}_{\alpha}$ is positive semidefinite and $\overset{\circ}{\Gamma}{}^{\alpha}\overset{\circ}{D}_{\alpha}$ cannot have zero modes for spaces where the internal curvature scalar is negative as for spheres and most homogeneous spaces. (However, compact spaces with positive scalar curvature admitting isometries and massless spinors are known.) No chiral spinors can be obtained for internal spaces with negative curvature scalar if the relevant kinetic operator is the Dirac operator.

This constraint becomes very powerful when combined with topological considerations[15]. If fermions are chiral with respect to some gauge group G, they must be chiral with respect to at least one U(1) subgroup of G. (In realistic theories, this U(1) subgroup can be identified with hypercharge.) Consider now continuous deformations of the internal space K^D leaving this U(1) subgroup invariant. (This means that two spaces K^D_1 and K^D_2 obtained form each other by an infinitesimal deformation have both the same U(1) isometry.) The number N_c of chiral fermions with respect to this U(1) group cannot be changed by such continuous deformations. As a consequence, no chiral fermions can be obtained on an internal space K^D if it can be continuously deformed, keeping the U(1) group unbroken, into a space with negative scalar curvature. This is a very strong constraint on candidates for an acceptable ground state. As I learned from E. Witten during the time between giving this talk and writing it down, it may even be too strong. Indeed, there is a theorem by Atiyah and Hirzebruch[16] that for the Dirac operator the index N_c vanishes identically for arbitrary compact spaces K^D. No realistic Kaluza-Klein theory can be constructed with our two assumptions stated above without a modification of gravity or of the standard approach to dimensional reduction.

ACKNOWLEDGEMENT

I would like to thank the organizers of the First Capri Symposium and of the 1983 Cargèse Summer Institute on Particles and Fields for my two interesting and enjoyable stays on Tyrrhenian islands this summer.

REFERENCES

1. Th. Kaluza, Sitzungsber. Preuss. Akad. Wiss., Berlin. Math. Phys. K1, (1921) 966; O. Klein, Z. Phys. 37 (1926) 895.
2. J. Rayski, Acta Physica Polonica 27 (1965) 89;
R. Kerner, Ann. Inst. Henri Poincaré 9 (1968) 143;
Y.M. Cho and P.G.O. Freund, Phys. Rev. D12 (1975) 1711;

J. Scherk and J.H. Schwarz, Nucl. Phys. B153 (1979) 61.
3. J.F. Luciani, Nucl. Phys. B135 (1978) 111;
E. Witten, Nucl. Phys. B186 (1981) 412;
A. Salam and J. Strathdee, Ann. Phys. 141 (1982) 316.
4. C. Wetterich, Phys. Lett. 110B (1982) 379.
5. S. Weinberg, Univ. of Texas at Austin preprint, to appear in Phys. Lett. B.
6. P.G.O Freund, Phys. Lett. 120B (1983) 335.
7. C. Wetterich, Phys. Lett. 113B (1982) 377.
8. A. Chodos and S. Detweiler, Phys. Rev. D21 (1980) 2167.
9. Q. Shafi and C. Wetterich, CERN preprint TH-3613, to appear in Phys. Lett. B.
10. A. Guth, Phys. Rev. D23 (1981) 347;
A.D. Linde, Phys. Lett. 108B (1982) 389;
A. Albrecht and P.J. Steinhardt, Phys. Rev. Lett. 48 (1982) 1220.
11. W. Nahm, Nucl. Phys. B135 (1978) 149.
12. C. Wetterich, Nucl. Phys. B211 (1983) 177.
13. C. Wetterich, Nucl. Phys. B222 (1983) 20.
14. A. Lichnerowicz, C.R. Acad. Sci. Paris, Série A-B 257 (1963) 7.
15. C. Wetterich, Nucl. Phys. B223 (1983) 109.
16. M.F. Atiyah and F. Hirzebruch, in Essays on topology and related topics, p. 18, ed. A. Haefliger and R. Narasimhan, Springer, Berlin 1970.

TOPICS IN ELEMENTARY PARTICLE PHYSICS

S. L. Glashow

Lyman Laboratory of Physics
Harvard University
Cambridge, MA 02138

INTRODUCTION

It is nearly thirty years since Yang and Mills invented non-Abelian gauge theory. We have come a long way since then. Today, all of the elementary particle interactions (strong, weak, and electromagnetic) are expressed in terms of such quantum field theories. The conjunction of electroweak SU(2)xU(1) and color SU(3) acting upon the new periodic table of quarks and leptons comprises what is known as the standard theory. In terms of a certain number of numerical parameters (E.g., the Cabibbo angle, the fine-structure constant, the tau-electron mass ratio), the standard theory offers a complete and correct description of all known elementary particle phenomena. What were the relevant problems of particle physics when I was a graduate student have, in large measure, been solved. The remaining problems are not so much new as they are profound and intractable :

The Einstein Problem

Einstein devoted the last three decades of his life to the construction of a unified theory of gravitation and electromagnetism. His failure was due in part to his reluctance to accept quantum mechanics ("God does not play dice.") and his aversion to the existence of peculiarly nuclear forces (He once argued that the nucleus is held together by gravitational forces). Today, we have a unified theory of all forces except gravity, but still no quantum theory of gravity. While many of today's physicists devote themselves to this problem, as yet they produce more heat than light : unobservable particles and forces, and extra dimensions of space-time.

The Dirac Problem

This concerns the existence of very large numbers in our theory of physics, such as the ratio of the gravitational attraction and the electric repulsion of two electrons. In a "correct" theory, Dirac argues that no large dimensionless numbers can appear as fundamental parameters. They are not merely unsightly, but unnatural. To keep a parameter small, in the context of quantum field theory, requires careful fine tuning of renormalized quantities. (Another possibility, advocated by Dirac, is the hypothesis that large numbers simply reflect the great age of our universe. Dirac conjectures that the gravitational constant is decreasing with time, $\dot{G}/G = 1/T$ where $T \cong 20$ billion years. Only very recently has experimental data ruled out this fascinating speculation.) Under the guise of "the gauge hierarchy problem", this question has the complete attention of many of today's practitioners.

The Rabi Problem

"Who ordered that?" said Isidore Rabi upon the identification of the muon as an obese electron. One family of quarks and leptons would be enough for the operation of spaceship Earth and its auxiliary solar power station. Why are there three families? Are there, perhaps, four? Why do all fermion families have exactly the same structure? Nanopoulos offers an "anthropic" partial answer to the first question. At least three families are needed if automatic CP conservation is to be avoided in the standard theory. CP violation was essential for the synthesis of baryons in the early universe : without it, intelligent life could never have evolved. Such arguments are satisfactory only to astrophysicists and followers of Velikovsky. Moreover, "anthropic" is not even an English word. The problem is not solved.

The Cabibbo Problem

Nicola showed how the introduction of a single new parameter could explain much about the pattern of weak interaction decay phenomena. It was a grand accomplishment. Today, the standard theory requires the specification of nineteen fundamental dimensionless parameters. Surely, these are too many arbitrary constants to describe a truly fundamental theory. Most attempts to compute some of these parameters have led to the introduction of even more. For example, the grand unified theory reduces the number of gauge coupling constants from three to one. However, the number of independent Yukawa couplings is vastly increased.

Permit me to conclude this section with a brief chronology of the development of the standard theory over the past thirty years :
1954 Yang and Mills invent non-Abelian gauge theories involving massless vector mesons.

1957 Schwinger suggests that these theories offer the promise of an electroweak unification.

1961 Goldstone discovers that spontaneous symmetry breaking leads to the appearance of massless scalar mesons. Glashow invents the SU(2) x U(1) electroweak model.

1962 Gell-Mann and Ne'eman propose the eightfold way.

1963 Gell-Mann and Zweig invent quarks in order to explain the successes of the eightfold way.

1964 The predicted Ω^- is discovered at Brookhaven. Higgs, and others, show that massless Goldstone bosons combine with massless gauge bosons to produce massive vector bosons in spontaneously broken gauge theories. Bjorken and Glashow propose charm.

1965 Greenberg, Han, and Nambu suggest the existence of quark color as a hidden variable.

1967 Weinberg and Salam utilize the Higgs mechanism to convert the SU(2) x U(1) electroweak model into a genuine theory of the interactions of leptons. Deep inelastic electron scattering at SLAC is interpreted by Bjorken as evidence for pointlike constituents of hadrons, i.e., quarks.

1970 By adopting charm, Iliopoulos, Maiani, and Glashow extend the electroweak theory to describe hadrons.

1971 Gerard 't Hooft establishes the renormalizability of the electroweak theory.

1972 Asymptotic freedom of non-Abelian gauge theory is established by Gross, Wilczek, and Politzer. Quantum chromodynamics emerges as a realistic theory of strong interactions.

1973 The neutral currents predicted by the electroweak theory are observed at CERN. Grand unified theories are proposed.

1974 The J/Ψ particle is discovered at Brookhaven and at SLAC.

1975 Quark jets seen at SLAC provide additional evidence of quarks.

1976 Particles with bare charm are observed at SLAC and at Brookhaven. The existence of a third family of quarks and leptons is established by experiments at Fermilab and at SLAC.

1978 Atomic parity violation is detected and measured at SLAC. It agrees with theoretical predictions.

1979 The Nobel Prize is awarded jointly to Glashow, Salam, and Weinberg, for their independent contributions to the electroweak synthesis.

1981 Further evidence confirming the standard theory comes from DESY : the observation of gluon jets and the measurement of the muon asymmetry.

1983 The intermediate vector bosons $W^{\pm}$ and Z^0 are produced and detected at the CERN collider. The results once again confirm the standard theory.

Clearly, our discipline has reached a plateau. We have in hand an excellent theory which is compatible with all experimental data. It is not at all clear what should be done next. We turn to a brief historical digression for guidance.

HISTORICAL DIGRESSION

The search for the ultimate constituents of matter and for the laws by which they combine has a somewhat periodic history, alternating between Chaos, the discovery of Order, and the revelation of a new and confusing level of Structure. Begin with the chemical elements. Lavoisier was aware of some two dozen of them. Subsequent technical developments (like spectroscopy) led to the discovery and the synthesis of many more. Today, there are 108 known chemical elements, of which some twenty must be artificially produced. The time evolution of the number of known elements is shown in fig. 1. Clearly, there are too many different varieties of atoms for them to be regarded as fundamental entities.

Order was established in 1869 with the development of the periodic table of the elements by Mendeleev. Not only did this system congently display the regularities in chemical and physical properties of the elements, but it was predictive. The table had holes corresponding to elements yet undiscovered. Their chemical and physical properties were predicted. The discoveries of Gallium (1875), Scandium (1879), and Germanium (1886) showed the power of the periodic table. Many scientists took the success of the periodic table as an indication of the composite nature of atoms.

On the other hand, the Table was seriously flawed. Chemists did not succeed in predicting the existence of the inert gases. The discovery of Argon in 1894 as a new element comprising 1 percent of the atmosphere was not initially accepted by Mendeleev. There was no room for it in his Table.

Spectroscopy led to further evidence of atomic structure. The regularity of the Balmer series was recognized in 1885. But it was the discovery of the electron in 1895 and the atomic nucleus in 1911 that led us to the essentially correct view of atomic structure : the nuclear atom of Rutherford, Bohr, and Sommerfeld.

Thus, we were led to the next level of chaos, that of atomic nuclei. With the discovery of isotopes, it became clear that there were more species of the nucleus than of the atom. Nuclei could not be fundamental. The first tantalizing hint of nuclear order is Prout's law : that all atomic weights should be integer multiples of that of Hydrogen. (Indeed, it was in the test of this law that Lord Rayleigh performed the precise density measurements that led to the almost serendipidous discovery of Argon). Prout's Law, though inexact, is remarkably well satisfied. However, it was Moseley's work that led to a more precise and predictive level of nuclear order. Moseley measured the characteristic X-rays of all the known elements from Aluminium to Gold. The systematic dependence of X-ray wavelength upon atomic weight revealed another integer law of the nuclei : atomic number, which Moseley correctly identified as the nuclear

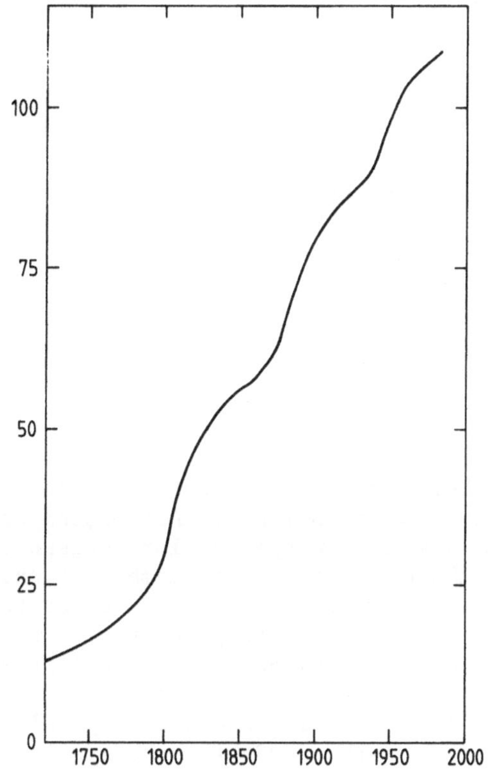

Fig. 1. The number of known chemical elements as a function of time.

charge in multiples of the electronic charge. Again, there were predictive holes in Moseley's table corresponding to the then undiscovered elements Rhenium and Technetium. The stage was set for the discovery of the nuclear constituents.

With the discovery of the neutron in 1932 a new plateau was reached. Nucleons and electrons were the components of all matter. A new nuclear force was needed to bind nucleons together. The hypothetical "mesotrons" (now called pions), which mediate this force were discovered in cosmic rays in 1948. Yukawa won his Nobel Prize and the physics community rejoiced in its discovery of the ultimate constituents of matter. But, only a few years later the nuclear particles were to be subjected to yet a new variety of population explosion.

With the development of large particle accelerators and of detection systems, like the bubble chamber, many cousins of the pions and nucleons were discovered, as well as a host of strange particles. The course of this development is shown in fig. 2. Many of the new particles were recognized to be excited states of the nucleon. Thus, energy level diagrams are common to atoms, to nuclei, and to the proton itself, as shown in fig. 3. This was a strong hint of the composite nature of hadrons. But, further progress to understand the structure of hadrons depended on the establishment of a new level of order. This emerged with the development of the 8-fold way of Gell-Mann and Ne'eman. Hadrons were seen to appear as complete representations of the group SU(3), which was supposed to be an approximate symmetry group of the strong interactions. Baryons transformed according to one or other of three representations : one-dimensional, eight-dimensional, or ten-dimensional. All mesons were either singlets or octets. One of the spectacular successes of this scheme was the prediction of the Ω^-, the tenth member of the $J = 3/2^+$ baryon decimet. With its discovery in 1964, it was clear to all that the eightfold-way was useful, valid, and predictive. But, why only certain representations, and why SU(3) among all groups?

Those questions were answered elegantly by Gell-Mann and Zweig with their invention of quarks, and with the mysterious quark rules : three quarks to a baryon, quark-antiquark to a meson, three antiquarks to an antibaryon, no other observable quark configurations, and spin-1/2 quarks to be treated as bosons. Only much later did a quantitative theory of strong interactions emerge in which these rules became comprehensible : today's gauge theory based on exact color SU(3) —quantum chromodynamics.

Many experiments have identified quarks and gluons as the most fundamental known nuclear constituents. First and foremost, perhaps, are the deep-inelastic lepton scattering experiments initiated at SLAC in the 1960's. These results were incontroversible evidence for the existence of pointlike charged particles (partons) within

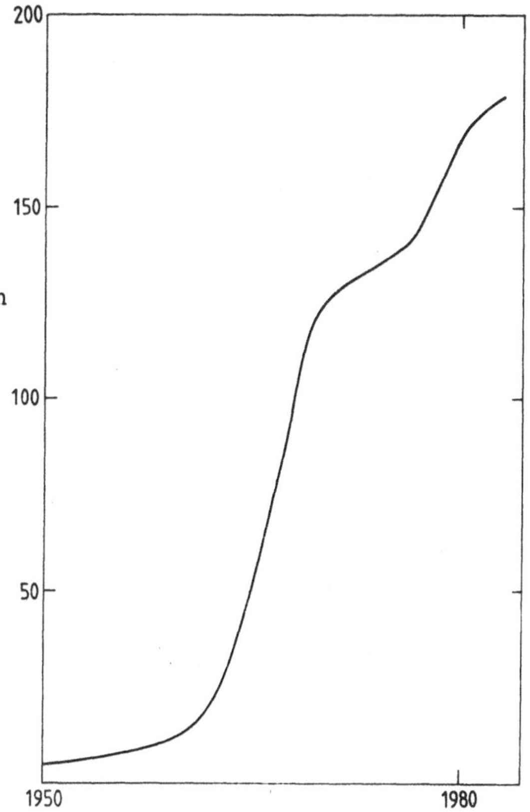

Fig. 2. The number of known hadrons as a function of time.

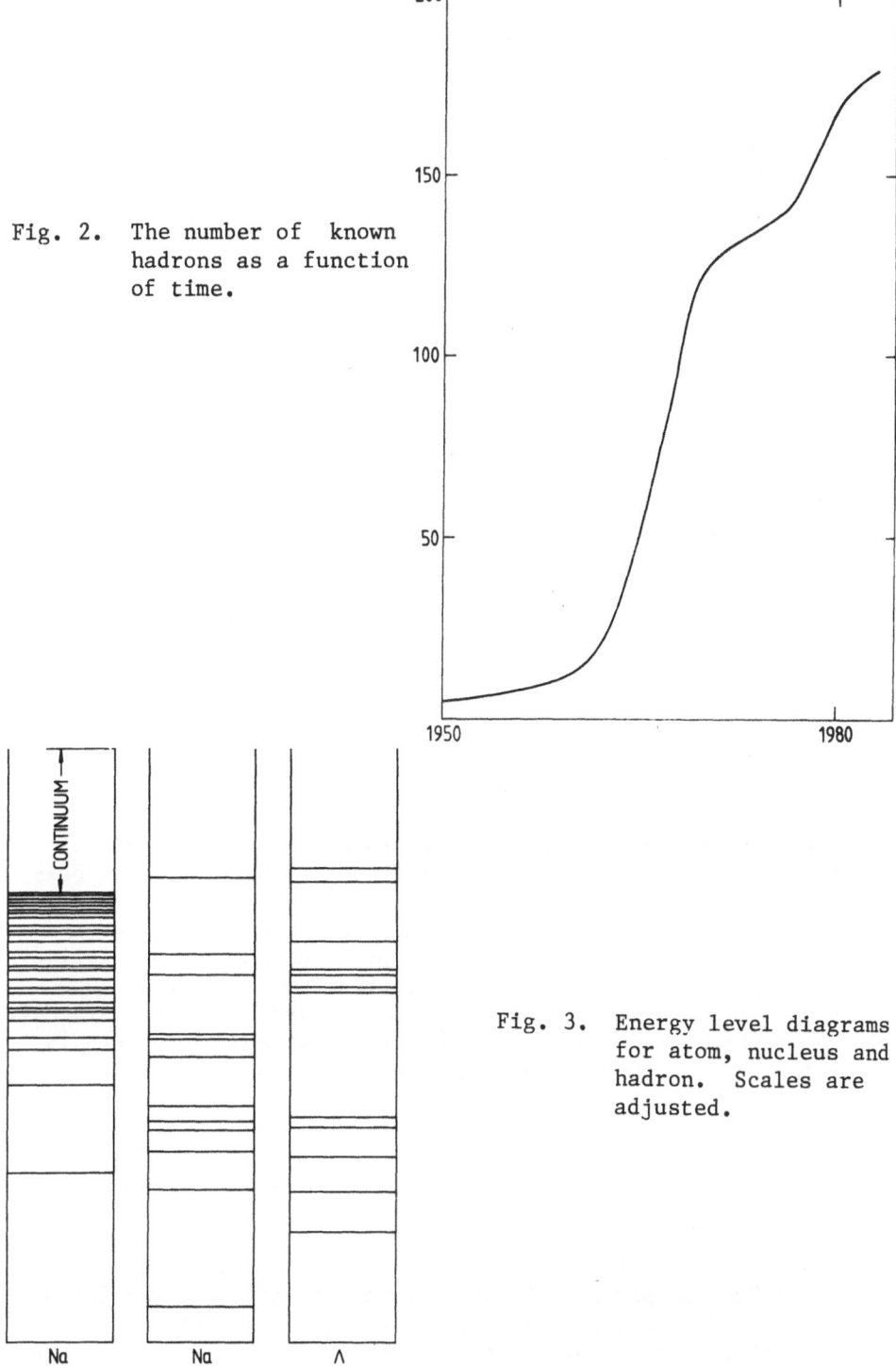

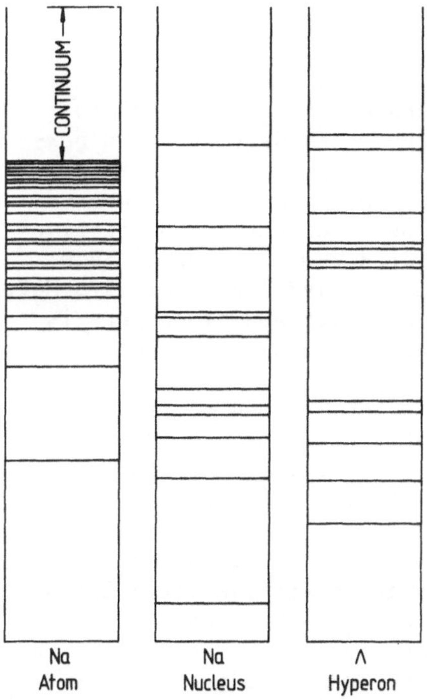

Fig. 3. Energy level diagrams for atom, nucleus and hadron. Scales are adjusted.

Electric Charge			
-1	-1/3	0	2/3
e	d	ν_e	u
μ	s	ν_μ	c
τ	b	ν_τ	

Fig. 4. Periodic table of quarks and leptons. Will there be a fourth row? Will there be new columns?

the nucleon. With the overwhelming success of the quark model in explaining hadron spectroscopy, it soon became evident that the parton is a quark. With the discovery of the J/Ψ particle, a cleaner spectroscopic system emerged in which the quark model (supplemented with the charmed quark) proved itself to be a correct and predictive system. For example, the recently discovered $F = c\bar{s}$ meson at 1970 MeV lies almost precisely at its predicted (in 1975) mass of 1975 MeV. Finally, the observation and measurement of quark and gluon jets is as close to "seeing" these fundamental particles as we may get.

The number of truly fundamental particles is none too small for comfort : seventeen. The twelve fundamental fermions are shown in today's periodic table of quarks and leptons, fig. 4. Of course, it may turn ourt that the number of fermion families (or, rows of the table) is larger than three, but not much larger since cosmology constrains the number of neutrinos to be not greater than four. Beyond these "matter particles" there are at least five varieties of "force particles" : photons, gluons, $W^{\pm}$, Z^0, and the elusive Higgs boson. To date, fourteen of the particles have been "seen". All that remain are the top quark, the tau neutrino, and the Higgs boson.

Our theoretical framework : a gauge theory based upon SU(3) x SU(2) x U(1), offers a complete and correct description of all confirmed phenomena at accessible energies. There are no generally established loose ends that require something extra in the theory. However, there are some intriguing indications. One is the existence of unexplained same-sign dileptons in high-energy neutrino interactions. This curious effect cannot be explained in terms of associated charm production, since the Ohio State triggered emulsion experiment shows neutrino production of $\sim$100 single charmed particles, but not one candidate for production of charmed particle pairs. A second potential problem is the sighting of apparent decays of Z^0 into a lepton pair and a photon. Such a process, if it is really there, admits no conventional explanation. It would mean that the desert must begin to bloom in the very near future.

WHERE IS THE TOP QUARK ?

The periodic table of quarks and leptons may be completed with the discovery of the tau neutrino and the top quark. Let us digress about the tau neutrino. While it is not yet directly established, indirect arguments for its existence are all but convincing. First of all, it is established that the tau lepton behaves, like the muon and the electron, as a left-handed doublet and a right-handed singlet. The tau production asymmetry has been measured at PETRA, and it agrees with standard predictions. Similarly, the Michel ρ parameter characterizing the decays $\tau \to e\nu\nu$ and $\tau \to \mu\nu\nu$ is observed to be 3/4, as it must be in the standard model. Alternatives such as a vector model (with both helicities of tau transforming as doublets), or models in which tau is primarily a weak singlet (with mixing) are clearly ruled out. Another possibility, wherein the tau neutrino is heavier than the tau, would require significant mixing for tau to decay. This possibility is ruled out by the "correct" observed tau lifetime, and the experimental absence of large neutrino mixing effects among ν_e and ν_μ. The only remaining possibility is that ν_τ does exist, and has a mass less than the current experimental upper bound of 250 MeV. Despite these indirect arguments, it would be desirable to have direct evidence for the existence of ν_τ. Dedicated experiments of this kind are in progress.

The discovery of the top quark is certainly overdue. With the discovery of upsilon at 9.5 GeV, it was generally believed that toponium would soon follow. However, it is now certain from research at PETRA that this state is heavier than 42 GeV, if indeed it exists. Is there any case for a topless universe, in which the b quark is primarly a weak singlet? In this model, the GIM mechanism is sacrified to the extent that the b quark is admixed into the weak current. However, the long observed B lifetime ($\sim 10^{-12}$s) ensures that there is no problem with GIM violating contributions to the K_L-K_S mass splitting. This topless theory fails for another reason : the unavoidable appearance of beauty-violating neutral currents. Indeed, we straightforwardly compute

$$R = \frac{\Gamma(B\to\ell^+ + \ell^- + \ldots)}{\Gamma(B\to\ell^\pm + \ldots)} = 0.13$$

in such topless theories. This is in flat contradiction with the CESR experimental upper limit $R < 0.03$. Thus, the current experimental situation demands the existence of the top quark.

Ginsparg, Wise, and I recently deduced a lower limit to the top quark mass[1]. The argument, framed in the conventional three-family model, is entirely straightforward. First, we obtain the following estimates of the Kobayashi-Maskawa parameters

$$s_2^2 + s_3^2 + 2s_2s_3c_\delta = 4.2 \times 10^{-3}\ R(b\to c)(10^{-12}s/\tau_B)$$

$$s_3^2 = 3.9 \times 10^{-2}\ R(b\to u)(10^{-12}s/\tau_B),$$

where R signifies the B-decay branching ratios and τ_B is the B lifetime. Recent SLAC experiments indicate that τ_B is near to 10^{-12}s, while recent CESR experiments show that $R(b\to u)$ is at most 5 percent. We may conclude that both s_2 and s_3 are quite small.

Next, we compute the imaginary part of the K^0 mass matrix from the box diagram. Unlike the real part of the mass matrix, the CP-violating part is reliably calculable by use of a short-distance expansion. Our constraint upon the t quark mass is obtained by requiring this computation to yield the observed strength of CP violation. We obtain a lower limit to the top quark mass as a function of the B lifetime. Unfortunately, the result is very sensitive to τ_B, whose exact value is not now known. For example, we obtain $M(t) > 30$ GeV if $\tau_B > 0.75 \times 10^{-12}$s, and the much stronger bound $M(t) > 50$ GeV if $\tau_B > 1.25 \times 10^{-12}$s. All of this is especially interesting in the light of rumors which are occasionally heard about the possible sighting of the top quark at the CERN Collider.

What will it mean if the values of τ_B and $M(t)$ are found to violate the bound? It may indicate that the Kobayashi-Maskawa model of CP violation is erroneous, or it may signify the existence of a fourth family of quarks and leptons. In either case, it would imply the existence of more interesting physics discoveries at or below the TeV scale.

Let me conclude this section with a new and somewhat fanciful (i.e., unpublishable) prediction for the value of the top quark mass. It is an outgrowth of an earlier prediction presented at a previous Cargèse school. (See also ref. 2) The essential hypothesis is that the mass matrices for up-like quarks, down-like quarks, and charged leptons are linearly dependent. In the approximation of small mixing angles, this yields the determinantal formula

$$\begin{vmatrix} e & d & u \\ \mu & s & c \\ \tau & b & t \end{vmatrix} = 0,$$

which predicts toponium to lie between 30 and 40 GeV. This prediction is proven to be wrong by experiments done at PETRA. However, there was a hidden and unstated hypothesis in the above analysis : that the leptons are assigned to families in order of their masses. This is not an obligatory hypothesis. Consider the alternative in which the tau lepton and the electron are interchanged. In this case we obtain a different determinantal mass formula :

$$\begin{vmatrix} \tau & d & u \\ \mu & s & c \\ e & b & t \end{vmatrix} = 0.$$

One way in which this formula can be satisfied is if

$$u = d = \mu = e = 0$$

$$ts = cb.$$

In this lowest-order approximation to an unspecified theory, the light quarks and leptons are massless, and all K.M. angles vanish. The "bare" mass of the top quark is $t = cb/s$, a familiar mass formula most recently come upon by Achiman[3]. When properly subject to radiative corrections, the mass of the top meson is predicted to be

$$m(t) \cong m_t(2m_t) \cong 33 \pm 6 \text{ GeV}.$$

If this prediction is satisfied, a result which can be established at the CERN Collider, several different groups of theorists, each touting a very different model, will claim vindication of their work.

Our model, in particular, makes one other startling prediction in the context of SU(5) grand unification. The interchange of electron and tau affects the couplings of the X boson responsible for proton decay. In our skewed picture, the $\pi^0 e^+$ decay mode is forbidden, while the $K^0 \mu^+$ decay mode proceeds at its conventional rate. This possibility is not excluded by published IMB data, since the upper limit on $p \to K^0 \mu^+$ is presently of order 10^{31} yrs, and the SU(5) prediction for this mode is $10^{30\pm1.7}$ yrs. In this scheme, the decay muon is expected to be unpolarized, in contradistinction to predictions based upon Higgs dominance or supersymmetry.

SHORT TAKES

In the remainder of this paper, I discuss subjects that were considered in my Cargèse lectures, but which have subsequently (or will soon) appear in print.

(1) "Photon Oscillations and the Cosmic Background Radiation"[4]

We consider the possible existence of a second species of photon which is uncoupled to known forms of matter. Explicit mass terms in the Lagrangian can give rise to small photon masses and to oscillations of photon identity, without sacrificing the renormalizability of the gauge theory. Current upper limits on the photon mass are $\sim 6 \times 10^{-16}$ eV/c^2. Photon oscillations corresponding to much smaller masses can significantly alter the spectral shape of the CBR. Indeed, we show that an apparent discrepancy between theorical and observed CBR spectra can be resolved in terms of photon oscillations, and a mass parameter of 5×10^{-18} eV/c^2. Our model predicts the appearance of measurable oscillations in the CBR spectrum in the centimeter wavelength band. Experiments have been initiated to search for the effect.

(2) "Geophysical and Cosmological Sources of Antineutrinos"[5]

The sun is an intense source of neutrinos, but not of antineutrinos. The decays of naturally occuring radioactive elements in the Earth (K, Rb, Th and U) provides a local source of antineutrinos, with a flux roughly four orders of magnitude less than the solar neutrino flux at the Earth's surface. These could be detected with appropriate radiochemical or RIS (resonance ionisation spectroscopy) procedures, wherein the antineutrinos induce β^+ decay or stimulate resonant electron capture. The experiment appears to be feasible with extreme difficulty. The result —a direct measurement of the surface antineutrion flux— is of considerable geological interest. It is a direct measure of the radiogenic production of heat in the Earth, and it depends upon the imprecisely known distribution of radioactive elements in the deep mantle.

We also consider the production of antineutrinos by distant supernovae. (Nearby supernovae produce an easily detectable signal, but they only occur every several decades). It is easy to estimate crudely the flux of antineutrinos from the sum of all supernovae. Each supernovae produces about 10^{58} neutrinos and antineutrinos of mean energy 10 MeV. At the present epoch, each galaxy has suffered $\sim 10^9$ such supernovae, and thus, has been a source of 10^{67} antineutrinos. Since there is about one galaxy per 10^{74} cm^3 in the visible universe, we conclude that the relict $\bar{\nu}$ density $\sim 10^{-7}$ cm^{-3}, corresponding to a flux of ~ 3000 $cm^{-2}s^{-1}$. The mean $\bar{\nu}$ energy is redshifted by a factor of 3/5. Such a relict $\bar{\nu}$ flux is possibly measurable, and would yield a more precise measure of supernovae frequency throughout universal history.

(3) "Staying Alive With SU(5)"[6]

In this paper, we consider twenty variations of naive SU(5) in which there appear various "split" SU(5) multiplets of fermions : some with observable masses (e.g. $\leq$ 1 TeV), and some with masses at the unification scale. Three of the variations are compatible with present observations of $\sin^2\Theta$ and limits on proton decay. As our results were incorrectly reproduced in the published article, we show a corrected illustration of our results in fig. 5.

(4) "Neutrino Exploration of the Earth"[7]

We show how the neutrinos produced by a multi-TeV proton synchrotron may be used for purposes of geological research. Project GENIUS (geological exploration by neutrino-induced underground sound) is designed to search for deposits of oil and gas at large distances from the accelerator. It depends upon the coherent sound signal produced at depth by millions of neutrino interactions along the underground neutrino beam. Surface measurements of the acoustic pulse provide a remote underground probe. Project GEMINI (geological exploration with muons produced by neutrino interactions) is designed to search for distant deposits of high-Z ores. It depends upon the surface measurement of neutrino-induced muons which were produced in the last few kilometers of the neutrinos'underground voyage. Project GEOSCAN is a flux-independent procedure to determine the vertical density profile of the Earth, and especially its core. It depends upon the angle and energy dependence of the attenuation as the neutrino beam traverses the whole Earth.

(5) "Anomalous Z Decays and the Extended Family"[8]

This work was not presented at Cargèse. Rather, it has been inspired by the remarks of C. Rubbia at the Summer School. Both the UA1 and UA2 collaborations have reported convincing observations of the production and decay into e^+e^- of the Z^0. Both groups also report the observation of anomalous events of the sort $Z^0 \rightarrow e^+e^-\gamma$. We can find no explanation of this phenomenon in terms of the SU(2)xU(1) electroweak theory. Thus, we argue that the actual anomalous final state may involve two photons. In the observed events, one of the photons may be approximately collinear with one of the other particles, thus mimicking an $e^+e^-\gamma$ final state. We have extended the usual electron system to include the electron, its neutrino, and three heavier leptons (neutral, singly charged (E) and doubly charged.) The model is unconventional in that the left-handed positron is assigned to the central location of an SU(2) triplet. In our model, the overwhelmingly dominant decay of the E is into an electron and photon, so that Z^0 decays according to

$$Z^0 \rightarrow E^+ + E^- \rightarrow e^+ + e^- + 2\gamma.$$

If our hypothesis is correct, the branching ratio of Z^0 into this channel must be comparable with the branching ratio into e^+e^-, and

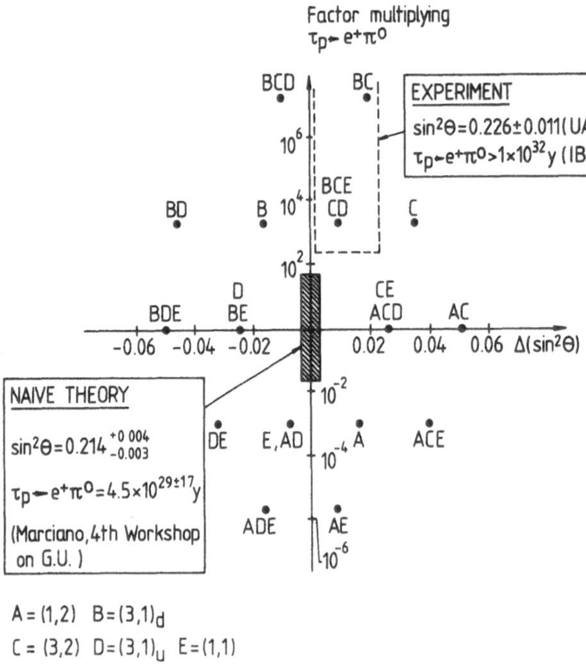

Fig. 5. Staying Alive With SU(5). Erratum : In figure 1, some points were incorrectly plotted with respect to $\Delta(\sin^2\Theta)$. The corrected figure 1 is shown here. The split fermion multiplets consistent with experiment have light fermions corresponding to BC, BCE and CD in the notation of the paper.

the observed "one-photon" events must be convincingly reinterpreted to involve two photons. We look forward to additional experimental data.

REFERENCES

1. P.H. Ginsparg, S.L. Glashow and M.B. Wise, Phys. Rev. Lett. 50, 1415 (1983).
2. S.L. Glashow, Phys. Rev. Lett. 45, 1914 (1980).
3. Y. Achiman, Phys. Lett. 131B, 366 (1983).
4. H. Georgi, P. Ginsparg and S.L. Glashow, to appear in Nature.
5. L. Krauss, D. Schramm and S.L. Glashow, submitted to Nature.
6. P. Frampton and S.L. Glashow, Phys. Lett. 131B, 340 (1983).
7. A. De Rujúla, S.L. Glashow, R.R. Wilson and G. Charpak, Phys. Rep., in press.
8. P. Frampton and S.L. Glashow, in preparation.

INDEX